Materials Horizons: From Nature to Nanomaterials

Series Editor

Vijay Kumar Thakur, School of Aerospace, Transport and Manufacturing, Cranfield University, Cranfield, UK

Materials are an indispensable part of human civilization since the inception of life on earth. With the passage of time, innumerable new materials have been explored as well as developed and the search for new innovative materials continues briskly. Keeping in mind the immense perspectives of various classes of materials, this series aims at providing a comprehensive collection of works across the breadth of materials research at cutting-edge interface of materials science with physics, chemistry, biology and engineering.

This series covers a galaxy of materials ranging from natural materials to nanomaterials. Some of the topics include but not limited to: biological materials, biomimetic materials, ceramics, composites, coatings, functional materials, glasses, inorganic materials, inorganic-organic hybrids, metals, membranes, magnetic materials, manufacturing of materials, nanomaterials, organic materials and pigments to name a few. The series provides most timely and comprehensive information on advanced synthesis, processing, characterization, manufacturing and applications in a broad range of interdisciplinary fields in science, engineering and technology.

This series accepts both authored and edited works, including textbooks, monographs, reference works, and professional books. The books in this series will provide a deep insight into the state-of-art of Materials Horizons and serve students, academic, government and industrial scientists involved in all aspects of materials research.

More information about this series at http://www.springer.com/series/16122

Gopinathan Anilkumar · Salim Saranya
Editors

Green Organic Reactions

Springer

Editors
Gopinathan Anilkumar
School of Chemical Sciences
Mahatma Gandhi University
Kottayam, Kerala, India

Salim Saranya
School of Chemical Sciences
Mahatma Gandhi University
Kottayam, Kerala, India

ISSN 2524-5384 ISSN 2524-5392 (electronic)
Materials Horizons: From Nature to Nanomaterials
ISBN 978-981-33-6896-5 ISBN 978-981-33-6897-2 (eBook)
https://doi.org/10.1007/978-981-33-6897-2

This Springer imprint is published by the registered company Springer Nature Singapore Pte Ltd.
The registered company address is: 152 Beach Road, #21-01/04 Gateway East, Singapore 189721, Singapore

Dedicated with great respect to Prof. Yasuyuki Kita for his outstanding contributions in Organic synthesis and for introducing one of us (GA) into the fascinating field of Green Chemistry.

Preface

A sustainable development in the scientific world is inevitable nowadays, especially in the field of chemistry. The great contributions that our scientific society has given to the world for its development are uncountable. But the major drawback that we were facing for centuries was the lack of proper method to avoid or minimize the pollution caused through this scientific evolution. In this context, the idea of green chemistry has emerged, which is to reduce or eliminate the use of hazardous substances and make it more earth-friendly. Hence, green chemistry is also named as sustainable chemistry. According to this concept, greener methods flourished in all areas of research including organic chemistry. We have tried to encompass major advancements in green organic reactions in this book.

For clarity and brevity, we have divided the whole book into 17 chapters, where the first two chapters are dedicated for the full interpretation of green chemistry and its 12 principles. Solvent plays a major role in organic reactions, and in greener aspect, if we are able to use an environment-friendly solvent or avoid the use of solvents, it would be a great contribution toward green chemistry. There are many reports under these categories, and hence, we have tried to include different chapters like green organic reactions in water, supercritical fluids as well as organic reactions under solvent-free conditions, solid-state reactions, etc. In addition, a chapter is devoted solely for environmentally benign organic solvents. Microwave-assisted reactions have contributed much in the discipline of green chemistry, and to explain the whole reactions in a single chapter is somewhat impossible. So we have focused on the recent advances in microwave-assisted amination reactions. The book also gives a clear vision on the applications of green chemistry in the field of analytical chemistry, pharmaceutical chemistry, etc. The impact of catalysis in organic chemistry under greener aspects has been discussed through chapters like green organocatalysis, green nanocatalysts in organic synthesis, visible light-catalyzed asymmetric synthesis, and green chemistry of recoverable catalysts. The book also comprehends green chemistry on C-H activation as well as atom economic green organic reactions. Flow chemistry is a flourishing area in the globe, and it has to do much in sustainable chemistry. So, a special mention of its role in the field of chemical synthesis is also discussed.

Hence, we hope that this book will help the scientific world including students, researchers, academicians, and scientists to get an overview of organic reactions in a greener aspect.

Kottayam, India
April 2020

Gopinathan Anilkumar
Salim Saranya

Contents

About the Editors

Dr. Gopinathan Anilkumar is currently Professor of Organic Chemistry at the School of Chemical Sciences of Mahatma Gandhi University, Kottayam, India. He obtained his Ph.D. in 1996 from the Regional Research Laboratory (later renamed the National Institute for Interdisciplinary Science and Technology, NIIST-CSIR), Trivandrum, with Dr. Vijay Nair. He undertook postdoctoral studies at the University of Nijmegen, The Netherlands (with Prof. BinneZwanenburg), Osaka University, Japan (with Prof. Yasuyuki Kita), Temple University, USA (with Prof. Franklin A. Davis), Leibniz-Institut fur OrganischeKatalyse (IfOK), Germany (with Prof. Matthias Beller), and Leibniz-Institut fur Katalyse (LIKAT), Germany (with Prof. Matthias Beller). Then, he was Senior Scientist at AstraZeneca, India. His research interests are in the areas of organic synthesis, medicinal chemistry, green chemistry and catalysis, in particular ruthenium-, iron-, zinc-, and copper-catalyzed reactions. He has 7 patents and has published more than 125 papers in international journals. He is a recipient of the Dr. S. Vasudev award from Kerala State Council for Science, Technology, and Environment, Government of Kerala, for best research in 2016. His h-index is 32.

Ms. Salim Saranya is Doctoral Student of Prof. G. Anilkumar at Mahatma Gandhi University, Kottayam, Kerala, India. Her area of work is organic synthesis and catalysis including exploration of greener protocols. She has published over 27 research papers and reviews. She has attended and presented papers in many national and international conferences and seminars. She received her B.Sc. and M.Sc. degrees from Christian College, Chengannur, in 2012 and 2014, respectively. She qualified the CSIR-UGC National Eligibility Test in 2015 with a research fellowship.

Abbreviations

[AcMim]X	Acylmethylimidazolium halide
[bmim][BF_4]	1-butyl-3-methylimidazolium tetrafluoroborate
[bmim][Br]	1-butyl-3-methylimidazolium bromide
[bmim][NTf_2]	1-butyl-3-methylimidazolium bis(trifluoromethylsulfonyl)imide
[bmim][PF_6]	1-butyl-3-methylimidazolium hexafluorophosphate
[bmim][TFA]	1-butyl-3-methylimidazolium trifluoroacetate
[$Cp^*RhCl_2]_2$	Pentamethylcyclopentadienyl rhodium dichloride
[TEA][HPS]	Triethanolammonium 3-hydroxypropane-1-sulfonate
2-MeTHF	2-methyltetrahydrofuran
7-ACA	7-aminocephalosporanic acid
Å	Angstrom
AC	Activated carbon
AcOH	Acetic acid
ACS	American chemical society
ACV NRPS synthatase	*L*-δ-(α-aminoadipoyl)-*L*-cysteinyl –*D*-valine nonribosomal peptide synthatase
AE	Atom economy
AgNps	Silver nanoparticles
AIBN	Azobis isobutyronitrile
ANS	8-anilino-1-naphthalenesulfonic acid
APCI	Atmospheric pressure chemical ionization
API	Active pharmaceutical intermediate
AuNps	Gold nanoparticles
AZT-TP	Azidothymidine triphosphate
BAIL	Brønsted acid ionic liquid
BDEs	Bond dissociation energies

BIES	Base-assisted intramolecular electrophilic-type substitution
Bmim	1-Butyl-3-methylimidazolium
BMOPs	Polybismaleimide-based microporous organic polymers
BOD	Biochemical oxygen demand
BQ	Benzoquinone
$BrCCl_3$	Bromotrichloromethane
Bu_4NI	Tetrabutylammonium iodide
CalB/CALB	*Candida antarctica* lipase B
CAN	Ceric ammonium nitrate
CDC	Cross-dehydrogenative couplings
CE	Capillary electrophoresis
CHS	Caffeine hydrogen sulphate
CMC	Critical micellar concentration
CMD	Concerted metallation deprotonation
COP-213	Covalent organic polymer
COX-2	Cyclooxygenase-2
CS-LDL-DOX NPs	Chitosan-low density lipoprotein doxorubicin nanoparticles
CTAB	Cetyltrimethylammonium bromide
Cu(II)-TD@nSiO_2	Copper immobilized on nanosilica triazine dendrimer
CuAAC	Copper(I)-catalyzed alkyne-azide cycloaddition
Cy	Cyclohexyl
d4T	Stavudine or 2’,3’-didehydro-2’,3’-dideoxythymidine
DABCO	1,4-Diazabicyclo[2.2.2]octane
DAOC synthetase	Deacetoxycephalosporin-C synthatase
DBU	1,8-Diazabicyclo[5.4.0]undec-7-ene
DCC	Dicyclohexylcarbodiimide
DCE	1,2-Dichloroethane
DCM	Dichloromethane
DDS	Drug delivery system
DDT	Dichlorodiphenyltrichloroethane
DES	Deep eutectic solvents
DG	Donor group
DIPEA	Diisopropylethyl amine
DMA	Dimethylacetamide
DMAP	4-(*N,N*-dimethylamino)pyridine
DME	Dimethoxyethane
DMF	Dimethylformamide
DMP	Dess-Martin-Periodinane

DMPSi	Dimethyl polysilane
DMSO	Dimethylsulfoxide
DNA	Deoxyribonucleic acid
DO	Dissolved oxygen
DOE	Design of experiments
DOX	Doxorubicin
DPEphos	Bis[(2-diphenylphosphino)phenyl] ether
DPPF	1,1'-Bis(diphenylphosphino)ferrocene
DPP-IV	Dipeptidyl peptidase-4
DPPP	1,3-Bis(diphenylphosphino)propane
DPZ	5,6-Bis(5-methoxythiophen-2-yl)pyrazine-2,3-dicarbonitrile
DTBP	Di-*tert*-butyl peroxide
EDA	Electron donor-acceptor
EDC	1-Ethyl-3-(3-dimethylaminopropyl) carbodiimide
EDS	Energy dispersive spectroscopy
EDX	Energy-dispersive X-ray spectroscopy
EL	Ethyl lactate
EPA	Environmental protection agency
ESI	Electrospray ionization
EtOH	Ethanol
e-waste	Electronic-waste
EWG	Electron withdrawing group
$Fe_3O_4@SiO_2$-imid-PMAn	Immobilization of $H_3PMo_{12}O_{40}$ nanoparticles (PMAn) on imidazole functionalized $Fe_3O_4@SiO_2$ nanoparticles
FEP	Fluorinated ethylene propylene
FT-IR	Fourier-transform infrared spectroscopy
GAAS	Gluconic acid aqueous solution
GC	Gas chromatography
GCI	Green chemistry institute
GLP	Good laboratory practice
GNS	Graphene nanosheets
GO	Graphene oxide
GOx	Glucose oxidase
GPS	Global positioning system
GVL	γ-Valerolactone
GXL	Gas expanded liquids
h	Hour
HALMD	Hydrophilic, aromatic, and low molecular-weight drugs
HAT	Hydrogen atom transfer
HCP	Hyper cross-linked polymers

HFIP	1,1,1,3,3,3 hexafluoropropan-2-ol
HIV	Human immunodeficiency virus
Hmc	Hollow mesoporous carbon
HNTs	Halloysite nanotubes
HOBt	Hydroxybenzotriazole
HPLC	High performance liquid chromatography
HSBM	High-speed ball milling
HSV	Herpes simplex virus
HSVM	High-speed vibration milling
IBA	Iodosobenzoic acid
IBX	Iodoxybenzoic acid
ICH	International Conference on Harmonization
IL	Ionic liquid
ILAG	Ion- and liquid-assisted grinding
IoT	Internet of things
IPA	Isopropyl alcohol
IR	Infrared radiation
IUPAC	International Union of Pure and Applied Chemistry
K_h	Hydrolysis constants
KSF	Acidic clay montmorillonite
LAG	Liquid-assisted grinding
LASC	Lewis acid–surfactant-combined
LASSC	Lewis acid–surfactant–SiO_2-combined
LC	Liquid chromatography
LCA	Life cycle analysis
LDI-MS	Laser-desorption/ionisation mass spectrometry
LDL	Low density lipoprotein
LED	Light emitting diode
LKADH	Lactobacillus kefir alcohol dehydrogenase
$Ln(OTf)_3$	Lanthanide triflates
MALDI-MS	Matrix-assisted laser-desorption/ionisation mass spectrometry
MCF-7	Michigan cancer foundation-7
m-CPBA	*Meta*-chloroperbenzioc acid
Me_4NCl	Tetramethylammonium chloride
MIL	Material of institut Lavoisier
min.	Minute
MIP	Molecularly imprinted polymer
$MnFeCaFe_2O_4$@starch@aspartic acid MNPs	Aspartic-acid-loaded starch-functionalized Mn–Fe–Ca ferrite magnetic nanoparticles
MNP	Metal nanoparticle
MOEMIM.Oms	1-Ethyl-3-methylimidazolium methanesulfonate

MOEMIM.TFA	1-Ethyl-3-methylimidazolium trifluoroacetic acid
MOEPy.Oms	*N*-Methoxyethylpyridinium methanesulfonate
MOF	Metal-organic framework
MOL	Molsidomine
MR	Merrifield resins
MS	Mass spectrometry
MS	Meglumine sulphate
MTBE	Methyl *tert*-butyl ether
MW	Molecular weight
MWCNTs@NHBut/PTA	Phosphotugstic acid immobilized on aminated multiwalled carbon nanotubes
Myo	Myoglobin
Na_2EDTA	Ethylenediaminetetraacetic acid, disodium salt
NaBARF	Sodium tetrakis[3,5-bis(trifluoromethyl)phenyl]borate
NADP	Nicotinamide adenine dinucleotide phosphate
NADPH	Reduced nicotinamide adenine dinucleotide phosphate
NBS	*N*-Bromosuccinimide
NHC	N-Heterocyclic Carbene
NHS	*N*- Hydroxyl succinimide
NIR	Near infrared
NMP	*N*-Methyl-2-pyrrolidone
NMP	Nucleoside-5'-monophosphates
NMR	Nuclear magnetic resonance
NMSM	(*E*)-*N*-Methyl-1-(methylthio)-2-nitroethenamine
Nps	Nanoparticles
NQ	Naphthoquinones
NRPS	Nonribosomal peptide synthetase
NZ	Acridinium salt derivative
OA	Oxidative addition
ODS	Octadecyl silica
OMS	Manganese octahedral molecular sieves
OpI	Overall process improvement
PAc-β-CD	Peracetylated-β-cyclodextrin
PAH	Polycyclic aromatic hydrocarbons
PASE	Pot atom economic and step economic
PCA	Principal component analysis
PCB	Polychlorinated biphenyl
PEEK	Polyether ether ketone
PEG	Polyethylene glycol

PEGMA-g-TEGBDIM	Polyethylene glycol methacrylate-grafted tetra-ethylene glycol-bridged dicationic imidazolium-based ionic liquid
PIB	Polyisobutylene
pKa	Acid dissociation constant
PLA	Polylactic acid
PLE	Pressurized liquid extraction
PMB	*p*-Methoxy benzyl
PMI	Product mass intensity
PMMA	Poly (methyl methacrylate)
PMO	Mesoporous organosilica
PMPsi	Poly (methyl-phenylsilane)
POLAG	Polymer-assisted grinding
POM	Polyoxometallate
POP	Persistent organic pollutants
PS	Polymer supported
PSAs	Polymer-supported acid catalysts
ps-BEMP	Polymer supported 2-*tert*-butylimino-2-diethylamino-1,3-dimethylperhydro-1,3,2-diazaphosphorine
PTC	Phase transfer catalyst
PTS	Polyoxyethanyl-α-tocopheryl sebacate
p-TSA	*p*-toluenesulfonic acid
PμZ	Paper micro-zone
QbD	Quality by design
QToF	Quadrupole time of flight
RasADH	Ralstonia ADH
rGO	Reduced graphene oxide
RME	Reaction mass efficiency
ROP	Ring opening polymerization
RPG	Relative process greenness
RPI	Relative process (waste) improvement
RPLC	Reversed phase liquid chromatography
RT/rt	Room temperature
RTILs	Room-temperature ionic liquids
SB-DABCO$^+$Cl$^-$	Silica-bonded *n*-propyl-4-aza-1-azoniabicyclo[2.2.2]octane chloride
SB-DBU$^+$Cl$^-$	Silica-bonded 5-*n*-propyl-octahydro-pyrimido[1,2-*a*]azepinium chloride
$scCO_2$	Supercritical carbon dioxide
SCF	Supercritical fluids
scH_2O	Supercritical water

SDS	Sodium dodecyl sulfate
S_E Ar	Electrophilic aromatic substitution
SEM	Scanning electron microscope
SET	Single-electron transfer
SFC	Supercritical fluid chromatography
SFCD	Supercritical fluid chemical deposition
SFD	Supercritical fluid deposition
SFE	Supercritical fluid extraction
SILM	Supported ionic liquid membrane
SPINOL	1,1'-Spirobiindane-7,7'-diol
ST1535	[2-Butyl-9-methyl-8-(2H-1,2,3-triazol 2-yl)-9H-purin-6-xylamine]
TBAB	Tetra-butylammonium bromide
TBDMS	*Tert*-butyldimethylsilyl
TBHP	*Tert*-butyl hydroperoxide
^{t}Bu	Tertiary butyl
TDS	*Tert*-hexyldimethylsilyl
TEA	Triethylamine
TEBSA	4-Triethylammonium butane-1-sulfonic acid
TEDA/IMIZ-BAIL@UiO-66	Triethylenediamine or imidazole Brönsted acidic, ionic-liquid-supported Zr metal-organic structure
TEM	Transmission electron microscopy
TEMPO	(2,2,6,6-tetramethylpiperidin-1-yl)oxyl
TFA	Trifluoroacetic acid
TFT	Trifluridine
THF	Tetrahydrofuran
THQ	Tetrahydroquinoline
TMS	Trimethylsilyl
TMSCN	Trimethylsilyl cyanide
TMSOTf	Trimethylsilyl trifluoromethanesulfonate
TOAB	Tetraoctylammonium bromide
TOF	Turn over frequency
TON	Turn over number
TPAB	Tetraptopylammonium bromide
TPGS	Polyoxyethanyl-α-tocopheryl succinate
$TropBF_4$	Tropylium BF_4
TSA	Tungstate sulfuric acid
UHP	Ultra high performance
UNCA	Urethane protected α-amino acid N-carboxyanhydrides
US	Ultrasound
UV-Vis	Ultraviolet-visible
VOC	Volatile organic compounds

WEP	Water-extractable pomelo
WERC	Water exchange rate constants
WHO	World Health Organization
XPS	X-ray photoelectron spectroscopy
XRD	X-ray diffraction
σ BM	σ-bond metathesis

Chapter 1
Introduction to Green Chemistry

Hosam M. Saleh and Amal I. Hassan

1 Introduction

Green chemistry is the design of chemical products and processes that lessen or take away the usage of the perilous substances. It is a highly effective means of preventing pollution because they apply innovative scientific solutions [1]. EPA has produced green chemistry principles, but there is no consensus on its concept [2]. The pollution is caused by the outcome of many chemical industries on the environment [3]. Efforts had to be made to design the manufacturing processes in a way that reduces this waste and has no negative impact on the environment, as well as to develop appropriate disposal methods.

In order to perform the reactions, it is necessary to verify several key factors, namely the choice of feedstocks, solvents and catalysts. For example, gasoline should be avoided as a solvent at all costs because it is carcinogenic by nature[4]. Thus, there is an imperative need to improve synthetic chemistry either by the starting materials which are not harmful to the environment or by designing new synthesis methods that reduce the utilization of toxic substances [5].

Moreover, green chemistry is looking for alternatives in the early stage of material design to remove the utilization or production of perilous substances in the manufacture of chemical products so that traditional treatment techniques will be obviated[6]. It is an extremely prosperous approach to pollution averting as a result of it stratifies innovative scientific solutions to real-world environmental issues.

H. M. Saleh (✉) · A. I. Hassan
Radioisotope Department, Nuclear Research Center, Atomic Energy Authority, Dokki 12311, Giza, Egypt
e-mail: hosam.saleh@eaea.org.eg; hosamsaleh70@yahoo.com

A. I. Hassan
e-mail: virtualaml@gmail.com

G. Anilkumar and S. Saranya (eds.), *Green Organic Reactions*, Materials Horizons: From Nature to Nanomaterials, https://doi.org/10.1007/978-981-33-6897-2_1

2 History and Origins of Green Chemistry

The toxic chemicals and their threat to humankind and ecosystems were an early ultimatum to the chemical society, and it also sought clean and environmentally friendly means to prevent further deterioration of the environmental situation. Green chemistry has gone through many stages, but before it actually appeared in the 1990 s, it was preceded by many studies that formed the solid foundations of this new science, which we believe were the early beginnings of green chemistry [7]. The first ideas and basic concepts of green chemistry have developed since the beginning of the 1990s. The name "green chemistry" did not appear in books and literature twenty years ago. The most powerful issue was finding optimal ways to reduce pollution in its various forms. This represents a marked difference between environmental chemistry and green chemistry, and in the early 1990s, specifically in 1991, the green chemistry program and policies were introduced in the USA. At the beginning of the first half of 1990, Italy and the UK launched their first programs, subsequently Japan joined the previous countries during the second half of the same decade. Specialists in this field have defined a comprehensive definition of this significant branch of chemistry, namely the investigation and design of compounds and chemical methods to mitigate or prevent the use and production of hazardous substances [8]. The chemical society's tendency to find chemical ways to treat waste and mitigate the environmental impact of pollutants has not always been the least economically viable solution. This problem suggested much balance in favor of green chemistry principles, which can achieve "perfect chemical manufacturing" [9].

In the 1980s and 1990s, there were several terms known to this field, such as clean chemistry, environmental chemistry, green chemistry, benign chemistry and sustainable chemistry. These terms are broadly defined and are subject to discussion among chemists. As Fig. 1 illustrates, the term has become, and remains, the most popular among a number of alternatives in chemical science.

In 1991, when the term "green chemistry" was officially defined, most of the environmental solutions were post facto. They were resolved in court decisions, mandating bulldozer cleanup of legacy waste [10]. Waste needs to be treated and filtered, and smokestack scrubbers need to be put on. It was an end-of-pipe solution. This was the mindset at the time. During the 1990s, the green chemistry movement began to gain global standing. About 175 countries ratified the Basel Convention in 1992 on controlling the transportation and disposal of hazardous wastes. The agreement assists countries in developing policies to prevent and combat hazardous waste and forging partnerships between industry and academia to encourage innovation in green material design [11].

The emergence of this field in the chemical arena led to the launch of the movement of benign chemistry and encouraged the utilization of techniques of chemistry that mitigate the usage of substances that pose intimidation to human health or the environment. This progressive movement urged to publish several texts in this field between 1994 and 1999 [12].

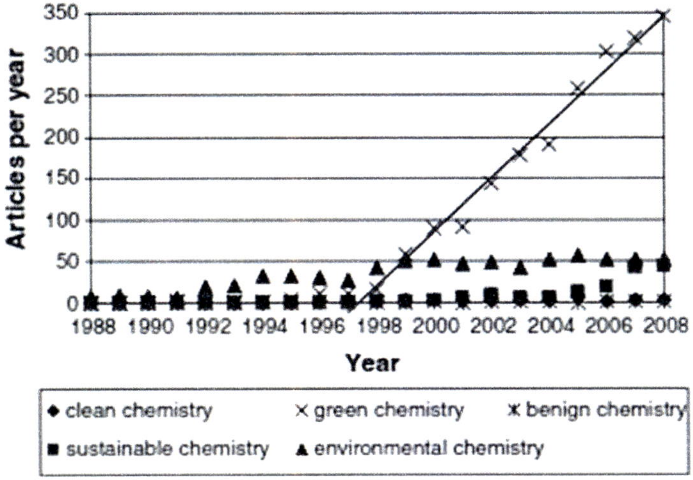

Fig. 1 Usage of the idiom "green chemistry" in research published from 1988 to 2008. *Source* ISI web of knowledge: web of science

Currently, this science is in a stage of growth and prosperity. We believe that it has become a recognized science in the whole world, and interest in it requires and works to develop new conceptual approaches to the development of twenty-first-century technologies to make them safer. In this regard, there are many studies that have been able to link this science with other fields of chemistry, as well as early achievements in this division of science [13].

The objective of this new science has been defined as the development of principles or synthesizing and applying the results of chemicals that minimize or expel hazardous (harmful) substances away and are not concentrated in the environment [14] as summarized in Fig. 2.

In recent years, many governments have taken measures to prevent the synthesis and manufacture of pollutants causing damage to our environment. For example, there are currently more than 70,000 chemical compounds in circulation in the USA. Of course, pollution control agencies have a limited ability to monitor the release of all of these chemicals, but these agencies focus on a few chemicals that threaten humans and the environment. Figure 3 shows the amount and percentage of chemicals released into the environment in 1992.

Bioproducts are split into two categories: bioenergy and non-bioenergy. Table 1 shows the classification of vital products.

Finally, between 1997 and 2011, the International Union of Pure and Applied Chemistry (IUPAC), with ACS and GCI, held four conferences on green chemistry. The conferences included adopting green policies and education in green chemistry.

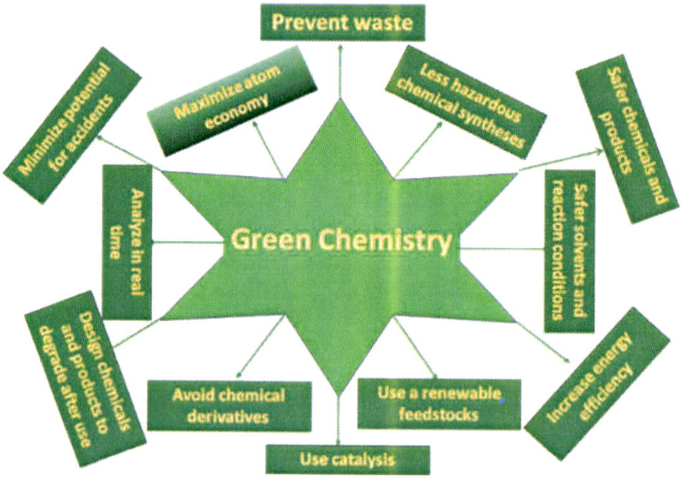

Fig. 2 Target and the concepts of green chemistry as represented in [14]

Fig. 3 Quantity and percentage of chemicals released to environment in 1992

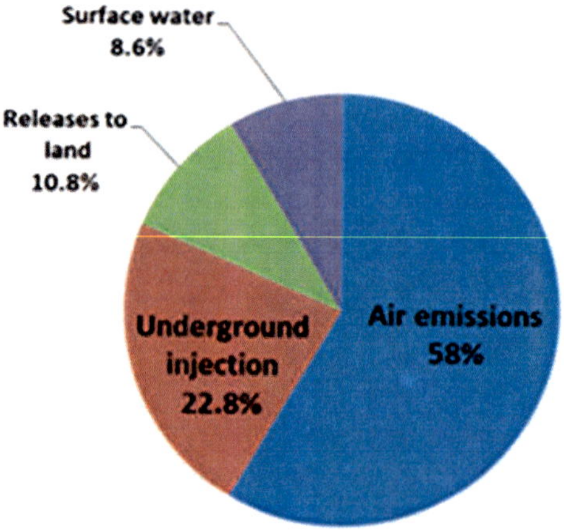

3 Green Chemistry for the Environment

Within the confines of the broad framework of chemistry and the environment, the chemistry of the surrounding environment must relate to the consequences and fate of chemicals and synthetic substances present in the surrounding [16]. Although facile and somewhat restricted, this picture allows us to think about environmental chemistry as a science that observes, understands and predicts measures. Consequently, monitoring, interpretation and prediction of pollution problems will be supplemented by scientific and technological data that attempt to reduce the degree of pollutant chemicals [17]. After that, the regression may be due to treatment and disability

Table 1 Categorization and segmentation of bioproducts

Class	Section	Products
Bioenergetic	Biofuels	Liquid fuels, such as ethanol, methanol, biobutanol, fuel oil and biodiesel
	Bioenergy	Solid/liquid biomass to generate heat and electricity
	Biogas	Gaseous fuel, such as biogas, biomethane and synthetic gas to create heat and electricity
Non-bioenergetic	Bioplastics/biopolymers	Bioplastics from vegetable oils and C6 and C5 sugars
	Biocompounds	Extracted from agricultural and some forest fibers (hemp, linen) which use in the production of some parts in the vehicles
	Biochemicals	Basic chemical products and industrial specialties, including grouts, paints, lubricants and solvents
	Biomedicine	Pharmaceutical products such as the antibodies and vaccines produced by genetically modified plants; Cosmetic product such as soaps, body creams and lotions

Source Gobina [15]

techniques, respectively (Fig. 4), however, the waste product may be viewed as still not present [17, 18].

Green chemistry chiefly needs to deal with the production of chemicals; these clean technologies must be inevitably hinged on chemical preparative conversions. According to this view, green chemistry appears to be constrained by preparatory chemistry [20, 21]. The correlation of the organic chemical nature of many pollutants explains why organic chemists tend to think of controls that harness green chemistry to serve organic chemistry. It is clear that environmental degradation is

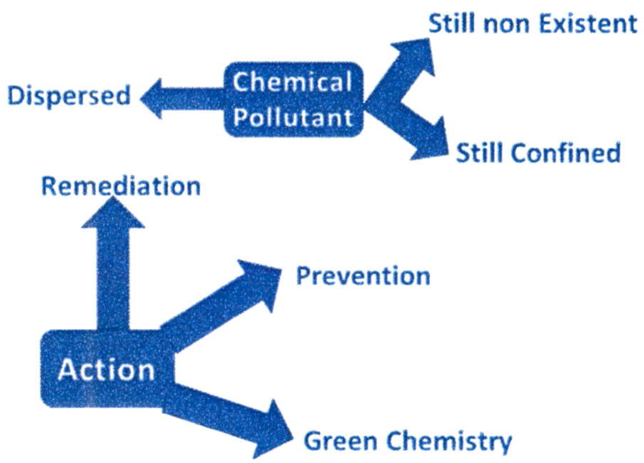

Fig. 4 Alternative chemistry and the lofty goal for its existence as represented [19]

because of chemicals and hence an understanding of the mechanism of this degradation and access to its structural fundamentals at a particular location should be taken into account when designing any innovative molecular structure if released into the system around us [22]. The apprehension of environmental chemistry should also prioritize the objectives of green chemistry for its total utilization as well as chemicals and chemically produced materials that must be replaced by other useful and not harmful substances. Moreover, the concept of green chemistry is not only about the development of clean and safe synthesis, but also applies to the reduction or suppression of the generation or introduction of any harmful, hazardous or chemically produced substances into the biosphere [23]. Green chemistry also reduces the chemical processes that ultimately lead to the production of millions of tons of harmful waste, which in turn require treatment or recycling for disposal [24]. Some giant companies of chemical and petrochemical industries spend annually about $ 1 billion on matters related to the preservation of the environment, which is spent on research and development, hence to give prominence and effectiveness of green chemistry in addressing the problem of pollution. Instead of treating and controlling pollution, we focus on preventing and stopping the formation of pollutants and harmful substances from the ground [25]. This will lead to good economic performance for poor and developing countries, and these countries can support their development projects. Sustainable development is achieved by having a convergence between the concept of this development and green chemistry and achieving economic and environmental prosperity. Also, increasing reliance on renewable energy sources is to obtain the raw materials needed in the chemical industry, and this supports biotechnology. This support is achieved by providing useful technologies in the use of energy for plants or in bioenergy transfers, such as biodiesel production.

Natural sciences are geared toward facing new challenges strongly as they provide the basic knowledge necessary to promote appropriate new technology, in order to promote sustainable development. This interest became particularly directed at chemistry, which in turn led to the emergence of green chemistry about 20 years ago [26]. Chemistry has a discriminatory role in science, for two basic reasons. First, industrial chemistry that was developed and practiced throughout the twentieth century was a strong contributor to environmental disasters due to the spread of toxic pollutants from manufacturing processes and their deposition in the environment. This requires checking the materials used and manufactured from all our vital and daily products. There is also a wide range of manufactured materials that support our daily life from household cleaners, cosmetic products, plastic materials, textiles and building materials, as well as chemicals used in medicine, agriculture and fuel for transportation, which requires scrutiny of the materials used and manufactured from all of our daily vital products [16]. Despite this technology that chemistry contributed to, it was a double-edged sword. The increased progress has contributed to the simultaneous degradation of the biosphere's environment including human health through pollution and unsustainable use of natural resources. Of course, chemistry is a branch of science that aims to search for innovations that contribute to processes of environmental balance, which improve the general quality of life and reduce environmental collapse and the atmosphere [16].

4 Green Chemistry and Other Chemistry

Certain chemicals are usually treated as "hazardous waste" (such as pesticides, oils, lubricants and minerals), whether they are produced individually at the production site or during wastewater treatment[27]. Although chemicals are important for development and are responsible for advances in health, some types of chemicals, such as persistent organic pollutants, can accumulate to dangerous levels in humans and the surrounding environment, with adverse effects on reproduction, growth, immunity and hormones. Thus, it can turn inside the organism or its surroundings into carcinogens and become disastrous. Substances that negatively affect the endocrine glands, such as dioxins, furans, polychlorinated biphenyls and DDT are considered the most dangerous chemicals to human health. As well as, the substances that are likely to cause endocrine disorders (phthalate and bisphenol-A) can occur during food, water, dust, air and skin contact with various substances. These materials, as mentioned above, can appear as chemical additives that were used daily in electronics, electrical equipment, household cleaning products, fabrics and furniture [28]. Although information is available on only a few aspects of chemical exposure, it is estimated that some 107,000 people die annually from asbestos exposure. About 674,000 people died from lead exposure in 2010 [29]. It is also well known to the scientific community that lung cancer kills 1.6 million people annually, and it is estimated that about 36 per cent of these 568,000 deaths are associated with occupational exposure to harmful chemicals and air pollution [30]. Overexposure and improper use of pesticides contribute to the poisoning of at least 3 million people, especially among poor rural workers. Activities such as mining are particularly detrimental to the health of fragile communities in Africa, Latin America and Asia [31].

In particular, children are vulnerable to these harmful effects from those chemicals. Products of these substances may precipitate in the brain and lead to severe consequences, as they may cause severe mental illness and mental retardation. For example, exposure to mercury and lead in the womb and during the early stages of a child's brain may lead to mental retardation, vision and hearing loss and developmental barriers [32] and at the same time, contaminate the agricultural soils with heavy metals such as lead, chromium and cadmium and interfere in farm operations by using sewage sludge as fertilizer and using pesticides containing minerals. The agricultural domain is additionally the most users of antibiotics within the world, exploiting 70% of all factory-made antibiotics. The extravagant use of pharmaceutical merchandise (antibiotics and antimicrobial agents), both in human medicine and in veterinary practices, can participate in the emergence of resistant strains of microbes in humans, thus posing a serious health threat [33].

Global solid waste production is predicted to double by 2025 at different rates by region and country; the upper the financial gain level and also the rate of urbanization, the larger the number of solid waste generated [34]. The fastest growing waste type is electronic waste (e-waste). The world produced approximately 42 million tons of these wastes in 2014, and this figure continues to rise. Although recycling is positive, improper dismantling and final disposal of that e-waste containing various hazardous contents are unsafe. These toxins emitted from these residues as well as deposited in the soil and suspended in air and water can have severe effects on the endocrine glands and thus deterioration of human life [35].

Continued reduction in the quantities of hazardous wastes and other transboundary movements is also critical, as this is a clear constraint especially for developing countries that have wasted but cannot manage or dispose it safely [36].

5 Benefits of Green Chemistry

Sustainable chemistry plays a prominent role in the real and actual association of biochemistry and medicinal plants. Not only do the chemical compounds extracted from plants provide the body with the necessary elements and treat damaged cells, but they are safer than their organic and laboratory chemical preparations and have no chemical effects and therefore are called green chemistry, i.e., inspired by nature.

It also plays key roles in antioxidants in many areas of current green chemistry research to understand its bioactivity as well as to seek inexpensive and renewable sources as well as to develop green technologies to isolate or synthesize powerful antioxidants that have a prominent effect in vivo without damage [37]. The results may lead to new antioxidant treatments that reduce the effects of environmental pollution and contribute to the prevention of many diseases that affect humans and bode well for the future of mankind [38]. There are different patterns of free radicals, and there is something in common with all of them; a single electron generates an unbalanced electrical (electrical) charge. Thus, the free root seeks to correct by finding another electron to restore its balance. During this process, the free radicals can steal electrons from stable molecules whose proximity coincides with them; this may lead to the initiation of a chain reaction and a cascade of molecular sabotage begins. Antioxidants add an enormous number of electrons to the blood vessels, thus balancing the free oxygen atoms, i.e., the cell being depleted returns the electron to its normal balance and again becomes able to function. Additionally, antioxidants remove free radicals and free nitrogen after they are formed, resist and convert them into other forms of oxidation inhibiting their formation according to transitional orders that stimulate the production of these cracks [38]. Stray oxygen atoms or free radicals are highly effective chemical compounds because they lose one of their electrons and look hard for the electron. Thus, they contain non-double electrons in their outer orbit [39]. This makes them move to try to recover the lost electron through the body from other body compounds [40]. Thus, they are causing stray oxygen atoms to damage tissues and cells by breaking down the protective cellular barrier by interaction with phospholipids in the cell membranes [41]. This leads to damage to the DNA as well as the collagen layer of the skin because these roots merge with the cells of the body and tissues to obtain the lost electron that is active in the body to a large extent during the functions of normal metabolism. The roots interact with the oxidizing cells in the body and hinder the ability of cells to do their work so that they play a big role in many diseases, including premature aging and cancer.

Conversely, taking antioxidants can reduce exposure to free radicals [42]. One of the most popular green antioxidants is polyphenols, which are widely available in plants and generally found in organic extracts [42]. Polyphenols are renowned for their antioxidant properties, are antioxidants that can stop the spread of oxidative reactions [43]. Moreover, they help prevent diseases such as inflammation, atherosclerosis and cancer development and are also widely used in cosmetics production for its role in protecting the skin and maintaining its freshness [44].

Studies have shown that tea polyphenols and other active substances such as flavanols, anthocyanins, flavonoids, flavones and phenolic acids have the effect of detoxifying the circumferential waste and anti-radiation and can effectively block radioactive substances in the bone marrow, which can rapidly secrete strontium 90 and cobalt 60 [45]. The results of studies on natural antibiotics such as grape polyphenols, a water-soluble plant substance, methanol, ethanol and other organic solvents have proven to have a strong effect on these updated toxins. Studies have also found that these polyphenols consist of phenolic acids (Epicatechin), tannin condensation along with the preceding substances [46].

By nature, more than 50 antioxidant species have been found in grape seed extracts, a product that contains the most antioxidant types that serve as natural biological treatments devoid of toxic chemicals [44]. Apple polyphenols (apple extract) have been realized through studies that immature apples contain polyphenols 10 times the content of ripe apples. So, the polyphenols extract of apples using immature fruit is very suitable for this purpose. Apple polyphenols have antioxidant and detoxification function, retain the freshness, protect color, prevent loss of vitamins, etc., so they can be used in food processing, improving product quality and in many good alternative remediation [47]. Simultaneously, apple polyphenols also contain an assortment of health functions, such as obviating tooth decay, averting of hypertension, preclusion of allergies and anti-tumor properties. Anti-mutations impede the absorption of ultraviolet radiation, so it can be used in the health food industry and cosmetics. Application of polyphenols in health food and functional food as additives need only 50–500 ppm to have a full effect [48].

Clean chemistry uses green nanotechnology, which aims to produce clean technology using nanotechnology applications to enhance the environmental sustainability of processes currently producing negative impacts on the environment. There have also been studies on the packaging of antioxidants in a biodegradable fatty cortex in nanoparticles. These particles are fashioned to preserve antioxidants from unwanted metabolic decomposition and also the arrival of active substances to the desired target [49].

One example is the encapsulation of curcumin in poly nanoparticles (lactic acid and glycolic) to increase their solubility and bio-efficacy by oral administration, allowing smaller doses of curcumin to be as effective as larger doses, thus overcoming undesirable effects [50]. Resveratrol was also loaded into nanoparticles to increase their bioactivity [51].

The pros of using green nanoparticles are the shelf life of those nanoparticles used in the manufacture of antioxidants should be considerably long. This makes the technique successful, performing its purpose and has found that nanoparticles are stable for several hours, giving them strength and superiority over others [52].

The traditional approach of society and industry through environmental policies is to reduce risks by reducing exposure time to chemicals when stabilizing the risk of substances and using data on toxicity of substances and knowing the effectiveness of controlling exposure time used can maneuver until the risk is reached to a certain extent as acceptable level [52]. Another disadvantage of controlling exposure time is to reduce the risk that the use or release of a chemical compound may affect individuals who do not use these means of control. For example, workers may wear gloves or goggles, etc., to protect themselves from exposure to high levels of a certain substance known to have a dangerous effect, but the situation is different for the exposure of safe individuals who are downstream of the wind which carries this hazardous substance and those who do not have methods of protection from exposure controls [53].

Given the uncertainty of chronic effects, the effects of bioaccumulation and the very high synergetic effects of a large number of compounds in our current information, the use of exposure control to minimize risks to society is now in question. Besides, the limited control in the time of exposure is that this control may not be successful [54]. No face protection system, respirators, glasses, gloves or protective clothing are perfect, whereas such equipment used to reduce exposure time may fail to perform its mission and therefore the personnel [55]. Therefore, the users are exposed to the maximum risk of handling these hazardous materials. Conversely, risk reduction through green chemistry using appropriate procedures cannot be countered by the potential for failure. Using various techniques and methods such as the use of safe starting materials, alternative catalysts and alternative solvents, the risk of compounds will be reduced. Therefore, the use of non-harmful substances will not have potential toxicity and thus will not counteractively affect human health or the environment [56].

The advantage of green chemistry is that by addressing the solution to reduce risks reap economic benefits by reducing the cost of the starting materials for production, reduce the time of chemical reactions, increase the proportion of chemical conversion required, high selectivity and speed of separation of compounds as well as reducing energy used [57].

6 Future of Green Chemistry

In the academic and professional context, there are many conferences and seminars that discuss the future of this clean chemistry, including the Green and Sustainable Chemistry Conference, which brings together academic and business representatives to showcase business, exchange ideas and learn. The realization of such a design comes if everyone plays a role, it does not matter if it is small, and when we join all the parts it will be a great thing. Finally, we must have positive views of the future of green chemistry because it encompasses the future of our world [58].

7 Conclusion

With the rising cognizance of the prominence of preventing pollution among scientists and the industry community, the need to reduce the sources of this pollution is by changing the design of preparation methods and production processes so as not to lead to the production of waste in the first place. Even the absence of residues is not enough to protect the ambience which is surrounding humans but chemical products must be designed so that hazardous substances are not used during their production. By excluding the use of such materials, the impact on the environmental gains will be multiplied as a result of these new design processes. These choices should be made after careful analysis of the commercial product required. It is only natural that only those chemists (specialists in the synthesis of materials) who can rely on the chemical industry to discover and update commercial technologies can implement the nature of these choices. These updates require not only the preservation of the product and the improvement of its quality, but the development of new ways to synthesize these compounds so that manufacturing them is less expensive and less harmful to the environment. The protocol for the production of any new chemical shall include these principles. It is clear from the above; the objectives of green chemistry mechanisms have reduced environmental risks and health damage by adopting new methods, both at the level of research laboratories and industrial processes. Another reason for the intensive application of green chemistry by the chemical community is that it is based on basic molecular science as a way of solving environmental problems and does not address problems by bandage or patchwork to reduce risk.

References

1. Anastas PT, Beach ES (2009) Changing the course of chemistry. In: hanging the course of chemistry, green chemistry education. ACS Symposium Series American Chemical Society, Washington DC
2. Anastas PT, Zimmerman JB (2016) The molecular basis of sustainability Chem 1(1):10–12
3. Lu J-G, Li X, Zhao Y-X et al (2019) CO_2 capture by ionic liquid membrane absorption for reduction of emissions of greenhouse gas. Environ Chem Lett 17(2):1031–1038
4. Li N, Tompsett GA, Zhang T, Shi J et al (2011) Renewable gasoline from aqueous phase hydrodeoxygenation of aqueous sugar solutions prepared by hydrolysis of maple wood. Green Chem 13(1):91–101
5. De Simone R, Chini MG, Riccio R et al. Structure-based discovery of inhibitors of microsomal prostaglandin E2 synthase – 1, 5-lipoxygenase and 5-lipoxygenase-activating protein: promising hits for the development of new anti-inflammatory agents. J Med Chem 54(6):1565–1575
6. Armor JN (1999) Striving for catalytically green processes in the twenty first century. Appl Catal A Gen 189(2):153–162
7. Contractor RM, Garnett DI, Horowitz HS et al (1994) A new commercial scale process for n-butane oxidation to maleic anhydride using a circulating fluidized bed reactor. Stud Surf Sci Cata 82:233–242
8. Clark JH (2016) Green and sustainable chemistry: an introduction. In: Summerton L, Sneddon HF, Jones LC, Clark JH (eds) Green and sustainable medicinal chemistry: methods, tools

and strategies for the twenty first century pharmaceutical industry. Royal Society Chemistry, London, pp 1–11
9. Anastas P, Eghbali N (2010) Green chemistry: principles and practice. Chem Soc Rev 39(1):301–312
10. Max M (2017) Advanced green chemistry-Part 1: Greener organic reactions and processes, vol 3. World Scientific
11. Lewis H, Gertsakis J, Grant T, Morelli N, Sweatman A (2017) Design + environment: a global guide to designing greener goods. Routledge, London
12. Woodhouse EJ, Breyman S (2005) Green chemistry as social movement? Sci Tech Hum Values 30(2):199–222
13. Mitarlis Ibnu S, Rahayu S, Sutrisno (2017) Environmental literacy with green chemistry oriented in twenty first century learning. AIP Conf Proc 1911(1):20020
14. Anastas PT, Warner JC (1998) Green chemistry. Frontiers 640:1998
15. Gobina E (2014) Biorefinery products: global markets. BCC
16. Scheringer M (2017) Environmental chemistry and ecotoxicology: in greater demand than ever. Environ Sci Eur 29(1):3
17. Stieger G, Scheringer M, Ng CA, Hungerbühler K (2014) Assessing the persistence, bioaccumulation potential and toxicity of brominated flame retardants: data availability and quality for 36 alternative brominated flame retardants. Chemosphere 116:118–123
18. Reif DM, Sypa M, Lock EF et al (2012) ToxPi GUI: an interactive visualization tool for transparent integration of data from diverse sources of evidence. Bioinformatics 29(3):402–403
19. Mestres R (2005) Green chemistry-views and strategies*(5 pp). Environ Sci Pollut Res 12(3):128–132
20. Sheldon RA (2012) Fundamentals of green chemistry: efficiency in reaction design. Chem Soc Rev 41(4):1437–1451
21. Sheldon RA (2016) Engineering a more sustainable world through catalysis and green chemistry. J R Soc Interf 13(116):20160087
22. Crawford SE, Hartung T, Hollert H et al (2017) Green toxicology: a strategy for sustainable chemical and material development. Environ Sci Eur 29(1):16
23. Maertens A, Anastas N, Spencer PJ et al (2014) Food for thought: green toxicology. Altern Anim Exp ALTEX 31(3):243–249
24. Abdel-Shafy HI, Mansour MSM (2018) Solid waste issue: sources, composition, disposal, recycling, and valorization. Egypt J Pet 27(4):1275–1290
25. Hjeresen DJ (2004) Green chemistry: the impact on water quality and supplies. In: Water and Sustainable Development: Opportunities for the Chemical Sciences: A Workshop Report to the Chemical Sciences Roundtable
26. Maxwell K, Benneworth P (2018) The construction of new scientific norms for solving grand challenges. Palgrave Commun 4(1):52
27. Abdel-Aziem SH, Abd El-Kader HAM, Ibrahim FM, Sharaf HA, El makawy AI (2018) Evaluation of the alleviative role of Chlorella vulgaris and Spirulina platensis extract against ovarian dysfunctions induced by monosodium glutamate in mice. J Genet Eng Biotechnol 16(2):653–660
28. Global status report on noncommunicable diseases (2014) World Health Organization
29. Lim SS, Vos T, Flaxman AD et al (2012) A comparative risk assessment of burden of disease and injury attributable to 67 risk factors and risk factor clusters in 21 regions, 1990–2010: a systematic analysis for the Global Burden of Disease Study 2010. Lancet 380(9859):2224–2260
30. Prüss-Ustün A, Wolf J, Corvalán C, Neville T, Bos R, Neira M (2016) Diseases due to unhealthy environments: an updated estimate of the global burden of disease attributable to environmental determinants of health. J Publ Health 39(3):464–475
31. Thundiyil JG, Stober J, Besbelli N, Pronczuk J (2008) Acute pesticide poisoning: a proposed classification tool. Bull World Health Organ 86:205–209
32. Mellon M, Fondriest S (2001) Union of Concerned Scientists. Hogging it: estimates of animal abuse in livestock. Nucleus 23:1–3

33. Anastas ND, Maertens A (2018) Integrating the principles of toxicology into a chemistry curriculum. In: Green Chemistry, Elsevier, pp 91–108
34. Gwinn M et al (2017) A public health perspective on twenty first century risk assessment. Am J Publ Heal 107(7):1032–1039
35. Torok B, Dransfield T (2017) Green chemistry: an inclusive approach. Elsevier
36. Attina TM, Trasande L (2013) Economic costs of childhood lead exposure in low-and middle-income countries. Environ Health Perspect 121(9):1097–1102
37. Beach ES, Cui Z, Anastas PT (2009) Green chemistry: a design framework for sustainability. Energy Environ Sci 2(10):1038–1049
38. Khan KA (2018) Natural products chemistry: the emerging trends and prospective goals. Saudi Pharm J 26(5):739–753
39. Halliwell B, Gutteridge JM (1984) Oxygen toxicity, oxygen radicals, transition metals and disease. Biochem J 219(1):1
40. Jianhui B (2010) Study on surface O_3 chemistry and photochemistry by UV energy conservation. Atmos Pollut Res 1(2):118–127
41. Ghosh N, Das A, Chaffee S, Roy S, Sen CK (2018) Reactive oxygen species, oxidative damage and cell death. In: Immunity and inflammation in health and disease, Elsevier
42. Altemimi A, Lakhssassi N, Baharlouei A, Watson D, Lightfoot D (2017) Phytochemicals: extraction, isolation, and identification of bioactive compounds from plant extracts. Plants 6(4):42
43. Zuin VG, Ramin LZ (2018) Green and sustainable separation of natural products from agro-industrial waste: challenges, potentialities, and perspectives on emerging approaches. In: Chemistry and chemical technologies in waste valorization, Springer, pp 229–282
44. Mojzer EB, Hrnčič MK, Škerget M, Knez Ž, Bren U (2016) Polyphenols: extraction methods, antioxidative action, bioavailability and anticarcinogenic effects. Molecules 21(7):901–938
45. Panche AN, Diwan AD, Chandra SR (2016) Flavonoids: an overview. J Nutr Sci 5:47–61
46. Suwal S, Marciniak A (2018) Technologies for the extraction, separation and purification of polyphenols–a review. Nepal J Biotechnol 6(1):74–91
47. Wilczyński K, Kobus Z, Dziki D (2019) Effect of press construction on yield and quality of apple juice. Sustainability 11(13):1–15
48. Hyun TK, Jang K-I (2016) Apple as a source of dietary phytonutrients: an update on the potential health benefits of apple. EXCLI J 15:565–569
49. Liguori I, Russo G, Curcio F et al (2018) Oxidative stress, aging, and diseases. Clin Interv Aging 13:757–772
50. Malik P, Ameta RK, Singh M (2014) Preparation and characterization of bionanoemulsions for improving and modulating the antioxidant efficacy of natural phenolic antioxidant curcumin. Chem Biol Interact 222:77–86
51. Richard PU, Duskey JT, Stolarov S, Spulber M, Palivan CG (2015) New concepts to fight oxidative stress: nanosized three-dimensional supramolecular antioxidant assemblies. Expert Opin Drug Deliv 12(9):1527–1545
52. Hecht F, Pessoa CF, Gentile LB, Rosenthal D, Carvalho DP, Fortunato RS (2016) The role of oxidative stress on breast cancer development and therapy. Tumor Biol 37(4):4281–4291
53. Little JC, Weschler CJ, Nazaroff WW, Liu Z, Hubal EAC (2012) Rapid methods to estimate potential exposure to semivolatile organic compounds in the indoor environment. Environ Sci Technol 46(20):11171–11178
54. Connon RE, Geist J, Werner I (2012) Effect-based tools for monitoring and predicting the ecotoxicological effects of chemicals in the aquatic environment. Sensors 12(9):12741–12771
55. Balbus JM, Boxall ABA, Fenske RA, McKone TE, Zeise L (2013) Implications of global climate change for the assessment and management of human health risks of chemicals in the natural environment. Environ Toxicol Chem 32(1):62–78
56. Welton T (2015) Solvents and sustainable chemistry. Proc R Soc A Math Phys Eng Sci 471(2183):20150502
57. Rogers L, Jensen KF (2019) Continuous manufacturing–the green chemistry promise? Green Chem 21(13):3481–3498

58. de Barros JO, Kogawa AC, Salgado HRN (2017) Short-term stability study of doxycycline tablets by high performance liquid chromatography, spectrophotometry in the ultraviolet region and turbidimetry. J Pharm Sci Exp Pharmacol 2018:43–49

Chapter 2
Principles of Green Chemistry

Amal I. Hassan and Hosam M. Saleh

1 Introduction

The green chemistry approach strives to achieve sustainability at the molecular level. Because of this goal, it is not surprising it has been applied to all industry sectors. From aerospace, automobile, cosmetic, electronics, energy, household products, pharmaceutical to agriculture, there are hundreds of examples of successful applications of award winning and economically competitive technologies [1]. The concept of green chemistry has had this large impact due to the fact that it goes beyond the research laboratory in isolation and has touched industry, education, environment, and the general public.

The green chemistry concept presents an attractive technology to chemists, researchers, and industrialists for innovative chemistry research and applications. Sustainable development is the development that "meets the needs of the present without compromising the ability of future generations to meet their own needs [2]. The imperative of sustainable development is leading to a wholesale rethinking of the way that humankind needs to approach its continued existence on earth. The green chemistry movement is bringing guidance as to how chemistry should be practiced now and in the future [3].

A. I. Hassan · H. M. Saleh (✉)
Radioisotope Department, Nuclear Research Center, Atomic Energy Authority, Dokki 12311, Giza, Egypt
e-mail: hosamsaleh70@yahoo.com

A. I. Hassan
e-mail: virtualaml@gmail.com

G. Anilkumar and S. Saranya (eds.), *Green Organic Reactions*, Materials Horizons: From Nature to Nanomaterials, https://doi.org/10.1007/978-981-33-6897-2_2

2 Basic Concepts of Green Chemistry

The basic principle, benign by design, emphasized that both the product and the process used to produce it should conform to the basic rules of sustainability. In their seminal book in 1998, Anastas and Warner established the major principles of green chemistry. Although since then several "new principles" have been added to the list, the original list is still applicable (Fig. 1). They are a guiding framework for the design of new chemical products and processes, applying to all aspects of the process life cycle from the raw materials used to the efficiency and safety of the transformation, the toxicity, and biodegradability of products and reagents used. They were summarized recently into the more convenient and memorable acronym productively [4].

These principles, which promote the clean production and green innovations, are already relatively widespread for industrial applications—as the production of agrochemicals—particularly, in countries with a well-developed chemical industry and with strict control over the emission of pollutants. They are based on the assumption that chemical processes with potential to impact negatively on the environment will be replaced by less polluting or non-polluting processes [5].

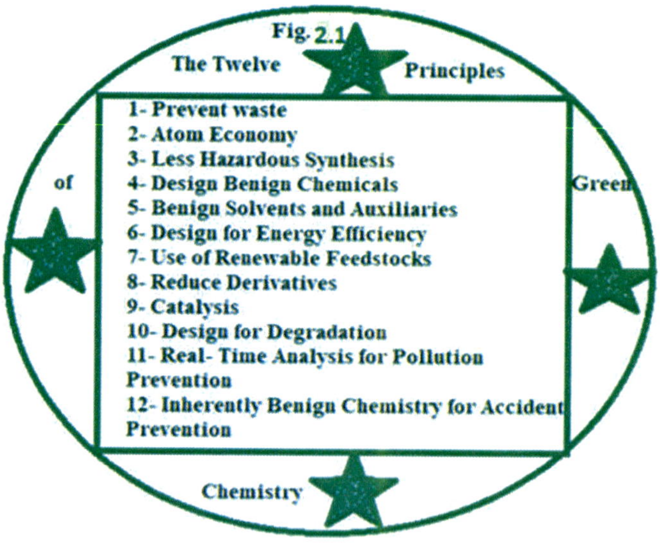

Fig. 1 Twelve principles of green chemistry [3]

2.1 *Preventing Waste Formation is Better Than Processing after it is Created*

In the past, the economics of chemical manufacturing was mainly concerned with the cost of raw materials and operating expenses to obtain the most products without taking into account any other aspects of human health and environmental safety [6]. But in the past 20 years, a major new factor has emerged—after a multitude of health and environmental disasters—added to the economics of industrialization: the cost of waste treatment and toxic waste disposal. The greater the risk of these wastes, the higher the cost of treatment and this applies to the large factory or small academic laboratory. Chemical industry men did not care about the problem of waste disposal or treatment and their logic was that their knowledge of hazardous materials enables them to deal with them and avoid their dangers [7]. It is an unrealistic logic similar to saying that a doctor's knowledge of treatment methods does not make him avoid diseases. These risks often cost more than addressing them. The criterion by which a particular chemical is used is the type of waste it produces and the extent of its damage. The burden of dealing with these wastes is that they cannot be properly treated, require high energy or cost in time or expenditure, or require high technology to separate them from the product and dispose of them or turn them into harmless forms [8]. Although these standards are important for waste, the most important assessment is their impact on human life and environmental safety. The mere fact is that the processes that produce residues need to be separated, processed, and disposed of. The use of hazardous materials requires special handling, means of protection, and precise precautions. In a broader sense, the type of material used and the method of manufacture should be assessed on the basis of the quality of the waste in light of the above considerations. In some cases, however, the damage may not be so great that we can completely change the process, but we cannot deny the existence of such damage and appreciate its limits [6]. The situation is different in universities and small research laboratories that suffer from the cost of disposing of wastes resulting from interactions—which limits scientific innovation and the only solution to reduce the cost of hazardous waste disposal in these laboratories is to follow green chemistry techniques, which reduce or limit spending and provide appropriate mechanisms to control the risks associated with chemical reactions [9].

2.2 *Achieve the Maximum Atom Economy*

Most of the twentieth-century references in organic chemistry did not write equilibrium chemical equations and rarely mentioned in these equations or hardly mention at all secondary products that accompany the basic output. The value and efficiency of the reaction are evaluated on the basis of the yield of the product of greatest interest, while by-products are worthless and often ignored and neglected. In some cases, the by-product is formed with larger masses and sizes than the desired product [10]. The chemical calculation is based on the moles of the reactants versus the moles of the products. If the mole of the reactant produces a mole of the resulting substance, the reaction yield is 100% or we consider it to be fully efficient despite the production of secondary substances in quantities that may exceed the primary product in the reaction equation.

Atom economy (AE) is currently used, where we measure the degree to which the reactants entered the final product [11–13]. Some known interactions can be assessed from the standpoint of AE as follows:

$$\%\text{ atom utilization} = \frac{\text{MW of desired product}}{\text{MW of desired product} + \text{waste products}} \times 100$$

Atom economy (atom efficiency is also used), first described in 1991 by Trost, is defined by the ratio of the molecular weight of the product and the sum of the molecular weights of all substances consumed in the stoichiometric equation of a reaction. Commonly it is expressed as a percentage. It is important to highlight that AE is based on the theoretical reaction (i.e., no unexpected by-products are factored in) and 100% theoretical yield. Therefore, the AE is the best possible scenario and can be used to assess a reaction at the theoretical level. For example, if a reaction scheme does not involve the formation of any expected by-product, the AE is 100% (Scheme 1). However, it is worth noting that it cannot be used exclusively to describe the environmental impact of a reaction: It may be that a reaction with 100% AE yields unexpected by-products (e.g., stereo- or regioisomers) that would decrease the actual AE [11]. Thus, a highly selective reaction with 80% theoretical AE and no unexpected by-products may have a less environmental impact than a 100% AE reaction that is accompanied by extensive unexpected by-product formation [14]. Although these are the two most important measures to describe the efficiency of a chemical process, several alternative metrics have been proposed, such as the reaction mass efficiency (RME) defined as the mass ratio of the obtained product to the total mass of the reactants, thus, incorporating the actual percent yield into traditional AE calculations. Carbon efficiency is similar to RME but only considers carbon as a part of the product or starting materials and reagents. Mass efficiency (total mass of the materials used divided by the mass of product obtained given as a percentage) and effective mass yield (the ratio of the mass of the desired product and the total mass of non-benign reactants) are other available ways to describe the environmental impact of processes. There is a general agreement in the literature that the AE and E-factor

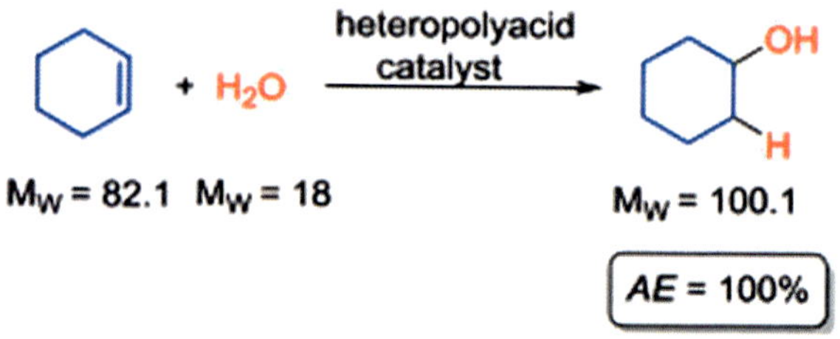

Scheme 1 Hydration of cyclohexene to cyclohexanol; a 100% atom economic process [15]

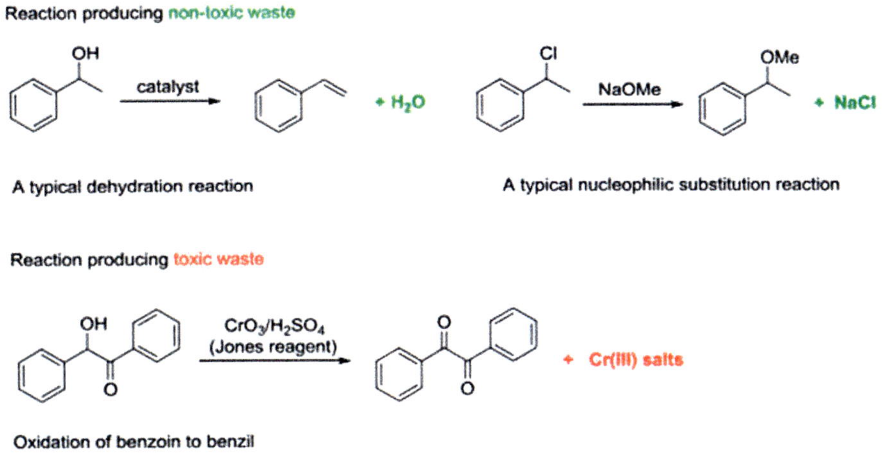

Scheme 2 Examples of chemical reactions producing toxic (red) and non-toxic (green) waste [15]

are the most applicable and widely used measures in many industries. Although the aforementioned metrics are able to estimate the impact of a process, there are other considerations to discuss, such as the chemical characteristics of the waste. Obviously, the ultimate process occurs with 100% AE and 0 E-factor; however, practical processes are different and mostly produce either expected or unexpected by-products that are considered waste. The nature of that waste is highly important. It is easy to realize that if the waste is water (e.g., dehydration reactions) or sodium chloride (nucleophilic substitutions) that are considered harmless, and the process is quite different from, to choose one example from many, the Jones oxidation of alcohols to ketones that generates a significant amount of chromium salt as waste (Scheme 2).

2.3 *Designing Less Dangerous Chemical Technologies*

The basic rule of green chemistry is to remove or minimize hazardous substances in all areas of chemistry without the need for laws to protect the environment. Green chemistry has also advanced through the skill and creativity that chemists currently possess with modern technology that preserves human health and the safety of the environment [16].

There are only two ways to reduce the risk of chemicals, by either reducing the concentration of the hazardous substance or reducing the exposure time of these substances [17]. Reducing chemical exposure takes many forms, including the use of protective clothing, gas masks, or techniques for controlling reactions. It is dangerous to disregard any kind of chemical hazards, no matter how small on the pretext that an expert chemist can deal with any substance no matter how dangerous it is (toxic or flammable) [18]. The realistic assessment of the hazardous substance is based on two factors: first, it is practically impossible to avoid exposure without increasing the cost (the cost of clothing, masks, etc.), thus, adding a cost that could have been dispensed with [19]. The second is a moral factor, meaning that as chemists we have a responsibility to innovate in the search for materials and pathways to produce environmentally friendly chemicals that are safe to human health. We must face hazardous substances from an environmental, legislative, and economic perspective and we have no choice. The environment has endured tremendous damage as a result of misuse and the short sight of some workers in the field of chemistry and in society in general and some have seen chemists as creative scientists and others are seen as polluters for the environment. Green chemistry reduces if not avoids hazardous substances [20]. One practical example, the use of alternative substitutes that have the least toxicity and are not dangerous to the environment, is the preparation of polycarbonate, a polymer resistant to high temperature and has many industrial uses. The old method: Phosgene (known for its high toxicity) was used as a starting material for the reaction with bisphenol-A and sodium hydroxide through condensation polymerization on the separating surface of two solvents, methylene chloride, and water. The risk of using a large amount of solvent CH_2Cl_2 (more than ten times more weight than the amount of polycarbonate produced) that it is a cancer-causing solvent, one of the 17 compounds to reduce their release into the environment, through the recommendations of the US Environmental Protection Agency EPA (Scheme 3) [21]. The process of recovering CH_2Cl_2 through this process requires a very high cost since the boiling point of this solvent is 40 °C and its solubility in water is very high.

Scheme 3 Phosgene process of polycarbonate showing its problems [22]

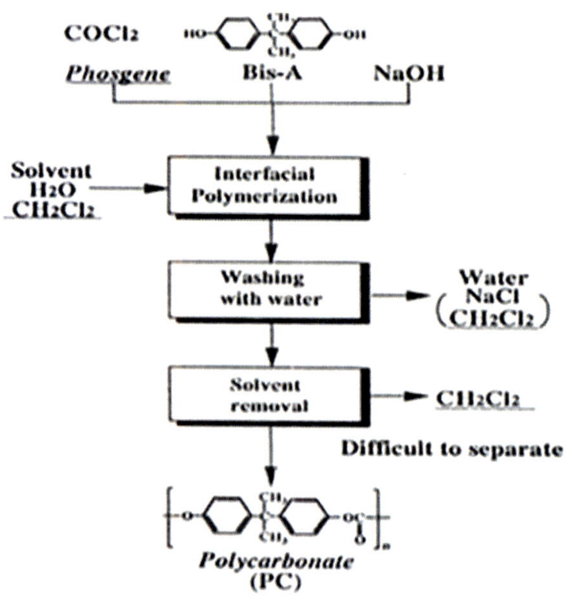

The improved method [22, 23] has avoided using phosgene and methylene chloride known as the melt process (Scheme 4). In this method, polycarbonate is obtained through the ester exchange reaction between bisphenol-A and diphenyl carbonate in the presence of a cofactor by excluding phenol.

The modern method of obtaining polycarbonate is an environmentally clean process based on polymerization in the solid state of amorphous polymers. The new method is called Asahi process and it consists of the following steps (Schemes 4 and 5).

Recent studies have shown that persistent exposure to polycarbonates can cause disruption of endocrine glands in humans. Bisphenol-A can significantly mimic the

Scheme 4 Melt process avoids using phosgene and methylene chloride in polycarbonate synthesis [22]

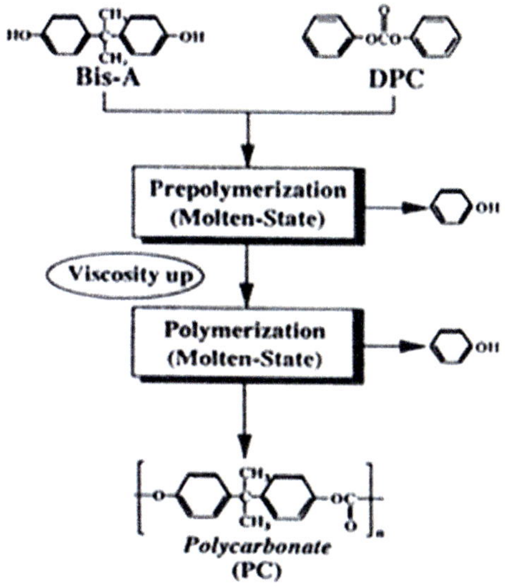

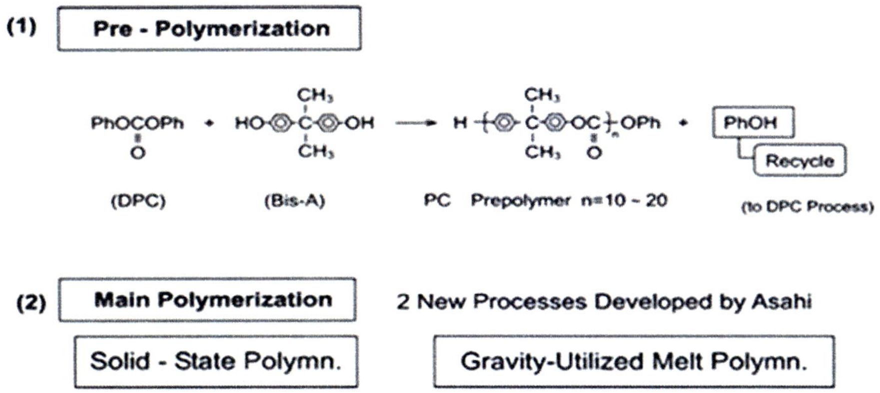

Scheme 5 Asahi polymerization process [22]

structure and function of estradiol with the ability to bind and activate estrogen receptors as a natural hormone. Other studies suggest that bisphenol-A may be a carcinogenic material. It can cause leukemia, breast cancer, and prostate tumor.

Polymers resulting from the use of the Asahi method are colorless, highly transparent and have other good properties such as thermal stability. It is resistant to solvents as well as water vapor. For example, after exposure to steam at 129 °C and for 20 h, this polymer retained its transparency. The new Asahi method succeeded in obtaining a high-quality polycarbonate compared to those resulting from the use of phosgene, and more importantly, the cost of production is competitive (much lower) than the method of phosgene use. In this example, the application of the basics of green chemistry is evident.

This new method complies with pollution prevention standards because it does not use highly toxic substances such as phosgene or carcinogenic solvents such as methylene chloride. Non-hazardous substances such as dimethyl carbonate were used with carbon dioxide as raw material and thus it became possible to obtain polycarbonates using carbon dioxide and bisphenols (Scheme 5) [24]. This process is more environmentally friendly because it does not use toxic phosgene and carcinogenic methylene chloride and use only carbon dioxide.

2.4 *Chemicals with High Functional Efficiency and Minimum Toxicity should be Designed*

More reliable chemicals can be designed simply by knowing the molecular structure of the material and through the advancement of analysis and measurement, mechanisms, or activity against tumors [25]. Scientists have been able to learn a lot about the relationship between the chemical compositions of biological effect [26]. The goal of designing safe chemicals is to balance the maximum functional efficiency with the least toxicity possible. Fortunately, these targets are attainable because chemists follow very closely the relationship between molecular structure and chemical properties [27].

2.4.1 Synthesis

Many of the new reactions that have been developed in the past decade add to the already existing green reactions that were discovered during the past century. Reactions based on cycloaddition, rearrangement, or multi-component reactions were already known and constitute one category of efficient reactions [28]. Cascade or tandem reactions, [29] enzymatic reactions, [30] C–H activation, [31] and metathesis [32] are rather new approaches and illustrate strong examples of cleaner, more efficient synthetic tools available to organic chemists.

This insistence in the preparation of safer chemicals in the present day is due to the progress in research that indicates the nature of chemical poisoning [33]. The performance of a chemical product was measured by the extent to which it achieved the function for which it was designed. It was not easy to know the stage at which toxicity occurs for the previous generation of chemicals, but now it is understood in detail the mechanism by which chemists can modify the composition of the substance to prevent or reduce the occurrence of adverse reactions [34]. Of course, the new reaction should not affect the functionality and performance of the desired product. This is by understanding the relationship between structure and function (such as the relationship between the presence of active groups and toxic effect), which facilitates control of the groups leading to the formation of harmful substances, bioavailability should also be reduced. If the substance has a detrimental effect on a particular organ of the body such as stomach, lung, or liver, it should not reach that organ by knowing its properties such as solubility or polarity, making it difficult to absorb through the biological walls of cells in the body. This organ will not be affected and thus is the best treatment [35].

2.5 *Usage of Benign Solvents and Auxiliaries*

There are many concerns about common solvents. Halogenated solvents such as methylene chloride, chloroform, ethylene perchloride, and carbon tetrachloride have long been found to be carcinogens. This applies, but to a lesser extent, to benzene and other aromatic hydrocarbons. Thus, despite the widespread use of solvents, the detrimental effect of most of them has been proven and should be minimized [36].

The harmful effect of solvent use in chemical reactions and even in daily life is not limited to human health in general but extends to the ecosystem in which we live [37]. Excessive use (especially fluorocarbons) has caused the erosion of the ozone layer in the stratosphere that protects the environment from impacts. Harmful to various rays such as cosmic rays, these substances have a high cleaning capacity and are used in the plastic industry as well as in refrigeration but have been shown to be primarily responsible for the erosion of the ozone layer. Volatile organic compounds are a large number of hydrocarbons and their derivatives and have been used as solvents in a large number of chemical applications that have led to the phenomenon of smog smoke [38]. This contributed indirectly to the respiratory diseases of many individuals, prompting many bodies in America and in many other countries of the world to enact legislation. This has cost many companies a lot of money to use alternative solvents or not to use solvents at all [39].

The most common method of separation and purification of materials is the recrystallization method, which requires energy or the addition of substances to change the solubility of the material. The other method used to separate the chemical compounds is by chromatography, which consumes fewer auxiliary substances but when used to separate large quantities require large amounts of solvents and accompanied by some of the substances that are hazardous. Therefore, we must take into account the waste and energy used when assessing the effect of additives on the entire chemical process [40].

Solvent-free reactions, systems that do not use solvents, are characterized by avoiding the risks that solvents cause to human health and the environment [41]. Many entities are involved in developing new methods to complete reactions without solvents such as smelting of reactants to ensure perfect mixing under ideal conditions for solvent interaction. All these methods avoid the use of solvents in chemical reactions [41].

Water systems have proven over the years their efficiency, selectivity, and environmental friendliness [42]. Water is undoubtedly the safest material on earth and therefore the safest solvent. Unlike conventional solvents, however, all cases using water as solvents should be examined separately [43].

The main problem in the use of solvents is their volatility, which negatively affects humans and the environment. One solution to this problem is the use of unrestricted solvents (non-volatile), which takes several forms, but they do the same purpose, which is the dissolution of materials without the use of volatile solvents, which protects human health and does not affect the environment [44]. This method is done by binding the solvent molecules to a solid support or solvent molecule directly on the core of a particular polymer. New polymers with solubility and less risk have been discovered [45]. One of the principles of green chemistry is not to resort to the use of auxiliary substances as much as possible, such as solvents and separation reagents or the use of harmless substances. Supercritical fluids are one of the alternatives adopted by green chemistry in order to reduce the dangers that occur during the traditional use of solvents and auxiliary substances [46]. Supercritical CO2 is an important example of these alternatives [47]. Whereas, CO_2 is present in the gaseous state under normal conditions and is converted to a solid state at high pressures above 5 atmospheres and temperatures below the triple point temperature of 65 °C.

Carbon dioxide has no toxicity and is not flammable and cheap and therefore can be used as an alternative to organic solvents [48]. Its use also provides a way to improve the way chemicals are prepared by accelerating the reaction and making the reaction more selective. As well as, the strength of the solvent in SC can be adjusted by the density of the medium in addition to the density-dependent properties such as the dielectric constant, viscosity, etc. Additionally, the pressure change leads to more control and selectivity of the reactions, which is formed by exposing the carbon dioxide molecules to a temperature [49]. One advantage of using carbon dioxide is that it cannot be oxidized and can therefore be typically used as a solvent in oxidation reactions.

SC carbon dioxide outperforms other solvents used in chemical processes. It can replace freons in the free radical polymerization process of fluorinated acrylates (free radical polymer) [50]. Jessop et al. [51] found that the use of carbon dioxide in the supercritical fluid state (SCF) accelerates the reaction of hydrogenation of carbon dioxide using homogeneous catalysts compared to other solvents to obtain formic acid or its derivatives in the presence of triethylamine or a mixture of triethylamine and methanol according to the following equations:

$$CO_2 + H_2 \xrightarrow[\text{SCF } CO_2]{\text{Ru }(PMe_3)4H_2} HCO_2H \xrightarrow[NEt_3]{} (HNEt_3+)HCO_2$$

$$CO_2 + H_2 \xrightarrow[\text{SCF } CO_2]{\text{Ru }(PMe_3)4Cl_2} HCO_2H \xrightarrow[NEt_3]{MeOH} HCO_2Me + H_2O$$

2.6 Increase in Energy Efficiency

In many chemical reactions, we need a certain amount of energy when the reagents and reactants are dissolved in a particular solvent [52]. Reactions often need thermal energy beyond the activation energy to complete the reaction [53]. This shows the importance of the catalysts because they reduce the activation energy and thus reduce the amount of energy needed to complete the reaction. Some reactions are exothermic and the amount of energy released is dangerous enough so the reaction must be controlled by intensive cooling and needed in high-speed reactions that take place in a fraction of a second. Refrigeration in chemical manufacturing processes is very important as it controls uncontrolled reactions that may lead to serious chemical accidents. It is noted that cooling, similar to heating is equally costly both environmentally and monetarily.

Purification and separation are the most energy-intensive chemical processes. Whether these processes are done by distillation or recrystallization, energy must be used to ensure that the impurities are separated from the products [54]. The chemist should design the method that requires the least amount of energy for the separation process and avoid methods that require large thermal or electrical energy to obtain the pure products [55]. Some reactions (cycloaddition and pericyclic) can be performed using ultrasonic energy. Through this technique, the state of the reactants changes in a way that stimulates chemical transformation. Like any other type of energy, the amount of energy needed for each reaction must be estimated to evaluate/calculate the optimal amount to achieve the goal. An important application of photovoltaic energy as an alternative to conventional chemical techniques is to obtain the products of the Friedel–Craft reaction, by using a new technology based on photochemical reactions. These reactions use Lewis acids such as aluminum chloride, tin chloride, or titanium tetrachloride. It also uses solvents such as nitrobenzene, carbon disulfide, carbon tetrachloride, and methane dichloride. These reactions produce corrosive by-products and chloride acids and most of these substances are polluting the environment. The method has been modified through some research to reduce the use of polluting solvents in this reaction, but this requires raising the temperature, which is an economic burden. The acyl hydroquinone compounds were prepared by the reaction of aldehydes with quinone using light energy in an alternative reaction to the traditional Friedel–Craft reaction.

2.7 *Utilization of the Renewable Raw Materials that Are not Exhaustible*

The difference between renewable and permeable sources is the time factor. The most accessible sources often refer to fossil fuels (petroleum and coal) that have been formed over millions of years [56]. By the same logic, the sun can be regarded as a non-renewable and draining source of energy. Renewable materials are often associated with biomaterials and plant origin. Carbon dioxide is generated continuously and in all places from human and natural sources. It can be considered as a renewable material, and methane can also be considered as a renewable material [57]. There is a concern in the world today about the depletion and inevitability of non-renewable materials affecting the economic and environmental dimensions [58]. It is known that how the use of fossil fuels has negatively affected human health and environmental safety. There is a historical environmental heritage that includes black lung, environmental destruction in coal mines, oil spills, and air pollution during oil refining [59]. There is an indirect impact on human health and environmental safety from petroleum and petrochemical industries that flourished in the latter half of the twentieth century. Petroleum hydrocarbons have highly reducing properties and therefore need oxidizing agents to employ and derive many products of various uses. Oxidation chemistry has been a major contaminant of human health and the environment mainly due to the use of heavy metals as oxidants such as chromium [60].

2.8 *Avoid Derivative Processes*

This means inhibitors, protection groups, or temporary modifications in physical and chemical processes. One of the most commonly used techniques today is the use of blocking or protecting groups—to protect a sensitive part of the molecule from the reaction conditions which, if left unprotected, may function as a whole [61]. One of these reactions is to protect alcohol by converting it into benzyl ether to conduct oxidation elsewhere in the molecule without the alcohol being affected. The derivation of this type is practically common in the preparation of high-end chemicals such as drugs, some dyes, and pesticides. It is common to formulate certain materials in such a way that they can mix with other materials [62].

2.9 *Catalysts are more Specialized than Selective Reagents*

The catalysts are distinguished in many cases from the use of equivalent quantities [63]. It is noted that the catalyst facilitates the transformation of the material without being consumed during the reaction and does not integrate with the final product, and the role of the catalyst in facilitating the reaction takes several forms:

2.9.1 Improve Selectivity

There is no doubt that there is a great deal of effort to select the catalyst with the specific selectivity. Selective catalysts achieve control over the reaction such as single versus multiple addition or preservation of reaction site, for example, C-methylation vs O-methylation as well as the control of stereochemistry. The selectivity of catalysts gives a great advantage as an important tool in green chemistry because it stimulates reactants and prevents or minimizes waste production [64, 65].

This input is usually impossible and therefore most operations are accompanied by compounds associated with the desired output as well as other residues, which must be within the environmentally acceptable limits. Some of these residues can be converted to ashes by burning. It can also be converted into stationary products. The catalysts play an important role in the conversion of industrial waste to fixed compounds. The catalysts are widely used commercially in petroleum distillation processes and in the chemical industry. More than 60% of industrial chemical products are made using catalysts [66]. The use of these substances is a burden on the economics of the production process as well as on the environment.

2.10 Designing the Chemical Products so that they do not Settle in the Environment (Biodegradable Materials)

The greatest concern is concentrated in chemicals that settle in the environment in the same form and enter plants and animals, stabilize and accumulate in their tissues. This accumulation affects these organisms by either direct or indirect poisoning [9].

The plastic is a type of chemical compound that has shown throughout its history great benefits in terms of diversity of uses and duration of survival, but these products stabilized after disposal either in places of burial or in the oceans or in the water environment, which caused environmental problems as a result of its physical properties as opposed to chemical properties [67]. Plastic materials are not digested, so there is a need to produce biodegradable plastics [68].

Many insecticides are organic halogen compounds. These compounds, although efficient, accumulate in plant and animal tissues in adipose tissue. DDT was one of the first to cause extensive destruction in this way [69].

When designing a chemical compound, green chemistry attaches importance to the effectiveness of the material for the purpose, it is designed and for its biodegradability [70]. If a plastic bag is made to collect the waste, after it is manufactured to the specifications required for the purpose, it must be taken into account the property of its ability to environmental degradation after use. Also, when designing biodegradable chemicals, decomposition products should be evaluated. Functional groups can be introduced into the molecular structures that facilitate their decomposition, such as hydrolyzable, biodegradable, or easily breakable groups to ensure its biodegradation. It should be borne in mind that the substances resulting from the decomposition do not pose any risk to human health or environment or the goal of green chemistry will not be achieved [5]. Consideration should be given to the impact of these substances on human health and the integrity of ecosystems, wildlife, and other pollution factors.

2.11 *Real-Time Analysis for Pollution Prevention*

The methodology of the analysis techniques should be further developed to allow real-time monitoring of preparation steps to control hazardous materials prior to their formation. Analytical chemistry has contributed to the detection and identification of environmental problems since the beginning of attention to environmental issues. The analytical chemistry effort is currently focused on developing methods and techniques to prevent or reduce the formation of hazardous substances during chemical reactions [71].

The development of analytical chemistry to serve the field of green chemistry is based on the premise that you cannot control what cannot be measured. To be able to influence the chemical process as it is conducted; there must be accurate and reliable sensors, observers, and analytical techniques to assess the risks in the process [72]. Using these features, the generation of hazardous by-products or side reactions can be controlled. When these toxic substances are detected during the course of the reaction, even in small amounts, the reaction conditions can be controlled to minimize or avoid the formation of these substances [71]. Sensors that monitor the reaction can automatically direct it to prevent the formation of these hazardous substances. Another example of using this method in analytical chemistry is to monitor the flow of the reaction to the point of completion [73].

2.12 *Reducing the Possibility of Accidents*

Storage of chemicals like oxidizers and combustible substances results in the chance of unleashing and reaction. If these compounds leak from their containers and react, they are going to cause an outsized hearth. This risk is simply avoided by storing these chemical species independently. Accident prevention begins with the identification and evaluation of risks. Many of the chemicals we still use are presenting serious current hazards and should be replaced with safer alternatives to prevent accidents whenever possible [74].

3 Conclusion

Green chemistry aims to increase the efficiency of the chemical process and reduce the risks to human health and the environment. In this chapter, the twelve principles of green chemistry are briefly discussed with appropriate examples, which give an overview of the importance of green chemistry and its influence on human health and environment. It also demonstrates that despite the absence of a (green) reaction completely, it is possible to minimize the total negative effects of research and chemical industries by following the twelve principles of green chemistry.

References

1. Leitner W, Jessop PG, Li CJ (2010) Handbook of green chemistry e green solvents. Wiley-VCH, Weinhheim
2. World Commission on Environment and Development (1987) Our common future. Oxford University Press, Oxford, United Kingdom
3. Anastas PT, Williamson TC (1996) Green chemistry: an overview. ACS Symp Ser 626:1–17
4. Tang S, Bourne R, Smith R et al (2008) The 24 principles of green engineering and green chemistry: improvements productively. Green Chem 10:268–269
5. Sheldon RA (2017) The E factor 25 years on: the rise of green chemistry and sustainability. Green Chem 19:18–43
6. Akadiri PO, Chinyio EA, Olomolaiye PO (2012) Design of a sustainable building: a conceptual framework for implementing sustainability in the building sector. Buildings 2:126–152
7. Abdel-Shafy HI, Mansour MS (2018) Solid waste issue: sources, composition, disposal, recycling, and valorization. Egyptian J Petrol 27(4):1275–1290
8. Elbeshbishy E, Okoye F (2019) Improper Disposal of Household Hazardous Waste: Landfill/Municipal Wastewater Treatment Plant. Intech Open
9. Marion P, Bernela B, Piccirilli A et al (2017) Sustainable chemistry: how to produce better and more from less? Green Chem 19:4973–4989
10. Mulvihill MJ, Beach ES, Zimmerman JB et al (2011) Green chemistry and green engineering: a framework for sustainable technology development. Annu Rev Environ Resour 36:271–293
11. Sheldon RA (2000) Atom efficiency and catalysis in organic synthesis. Pure Appl Chem 72(7):1233–1246
12. Eynde JJV (2016) How efficient is my (medicinal) chemistry? Pharmaceuticals (Basel) 9(2):26
13. Winterton N (2016) Green chemistry: deliverance or distraction? Clean Techn Environ Policy 18:991–1001
14. Patel KR, Sen DJ, Jatakiya VP (2013) Atom economy in drug synthesis is a playground of functional groups. Am J Adv Drug Del 1:73–83
15. Török B, Dransfield T (2018) Green chemistry: historical perspectives and basic concepts. In: Green Chemistry, p 8
16. Anastas P, Eghbali N (2010) Green chemistry: principles and practice. Chem Soc Rev 39:301–312
17. Cherrie JW, Semple S, Christopher Y et al (2006) How important is inadvertent ingestion of hazardous substances at work? Ann Occup Hygiene 50(7):693–704
18. Ganesan K, Raza SK, Vijayaraghavan R (2010) Chemical warfare agents. J Pharm Bioallied Sci 2(3):166–178
19. Curzons AD, Constable DJC, Mortimer DN et al (2001) So you think your process is green, how do you know?—Using principles of sustainability to determine what is green—a corporate perspective. Green Chem 3:1–6

20. Marco BA, Rechelo BS, Tótoli EG et al (2019) Evolution of green chemistry and its multidimensional impacts: a review. Saudi Pharmaceut J 27(1):1–8
21. Serini V (2000) Polycarbonates in Ullmann's encyclopedia of industrial chemistry. Wiley-VCH, Weinheim
22. Kim WB, Joshi UA, Lee JS (2004) Making polycarbonates without employing phosgene: an overview on catalytic chemistry of intermediate and precursor syntheses for polycarbonate. Ind Eng Chem Res 43(9):1897–1914
23. Fukuoka SJ, Patent 4-257646 A1 (Sep. 11, 1992); 7-91237 B1 (Oct. 4, 1995), assigned to Asahi Kasei Chemicals Corp
24. Komiya K, Fukuoka S, Aminaka M et al (1996) New process for producing polycarbonate without phosgene and methylene chloride. Green Chem Des Chem Environ 26:20–32
25. Hansson SO, Molander L, Ruden C (2011) The substitution principle. Regul Toxicol Pharmacol 59:454–460
26. Vallamkondu J, Corgiat EB, Buchaiah G et al (2018) Liquid crystals: a novel approach for cancer detection and treatment. Cancers 10(11):462–482
27. Dichiarante V, Ravelli D, Albini A (2010) Green chemistry: state of the art through an analysis of the literature. Green Chem Lett Rev 3:105–113
28. Trost BM (1991) The atom economya search for synthetic efficiency. Science 254:1471–1477
29. Zhu J, Bienayme H (2005) Multicomponent reactions. Wiley-VCH Verlag, GmbH & Co
30. Murai S (1999) activation of unreactive bonds and organic synthesis, topics in organometallic chemistry (vol 3). Springer, Berlin Heidelberg
31. Grubbs RH (2004) Olefin Metathesis. Tetrahedron 60:7117–7140
32. Silverman RB (2002) The organic chemistry of enzyme-catalyzed reactions. In: Bommarius AS, Riebel BR (eds) Biocatalysis, Academic Press, New York. Wiley-VCH Verlag GmbH & Co KGaA
33. Rani B, Maheshwari R, Chauhan AK et al (2012) Potentiality of green chemistry for future perspectives. Int J Pharm Chem Sci 1(1):97–194
34. Crawford SE, Hartung T, Hollert H et al (2017) Green toxicology: a strategy for sustainable chemical and material development. Environ Sci Eur 29(1):16
35. Chen FM, Liu X (2016) Advancing biomaterials of human origin for tissue engineering. Prog Polym Sci 1(53):86–168
36. Welton T (2015) Solvents and sustainable chemistry. Proc R Soc 471(2183):20150502
37. Clarke CJ, Tu WC, Levers O et al (2018) Green and sustainable solvents in chemical processes. Chem Rev 118(2):747–800
38. Yang GP, Wu X, Yu B et al (2019) Ionic liquid from vitamin B1 analogue and heteropolyacid: a recyclable heterogeneous catalyst for dehydrative coupling in organic carbonate. ACS Sustain Chem Eng 7(4):3727–3732
39. Giorgianni G, Abate S, Centi G et al (2018) Effect of the solvent in enhancing the selectivity to furan derivatives in the catalytic hydrogenation of furfural. ACS Sustain Chem Eng 6(12):16235–16247
40. Coskun O (2016) Separation techniques: chromatography. North Clin Istanb 3(2):156–160
41. Himaja M, Poppy D, Asif K (2011) Green technique-solvent free synthesis and its advantages. Int J Res Ayurv Pharm 2(4):1079–1086
42. Cornils B, Herrmann WA (eds) (1998) Aqueous phase organometallic catalysis—concepts and applications. Wiley-VCH, Weinheim, Germany
43. Samorì C, Basaglia M, Casella S et al (2015) Dimethyl carbonate and switchable anionic surfactants: two effective tools for the extraction of polyhydroxyalkanoates from microbial biomass. Green Chem 17(2):1047–1056
44. Reinhardt D, Ilgen F, Kralisch D et al (2008) Evaluating the greenness of alternative reaction media. Green Chem 10:1170–1181
45. Breslow R, Anastas PT, Williamson TC (1998) Green Chem. Oxford University Press, New York p, p 225
46. Khaw KY, Parat MO, Shaw PN et al (2017) Solvent supercritical fluid technologies to extract bioactive compounds from natural sources: a review. Molecules 22(7):1186–1207

47. Peach J, Eastoe J (2014) Supercritical carbon dioxide: a solvent like no other. Beilstein J Org Chem 10:1878–1895
48. Nautiyal OH (2016) Food processing by supercritical carbon dioxide-review. EC Chem 2(1):111–135
49. Darani KK, Farahani VE (2005) Application of supercritical fluid extraction in biotechnology. Crit Rev Biotechnol 25(4):1–12
50. De Simone JM, Guan Z, Elsbernd CS (1992) Synthesis of fluoropolymers in supercritical carbon dioxide. Science 257(5072):945–947
51. Jessop PG, Hsiao Y, Ikariya T et al (1994) Catalytic production of dimethylformamide from supercritical carbon dioxide. J Am Chem Soc 116(19):8851–8852
52. Herron JA, Kim J, Upadhye AA (2015) A general framework for the assessment of solar fuel technologies. Energy Environ Sci 8:126–157
53. Gao J, Ma S, Major DT (2006) Mechanisms and free energies of enzymatic reactions. Chem Rev 106(8):3188–3209
54. Frank L (1996) Loss prevention in the process industries, vol 3, 2nd edn. Butterworth-Heinemann, Boston
55. Barton J, Rogers R (1997) Chemical reaction hazards: a guide to safety (2nd edn). Institution of Chemical Engineers, Davis Building, Rugby, Warwickshire UK
56. Alalwan HA, Alminshid AH, Aljaafari HAS (2019) Promising evolution of biofuel generations. Subject Rev Renew Energy Focus 28:127–139
57. Vasconcelos BR, Lavoie JM (2019) Recent advances in power-to-X technology for the production of fuels and chemicals. Front Chem 7:392
58. Miller RG, Sorrell SR (2014) The future of oil supply. Phil Trans R Soc A 372:20130179
59. Kotcher J, Maibach E, Choi WT (2019) Fossil fuels are harming our brains: identifying key messages about the health effects of air pollution from fossil fuels. BMC Public Health 19(1):1079
60. Rai PK, Lee S, Zhang M (2019) Heavy metals in food crops: health risks, fate, mechanisms, and management. Environ Int 125:365–385
61. Peterson EA, Manley JB (2015) Green chemistry strategies for drug discovery. Royal Soc Chemi UK. RSC Drug Disc Ser 46:19
62. Manahan SE (2006) Green chemistry and the ten commandments of sustainability (2nd edn) Chem Char Research, Inc Publishers Columbia, Missouri U.S.A pp 183
63. Kalz KF, Kraehnert R, Dvoyashkin M et al (2017) Future challenges in heterogeneous catalysis: understanding catalysts under dynamic reaction conditions. Chem Cat Chem 9(1):17–29
64. Davis HJ, Phipps RJ (2017) Harnessing non-covalent interactions to exert control over regioselectivity and site-selectivity in catalytic reactions. Chem Sci 8:864–877
65. Argyle MD, Bartholomew CH (2015) Heterogeneous catalyst deactivation and regeneration: a review. Catalysts 5(1):145–269
66. Emam EA (2013) Clays as catalysts in petroleum refining industry. J Sci Tech 3(4):356–375
67. Eriksen M, Thiel M, Prindiville M et al (2017) Microplastic: what are the solutions? Freshwater microplastics part of the handbook of environmental chemistry, Springer, pp 273–298
68. Hammer J, Kraak MH, Parsons JR (2012) Plastics in the marine environment: the dark side of a modern gift. Rev Environ Contam Toxicol 220:1–44
69. Jayaraj R, Megha P, Sreedev P (2016) Organochlorine pesticides, their toxic effects on living organisms and their fate in the environment. Interdiscip Toxicol 9(3–4):90–100
70. Jordan A, Gathergood N (2013) Designing safer and greener antibiotics. Antibiotics 2(3):419–438
71. Ivanković A, Dronjić A, Bevanda AM et al (2017) Review of 12 principles of green chemistry in practice. Int J Sustain Green Energy 6(3):39–48
72. Riđanović L, Ćatović F, Riđanović S The green chemistry-ecological revolution in the classroom. In: 8th research/expert conference with international participations "QUALITY 2013", Neum, B&H, June 06–08, pp 447–452. In Bosnian
73. Bharadwaj M, Neelam M (2015) The advantages and disadvantages of green technology. J Basic Appl Eng Res 2(22):1957–1960

74. Collins TJ (2017) Review of the twenty-three-year evolution of the first university course in green chemistry: teaching future leaders how to create sustainable societies. J Clean Prod 140:93–110

Chapter 3
Organic Reactions in Water

Nissy Ann Harry, K. R. Rohit, and Gopinathan Anilkumar

1 Introduction

Scientific community has started focusing on aspects of green chemistry due to severe environmental issues caused by waste produced as a result of chemical processes. Among the twelve principles of green chemistry, the selection of green solvents to minimise toxicity and pollution is a matter of great interest [1]. In a chemical reaction, solvents account for a major fraction of total mass handled and hence its choice is very important. Among all the possible solvents, water is the greenest solvent with smallest environmental impact [2]. Due to the intrinsic nature of the insolubility of organic substances in water, the scientific world was reluctant to use water as a medium in organic reactions in early days. But the natures call for safer chemistry demands us to consider water as a solvent for chemical processes since water is the greenest solvent that we ever have and the nature's choice in biological systems. In addition to that, there are many more good reason to choose water as a solvent since it is the most abundant liquid in our reach with great thermal stabilities and properties such as non-flammability, moderate viscosity, pH variability, etc. [3, 4]. Recently, water-based reactions were divided as "in water" or "on water" by Butler, Coyne and Fokin [5, 6]. The "in water" reactions are those where the reactants form a homogeneous mixture in water medium, whereas "on water" forms a heterogeneous mixture in water, which has been discussed here [7].

N. A. Harry · K. R. Rohit · G. Anilkumar (✉)
School of Chemical Sciences, Mahatma Gandhi University, Priyadarsini Hills, Kottayam, Kerala 686560, India
e-mail: anilgi1@yahoo.com

N. A. Harry
e-mail: nissyannharry@gmail.com

K. R. Rohit
e-mail: rkrnovachemia@gmail.com

G. Anilkumar and S. Saranya (eds.), *Green Organic Reactions*, Materials Horizons: From Nature to Nanomaterials, https://doi.org/10.1007/978-981-33-6897-2_3

2 Reactions on Water

The chemistry of water reactions was undisclosed until the inferences that water can speed up and give selectivity for Diels–Alder reaction by Breslow and co-workers [8] and shed light on the misconception that "high solubility of the organic reactants in solvent is required for better yields". The contributions of Sharpless accelerated the wide acceptance of water chemistry and he put forward the term "on water" reactions [9]. The terminology simply suggests that the reaction takes place heterogeneously or the organic reactants form an emulsion in the water solvent. The trans-phase hydrogen bonding at organic-water interface activates the reactants and sometimes stabilises the transition states. The hydrophobic nature of the reactants brings them closer and thereby increases the reaction rates. Nowadays, on water method is used as a prominent methodology for various catalysed as well as non-catalysed organic reactions, especially aldol reaction, Claisen rearrangement, allylation, functionalisation of olefins, multicomponent reactions, cross-coupling, C–H functionalisation, etc. [10–13].

2.1 *Catalyst-Free Reactions on Water*

In the modern scenario, the catalyst-free reactions represent that the reactions take place in the absence of a catalyst without compromising its yield and speed. Sometimes, they are faster than the catalytic reactions. In these methodologies, water enhances the reaction by better stabilisation of the transition state than the reactants through hydrogen bonding (Fig. 1) [14].

In 2015, Dada and co-workers demonstrated the application of on water chemistry in a multicomponent C–H activation reaction in which the sp^3 C–H of methyl azaarenes was functionalised with isatins and malononitrile (Scheme 1). This protocol exhibits moderate to good yields for various azaarenes and isatins [15].

Later Legros used "on water" strategy for the synthesis of (*Z*)-enaminones with *trans*-4-methoxy-3-buten-2-one as a useful surrogate of 3-butyn-2-one and anilines [16]. It was observed that water enhances the reaction rate by a 45–200-fold factor in comparison with other solvents (Scheme 2).

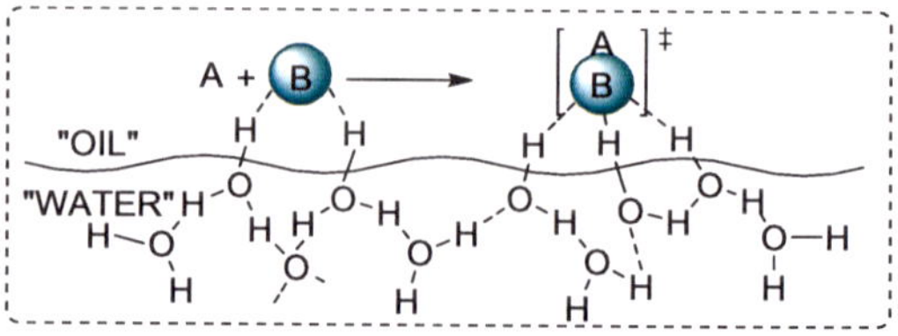

Fig. 1 Stabilisation of transition state via hydrogen bonding interactions

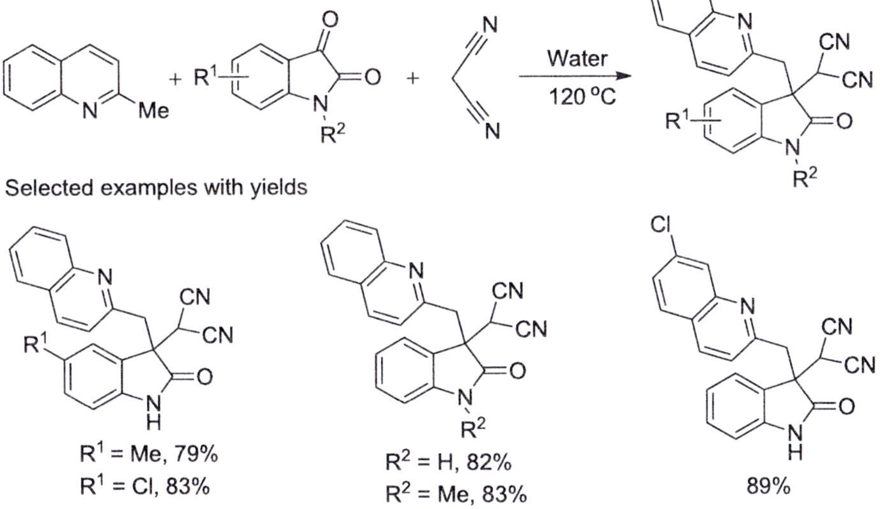

Scheme 1 Synthesis of 3,3-disubstituted oxindole derivatives via on water protocol

Scheme 2 Synthesis of (*Z*)-enaminones using on water methodology

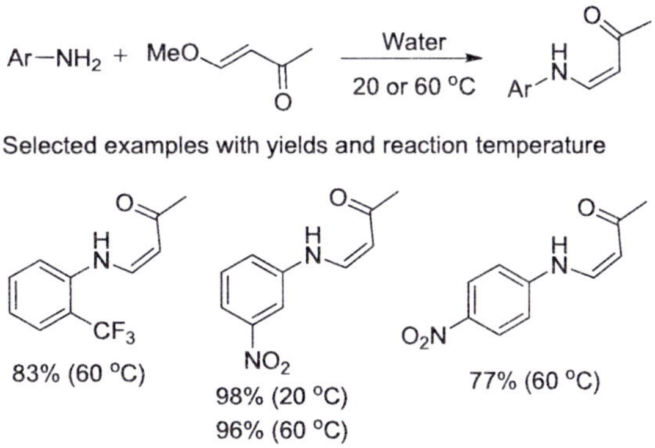

A transition metal-free cycloaddition reaction of tosylazide and enaminones was performed successfully on water by Wan et al. for the synthesis of 4-acyl-NH-1,2,3-triazoles with broad substrate scope under very mild reaction condition (Scheme 3) [17]. The mechanism involves the cycloaddition between enaminones and tosylazide to form 1,2,3-triazoline. The hydrogen bonds with water favour the elimination of amino group and the acidic proton at α-position of the acyl group to give N-tosyl-1,2,3-triazole intermediate which then undergoes aminolysis and/or hydrolysis to form the products.

As part of green convergence, the incorporation of microwave (MW) heating in a reaction is an admirable step which was achieved by Wacharasindhu et al. through the catalyst-free amination of 2-mercaptobenzoxazoles (Scheme 4) [18]. Later, they

Scheme 3 Synthesis of 4-acyl-NH-1,2,3-triazoles on water

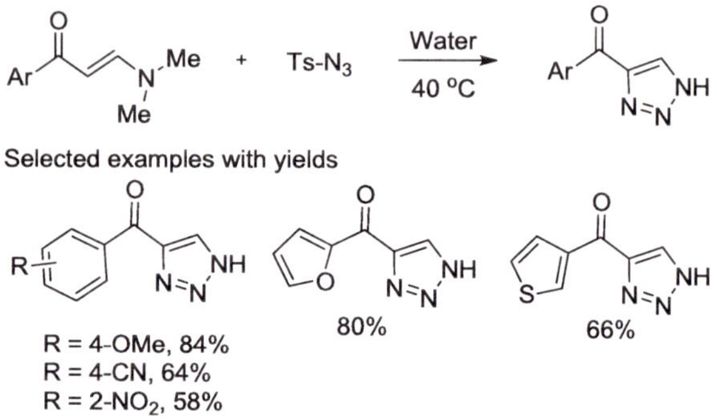

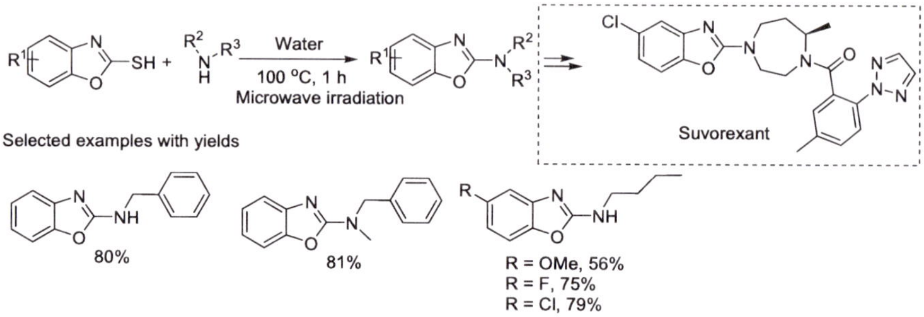

Scheme 4 Amination of 2-mercaptobenzoxazoles using microwave assisted on water protocol

extended the strategy to synthesise a useful drug suvorexant which is used to cure insomnia.

2.2 *Reactions on Water Using Metal Catalyst*

Catalysis is also inevitable in the modern scenario of green convergence because of its abilities to guide the reaction in a selective manner in very short reaction time. So, the combination of these two safer adaptations fascinates modern chemists in a great extent and there are many reports in various kinds of reactions on water [19].

Arvidsson and co-workers introduced a palladium catalysed synthesis of biaryl sulfonyl fluorides by the ligand-free Suzuki–Miyaura cross-coupling strategy [20]. 1 mol% of $Pd(OAc)_2$ along with triethylamine empowers the reaction of variously substituted boronic acids and halo phenyl sulfonyl fluorides on water with good to excellent yields (Scheme 5).

Metal catalysed C–N bond formations on water can be achieved via direct C–H amination reactions as well as cross-coupling reactions.

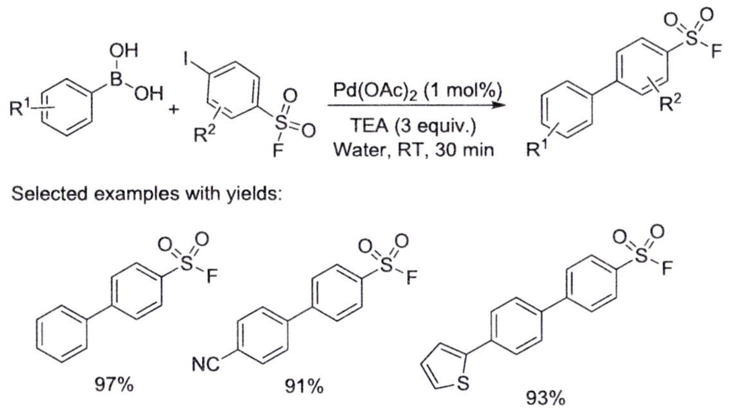

Scheme 5 Palladium catalysed Suzuki–Miyaura cross-coupling carried out on water

Water promoted Rh catalysed C–H amination was reported by Lu et al. in which pyridine was used as the directing group [21]. The reaction shows good substrate scope with electron deficient azides. N-Arylation of various amines through copper catalysed Ullmann coupling was reported [22]. Specially synthesised proline amide ligand along with CuI furnish the catalytic system. In the substrate scope analysis, various amines including nitrogen bases were used showing promising yields (Scheme 6).

Synthesis of heterocyclic compounds can be easily achieved through on water protocols.

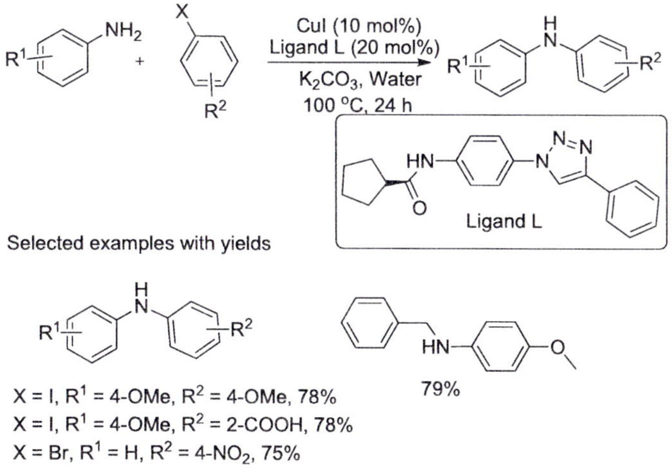

Scheme 6 Copper catalysed *N*-arylation via Ullmann coupling reactions

NHC-copper catalysed sequential *N*-alkynylation/condensation of *o*-aminofluoromethylketones with alkynes resulting the synthesis of 4-trifluoromethylquinolines and naphthyridines was reported. This reaction works well with very low catalytic loading at 100 °C on water [23]. In 2017, Shao and co-workers developed 7-azaindole synthesis through silver catalysed cycloisomerisation of amines (Scheme 7) [24]. Due to the success of this protocol in various substrates they extended this protocol to the synthesis of isoquinolones using amides instead of amines which was not that much promising in comparison with the azaindole synthesis.

C–H activation reactions for C–C bond formations are also reported on water.

Keesara et al. reported a polymer supported Pd(II) catalysed aryl ketone synthesis by the ortho-acylation of 2-aryl pyridines with toluene derivatives through sp^2 C–H activation [25]. TBHP was used as the oxidant in this on water protocol. Moderate to good yields were observed for various substituted substrates in which electron-donating groups were found to be better performing. Palladium catalysed direct C-3 arylation of imidazo[1,2-*a*]pyridines was reported by Rode et al. [26]. In this ligand-free approach, aryl iodides were found to be better alkylating agents in comparison with bromides and chlorides. Direct C-3 (hetero)arylation of 2H-pyrazolo[3,4-*b*]pyridines was reported by Guillaumet et al. with aryl iodides as the coupling partner [27]. This palladium catalysed strategy works well in water at 70 °C and showed good to excellent yields for various substituted aryl iodides (Scheme 8).

Click reactions can be easily performed on water. Various biologically important coumarins were synthesised by Shi and co-workers by introducing an ultrasound-assisted CuCl catalysed 1,3-dipolar cycloaddition of azides and alkynes at room temperature (Scheme 9) [28]. 1,4-Disubstituted 1,2,3-triazoles were synthesised successfully by Chandak and co-workers using Cu(II)/DABCO/AcOH catalytic system by the cycloaddition of alkynes with azides [29].

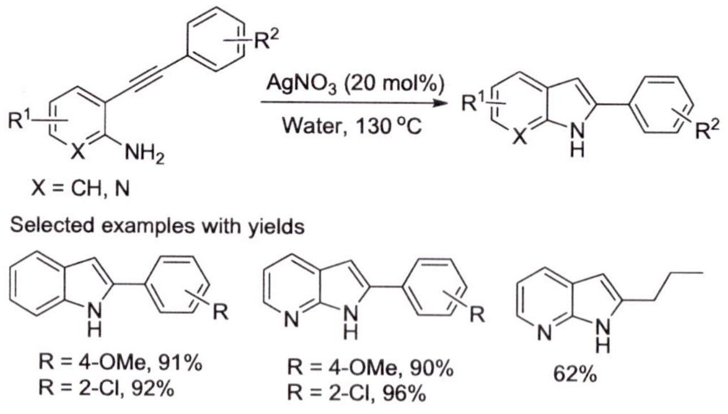

Scheme 7 On water synthesis of indoles and azaindoles

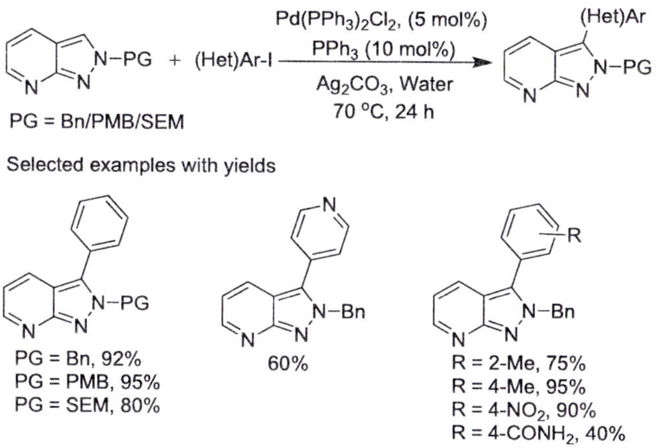

Scheme 8 C-3 (hetero)arylation of 2*H*-pyrazolo[3,4-*b*]pyridines

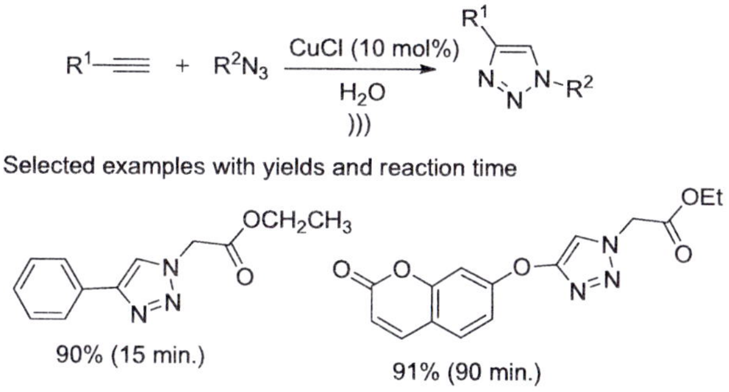

Scheme 9 Ultrasound-assisted CuCl catalysed 1,3-dipolar cycloaddition of azides and alkynes

3 Organic Reactions in Water: Reactions in Micellar Media

Surfactants are surface active agents that in aqueous solution form spontaneous micellar aggregates above a critical micelle concentration with hydrophilic head in contact with the surrounding solvent with its hydrophobic tails sequestering in the micelle centre. The use of surfactants under micellar conditions is one of the simplest ways to achieve catalysis in water, where the micelle behaves as nano-reactors [30–40].

Most of the commercially available surfactants are obtained from petroleum feed-stock. Depending upon the nature of head groups of surfactants, it is classified as cationic [41], anionic [42], neutral [43], zwitterionic [44] and Gemini surfactants [45] (Fig. 2). Biosurfactants are prepared from natural resources such as bacteria and yeast, and thus it is biodegradable and eco-friendly. Such anionic and neutral surfactants have hydrophobic part based on long-chain fatty acids, α-alkyl-β-hydroxy fatty acids

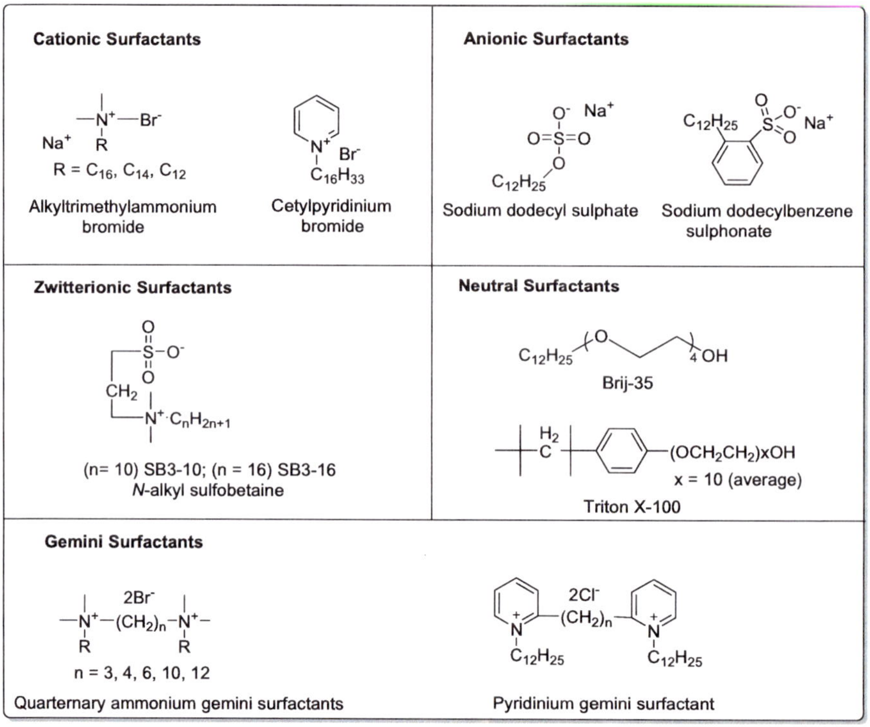

Fig. 2 Representative examples for a different class of surfactants

and the hydrophilic portion based on amino acids, cyclic peptides, carbohydrates, phosphates, alcohols or carboxylic acids [46, 47].

The type of aggregates formed depends upon the type of molecular structure of amphiphile, geometry of the molecule, reaction conditions such as pH and temperature [48]. Hence, surfactants are specially designed to carry out the reactions, for example, designing of surfactants that enables oxygen solvation within the micelle core for oxidation reactions, whereas some other surfactants prevent the oxygen solubility so that oxygen-sensitive phosphine ligand-chelated metal catalysts get more lifetime inside the micelle core. Surfactants play different roles, they can act either as a medium to disperse water insoluble organic catalysts, transition and non-transition metals, or can directly promote a reaction. Organic reactions in water using surfactants are recently reviewed [49, 50].

3.1 Reactions in Water Directly Promoted by Surfactants

Some reactions are directly promoted by surfactants in which hydrophobic interior of micelles favour dehydration reactions. Micellar systems facilitate condensation reactions by expulsion of water thereby causing dehydration.

Chakraborthi and co-workers investigated the effect of about 50 surfactants for the synthesis of quinoxalines [51]. 1,2-diamines and 1,2-dicarbonyls were reacted at room temperature using different surfactants and found that non-ionic surfactant Tween 40 gave the highest yield (Scheme 10). The order of catalytic potential was found to be neutral surfactants > anionic surfactants > Brønsted acid surfactants > cationic surfactants. The reactions in aqueous medium showed an enhanced reaction rate in comparison with that in organic solvents.

A one-pot three-component reaction for the synthesis of spirooxindoles with fused chromenes using weakly basic surfactants like sodium stearate in water was reported [52]. The basic property of the surfactants enables the deprotonation of malononitrile which reacts with isatin and further with 4-hydroxy-coumarin resulting in the formation of the corresponding spirooxindole derivatives in excellent yields (Scheme 11).

Kumar and co-workers reported a Mannich-type one-pot three-component reaction of secondary amines, aldehydes and indoles for the synthesis of 3-amino alkylated indoles [53]. 3-Amino alkylated indoles were selectively formed when the reaction was carried out in water using sodium dodecyl sulphate (SDS) as surfactant, whereas usage of other Brønsted and Lewis acids led to the formation of bis-indole derivative exclusively (Scheme 12). Water expulsion results in the initial formation of iminium cation and it is favoured by hydrophobic core of the micelle, which then reacts with indole to furnish the product.

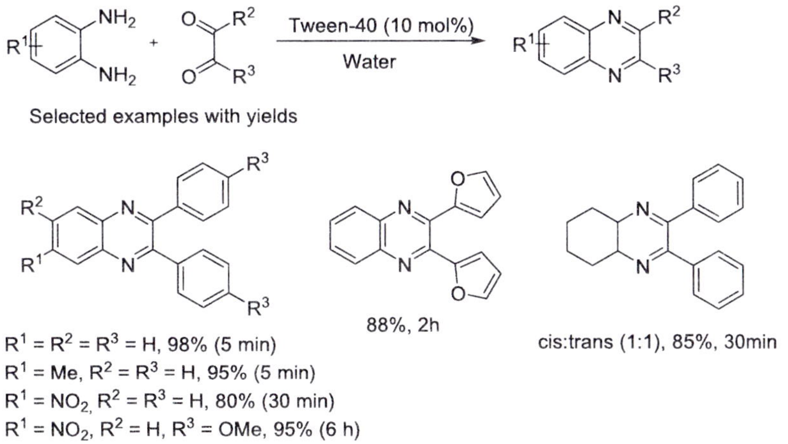

Scheme 10 Synthesis of quinoxalines from 1,2-diamines and 1,2-dicarbonyl using Tween 40 as catalyst

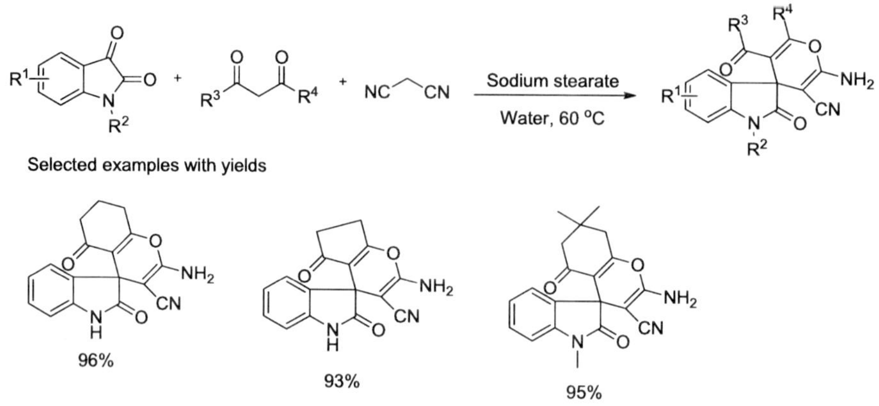

Scheme 11 Synthesis of spirooxindoles with fused chromenes using sodium stearate in water

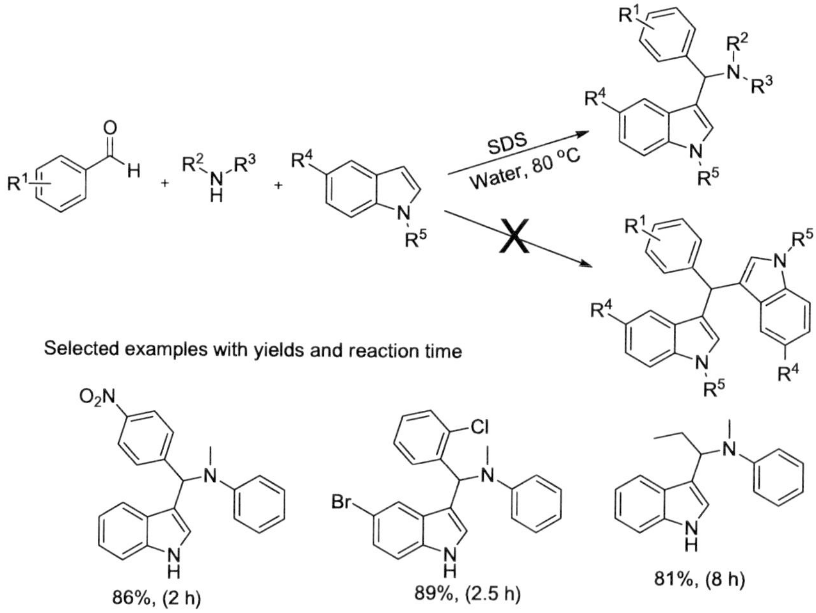

Scheme 12 Selective synthesis of 3-amino alkylated indoles via Mannich-type one-pot three-component reaction using SDS as catalyst

3.2 Reactions Promoted by Metals Along with Surfactants

Phosphene-ligated palladium catalysts were found to have longer lifespan in the oxygen-free hydrophobic core of micelle. Beverina and co-workers reported a

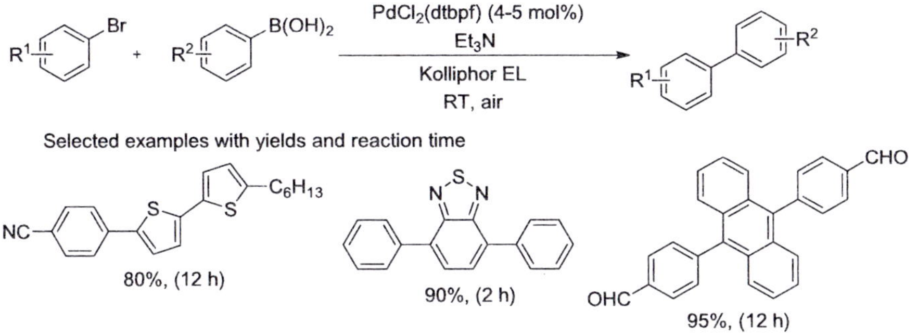

Scheme 13 Suzuki–Miyaura coupling catalysed by $PdCl_2$(dtbpf) using Kolliphor EL as surfactant

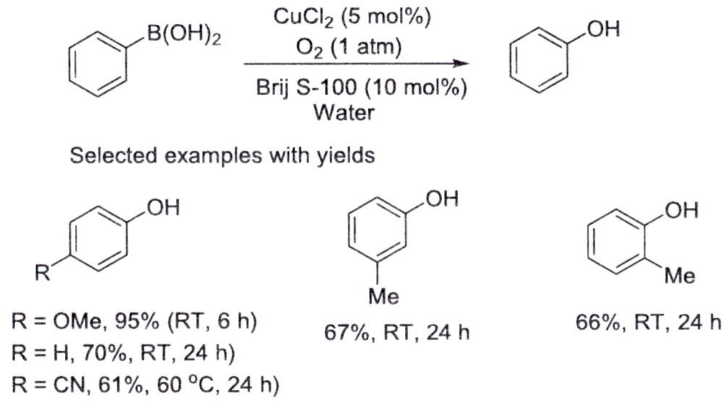

Scheme 14 Oxidative hydroxylation of aromatic boronic acids to phenols

Suzuki–Miyaura coupling catalysed by 4 mol% $PdCl_2$(dtbpf) using Kolliphor EL as surfactant (Scheme 13) [54].

$CuCl_2$ catalysed oxidative hydroxylation of aromatic boronic acids to phenols was accomplished in water using amphiphilic surfactants Brij S-100 without using any bases and ligands [55]. The reaction proceeded well with several electron-donating and electron-withdrawing boronic acids to the corresponding phenols in moderate to excellent yields (Scheme 14).

4 Design and Synthesis of New Surfactants

Recently, researchers focus on the synthesis, design and application of new surfactants. A newly designed surfactant includes polyoxyethanyl-α-tocopheryl sebacate (PTS), a non-ionic surfactant containing racemic vitamin E as hydrophobic part and sebacic acid and PEG-600 as hydrophilic portion. A modification of PTS is

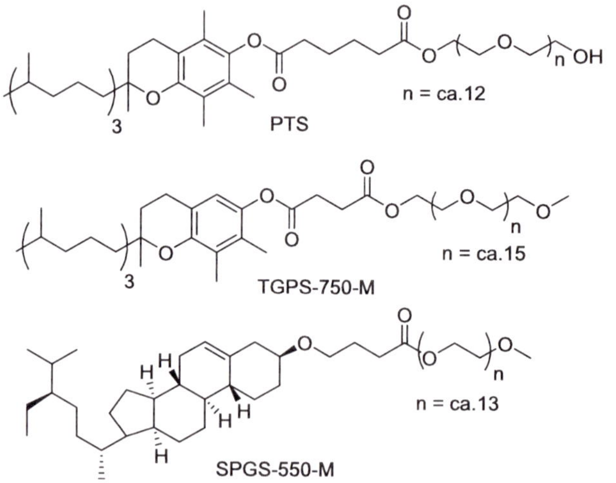

Fig. 3 Representative examples for "designer surfactants"

polyoxyethanyl-α-tocopheryl succinate (TPGS) containing hydrophobic part as α-tocopherol moiety, a succinic spacer and hydrophilic part as poly(ethyleneglycol) methyl ether chain (PEG-750-M). Even though this two surfactants share similar hydrophobic part they catalysed different kinds of reactions since the shapes of micellar aggregates formed are different. Lipshutz and co-workers have studied synthesis and applications of the new surfactants and some of them are included here [56–58] (Fig. 3).

Lipshutz and co-workers uncovered a Suzuki–Miyaura coupling in water using 300 ppm level of palladium catalyst, HandaPhos as ligand, using TPGS-750-M/H_2O as the reaction medium (Scheme 15) [59]. The nature of nano-micelles formed was studied via cryo-TEM measurements. It was observed that the presence of a co-solvent like THF, toluene or acetone enlarged the micellar size by expanding the inner volume available for reaction thereby increasing the reaction rate. Electron-donating, electron-withdrawing as well as heterocyclic educt was tolerated to furnish the coupled products in good to excellent yields.

Selective C–H arylation of indoles was achieved using surfactant SPGS-550-M in presence of 1-mol% of $[(Cinnamyl)PdCl]_2$ under mild conditions [60]. The nature of site selective ligand was found to be crucial for the site selectivity, where the DPPF and DPPP ligands were most effective in promoting the arylation at C3–H and C2–H, respectively (Scheme 16).

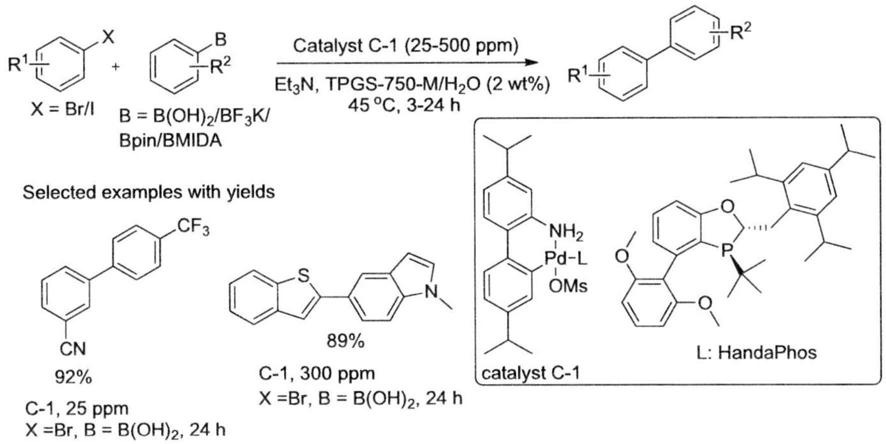

Scheme 15 Pd catalysed Suzuki–Miyaura coupling in water

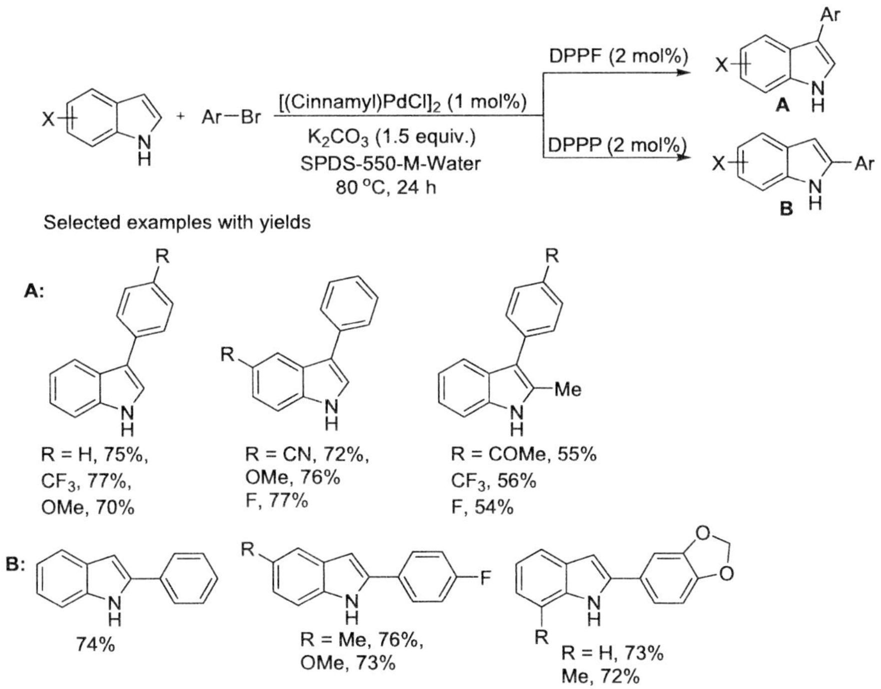

Scheme 16 Palladium catalysed selective C–H arylation of indoles was achieved using surfactant SPGS-550-M

5 Conclusion

It is high time to choose some eco-friendly solvents for organic reactions, since solvents contribute towards the major portion of total mass of a chemical reaction. Water is the nature's reaction medium. Hydrophobic pockets formed in water for lipophilic substrates act as reaction vessels resembling the chemistry occurring in nature and that guide us a clear path to go forward by widening this area of research. It is clear from the forgone discussion that water can be a suitable solvent for many reactions if we finely tune the reaction conditions. On water reactions where the trans-phase hydrogen bonding of water with the reactants and the closeness of the reactants due to its hydrophobic nature are the main reasons for the rate enhancements. Many organic reactions like C–H-activation, cycloaddition, Michael addition and even highly sensitive radical reactions could be effectively carried out on water. The use of surfactants under micellar conditions is another easiest method to achieve catalysis in water, where the micelle behaves as nano-reactors. Chemists are on the way to design new surfactants that can lead towards greener reaction pathways.

References

1. Anastas PT, Warner JC (1988) Green chemistry—theory and practice. Oxford University Press, New York
2. Kitanosono T, Masuda K, Xu P, Kobayashi S (2018) Catalytic organic reactions in water toward sustainable society. Chem Rev 118:679–746
3. Lindstrom UM (2008) Organic reactions in water: principles strategies and applications. Wiley, Hoboken
4. Li C-J, Chan T-H (2007) Comprehensive organic reactions in aqueous media. Wiley, Hoboken
5. Butler RN, Coyne AG (2010) Water: nature's reaction enforcer—comparative effects for organic synthesis "in-water" and "on-water". Chem Rev 110:6302–6337
6. Butler RN, Coyne AG (2016) Organic synthesis reactions on-water at the organic–liquid water interface. Org Biomol Chem 14:9945–9960
7. Harry NA, Radhika S, Neetha M, Anilkumar G (2019) Recent advances and prospects of organic reactions "on water". Chem Sel 4:12337–12355
8. Breslow R (2004) Determining the geometries of transition states by use of antihydrophobic additives in water. Acc Chem Res 37:471–478
9. Narayan S, Muldoon J, Finn MG, Fokin VV, Kolb HC, Sharpless KB (2005) "On Water": unique reactivity of organic compounds in aqueous suspension. Angew Chem Int Ed 44:3275–3279
10. Li C-J, Chan T-K (1997) Organic. Reactions in aqueous media. New York, Wiley
11. Li C-J, Chen L (2006) Organic chemistry in water. Chem Soc Rev 35:68–82
12. Engberts JBFN, Blandamer MJ (2001) Understanding organic reactions in water: from hydrophobic encounters to surfactant aggregates. Chem Commun 1701–1708
13. Sinou D (2002) Asymmetric organometallic-catalyzed reactions in aqueous media. Adv Synth Catal 344:221–237
14. Harry NA, Cherian RM, Radhika S, Anilkumar G (2019) A novel catalyst-free, eco-friendly, on water protocol for the synthesis of 2,3-dihydro-1*H*-perimidines. Tetrahedron Lett 60:150946–150949
15. Yaragorla S, Singh G, Dada R (2016) 'On water synthesis' of oxindoles bearing quaternary carbon center through C–H (sp3) functionalization of methyl azaarenes. Tetrahedron Lett 57:591–594

16. Jebari M, Pasturaud K, Picard B, Maddaluno J, Rezgui F, Chataigner I, Legros J (2016) "On water" reaction of deactivated anilines with 4-methoxy-3-buten-2-one, an effective butynone surrogate. Org Biomol Chem 14:11085–11087
17. Yang L, Wu Y, Yang Y, Wen C, Wan JP (2018) Catalyst-free synthesis of 4-acyl-*NH*-1,2,3-triazoles by water-mediated cycloaddition reactions of enaminones and tosyl azide. Beilstein J Org Chem 14:2348–2353
18. Tankam T, Srisa J, Sukwattanasinitt M, Wacharasindhu S (2018) Microwave-enhanced on-water amination of 2-mercaptobenzoxazoles to prepare 2-aminobenzoxazoles. J Org Chem 83:11936–11943
19. Zheng Q, Dong J, Sharpless KB (2016) Ethenesulfonyl fluoride (ESF): an on-water procedure for the kilogram-scale preparation. J Org Chem 81:11360–11362
20. Chinthakindi PK, Kruger HG, Govender T, Naicker T, Arvidsson PI (2016) On-water synthesis of biaryl sulfonyl fluorides. J Org Chem 81:2618–2623
21. Ali MA, Yao X, Li G, Lu H (2016) Rhodium-catalyzed selective mono- and diamination of arenes with single directing site "on water". Org Lett 18:1386–1389
22. Chakraborti G, Paladhi S, Mandal T, Dash J (2018) "On water" promoted ullmann-type C–N bond-forming reactions: application to carbazole alkaloids by selective N-arylation of aminophenols. J Org Chem 83:7347–7359
23. Czerwiński P, Michalak M (2017) NHC-Cu(I)-catalyzed friedländer-type annulation of fluorinated o-aminophenones with alkynes on water: competitive base-catalyzed dibenzo[b, f][1,5]diazocine formation. J Org Chem 82:7980–7997
24. Sun H, Xiao L, Li W, Xie Q, Shao L (2017) On-water silver(I)-catalyzed cycloisomerization of acetylenic free amines/amides towards 7-azaindole/indole/isoquinolone derivatives. Synthesis 49:4845–4852
25. Perumgani PC, Parvathaneni SP, Keesara S, Mandapati MR (2016) Recyclable Pd(II)complex catalyzed oxidative *sp2* C–H bond acylation of 2-aryl pyridines with toluene derivatives. J Organomet Chem 822:189–195
26. Kalari S, Babar DA, Karale UB, Makane VB, Rode HB (2017) On water direct arylation of imidazo[1,2-*a*]pyridines with aryl halides. Tetrahedron Lett 58:2818–2821
27. Faarasse S, Kazzouli SE, Naas M, Jouha J, Suzenet F, Guillaumet G (2017) "On water" direct C-3 arylation of 2*H*-Pyrazolo[3,4-*b*]pyridines. J Org Chem 82:12300–12306
28. Li X, Chen X, Jiang Y, Chen S, Qu L, Qu Z, Yuan J, Shi H (2016) Highly efficient ultrasonic-assisted cucl-catalyzed 1,3-dipolar cycloaddition reactions in water: synthesis of coumarin derivatives linked with 1,2,3-triazole moiety. J Heterocyclic Chem 53:1402–1411
29. Sarode PB, Bahekar SP, Chandak HS (2016) DABCO/AcOH jointly accelerated copper(I)-catalysed cycloaddition of azides and alkynes on water at room temperature. Synlett 27:2681–2684
30. Sorrenti A, Illa O, Ortuño RM (2013) Amphiphiles in aqueous solution: well beyond a soap bubble. Chem Soc Rev 42:8200–8219
31. Dwars T, Paetzold E, Boehme G (2005) Reactions in micellar systems. Angew Chem Int Ed 44:7174–7199
32. Yu X, Jiang B, Cheng S, Huang Z, Zeng X (2003) Comparative reactivities of metal cation-catalyzed hydrolysis of p-nitrophenyl picolinate in micellar solutions. J Dispers Sci Technol 24:761–765
33. Gaemers S, Keune K, Kluwer AM, Elsevier CJ (2000) Selective C−H bond activation with Na_2PtCl_4 in inverted micelles. Eur J Inorg Chem 2000:1139–1141
34. Tohma H, Takizawa S, Watanabe H, Fukuoka Y, Maegawa T, Kita Y (1999) Hypervalent iodine(V)-induced asymmetric oxidation of sulfides to sulfoxides mediated by reversed micelles: novel nonmetallic catalytic system. J Org Chem 64:3519–3523
35. Dadyburjor DB, Fout TE, Zondlo JW (2000) Ferric-sulfide-based catalysts made using reverse micelles: effect of preparation on performance in coal liquefaction. Catal Today 63:33–41
36. Grassert I, Paetzold E, Oehme G (1993) Influence of different types of amphiphiles on the rhodium(I) complex-catalyzed asymmetric hydrogenation of (Z)-methyl-α-acetamidocinnamate in aqueous medium. Tetrahedron 49:6605–6612

37. Rodenas E, Dolcet C, Valiente M, Valeron EC (1994) Physical properties of dodecyltrimethylammonium bromide (DTAB) micelles in aqueous solution and their behavior as the reaction medium. Langmuir 10:2088–2094
38. Bunton CA (1979) Reaction kinetics in aqueous surfactant solutions. Catal Rev Sci Eng 20:1–50
39. Martinek KM, Yatimirski AK, Osipov AP, Berezin IV (1973) Micellar effects on kinetics and equilibrium of synthesis and hydrolysis of benzylideneaniline: a general kinetic conception of micellar catalysis. Tetrahedron 29:963–969
40. Scrimin P, Tecilla P, Tonellato U (1994) Chiral lipophilic ligands. 1. enantioselective cleavage of alpha-amino acid esters in metallomicellar aggregates. J Org Chem 59:4194–4201
41. Müller W, Déjugnat C, Zemb T, Dufrêche J, Diat O (2013) How do anions affect self-assembly and solubility of cetylpyridinium surfactants in water. J Phys Chem B 117:1345–1356
42. Bergstrom M (2001) Synergistic effects in mixtures of an anionic and a cationic surfactant. Langmuir 17:993–998
43. Lai DT, Connor CJO (2000) Synergistic effects of surfactants on kid pregastric lipase catalyzed hydrolysis reactions. Langmuir 16:115–121
44. Mulqueen M, Blankschtein D (2000) Prediction of equilibrium surface tension and surface adsorption of aqueous surfactant mixtures containing zwitterionic surfactants. Langmuir 16:7640–7654
45. Manet S, Karpichev Y, Dedovets D, Oda R (2013) Effect of hofmeister and alkylcarboxylate anionic counterions on the krafft temperature and melting temperature of cationic gemini surfactants. Langmuir 29:3518–3526
46. Lu JR, Zhao XB, Yaseen M (2007) Biomimetic amphiphiles: biosurfactants. Curr Opin Colloid Interf Sci 12:60–67
47. Berti D (2006) Self assembly of biologically inspired amphiphiles. Curr Opin Colloid Interf Sci 11:74–78
48. Bagchi B (2005) Water dynamics in the hydration layer around proteins and micelles. Chem Rev 105:3197–3219
49. Lipshutz BH, Ghorai S, Clerget MC- (2018) The hydrophobic effect applied to organic synthesis: recent synthetic chemistry “in water”. Chem Eur J 24:6672–6695
50. Sorella GL, Strukul G, Scarso A (2015) Recent advances in catalysis in micellar media. Green Chem 17:644–683
51. Kumar D, Seth K, Kommi DN, Bhagat S, Chakraborti AK (2013) Surfactant micelles as microreactors for the synthesis of quinoxalines in water: scope and limitations of surfactant catalysis. RSC Adv 3:15157–15168
52. Wang L-M, Jiao N, Qiu J, Yu J-J, Liu J-Q, Guo F-L, Liu Y (2010) Sodium stearate-catalyzed multicomponent reactions for efficient synthesis of spirooxindoles in aqueous micellar media. Tetrahedron 66:339–343
53. Kumar A, Gupta MK, Kumar M, Saxena D (2013) Micelle promoted multicomponent synthesis of 3-amino alkylated indoles *via* a Mannich-type reaction in water. RSC Adv 3:1673–1678
54. Mattiello S, Rooney M, Sanzone A, Brazzo P, Sassi M, Beverina L (2017) Suzuki-Miyaura micellar cross-coupling in water, at room temperature, and under aerobic atmosphere. Org Lett 19:654–657
55. Inamoto K, Nozawa K, Yonemoto M, Kondo Y (2011) Micellar system in copper-catalysed hydroxylation of arylboronic acids: facile access to phenols. Chem Commun 47:11775–11777
56. Lipshutz BH, Ghorai S, Leong WWY, Taft BR (2011) Manipulating micellar environments for enhancing transition metal-catalyzed cross-couplings in water at room temperature. J Org Chem 76:5061–5073
57. Lipshutz BH, Abela AR (2008) Micellar catalysis of Suzuki–Miyaura cross-couplings with heteroaromatics in water. Org Lett 10:5329–5332
58. Krasovskiy A, Duplais C, Lipshutz BH (2009) Zn-mediated, Pd-catalyzed cross-couplings in water at room temperature without prior formation of organozinc reagents. J Am Chem Soc 131:15592–15593
59. Takale BS, Thakore RR, Handa S, Gallou F, Reillyd J, Lipshutz BH (2019) A new, substtuted palladacycle for ppm level Pd-catalyzed Suzuki-Miyaura cross couplings in water. Chem Sci 10:8825–8831

60. Vaidya GN, Fiske S, Verma H, Lokhande S, Kumar D (2019) A micellar catalysis strategy applied to the Pd-catalyzed C–H arylation of indoles in water. Green Chem 21:1448–1454

Chapter 4
Microwave-Assisted Amination Reaction: A Green Approach

Salim Saranya, Thaipparambil Aneeja, and Gopinathan Anilkumar

1 Introduction

The non-conventional method of heating has gained much importance in synthetic organic chemistry due to its advantages over conventional heating methods for organic transformations. The inherent attributes of microwave are being less hazardous and economical providing quick conversion at a faster rate in fewer steps and have attracted more attention toward it in the recent years. Another important advantage of microwave reaction is that it avoids or lowers the formation of by-products. Compared to conventional heating methods in which heating occurs slowly because the reactants get activated by transferring heat through the walls of the vessel, in MW assisted heating, microwaves directly come into contact with molecules and activate the reactants in less than a nano-second and hence making faster heating. MW assisted heating could also be applied for the formation of bioactive molecules as well as in the synthesis of compounds used for drug delivery. Hence, MW assisted reactions have great importance in modern synthetic organic chemistry.

MW assisted reactions are widely used in metal-catalyzed reactions and cross-coupling chemistry; but a collective account of the same is not discussed till date. Since the area is very vast, a comprehensive review covering all aspects of it in a single chapter is impossible. So considering the importance of C–N coupling in

S. Saranya · T. Aneeja · G. Anilkumar (✉)
School of Chemical Sciences, Mahatma Gandhi University, Priyadarsini Hills, Kottayam, Kerala 686560, India
e-mail: anilgi1@yahoo.com

S. Saranya
e-mail: saranyasalim123@gmail.com

T. Aneeja
e-mail: aneejat1995@gmail.com

G. Anilkumar and S. Saranya (eds.), *Green Organic Reactions*, Materials Horizons: From Nature to Nanomaterials, https://doi.org/10.1007/978-981-33-6897-2_4

organic synthesis, we discuss here a brief account of MW assisted amination reactions. Among the various bond formations, C–N bond formation has great significance due to its inevitable role in the formation of biologically and pharmaceutically active aliphatic, cyclic, aromatic, and heterocyclic compounds. The amination is a C–N bond formation in which the amino group is introduced by an amine source into an organic moiety by replacing a halogen, hydroxyl, or triflate group. Amination reaction is mainly classified into four types as (i) electrophilic amination, (ii) reductive amination, (iii) hydro-amination, and (iv) nucleophilic amination.

In this chapter, we focus on MW assisted amination reaction as a greener method in synthetic organic chemistry. For better understanding, clarity, and brevity, we have categorized the review as (i) metal-catalyzed amination reactions, (ii) metal-free amination reactions, and (iii) miscellaneous reactions.

2 Classification

2.1 Metal-Catalyzed Amination Reactions

A novel protocol for the synthesis of the dye 8-anilino-1-naphthalene sulfonic acid (ANS) has been developed by Georg and co-workers using microwave-assisted Ullmann coupling [1]. The microwave-assisted reaction of 8-chloro naphthalene sulfonic acid with differently substituted aryl amines resulted in ANS derivatives in low to good yields (Scheme 1).

MacArthur et al. reported a MW assisted amidation of aryl chlorides using 20 mol% of CuI, 60 mol% of *N,N'*-cyclohexane-1,2-diamine, 2.2 equiv. of K_2CO_3 and amide in acetonitrile at 200 °C for 45 min −1 h affording the desired product in moderate to good yields (Scheme 2) [2]. Various primary and secondary amides were reacted with several aryl chlorides. Aryl chlorides with substituents at para-position, except *para*-chloro, gave good yields while *ortho*-substituted ones gave only moderate yields, the reason for which is attributable to steric hindrance.

In 2003, Hamann and co-workers reported the microwave-assisted synthesis of 1-aminonaphthalenes and 5- and 8-aminoquinolines by amination of the respective

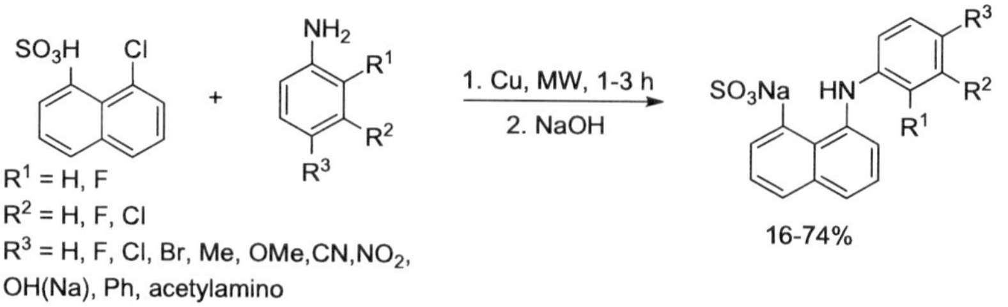

Scheme 1 Microwave-assisted synthesis of ANS derivatives

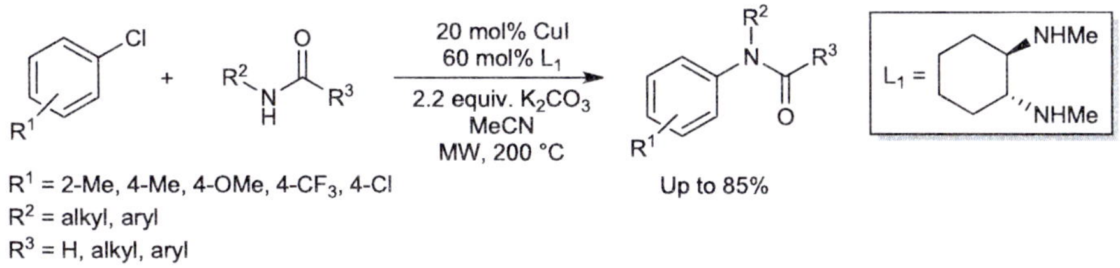

Scheme 2 MW assisted amidation of aryl chlorides

Scheme 3 Pd-catalyzed amination of 1-bromonaphthalenes and 5- and 8-bromoquinolines under microwave conditions

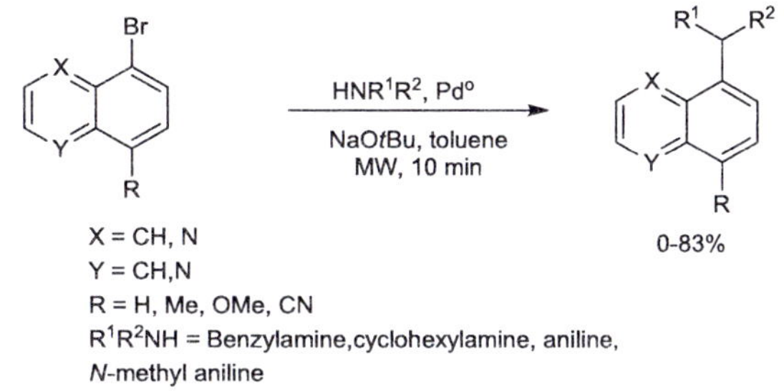

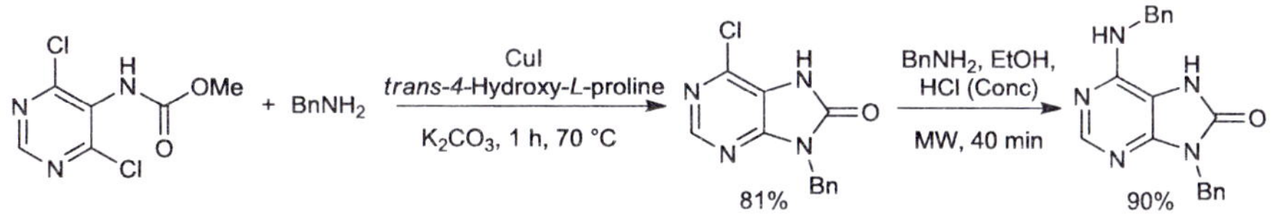

Scheme 4 Synthesis of 6,9-disubstituted purin-8-ones

aryl bromides using Pd as catalyst (Scheme 3) [3]. The optimized condition of the reaction is 2 mol% $Pd_2(dba)_3$, 6 mol% $PPFNMe_3$, and 1.5 equiv. of NaO^tBu under argon.

A novel strategy for the synthesis of 6,9-disubstituted purin-8-ones using 9-substituted 6-chloropurin-8-ones and amines has been disclosed by Sun and Zhong in 2010 (Scheme 4) [4]. The reaction proceeded via a Cu/4-hydroxy-*L*-proline–catalyzed coupling of dichloropyrimidin-5-ylcarbamate and cyclization is followed by microwave-assisted amination.

2.1.1 Catalyst-Free Amination Reactions

Zhou and co-workers reported the amination of mono-, di-, and tri-substituted aryl halides under catalyst-free conditions in presence of K_2CO_3 as base NMP as solvent at 250 °C for 20 h. [5] Here, the increase in the number of fluoro substituents was found to increase the yield of the products, and the *meta*-substituted halides showed higher activation compared to the corresponding *ortho*-derivative.

A similar type of amination was reported by Muller's group in 2007 under solvent- and catalyst-free conditions (Scheme 5) [6]. The microwave-assisted protocol provides a milder way toward the synthesis of N-substituted 5-nitroanthranilic acid derivatives in moderate to good yields. The initial reaction of 2-chloro-5-nitrobenzoic acid with aniline under solvent- and catalyst-free conditions afforded the desired product in 80% yield.

Baskaran and Reddy reported the microwave-assisted synthesis of fluoro-quinolones by amination of 7-halo-6-fluoroquinolone-3-carboxylic acids [7]. The reaction of 7-halo-6-fluoroquinolone-3-carboxylic acid with excess of amine in DMSO afforded the products in good to excellent yields.

A microwave-irradiated protocol for the synthesis of pyrrolidine derivatives via amination of dimesylate was discussed by Wu et al. [8] Under the optimized condition using 2-amino alcohol as solvent at 120 °C for 10 min afforded the desired pyrrolidine derivatives in moderate to excellent yields. One of the major advantages of the strategy is that the aliphatic amines also reacted well under this condition.

The amination of 2-bromo-3-chloropyridines using various aminoethanol could be achieved under MW irradiation in pyridine at 100 °C. [9] The initial reaction under conventional heating for 24 h at 180 °C in *i*-Pr_2EtN gave the product in 65% yield only. Further changing the conventional heating conditions to microwave irradiation for 1 h in 2-aminoethanol increased the yield to 87% (Scheme 6). Exploration of the substrate scope for aminoethanols revealed that un-substituted as well as cyclic secondary amines gave good yields. While substituted ones with either methyl or phenyl group and acyclic one afforded the desired products in moderate yield only.

Hu *et al.* disclosed the amination of 3-chloro-6-(4-methoxyphenyl)pyridazine using different amines in presence of 1.5 equiv. K_2CO_3 as base in DMF under MW irradiation for 15 min-18 h at 120 °C, which afforded the desired products in excellent yields (Scheme 7) [10]. The initial reaction of 3-chloro-6-(4-methoxyphenyl)pyridazine with benzylamine in presence of K_2CO_3 in DMF at 195 °C under microwave irradiation gave 97% of the product. Aliphatic amines, arylamines, and cyclic amines reacted well under this condition.

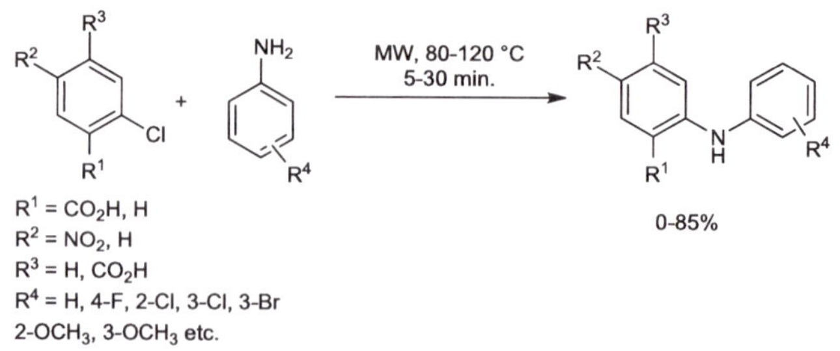

Scheme 5 Synthesis of *N*-arylanthranilic acid derivatives

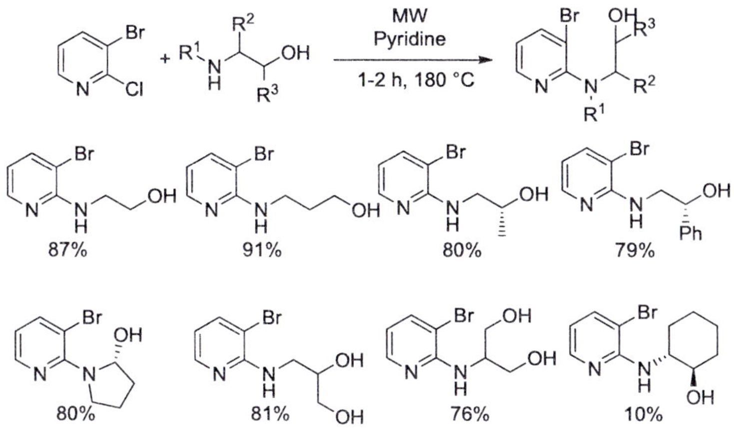

Scheme 6 Microwave-assisted amination of 3-bromo-2-chloropyridine with various aminoalcohols

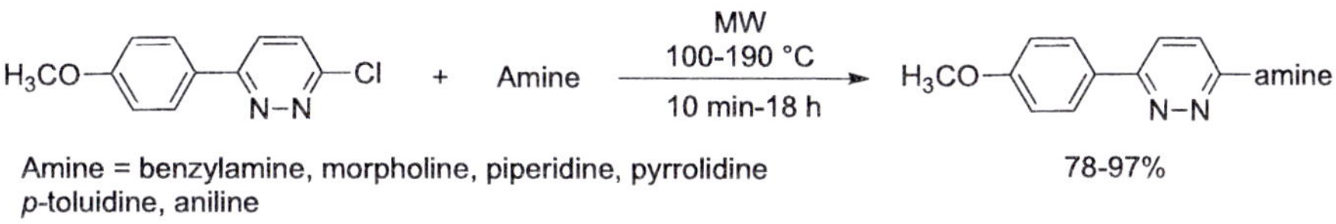

Scheme 7 Amination of 3-chloro-6-(4-methoxyphenyl)pyridazine under microwave irradiation

The amination of tosyl solketal with *n*-butylamine under solvent-free condition was studied by Bouquillon [11]. The amination reaction of different amines including aliphatic and heterocyclic reacted well.

Hranjec's group achieved the synthesis of 5-amino substituted benzimidazo[1,2-*a*]quinoline-6-carbonitriles by the amination of its halogen derivative. [12] The catalyst-free protocol provided the expected product in low to moderate yields. The same group described the synthesis of 2-amino substituted benzimidazo[1,2-*a*]quinolones in 33–99% yield [13, 14]. They have also reported the synthesis of amino and diamino derivative benzo[*b*]thieno[2,3-*b*]pyrido[1,2-*a*]benzimidazoles using the same protocol [15].

The optimization studies for the amination of periodic mesoporous organosilica (PMO) were carried out by Ferreira et al. in 2015 [16]. The microwave-assisted protocol could considerably reduce the reaction time from six days (conventional heating) to 15 min. without changing the yield.

A greener protocol for the synthesis of 6-substituted aminopurine was disclosed by Qu et al. (Scheme 8) [17]. A library of 6-substituted aminopurine analogs in moderate to good yield were synthesized from 6-chloropurine and various amines in water at 72 °C for 10 min. About 65–92% of the target products were obtained with moderate to high purity in 10 min at 200 W in presence of water.

Scheme 8 Synthesis of 6-substituted aminopurine analogs

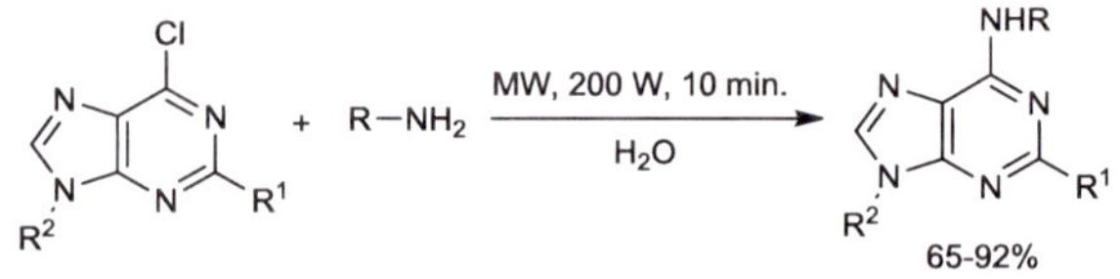

A similar work was reported by Schmalz's group where they discussed the amination of chloropurine derivative in the synthesis of acyclic nucleoside analogs (Scheme 9) [18]. The reaction of chloro-substituted precursor with alkyl or aryl amine using 1.1 equiv of diisopropylethylamine (DIPEA) in EtOH under microwave at 120 °C for 10 min afforded the desired product in good yield. The same reaction under conventional heating afforded the product in lower yield.

Ding developed a one-pot two-step protocol for the synthesis of 4,5-disubstituted pyrazolopyrimidines [19]. The initial step involves the displacement of chloro-substituents with different anilines, and the second step involves a Suzuki coupling (Scheme 10). Anilines with electron-releasing groups reacted well in presence of acetic acid and dioxane at 150 °C for 10 min under MW irradiation, while electron-withdrawing groups gave only comparatively lower yields.

The synthesis of 2-aminonicotinic acid by S_N2 displacement of chloro substituent of 2-chloronicotinic acid was described by McDonald et al. (Scheme 11) [20]. A range of amines like aliphatic amine, anilines, benzyl amines, etc., were studied, and the corresponding products were obtained in good yields.

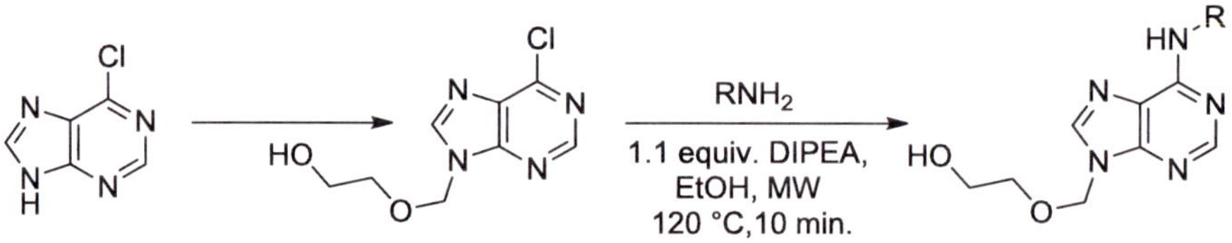

Scheme 9 Amination of chloropurine derivative in the synthesis of acyclic nucleoside analogs

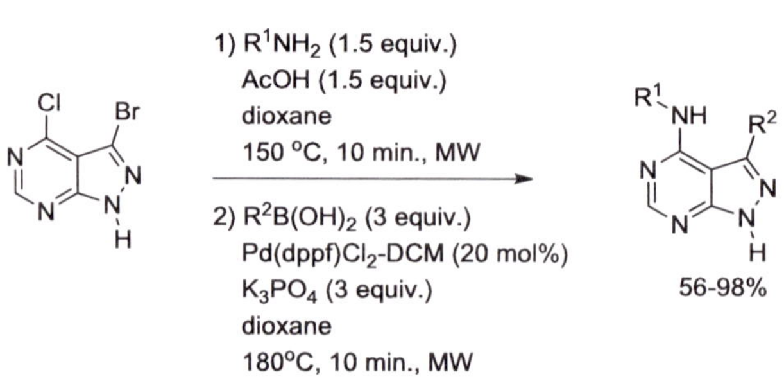

Scheme 10 Synthesis of 4,5-disubstituted pyrazolopyrimidines by microwave irradiation

Scheme 11 Synthesis of 2-aminonicotinic acids by reaction of 2-chloronicotinic acid with amines

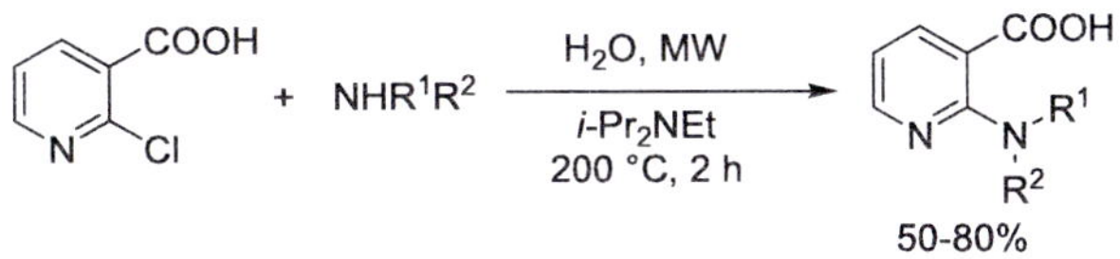

Reddy and co-workers put forward a metal- and solvent-free, atom economic procedure for the amination of 2-chloro azoles [21]. A wide variety of amines was studied and in all the cases, quantitative yields of the expected products were obtained within a few minutes.

In 2004, Turner and Poondra accomplished a novel route for *N*-substituted oxindoles through a two-step process in which initially, an amide bond formation occurs between anilines, 2-haloarylacetic acids and various amines in microwave conditions followed by intramolecular amidation catalyzed by palladium in presence of water (Scheme 12) [22]. The desired product was obtained in moderate to excellent yields. While examining the scope of the reaction, it was noted that using alkylamines, the reaction could be performed without isolation of the amide intermediate.

Samadi and co-workers in 2011 reported microwave-assisted transition metal-free synthesis of 2-(*N,N*-dimethyl)amine and 2-aminopyridine derivatives by the reaction amide solvents like formamide and dimethylformamide with 2-chloro-substituted pyridines (Scheme 13) [23]. It was observed that the yield of the reaction mainly depended on the substituents on the aromatic ring, and dimethyl formamide was found to the better solvent than formamide. This is an efficient method since the yields and reaction times are better than those in the conventional methods.

Saulnier *et al.* developed a synthetic pathway for the production of hydrogen halide salts of primary amines (Scheme 14) [24]. Under microwave irradiation, HX salts of primary amines were produced from the corresponding halides in methanol and 7 M ammonia at 130 °C from 0.5 to 2.5 h. This method helps in avoiding the production

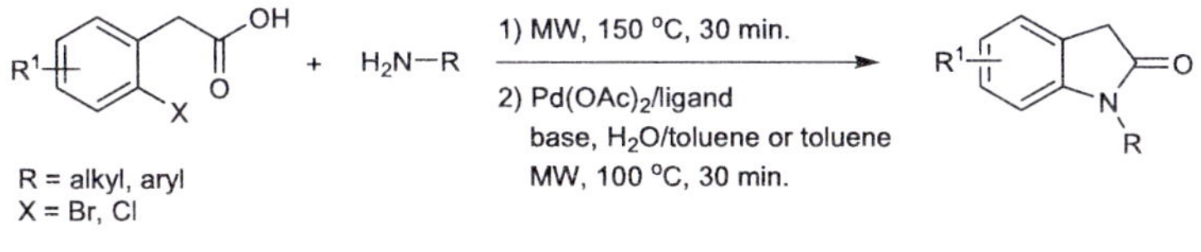

Scheme 12 Synthesis of substituted oxindoles

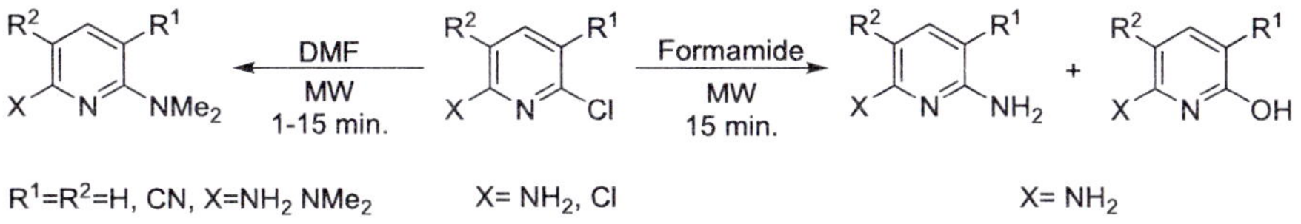

Scheme 13 Amination of 2-chloro-substituted pyridines with DMF and formamides under microwave irradiation

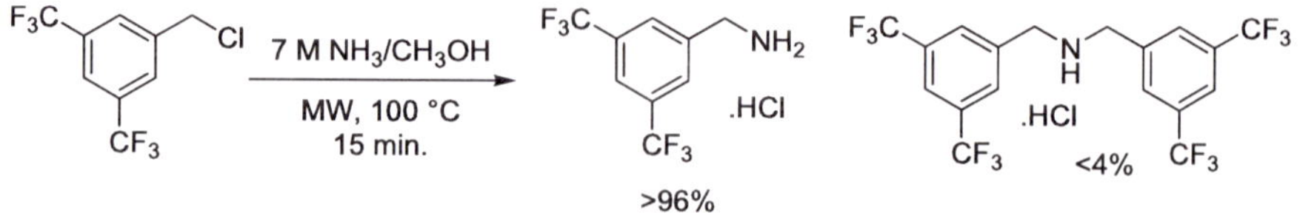

Scheme 14 Synthesis of benzyl amines from benzyl halides using microwave irradiation

of secondary amine as the side product, and about 90% of the yield was obtained by simple evaporation of the solvent. Thus, this method helps in the parallel synthesis since primary amine products are obtained by simple evaporation technique.

2.2 *Miscellaneous Reactions*

Various groups have studied base-catalyzed amination of aryl halides (Scheme 15, condition A), [25] potassium trifluoroborates [26] (Scheme 15, condition B), etc., using different amines, while Wang's group reported a base- and catalyst-free protocol for the amination of aryl triflates (Scheme 15, condition C) [27].

Turner et al. reported the synthesis of triarylamines through a microwave-assisted amination of aryldibromides [28]. Another group studied the amino functionalization of polymer under microwave irradiation. [29] The reaction goes through the formation of an intermediate from polymer-bound halide using a phthalimide, which in turn on amination with methyl amine in THF/H_2O under MW irradiation results in polymer-bound amine.

The direct amination of protected inosine using PyBroP under microwave irradiation has been developed by Scammells for the synthesis of functionalized N^6-substituted adenosines [30]. The initial reaction commenced with the TBS-protected inosine and *endo*-norbornylamine, which afforded low yields. Further optimization of the reaction conditions revealed that good yields of N^6-substituted adenosines

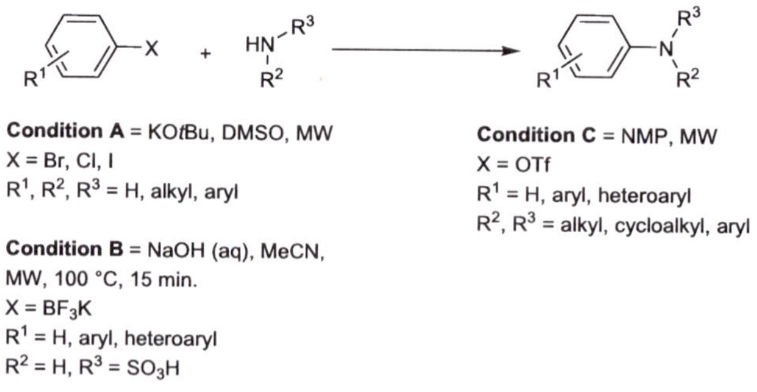

Scheme 15 Microwave-assisted amination of aryl halides, potassium trifluoroborates, and triflates

Scheme 16 Amination of 4,6-dichloropyrimidine with anilines

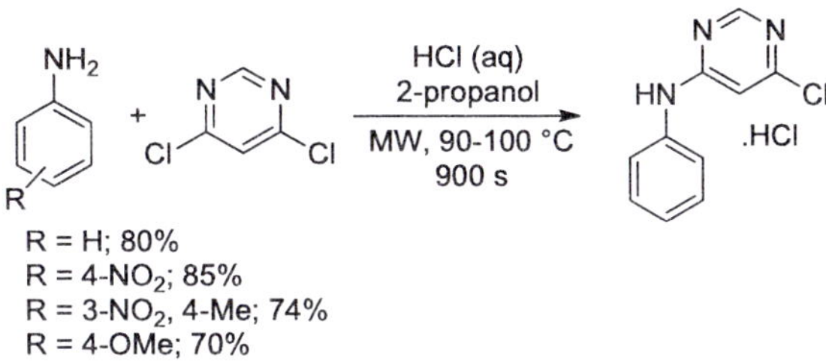

could be obtained when the reaction was carried out in presence of PyBrop, (±)-*endo*-norbornylamine hydrochloride for 15 min at 120 °C under microwave irradiation.

Moore and co-workers reported 2-*tert*-butylimino-2-diethylamino-1,3-dimethylperhydro-1,3,2-diazaphosphorine (BEMP)/polymer supported BEMP (ps-BEMP) mediated N-alkylation of weakly acidic amines [31]. To check the feasibility of the reaction, the scope of a few cyclic amines was also studied which offered moderate yield of the products.

Pyrimidines are important moiety present in nature as its derivatives [32]. Hence, the synthesis and functionalization of various pyrimidines have great importance in the scientific world. In 2006, Backes et al. discussed a one-pot rapid protocol for the functionalization of pyrimidines (Scheme 16) [33]. This method is an easy way for the functionalization of pyrimidines by microwave-assisted amination reaction followed by MW assisted Suzuki coupling.

3 Conclusion

MW assisted amination reactions are extensively studied in the past few years. Cu and Pd are the most commonly used metals in microwave-assisted amination reactions. From the foregoing discussions, it is clear that MW assisted reactions promote metal-free protocols making it more greener. The MW assisted strategy was also successfully applied for the synthesis of polymers. Moreover, the MW promoted protocol is also used in the synthesis of aminopurine derivatives for the synthesis of acyclic nucleoside analogs having tremendous application in pharmaceutical chemistry. The choice of catalysts, solvents, and reagents in organic reactions make the microwave-assisted technique a more sustainable, economic, and greener protocol.

Acknowledgements SS and TA thank the Council of Scientific and Industrial Research (CSIR), New Delhi, for the award of junior research fellowships. GA thanks the Kerala State Council for Science, Technology, and Environment (KSCSTE), Trivandrum, for a research grant (341/2013/KSCSTE dated 15. 03. 2013). The authors are also thankful to EVONIK Industries, Germany, for financial support (ECRP 2016 dated 6. 10. 2016).

References

1. Wang N, Faber EB, Georg GI (2019) Synthesis and spectral properties of 8–anilinonaphthalene-1-sulfonic acid (ANS) derivatives prepared by microwave-assisted copper(0)-catalyzed ullmann reaction. ACS Omega 4(19):18472–18477
2. Chang RK, Clairmont BP, Lin S, MacArthur AHR (2019) Amidation of aryl chlorides using a microwave-assisted, copper-catalyzed concurrent tandem catalytic methodology. Organometallics 38(22):4448–4454
3. Wang T, Magnin DR, Hamann LG (2003) Palladium-catalyzed microwave-assisted amination of 1-bromonaphthalenes and 5- and 8-bromoquinolines. Org Lett 5(6):897–900
4. Zhong Q-F, Sun L-P (2010) An efficient synthesis of 6,9-disubstituted purin-8-ones via copper-catalyzed coupling/cyclization. Tetrahedron 66:5107–5111
5. Meng X, Cai Z, Xiao S, Zhou W (2013) Microwave-assisted amination from fluorobenzenes without catalyst and strong base. J Fluor Chem 146:70–75
6. Baqi Y, Müller CE (2007) Catalyst-free microwave-assisted amination of 2-chloro-5-nitrobenzoic acid. J Org Chem 72:5908–5911
7. Reddy PG, Baskaran S (2001) Microwave assisted amination of quinolone carboxylic acids:an expeditious synthesis of fluoroquinolone antibacterials. Tetrahedron Lett 42:6775–6777
8. Chang Y-C, Chir J-L, Wu A-T et al (2009) Microwave-assisted synthesis of pyrrolidine derivatives. Tetrahedron Lett 50:4925–4929
9. Kim JG, Yang EH, Ha JD et al (2010) Microwave-assisted amination of 3-bromo-2-chloropyridine with various substituted aminoethanols. Tetrahedron Lett 51:3886–3889
10. Lin S, Liu Z, Hu Y (2007) Microwave-enhanced efficient synthesis of diversified 3,6-disubstituted pyridazines. J Comb Chem 9:742–744
11. Mbakidi J-P, Bouquillon S (2018) Glycerol-bsed ionic liquids: crucial microwaves-assisted synthetic step for solketal amines. J Mol Liq 252:218–222
12. Perin N, Nhili R, Ester K, Hranjec M et al (2014) Synthesis, antiproliferative activity and DNA binding properties of novel 5-Aminobenzimidazo[1,2-a]quinoline-6-carbonitriles. Eur J Med Chem 80:218–227
13. Perin N, Martin-Kleiner I, Hranjec M et al (2013) Biological activity and DNA binding studies of 2-substituted benzimidazo[1,2-a]quinolines bearing different amino side chains. Med Chem Commun 4:1537–1550
14. Vušak D, Perin N, Bertoša B (2017) Synthesis and antiproliferative activity of amino-substituted benzimidazo[1,2-a]quinolines as mesylate salts designed by 3D-QSAR analysis. Mol Divers 21(3):621–636
15. Perin N, Bobanović K, Hranjec M et al (2017) Antiproliferative activity of amino substituted benzo[b]thieno[2,3-b]pyrido[1,2-a]benzimidazoles explored by 2D and 3D cell culture system. Eur J Med Chem 125:722–735
16. Lourenço MAO, Gomes JRB, Ferreira P (2015) Optimization of the time and temperature of the microwave-assisted amination of phenylene-PMO. RSC Adv 5:9208–9216
17. Qu G, Han S, Zhang Z et al (2006) Microwave assisted synthesis of 6-substituted aminopurine analogs in water. J Braz Chem Soc 17(5):915–922
18. Lanver A, Schmalz H-G (2005) Microwave-assisted amination of a chloropurine derivative in the synthesis of acyclic nucleoside analogues. Molecules 10:508–515
19. Wu TYH, Schultz PG, Ding S (2003) One-pot two-step microwave-assisted reaction in constructing 4,5-disubstituted pyrazolopyrimidines. Org Lett 5(20):3587–3590
20. Quevedo CE, Bavetsias E, McDonald E (2009) Microwave-assisted synthesis of 2-aminonicotinic acids by reacting 2-chloronicotinic acid with amines. Tetrahedron Lett 50:2481–2483
21. Kumar KU, Dileep K, Reddy KH et al (2018) Microwave assisted amination of 2-chloro azoles with various substituted aryl piperazines and aryl sulfonylpiperazines under neat conditions. Curr Microw Chem 5:62–72
22. Poondra RR, Turner NJ (2005) Microwave-assisted sequential amide bond formation and intramolecular amidation: a rapid entry to functionalized oxindoles. Org Lett 7(5):863–866

23. Samadi A, Silvia D, Chioua M et al (2011) microwave irradiation-assisted amination of 2-chloropyridine derivatives with amide solvents. Synth Commun 41:2859–2869
24. Saulnier MG, Saulnier K, Struzynski CP et al (2004) Microwave-assisted synthesis of primary amine HX salts from halides and 7M ammonia in methanol. Tetrahedron Lett 45:397–399
25. Shi L, Wang M, Fan C-A, Zhang F-M, Tu Y-Q (2003) Rapid and efficient microwave-assisted amination of electron-rich aryl halides without a transition-metal catalyst. Org Lett 5(19):3515–3517
26. Kuik D, McCubbin JA, Tranmer GK (2017) Sonication and microwave-assisted primary amination of potassium aryltrifluoroborates and phenylboronic acids under metal-free conditions. Synthesis 49:1–7
27. Xu G, Wang Y-G (2019) Microwave-Assisted Amination from Aryl Triflates without Base and Catalyst. Org Lett 6(6):985–987
28. Horie M, Luo Y, Turner ML et al (2008) Triarylamine polymers by microwave-assisted polycondensation for use in organic field-effect transistors. J Mater Chem 18:5230–5236
29. Ljungdahl N, Martikainen L, Kann N (2008) Rapid microwave-assisted preparation of amino-functionalized polymers. Tetrahedron Lett 49:6108–6110
30. Ashton TD, Scammells PJ (2008) Microwave-assisted direct amination: rapid access to multi-functionalized n6-substituted adenosines aust. J Chem 61:49–58
31. Moore JE, Spinks D, Harrity JPA (2004) Microwave promoted amination of 3-bromoisoxazoles. Tetrahedron Lett 45:3189–3191
32. Lagoja IM (2005) Pyrimidine as constituent of natural biologically active compounds. Chem Biodivers 2(1):1–50
33. Hartung CG, Backes AC, Felber B (2006) Efficient microwave-assisted synthesis of highly functionalized pyrimidine derivatives. Tetrahedron 62:10055–10064

Chapter 5
Green Reactions Under Solvent-Free Conditions

Ghodsi Mohammadi Ziarani, Fatemeh Mohajer, and Razieh Moradi

1 Introduction

The green chemistry movement has promoted the industry to be much cleaner. As it was mentioned, the twelve principles of "green chemistry" were introduced in the 1990s by Paul Anastas and John Warner [1]. These points were implemented by chemists, academia, and even industries based on the synthetic strategy to minimize the production of toxic materials in different reaction conditions [1].

Green chemistry is an increasing agreement to accomplish organic reactions without damaging our environment and the most important way to achieve this goal is by conducting reactions under solvent-free conditions [2–6]. These trends are pretty more widespread due to its many advantages like cost, energy, and time-saving which create cleaner, safer, and more facile reactions. In this scenario, it is worthwhile to design reactions under favorable conditions in water or without any solvent [7].

In this area, there are several reactions to be highlighted through solvent-free approaches, which were accomplished in the presence of different catalysts and functionalized supports such as magnetic types [8, 9] mesoporous [10] graphene oxide [11–15], and other metal oxides [16–19].

Synthesis of various biologically active compounds [20] was reported under solvent-free conditions via different reagents such as pyrimidines [18, 21] which are significant heterocycles present in the core of various biologically active materials [22]. These compounds provided numerous medicinal properties such as antibacterial [23] antidiabetic [24] anticancer [21] antileishmanial [25] antiallergic [26]

G. Mohammadi Ziarani (✉) · F. Mohajer · R. Moradi
Department of Chemistry, Alzahra University, Vanak Square, P.O. Box 1993893973, Tehran, Iran
e-mail: gmohammadi@alzahra.ac.ir

F. Mohajer
e-mail: F.mohajer@alzahra.ac.ir

R. Moradi
e-mail: moradi3452@yahoo.com

G. Anilkumar and S. Saranya (eds.), *Green Organic Reactions*, Materials Horizons: From Nature to Nanomaterials, https://doi.org/10.1007/978-981-33-6897-2_5

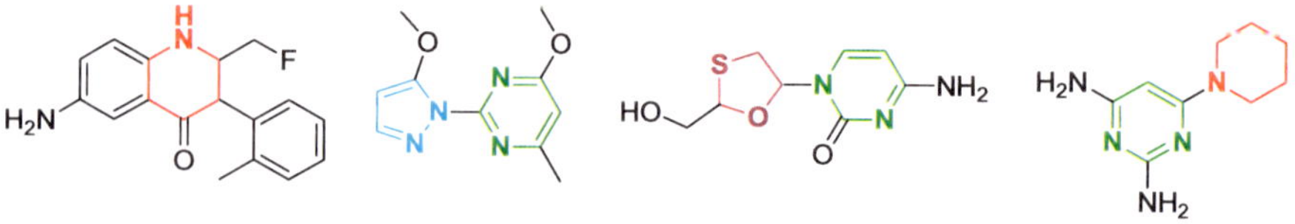

Fig. 1 Some biologically active compounds containing heterocyclic cores

antifungal [27] antipyretic [28] analgesic [29], and antidepressant [30]. Some of the important biologically active compounds such as afloqualone, epirizole, lamivudine, and minoxidil which are applied as anti-inflammatory [31] analgesic [31] anti-HIV [32] and antihypertensive [33] drugs are shown in Fig. 1. Several methods of green synthesis of pyrimidine derivatives under solvent-free conditions are reported.

Another important biologically active heterocyclic core is pyran, which is also synthesized under solvent-free conditions through various methods [34–37]. In this arena, pyrroles [38] quinazolinones [39, 40] benzodiazepines [41], and imidazoles [42, 43] also show biological activities and are synthesized under neat conditions.

A large number of reports dealing with the synthesis of compounds under solvent-free conditions are available. This chapter highlights and summarizes the recent (2016–2020) achievements in the green synthesis of important scaffolds under neat conditions and is classified as shown below [44].

2 Synthesis of Different Scaffolds Under Solvent-Free Conditions

Here, the green chemistry conditions applied in the synthesis of different scaffolds are highlighted.

2.1 The Synthesis of α-Aminophosphonate Derivatives

The α-aminophosphonate derivatives **4**, which are essential from medicinal perspectives [45] are the structural analogues of amino acids wherein a carboxylic moiety is replaced by phosphonic group. Cirandur et al., in 2019, synthesized the anti-oxidant and anti-inflammatory α-aminophosphonate derivatives **4** through a one-pot, three-component coupling reaction from 2-aminothiazole **1** with numerous aldehydes **2** and dialkyl phosphites **3** using caffeine hydrogen sulfate (CHS) as a recyclable catalyst under microwave (MW) irradiation and solvent-free conditions at room temperature in excellent yields as shown in Scheme 1 [46].

Cirandur and co-workers [47] also reported various synthetic α-aminophosphonates **6** from 2-methoxy-5-trifluoromethyl aniline **5**, numerous

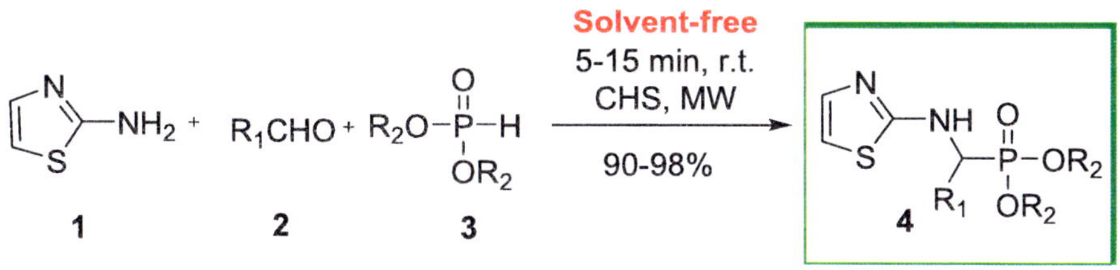

R_1= 4-$O_2NC_6H_4$, 4-ClC_6H_4, 4-BrC_6H_4,3-OCH_3-4-HOC_6H_3, furfuryl
R_2= CH_3,C_2H_5

Scheme 1 Synthesis of α-aminophosphonates under neat condition by Cirandur and co-workers

aldehyde derivatives **2**, and diethyl phosphite **3** in the presence of meglumine sulfate (MS) as an eco-friendly catalyst at room temperature under solvent-free conditions in high yields and short reaction time (Scheme 2).

In another report, Cirandur, and co-workers [48] accomplished the above reaction through Kabachnik–Fields method [49] in the presence of the nano-CuO-Au catalyst to yield α-amino phosphonates **7**, which act as α-glucosidase inhibitor and antioxidant, via three-component reaction of diverse aromatic aldehydes 2, 2-aminophenol **5**, and dimethyl phosphite **3** in solvent-free conditions at 60 °C (Scheme 3).

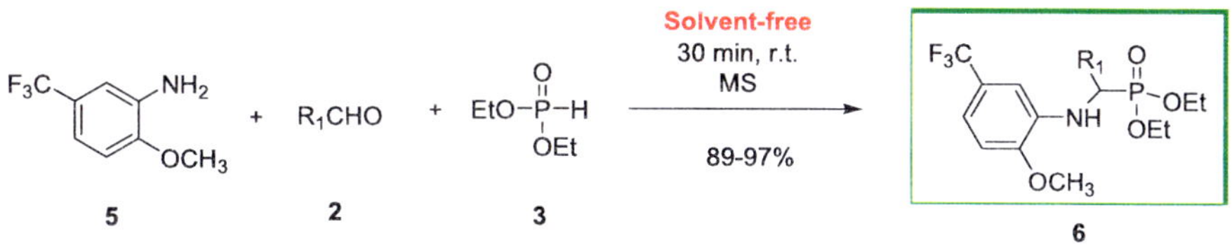

R_1= 4-$O_2NC_6H_4$ 4-ClC_6H_4 4-BrC_6H_4,3-OCH_3-4-HOC_6H_3, 4-HOC_6H_4

Scheme 2 Synthesis of α-aminophosphonates under neat conditions by Cirandur and co-workers

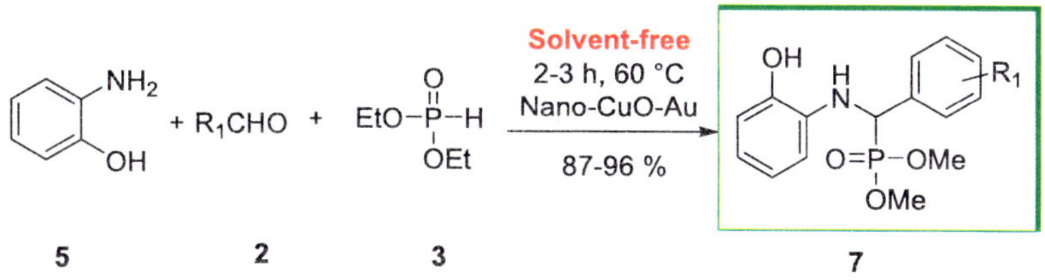

R_1= 4-$O_2NC_6H_4$, 2-HO-3,4-$(H_3CO)_2C_6H_2$, 4-$H_3COC_6H_4$,3-OCH_3-4-HO-4-$HC_3C_6H_2$, 4-ClC_6H_4, 4-$C_2H_5OC_6H_4$, 3,4-$(CH_3O)_2C_6H_3$, 4-$HCOC_6H_4$,2-F-5-BrC_6H_3, 3-Br-4-OH-5-$CH_3OC_6H_2$

Scheme 3 Synthesis of α-aminophosphonates via Kabachnik–Fields reaction under neat conditions by Cirandur and co-workers

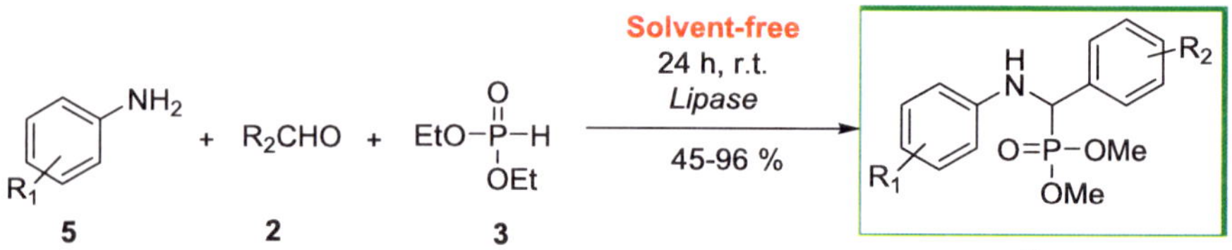

Scheme 4 Synthesis of α-aminophosphonates using lipase under neat conditions by Aribi-Zouioueche and co-workers

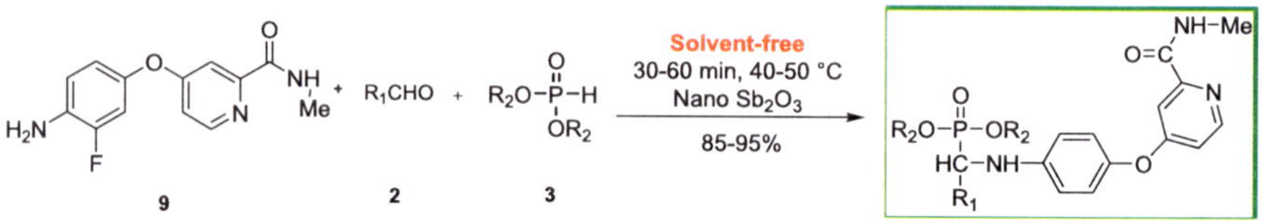

Scheme 5 Synthesis of α-aminophosphonates under neat conditions using nano-Sb_2O_3 by Cirandur and co-workers

In another report, Aribi-Zouioueche and co-workers reacted different aromatic aldehydes **2**, diverse compounds of aniline **5**, and dimethyl phosphite **3** under solvent-free conditions at room temperature using *Candida Antarctica lipase* as a biocatalyst to produce α-aminophosphonate compounds **8** (Scheme 4) [50]. Also, Esmaeilpour and co-workers [51] disclosed the mentioned process in the presence of Fe_3O_4@SiO_2imid-PMAn without any solvent, at room temperature or by ultrasonic irradiation.

In 2019, Cirandur et al. [52] developed a process for the creation of cytotoxic α-aminophosphonates **10** by a simple and proficient one-pot three-component reaction of 3-(4-amino-3-fluorobenzyl)-*N*-methylbenzamide **9** with diverse aromatic aldehydes **2** and dialkyl phosphite **3** using nano-Sb_2O_3 catalyst under solvent-free conditions in high yields at 40–50 °C as shown in Scheme 5.

2.2 *The Synthesis of Pyrimidine Derivatives*

In 2019, Esnaashari et al. [53] outlined the preparation of pyrimidines through multi-component reactions using triethylenediamine or imidazole Brönsted acidic, ionic liquid-supported Zr metal–organic structure (TEDA/IMIZ-BAIL@UiO-66). The dihydropyrido[2,3-*d*]pyrimidine compounds **13** were yielded from the reaction of 6-amino-1,3-dimethyl uracil **11**, several aromatic aldehydes **2**, and acetylacetone **12** in solvent-free conditions (Scheme 6).

The metal–organic structure called MIL-53(Fe) was used as a catalyst to provide the pyrimido[4,5-*d*]pyrimidine compounds **15** in one-step three-component reaction from isothiocyanate **14**, aromatic aldehydes **2**, and 6-aminouracil or *N,N*-dimethyl-6-aminouracil **11** in solvent-free condition at 110 °C in high yield. The other valuable advantage of this study is the recoverable nature of the catalyst supporting green chemistry (Scheme 7) [54].

Foroughifar and co-workers [55] reported the synthesis of dihydropyrimidine derivatives **17** as potent antibacterial agents from thiourea or urea **16**, acetylacetone **12**, and different aryl aldehydes **2** under solvent-free conditions in the presence of the $Mn_{0.5}Fe_{0.25}Ca_{0.25}Fe_2O_4$@starch@aspartic acid magnetic nanoparticles ($MnFeCaFe_2O_4$@starch@aspartic acid MNPs) as a catalyst. The merits of this reaction are easy workup, high-yield, and easy separation of catalyst (Scheme 8).

In another study, Bordoloi and co-workers [56] in 2019 demonstrated the synthesis of dihydropyrimidine derivatives **20** from various aldehydes **2**, the benzil **18,** and ammonium acetate **19** under the solvent-free condition at room temperature during 6–8 h in the presence of the water-extractable pomelo (WEP) as citrus fruit (Scheme 9).

Moradi and co-workers [57], in 2019, demonstrated the synthesis of benzochromenopyrimidines **23** from aldehydes **2**, β-naphthol **21,** and barbituric acids **22** by a green approach without any solvent through aminated multi-walled carbon

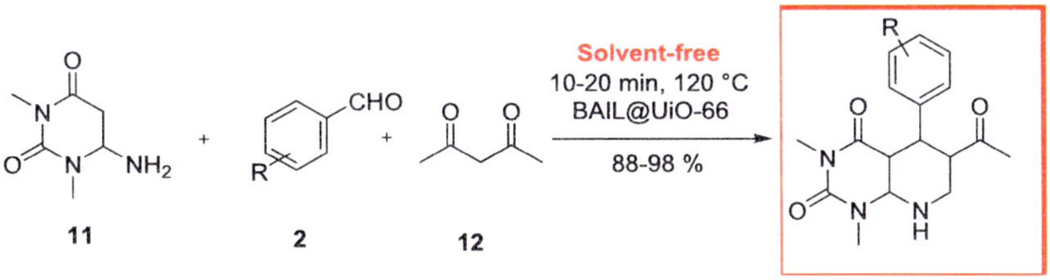

R= H, 4-NO$_2$,3 4-Cl,3,4-(CH$_3$)$_2$, 3-NO$_2$, 4-OCH$_3$, 4-CH$_3$,3,4-(OCH$_3$)$_2$, 4-CN

Scheme 6 Synthesis of dihydropyrimidines under neat conditions

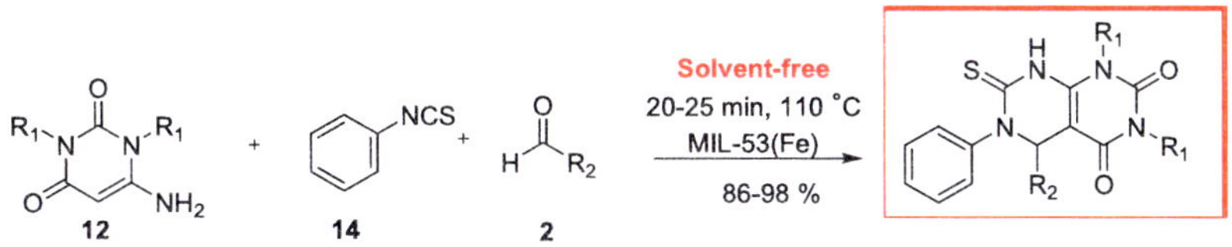

R$_1$= H, CH$_3$

R2= C$_6$H$_5$, 4-FC$_6$H$_4$, 4-BrC$_6$H$_4$,4-H$_3$CC$_6$H$_4$,4-O$_2$NC$_6$H$_4$,4-H$_3$COC$_6$H$_4$, 2-BrC$_6$H$_4$, 2-H$_3$CC$_6$H$_4$, 2-O$_2$NC$_6$H$_4$, 2-ClC$_6$H$_4$

Scheme 7 Synthesis of pyrimido[4,5-*d*]pyrimidines under neat condition by Shirirni and co-workers

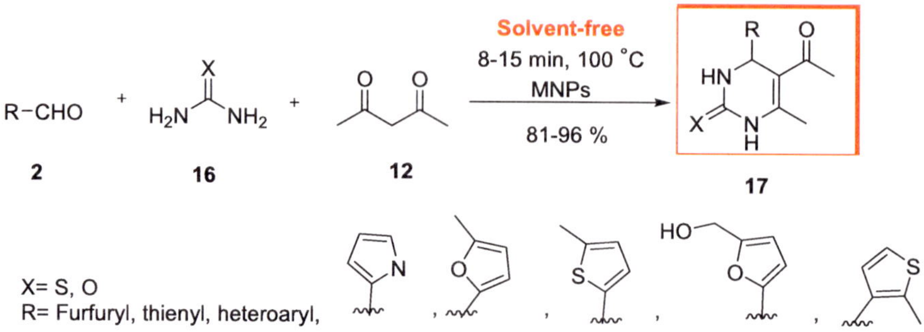

Scheme 8 Synthesis of dihydropyrimidines under neat conditions by Foroughifar and co-workers

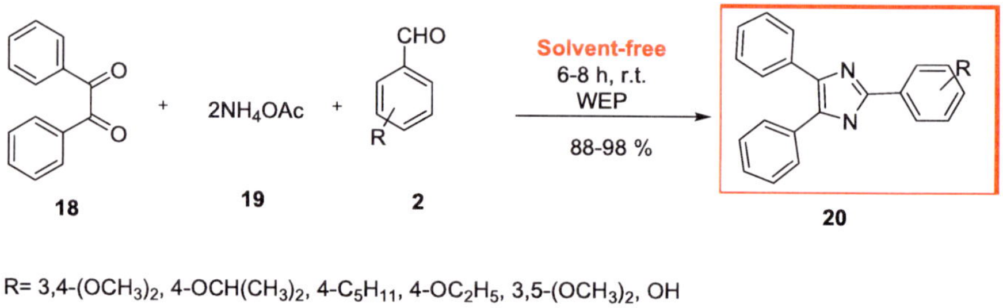

Scheme 9 Synthesis of dihydropyrimidines under neat conditions by Bordoloi and co-workers

nanotubes, which were functionalized by phosphotungstic acid and tungsten called (MWCNTs@NHBut/PTA) (Scheme 10).

Saleh and co-workers reported the synthesis of fused pyrimidine compounds **26** from enaminone **24** and various heterocyclic amines **25** under solvent-free conditions in the presence of the nano-like magnesium oxide (MgO) using ball-mill process. The merits of the reaction are easy workup, high-yield, and reusable catalyst. It is

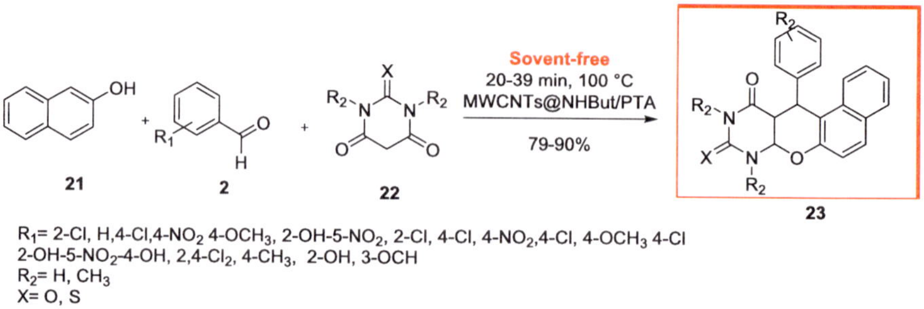

Scheme 10 Synthesis of benzochromenopyrimidines under neat conditions by Moradi and co-workers

important to mention that the aromatic aldehyde with electron-withdrawing groups reacted faster in comparison with electron-releasing groups (Scheme 11) [58].

A series of biologically active pyrimidine derivatives were successfully synthesized by Pal et al. [22] using graphite oxide as a green metal-free carbon catalyst (Scheme 12). In this method, cyanobenzaldehyde **2**, dimedone **27**, and 2-aminobenzothiazole **1** were reacted in the presence of graphite oxide tunder solvent-free reaction conditions (SFRC) at 60 °C for 20 min to provide the product **28** in 88% yield.

The synthesis of the pyrimidine derivatives **30** and **31** was developed through one-step multi-component reactions from 6-amino-1,3-dimethyl uracil **11**, 3,4-methylenedioxyphenol **29** or naphthalen-2-ol **21** and the suitable aryl aldehyde **2** in solvent-free conditions in the presence of SMA/Py/ZnO, which was prepared by the reaction of poly(styrene-co-maleic anhydride) with 3-aminopyridine and zinc oxide (Scheme 13) [59].

Pyrimidine scaffold was fused with triazole and pyrazole via a three-component reaction between 3-methyl-1-phenyl-2-pyrazolin-5-one **33**, various aromatic aldehydes **2,** and 3-amino-1,2,4-triazole **32** or pyrazole to yield the 4-aryl-substituted dihydropyrimidine derivatives **34** under solvent-free condition using tungstate sulfuric acid (TSA) as the green catalyst (Scheme 14) [60].

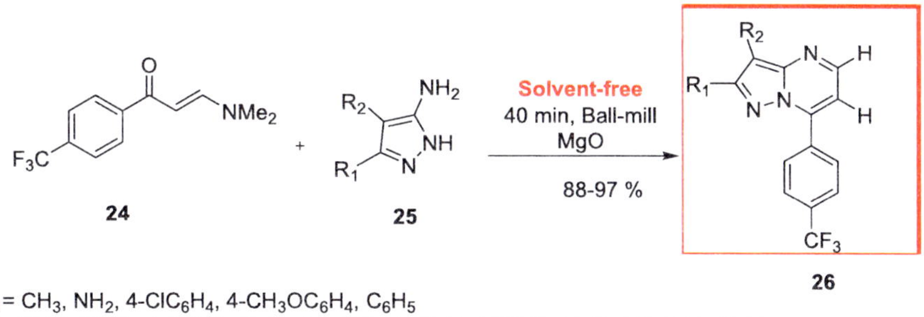

R_1= CH_3, NH_2, 4-ClC_6H_4, 4-$CH_3OC_6H_4$, C_6H_5
R_2= H, Ph-N=N-, 4-ClC_6H_4-N=N, 4-$O_2NC_6H_4$-N=N, 4-FC_6H_4-N=N, 3-$F_3C_6H_4$-N=N

Scheme 11 Synthesis of fused pyrimidines under neat conditions by Saleh and co-workers

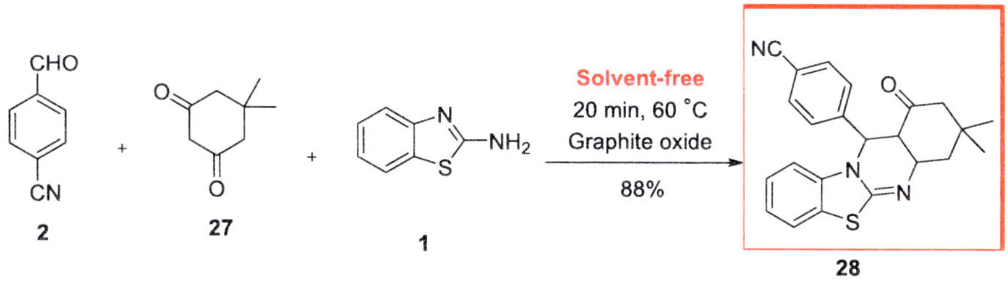

Scheme 12 Synthesis of fused pyrimidines under neat conditions by Pal and co-workers

R= C_6H_5, 4-FC_6H_4, 4-ClC_6H_4, 4-BrC_6H_4, 4-$NO_2C_6H_4$, 4-$CH_3C_6H_4$, C_6H_5, 4-FC_6H_4, 4-BrC_6H_4, 4-$NO_2C_6H_4$, 3-$NO_2C_6H_4$, 4-$CH_3C_6H_4$, 3,4-Cl_2-C_6H_3, 2,4-ClC_6H_3, 2,3-ClC_6H_3, 2,4-$H_3COC_6H_3$, 2-ClC_6H_4, 2-BrC_6H_4, 3-BrC_6H_4, 3-$H_3COC_6H_4$, 4-ClC_6H_4

Scheme 13 Synthesis of chromeno[2,3-*d*]pyrimidne-diones under neat conditions by Heravi and co-workers

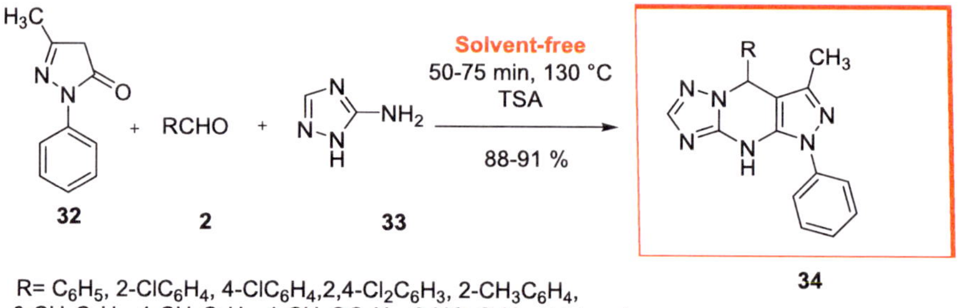

R= C_6H_5, 2-ClC_6H_4, 4-ClC_6H_4,2,4-$Cl_2C_6H_3$, 2-$CH_3C_6H_4$, 3-$CH_3C_6H_4$, 4-$CH_3C_6H_4$, 4-$CH_3OC_6H_4$, 3-$NO_2C_6H_4$, 2-thienyl

Scheme 14 Synthesis of fused dihydropyrimidines under neat conditions by Farahi and co-workers

This attempt reported the synthesis of the pyrimidine-triones 36 from various benzaldehydes **2**, barbituric acid **22**, and 4-amino-2*H*-chromene-2-one **35** using *p*-toluenesulfonic acid as catalyst under microwave irradiation and solvent-free conditions in high yields (Scheme 15) [61].

The heterogeneous copper ferrite ($CuFe_2O_4$) was used as a nanocatalyst to synthesize benzylpyrazolyl pyrido[1,2-*a*]pyrimidine derivatives **38** and pyrazole **39** through a three-component reactions from various aryl aldehydes **2**, 2-hydroxy-4*H*-pyrido[1,2-*a*]pyrimidine-4-ones **37**, and 3-methyl-1-phenyl-1*H*-pyrazol-5-one **32** in solvent-free conditions, and here, the catalyst could be easily removed magnetically from the mixture of reaction (Scheme 16) [62].

Through one-pot multi-component reactions, chromeno[2,3-*d*]pyrimidinetrione derivatives **40** were yielded from numerous aromatic aldehydes **2**, dimedone **27,** or cyclohexane-1,3-dione and barbituric acid **22** in the presence of $Sc(OTf)_3$ as catalyst under solvent-free condition at 100° C for 2 h (Scheme 17) [63].

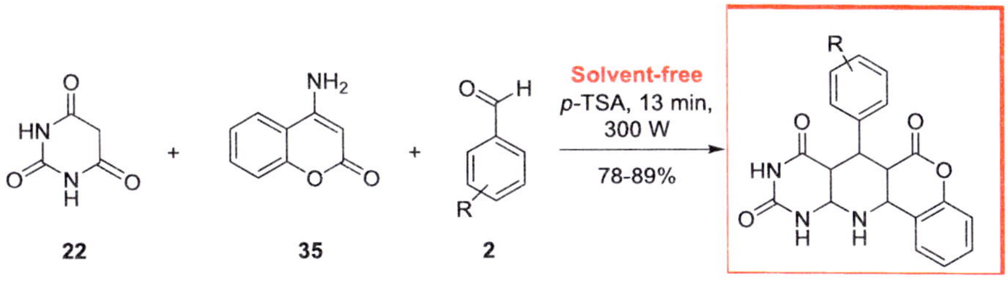

R = H, 2-CH_3, 3-CH_3, 4-CH_3, 4-F, 2-Cl, 3-Cl, 4-NO_2

Scheme 15 Synthesis of pyrimidines under neat conditions by Foroumadi and co-workers

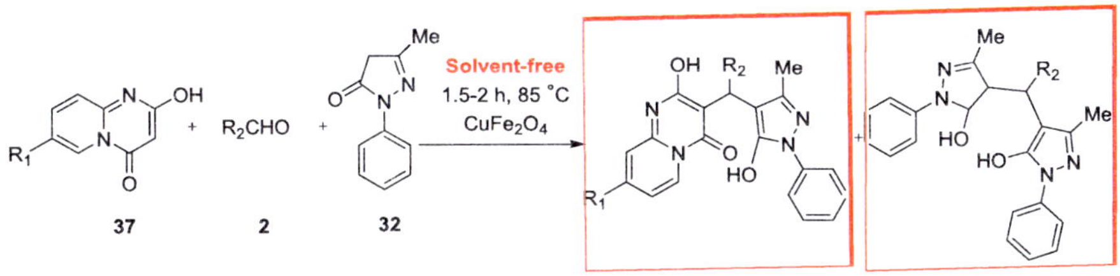

38 62-89 %

39 5-30 %

R_1= H, CH_3
R_2= Ar = C_6H_5, 4-ClC_6H_4, 4-BrC_6H_4, 4-CNC_6H_4, 2-$O_2NC_6H_4$, 3-$O_2NC_6H_4$, 3-$H_3COC_6H_4$,4-ClC_6H_4, 4-BrC_6H_4 2-naphthyl

Scheme 16 Synthesis of pyrimidines under neat conditions by Esmaeili and co-workers

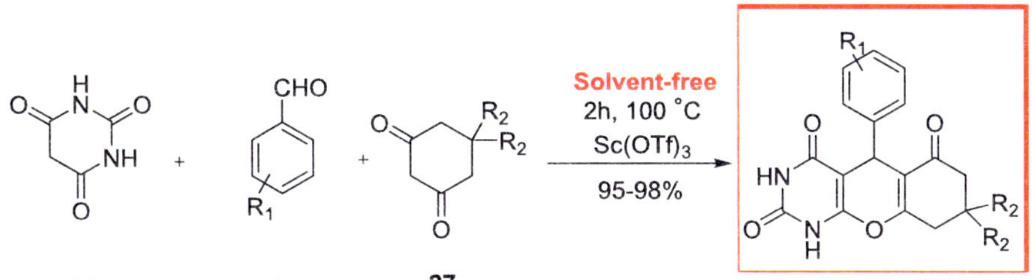

22 2 27 40

R_1= 4-$O_2NC_6H_4$, 4-ClC_6H_4, 4-$H_3CC_6H_4$, 4-FC_6H_4, 4-BrC_6H_4, 3-ClC_6H_4, naphtyl
R_2= CH_3, H

Scheme 17 Synthesis of chromeno[2,3-*d*]pyrimidinetriones under neat conditions by Kumari and co-workers

Pyrimidine-3-carbonitriles **43** were produced through multi-component condensation reaction of various aromatic aldehydes **2**, 3*H*-pyrido[1,2-*a*]pyrimidine-2,4-dione **41**, and malononitrile **42** using ZnO nanoparticles as suitable catalyst under solvent-free conditions (Scheme 18) [64].

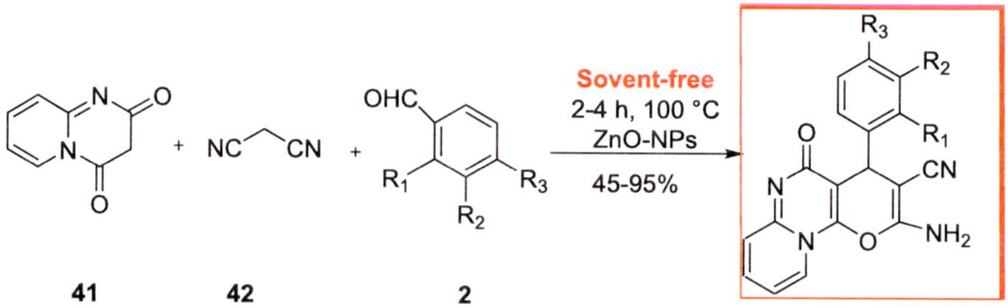

41 42 2 43

R_1= H, CH_3, NO_2, OCH_3, NH_2
R_2= H,OCH_3
R_3= F, Cl, CH_3, OCH_3, NO_2, H,

Scheme 18 Synthesis of pyranopyridopyrimidines under neat conditions by Mahmoud and co-workers

In 2018, Rawat and co-workers [18] synthesized the biologically active fused imidazo[1,2-*a*]pyrimidines **46** through coupling reaction involving 2-aminobenzimidazole **44**, aldehyde **2,** and terminal alkyne **45** in the presence of copper oxide nanoparticles as catalyst under solvent-free condition (Scheme 19).

In a similar study, Kim and co-workers [65] outlined a process to synthesize pyrimidine amine scaffolds **48** from 1*H*-benzo[*d*]imidazol-2-amine **44** and (*E*)-*N*-methyl-1-(methylthio)-2-nitroethenamine **47** with various aldehydes **2** under solvent-free condition at 80 °C using catalytic amounts of PEGMA-g-TEGBDIM (Scheme 20).

A series of pyrazolopyranopyrimidines **51** were obtained from four-component reaction of different aromatic aldehydes **2**, substituted barbituric acids **22**, hydrazine monohydrate **49**, and ethyl acetoacetate **50** in three methods (Method A: using SB-DABCO$^+$Cl$^-$; Method B: using SB-DBU$^+$Cl$^-$; and Method C: using NSB-DBU$^+$Cl$^-$) under solvent-free conditions (Scheme 21) [66].

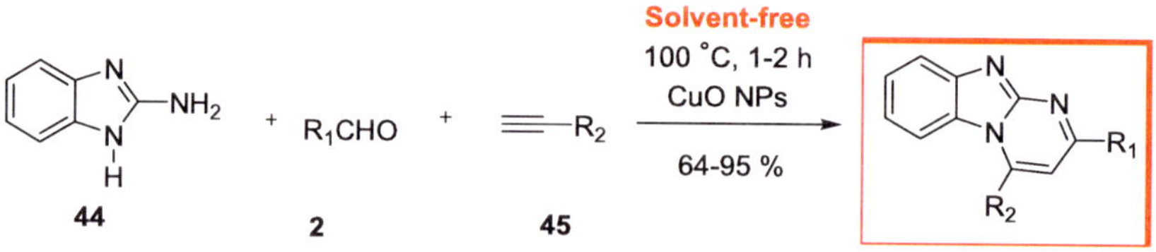

44 2 45 46

R_1= C_6H_5, 4-ClC_6H_4, 4-$O_2NC_6H_4$, 4F-C_6H_4, 4-$H_3CC_6H_4$, thienyl, 3-pyridyle, methylendioxyphenyl
R_2= C_6H_5, $CH_3OC_6H_4$, 4F-C_6H_4, 4-$H_3CC_6H_4$

Scheme 19 Synthesis of imidazo[1,2-a] pyrimidines under neat conditions by Rawat and co-workers

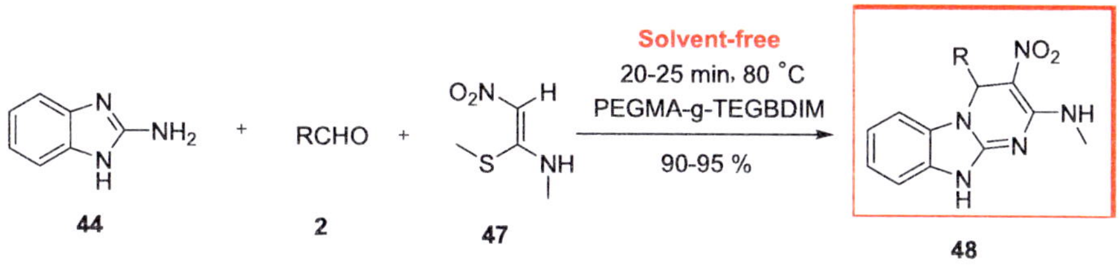

R= 4-BrC6H4, 4-ClC6H4, 4-FC6H4, 4H3CC6H4, 4-CH3OC6H5, 4-HOC6H4, 3-BrC6H4, 3-ClC6H4, 3-FC6H4, 3CH3C6H4, 2-BrC6H4, 2-ClC6H4, 2-FC6H4, 2-H3CC6H4, 2-HOC6H4, pyridyl, thionyl, pyridyl, 2,3,4-(CH3O)3C6H2

Scheme 20 Synthesis of pyrimidine amines under neat conditions by Kim and co-workers

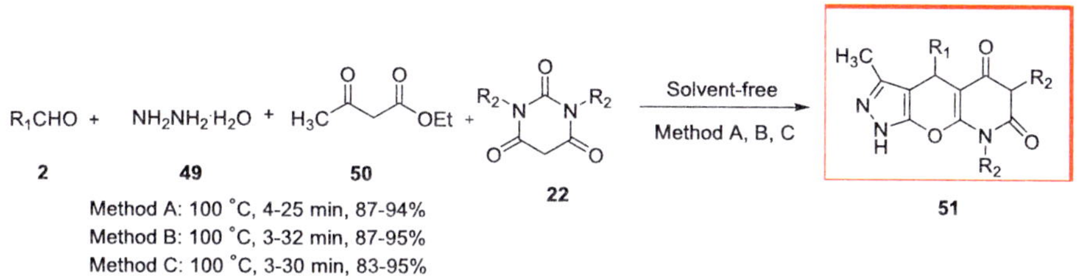

R1= 4-H3CC6H4, 3-O2NC6H4, 4-O2NC6H4, 4-ClC6H4, 2-CH3OC6H4, 4-ClC6H4, 2-CH3OC6H4, 3-ClC6H4, 2-ClC6H4, 3(CH3O)3C6H2, 3-O2NC6H4, 4-CH3OC6H4, 3-CH3OC6H4, 2-HOC6H4,

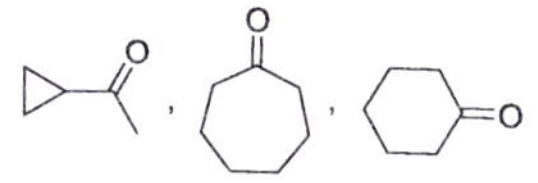

R2= CH3, H

Scheme 21 Synthesis of pyrazolopyranopyrimidines under neat conditions by Shaterian and co-workers

2.3 *The Synthesis of the Pyran Derivatives*

Ghasemzadeh and co-workers [36] synthesized tetrahydrobenzopyran derivatives **52** from different aromatic aldehydes **2**, dimedone **27**, ammonium acetate **19,** and ethyl acetoacetate **50** through a four-component coupling reaction in the presence of Co_3O_4 as nanocatalyst under solvent-free condition at 100 °C (Scheme 22).

Maghsoodlou and co-workers [37] reported the synthesis of the spiro–2–amino–4*H*–pyran derivatives **55** through a one-step three-component condensation reaction from malononitrile **45**, CH-acids **53**, and isatin **54** using Na_2EDTA as catalyst under solvent-free conditions at 70 °C (Scheme 23).

Siddiqui and co-workers [67] performed the synthesis of pyranopyrazole moieties **58** in the presence of the functionalized mesoporous silica and $NdCl_3$ called Nd-SM at 70 °C under solvent-free condition by Knoevenagel condensation [68] between various aromatic aldehydes **2** and ethyl cyanoacetate **56** [69]. The resulting

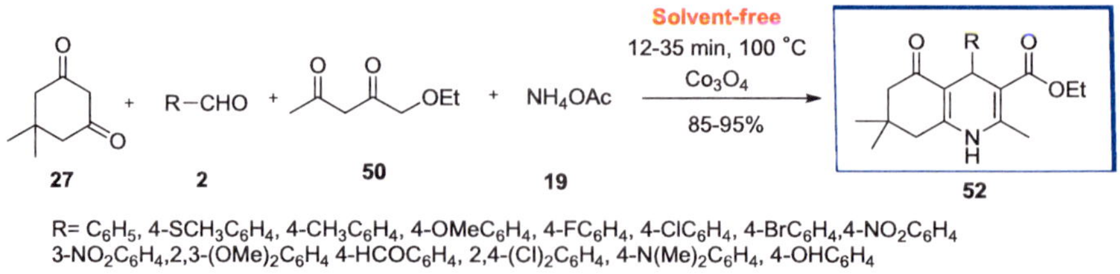

Scheme 22 Synthesis of pyrans under neat conditions by Ghasemzadeh and co-workers

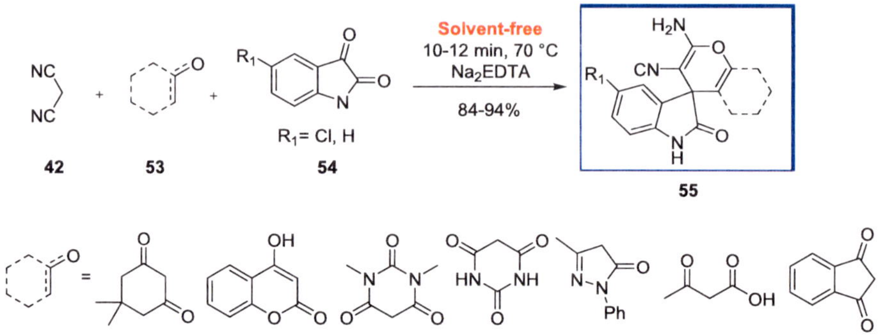

Scheme 23 Synthesis of spiro–2–amino–4*H*–pyrans under neat conditions by Maghsoodlou and co-workers

product reacts with substituted pyrazoline **57** through cyclization reaction to provide pyranopyrazole scaffolds (Scheme 24).

2.4 *The Synthesis of the Pyrrole Derivatives*

Pyrroles are one of the most significant moieties found in many natural compounds and biologically activate compounds [70]. Atar and co-workers disclosed the synthesis of tetrasubstituted pyrroles **61** which by the reaction of various types of amines **5**, substituted dialkyl acetylenedicarboxylates **60,** and β-nitrostyrene **59** in presence of the imidazolium Brønsted acidic ionic liquid as a metal-free catalyst under solvent-free condition. In this attempt, functionalized tetrasubstituted pyrroles were produced in acceptable yields (Scheme 25) [71].

After finding optimized conditions, nitromethane **62**, various aryl aldehydes **2**, 1,3-dicarbonyl derivatives **51,** and amine **5** were treated at room temperature through a one-step four-component reaction to afford polysubstituted pyrrole scaffolds **63** using functionalized Fe_3O_4 as the magnetic nanoparticle (Scheme 26) [72]. In the same study, the biologically active substituted pyrrole derivatives were formed from various ethyl acetoacetate, nitromethane, different benzaldehydes, and a variety of

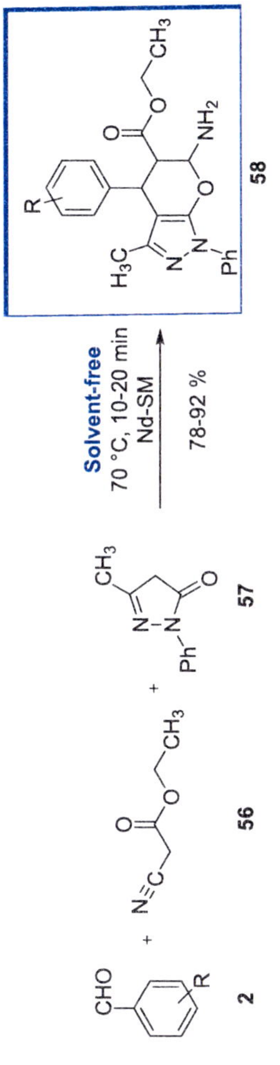

R= 4-Br, 3-Br, 3-OCH_3, thienyl, 3-NO_2, H, 4-OCH_3

Scheme 24 Synthesis of pyranopyrazoles under solvent-free conditions by Siddiqui and co-workers

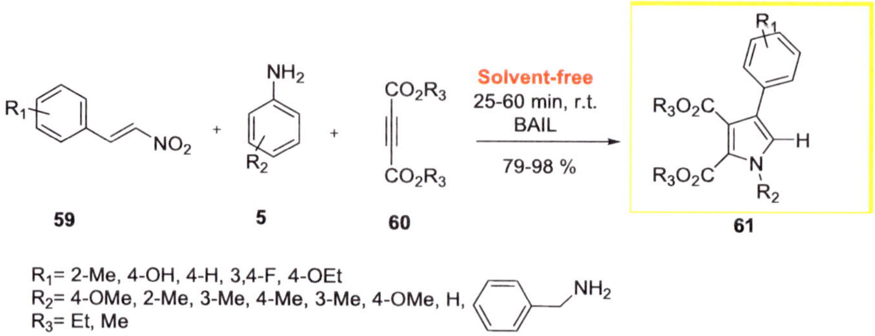

Scheme 25 Synthesis of pyrroles under neat conditions by Atar and co-workers

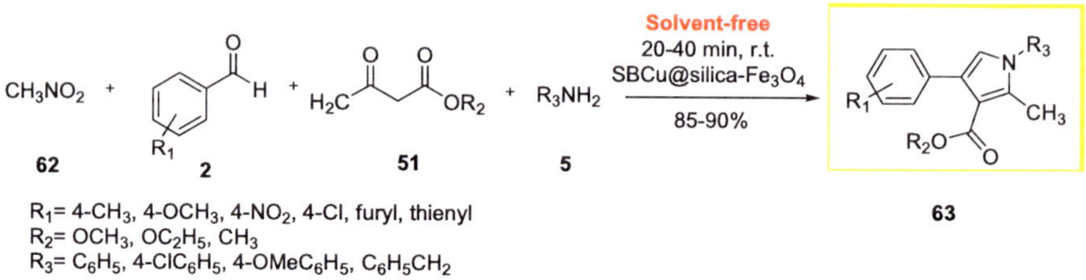

Scheme 26 Synthesis of pyrroles under neat conditions by Salehzadeh and co-workers

anilines in the presence of Cu@imine/Fe_3O_4 MNPs at 100 °C under solvent-free conditions in 15-25 min in 89–97% yield [38].

2.5 *The Synthesis of the Imidazole Derivatives*

Imidazole is a beneficial heterocyclic moiety found in many synthetic or natural compounds that attracted much attention through diverse and multi-purpose biological activity [73]. These properties marked them as valuable scaffolds for further study. Thus, Khandan-Barani and co-workers [74] designed the synthesis of 1,2,4,5-tetrasubstituted imidazoles **64** through multi-component reaction between various aryl aldehydes **2**, 1,2-dicarbonyl derivatives **11**, different amines **2,** and ammonium acetate **18** using glutamic acid as catalyst under solvent-free condition at 60 °C during 2–4 h in acceptable yields (Scheme 27).

In a study, the functionalized catalyst called γ-Fe_2O_3@TiO_2 (g-Fe_2O_3@TiO_2-EG-Cu(II)) yielded the tetrasubstituted imidazole scaffolds **65** from different types of aldehydes **2**, benzil **11,** and ammonium acetate **18** and substituted amines **5** under solvent-free conditions at 100 °C [75]. The mentioned catalyst was made

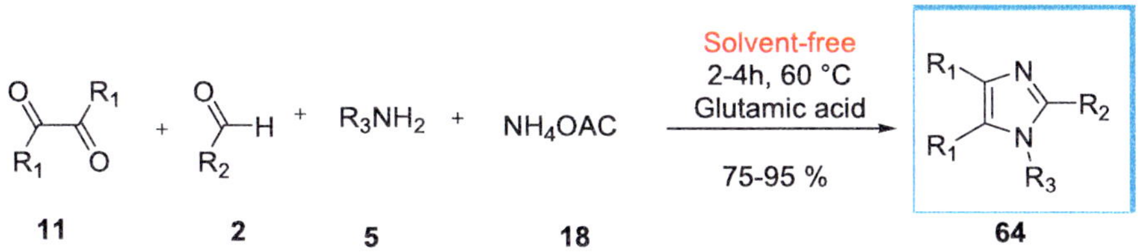

R_1= 4-ClC_6H_4, 4-MeC_6H_4, C_6H_5
R_2=C_6H_5, 4-Cl-C_6H_4, 4-NO_2-C_6H_4, 4-Br-C_6H_4,4-OMe-C_6H_4, 4-Me-C_6H_4, 4-OH-C_6H_4
3-OMe-C_6H_4, 4-Br-C_6H_4, 2-OH-C_6H_4, 4-NO_2-C_6H_4
R_3= C_6H_5, $C_6H_5CH_2$

Scheme 27 Synthesis of imidazoles under neat conditions by Khandan-Barani and co-workers

up of γ-Fe_2O_3 core and TiO_2 as a shell which was functionalized with guanidinated epibromohydrin and Cu (II) (Scheme 28).

In 2018, Sakram's group [42] developed a process by the condensation reaction of the benzil **11**, different aldehydes **2,** and ammonium acetate **18** under solvent-free condition to yield 2,4,5-trisubstituted imidazoles **66**. The best result was obtained at 300 W within 2–4 min in the presence of the poly(4-vinylpyridinium) bromide APVPB as an ionic liquid catalyst (Scheme 29).

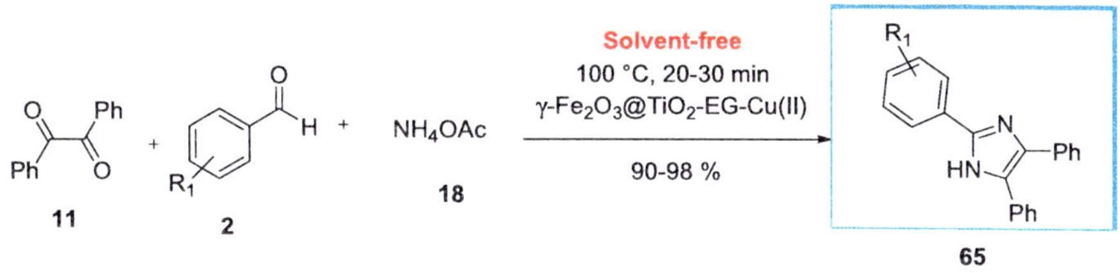

R_1= 4-NO_2, 4-CN, 4-F, 4-Cl, 4-Br, 4-CH_3, 2-CH_3, 4-CH_3

Scheme 28 Synthesis of imidazoles under neat conditions by Nejatianfar and co-workers

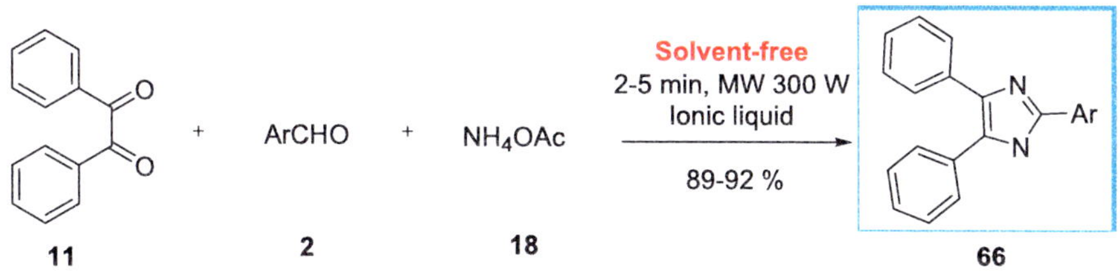

Ar= C_6H_5, C_9H_6N, 4-$H_3CC_6H_4$, 4-ClC_6H_4, C_6H_5CH=CH,C_9H_5ClN, 4-FC_6H_4,4-OHC_6H_4
2-ClC_6H_4, 3-$NO_2C_6H_4$, 4-Cl-2-OHC_6H_3, 5-$H_3CC_5H_5N$, 5-ClC_5H_4N

Scheme 29 Synthesis of imidazoles by MW irradiation under neat conditions by Sakram and co-workers

2.6 The Synthesis of the Propargylamine Derivatives

Negrón-Silva and co-workers [76] demonstrated the synthesis of diastereoselective propargylamine **69** using Cu functionalized mesoporous catalyst called Cu-MCM-41 [77]. The pyrrolidine **67**, phenylacetylene **68,** and different aldehydes **2** reacted through C–H activation to obtain various propargylamines using Cu-MCM-41 (Scheme 30).

Layek and co-workers [78] disclosed the synthesis of propargylamines **71** through one-pot multi-component reaction of different types of aldehydes **2** with various alkynes **70** and amines **5** using [Zn(*L*-proline)$_2$] as the catalyst (Scheme 31).

Zhang and co-workers [79] reported the synthesis of various propargylamines **73** using Cu(0)NPs@CMC as copper nanoparticle catalyst assembled on carboxymethylcellulose in solvent-free condition at 120 °C. To simplify the reaction process, several amines **5** were reacted with phenylpropiolic acid **72** and various aldehydes **2** in presence of the catalyst under neat conditions (Scheme 32).

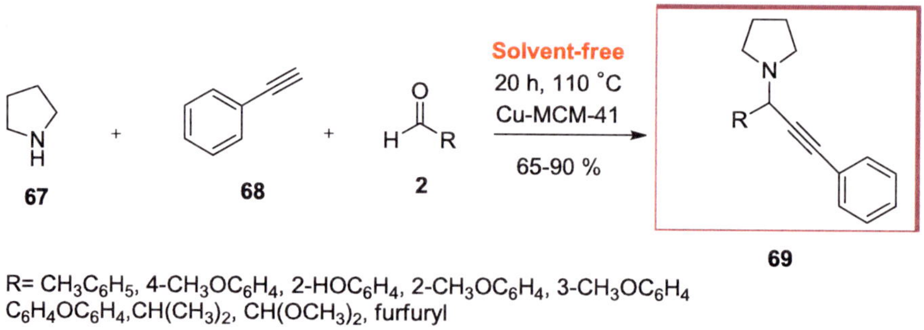

Scheme 30 Synthesis of propargylamines under neat conditions by Negrón-Silva and co-workers

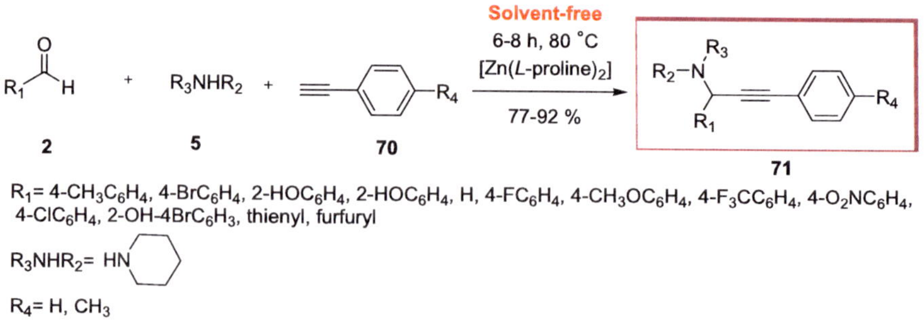

Scheme 31 Synthesis of imidazoles by under neat conditions by Layek and co-workers

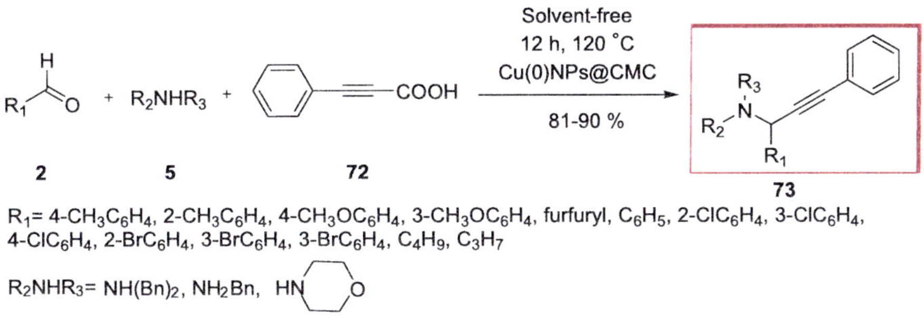

Scheme 32 Synthesis of propargylamines under neat conditions by Zhang and co-workers

3 Conclusion

The synthesis of different compounds through green reactions under solvent-free conditions is highly demanding in chemistry due to the great concern on our environment. There has been a great interest in developing environmentally benign reactions using green solvents and protocols, which lead to a series of reports using solvent-free green chemistry. A significant merit of this reaction is that many of these methods are simple and well organized. It is seen that α-amino phosphonates, pyrimidines, pyrans, pyrroles, imidazoles, and propargylamines are easily accessible by green chemistry process. The many successful models reported are applied in medicinal chemistry, drug discovery, organic synthesis, and material science. There is a future study to synthesis vast amounts of reactions in green chemistry under solvent-free conditions which reduce the cost of designing reactions.

Acknowledgements We highly appreciate the support of the Research Council of Alzahra University.

References

1. Sanderson K (2011) It's not easy being green. Nature 469(7328):18
2. Zeng J-W, Liu Y-C Hsieh P-A et al (2014) Metal-free cross-coupling reaction of aldehydes with disulfides by using DTBP as an oxidant under solvent-free conditions. Green Chem 16(5):2644–2652
3. Al-Shaal MG, Hausoul PJC, Palkovits R (2014) Efficient, solvent-free hydrogenation of α-angelica lactone catalysed by Ru/C at atmospheric pressure and room temperature. Chem Commun 50(71):10206–10209
4. Chen L, Lemma BE, Rich JS et al (2014) Freedom: a copper-free, oxidant-free and solvent-free palladium catalysed homocoupling reaction. Green Chem 16(3):1101–1103
5. Vidyacharan S, Shinde AH, Satpathi B et al (2014) A facile protocol for the synthesis of 3-aminoimidazo-fused heterocycles via the Groebke–Blackburn–Bienayme reaction under catalyst-free and solvent-free conditions. Green Chem 16(3):1168–1175

6. Zhou WJ, Fang L et al (2014) Tunable catalysts for solvent-free biphasic systems: pickering interfacial catalysts over amphiphilic silica nanoparticles. J Am Chem Soc 136(13):4869–4872
7. Sheldon RA (1996) Selective catalytic synthesis of fine chemicals: opportunities and trends. J Mol Catal A 107(1):75–83
8. Eidi E, Kassaee MZ, Nasresfahani Z et al (2018) Synthesis of quinazolines over recyclable Fe_3O_4@SiO_2-PrNH2-Fe^{3+} nanoparticles: a green, efficient, and solvent-free protocol. Appl Organomet Chem 32(12):e4573–e4582
9. Mirjalili BBF, Imani M (2019) Fe_3O_4@NCs/BF0.2: a magnetic bio-based nanocatalyst for the synthesis of 2,3-dihydro-1H-perimidines. J Chin Chem Soc 66(11):1542–1549
10. Mouradzadegun A, Mostafavi MA, Ganjali MR (2019) A facile and green synthesis of 2,4,6-triarylpyridine derivatives using the modified mesoporous organic polymer based on calix [4]resorcinarene: as an efficient and reusable heterogeneous acidic catalyst. Kinet Catal 60(2):187–195
11. Kooti M, Karimi M, Nasiri E (2018) A novel copper complex supported on magnetic reduced graphene oxide: an efficient and green nanocatalyst for the synthesis of 1-amidoalkyl-2-naphthol derivatives. J Nanopart Res 20(2):16–30
12. Shaikh M, Singh SK, Khilari S et al (2018) Graphene oxide as a sustainable metal and solvent free catalyst for dehydration of fructose to 5-HMF: a new and green protocol. Catal Commun 106:64–67
13. Hasanzadeh Banakar S, Dekamin MG, Yaghoubi A (2018) Selective and highly efficient synthesis of xanthenedione or tetraketone derivatives catalyzed by ZnO nanorod-decorated graphene oxide. New J Chem 42(17):14246–14262
14. Mousavi SR, Rashidi Nodeh H, Zamiri Afshari E et al (2019) Graphene oxide incorporated strontium nanoparticles as a highly efficient and green acid catalyst for one-pot synthesis of tetramethyl-9-aryl-hexahydroxanthenes and 13-Aryl-5H-dibenzo[b, i]xanthene-5,7,12,14(13H)-tetraones under solvent-free conditions. Catal Lett 149(4):1075–1086
15. Naeimi H, Zarabi MF (2018) One pot synthesis of aminonaphthoquinone derivatives using Cu(II) immobilized on hyperbranched polyglycerol functionalized graphene oxide as a reusable catalyst under solvent-free conditions. Tetrahedron 74(19):2314–2323
16. Eisavi R, Ahmadi F, Ebadzade B et al (2017) A green method for solvent-free conversion of epoxides to thiiranes using NH_4SCN in the presence of $NiFe_2O_4$ and $MgFe_2O_4$ magnetic nanocatalysts. J Sulphur Chem 38(6):614–624
17. Dhakar A, Goyal R, Rajput A et al (2019) Multicomponent synthesis of 4H-pyran derivatives using KOH loaded calcium oxide as catalyst in solvent free condition. Curr Chem Lett 8(3):125–136
18. Rawat M, Rawat DS (2018) Copper oxide nanoparticle catalysed synthesis of imidazo[1,2-*a*]pyrimidine derivatives, their optical properties and selective fluorescent sensor towards zinc ion. Tetrahedron Lett 59(24):2341–2346
19. Eskandari K, Khodabakhshi S (2018) An eco-friendly solvent-free synthesis of trisubstituted methane derivatives catalyzed by magnetic iron oxide nanoparticles as a highly efficient and recyclable catalyst. Lett Org Chem 15(6):463–471
20. Singaram K, Marimuthu D, Baskaran S et al (2017) Molecular dynamics and biological evaluation of 2-chloro-7-cyclopentyl-7H-pyrrolo[2,3-*d*]pyrimidine derivatives against breast cancer. Comb Chem High Throughput Screen 20(8):703–712
21. Xie F, Zhao H, Zhao L et al (2009) Synthesis and biological evaluation of novel 2,4,5-substituted pyrimidine derivatives for anticancer activity. Bioorg Med Chem Lett 19(1):275–278
22. Dam B, Jamatia R, Gupta A et al (2017) Metal-free greener syntheses of pyrimidine derivatives using a highly efficient and reusable graphite oxide carbocatalyst under solvent-free reaction conditions. ACS Sustain Chem Eng 12:11459–11469
23. Prakash O, Bhardwaj V, Kumar R et al (2004) Organoiodine (III) mediated synthesis of 3-aryl/hetryl-5,7-dimethyl-1,2,4-triazolo[4,3-*a*]pyrimidines as antibacterial agents. Eur J Med Chem 12:1073–1077
24. Lee HW, Kim BY, Ahn JB et al (2005) Molecular design, synthesis, and hypoglycemic and hypolipidemic activities of novel pyrimidine derivatives having thiazolidinedione. Eur J Med Chem 40(9):862–874

25. Ram V, Haque N, Guru P (1992) Chemotherapeutic agents XXV: synthesis and leishmanicidal activity of carbazolylpyrimidines. Elsevier Masson
26. Juby PF, Hudyma TW, Brown M et al (1979) Antiallergy agents. 1. 1,6-Dihydro-6-oxo-2-phenylpyrimidine-5-carboxylic acids and esters. J Med Chem 22(3):263–269
27. Agarwal N, Raghuwanshi SK, Upadhyay DN et al (2000) Suitably functionalised pyrimidines as potential antimycotic agents. Bioorg Med Chem Lett 10(8):703–706
28. Smith PAS, Kan RO (1964) Cyclization of isothiocyanates as a route to phthalic and homophthalic acid derivatives. J Org Chem 29(8):2261–2265
29. Vega S, Alonso J, Diaz JA et al (1990) Synthesis of 3-substituted-4-phenyl-2-thioxo-1,2,3,4,5,6,7,8-octahydrobenzo[4,5]thieno[2,3-*á*]pyrimidines. J Heterocycl Chem 27(2):269–273
30. Tani J, Yamada Y, Oine T et al (1979) Studies on biologically active halogenated compounds. 1. Synthesis and central nervous system depressant activity of 2-(fluoromethyl)-3-aryl-4(3H)-quinazolinone derivatives. J Med Chem 22(1):95–99
31. Vanderhaeghe H, Claesen M (1959) Pyrimidines IV—Hydrazinopyrimidines 68(1–3):30–46
32. Van Leeuwen R, Katlama C, Kitchen V et al (1995) Evaluation of safety and efficacy of 3TC (lamivudine) in patients with asymptomatic or mildly symptomatic human immunodeficiency virus infection: a phase I/II study. J Infect Dis 171(5):1166–1171
33. Devine BL, Fife R, Trust PM (1977) Minoxidil for severe hypertension after failure of other hypotensive drugs. Be Med J 2(6088):667
34. Waghmare AS, Pandit SS, Suryawanshi DM (2018) DABCO catalyzed green and efficient synthesis of 2-Amino-4H-Pyrans and their biological evaluation as antimicrobial and anticancer agents. Comb Chem High Throughput Screen 21(4):254–261
35. Rakhtshah J, Salehzadeh S, Zolfigol MA et al (2017) Mn(III)–pentadentate schiff base complex supported on multi-walled carbon nanotubes as a green, mild and heterogeneous catalyst for the synthesis of tetrahydrobenzo[b]pyrans via tandem Knoevenagel-Michael cyclocondensation reaction. Appl Organomet Chem 31(9):e3690–e3699
36. Ghasemzadeh MA, Elyasi Z (2017) Co_3O_4 nanoparticles as a robust and recoverable catalyst for one-pot synthesis of polyhydroquinolines and tetrahydrobenzopyrans. Iran J Catal 7(1):75–83
37. Jazinizadeh T, Maghsoodlou MT, Heydari R et al (2017) Na_2EDTA: an efficient, green and reusable catalyst for the synthesis of biologically important spirooxindoles, spiroacenaphthylenes and spiro-2-amino-4H-pyrans under solvent-free conditions. J Iran Chem Soc 14(10):2117–2125
38. Thwin M, Mahmoudi B, Ivaschuk OA et al (2014) An efficient and recyclable nanocatalyst for the green and rapid synthesis of biologically active polysubstituted pyrroles and 1,2,4,5-tetrasubstituted imidazole derivatives. RSC Adv 9(28):15966–15975
39. Novanna M, Kannadasan S, Shanmugam P (2019) Phosphotungstic acid mediated, microwave assisted solvent-free green synthesis of highly functionalized 2′-spiro and 2, 3-dihydro quinazolinone and 2-methylamino benzamide derivatives from aryl and heteroaryl 2-amino amides. Tetrahedron Lett 60(2):201–206
40. Gajaganti S, Kumari S, Kumar D et al (2018) An efficient, green, and solvent-free multi-component synthesis of benzimidazolo/benzothiazolo quinazolinone derivatives using Sc $(OTf)_3$ catalyst under controlled microwave irradiation. J Heterocycl Chem 55(11):2578–2584
41. Sibous S, Ghailane T, Houda S (2017) Green and efficient method for the synthesis of 1,5-benzodiazepines using phosphate fertilizers as catalysts under solvent-free conditions. Mediterr J Chem 6(3):53–59
42. Sonyanaik B, Ashok K, Rambabu S et al (2018) Facile one pot multi-component solvent-free synthesis of 2,4,5-trisubstituted imidazoles using “green” and expeditious ionic liquid catalyst under microwave irradiation. Russ J Gen Chem 88(3):537–540
43. Selvakumar K, Shanmugaprabha T, Kumaresan M et al (2017) One-pot multi-component synthesis of N, N′-alkylidene bisamides and imidazoles using heteropoly-11-tungsto-1-vanadophosphoric acid supported on natural clay as catalyst: a green approach. Synth Commun 47(22):2115–2126

44. Sarkar A, Santra S, Kundu SK et al (2016) A decade update on solvent and catalyst-free neat organic reactions: a step forward towards sustainability. Green Chem 18(16):4475–4525
45. Kukhar VP, Hudson HJA (eds.) (2000) Aminophosphonic and aminophosphinic acids and their derivatives as agrochemicals. Wiley, VCH Wieheim Germany, pp 443–482
46. Sudileti M, Chintha V, Nagaripati S et al (2019) Green synthesis, molecular docking, anti-oxidant and anti-inflammatory activities of α-aminophosphonates. Med Chem Res 28(10):1740–1754
47. Tellamekala S, Gundluru M, Sarva S et al (2019) Meglumine sulfate-catalyzed one-pot green synthesis and antioxidant activity of α-aminophosphonates. Synth Commun 49(4):563–575
48. Sreelakshmi P, Santhisudha S, Reddy GR et al (2016) Nano-Cuo–Au-catalyzed solvent-free synthesis of α-aminophosphonates and evaluation of their antioxidant and α-glucosidase enzyme inhibition activities. Synth Commun 48(10):1148–1163
49. Keglevich G, Bálint EJM (2012) Kabachnik-Fields React Mech Synth Use 17(11):12821–12835
50. Guezane-Lakoud S, Toffano M, Aribi-Zouioueche L (2017) Promiscuous lipase catalyzed a new P-C bond formation: green and efficient protocol for one-pot synthesis of α-aminophosphonates. Heteroatom Chem 28(6):e21408–e21418
51. Esmaeilpour M, Zahmatkesh S, Javidi J et al (2017) A green one-pot three-component synthesis of α-aminophosphonates under solvent-free conditions and ultrasonic irradiation using Fe3O4@SiO2-imid-PMAn as magnetic catalyst. Phosphorus, Sulfur Silicon Relat Elem 192(5):530–537
52. Poola S, Nadiveedhi MR, Sarva S et al (2019) Nano Sb_2O_3 catalyzed green synthesis, cytotoxic activity, and molecular docking study of novel α-aminophosphonates. Med Chem Res 28(4):528–544
53. Mirhosseini-Eshkevari B, Esnaashari M, Ghasemzadeh MA (2019) Novel brönsted acidic ionic liquids confined in UiO-66 nanocages for the synthesis of dihydropyrido[2,3-*d*]pyrimidine derivatives under solvent-free conditions. ACS Omega 4(6):10548–10557
54. Abdollahi-Basir MH, Shirini F, Tajik H et al (2019) MIL-53(Fe): introduction of a new catalyst for the synthesis of Pyrimido[4,5-*d*]pyrimidine derivatives under solvent-free conditions. J Mol Struct 1197:318–325
55. Afradi N, Foroughifar N, Pasdar H et al (2019) Aspartic-acid-loaded starch-functionalized Mn–Fe–Ca ferrite magnetic nanoparticles as novel green heterogeneous nanomagnetic catalyst for solvent-free synthesis of dihydropyrimidine derivatives as potent antibacterial agents. Res Chem Intermed 45(5):3251–3271
56. Tamuli KJ, Dutta D, Nath S et al (2017) A greener and facile synthesis of imidazole and dihydropyrimidine derivatives under solvent-free condition using nature-derived catalyst. Chem Sel 2(26):7787–7791
57. Ahmadi M, Moradi L, Sadeghzadeh M (2019) MWCNTs@NHBut/PTA: new efficient solid acid catalyst for solvent free synthesis of benzochromenopyrimidines. Appl Organomet Chem 33(8)
58. Al-Bogami AS, Saleh TS, Moussa TA (2018) A green synthesis, antimicrobial activity and cytotoxicity of novel fused pyrimidine derivatives possessing a trifluoromethyl moiety. Chem Sel 3(28):8306–8311
59. Daraie M, Heravi MM (2019) as a new biocompatible polymer supported nanocatalyst for the synthesis of chromeno[2,3-*d*] pyrimidine-diones through a novel and efficient pathway. Can J Chem 97(11):772–779
60. Farahi M, Karami B, Banaki Z et al (2017) TSA-catalyzed regioselective synthesis of medicinally important 4-aryl-substituted dihydropyrimidine derivatives fused to pyrazole and triazole scaffolds via an efficient and green domino reaction. Monatsh Chem 148(8):1469–1475
61. Firoozpour L, Yahyavi H, Ejtemaei R (2018) A one-pot, three-component, solvent-free synthesis of novel 6 h-chromeno[3',4':5,6]pyrido[2,3-*d*]pyrimidine-triones under microwave irradiation. J Chem Res 42(12):604–607
62. Jannati S, Esmaeili AA (2017) An efficient one-pot synthesis of highly functionalized benzylpyrazolyl pyrido[1,2-a]pyrimidine derivatives using $CuFe_2O_4$ nanoparticles under solvent-free conditions. Res Chem Intermed 43(12):6817–6833

63. Kumari S, Kumar D, Gajaganti S et al (2019) $Sc(OTf)_3$ catalysed multicomponent synthesis of chromeno[2,3-*d*]pyrimidinetriones under solvent-free condition. Synth Commun 49(3):431–443
64. Mahmoud NFH, El-Saghier AM (2019) Multi-component reactions, solvent-free synthesis of substituted pyrano-pyridopyrimidine under different conditions using ZnO nanoparticles. J Heterocycl Chem 56(6):1820–1824
65. Reddy MV, Byeon KR, Park SH et al (2017) Polyethylene glycol methacrylate-grafted dicationic imidazolium-based ionic liquid: Heterogeneous catalyst for the synthesis of aryl-benzo[4,5]imidazo[1,2-*a*]pyrimidine amines under solvent-free conditions. Tetrahedron 73(35):5289–5296
66. Rigi F, Shaterian HR (2017) Silica-supported ionic liquids prompted one-pot four-component synthesis of pyrazolopyranopyrimidines, 3-methyl-4-aryl-4,5-dihydro-1*H*-pyrano[2,3-c]pyrazol-6-ones, and 1,6-diamino-2-oxo-1,2,3,4-tetrahydropyridine-3,5-dicarbonitriles. Polycyclic Aromat Compd 37(4):314–326
67. Rather RA, Siddiqui ZN (2018) Synthesis, characterization and application of Nd-Salen schiff base complex Immobilized Mesoporous Silica in solvent free synthesis of pyranopyrazoles. J Organomet Chem 868:164–174
68. Yue C, Mao A, Wei Y et al (2008) Knoevenagel condensation reaction catalyzed by task-specific ionic liquid under solvent-free conditions. Catalysis Comm 9(7):1571–1574
69. Mather BD, Viswanathan K, Miller KM et al (2006) Michael addition reactions in macromolecular design for emerging technologies. Prog Ploy Sci 31(5):487–531
70. Estévez V, Villacampa M, Menéndez JC (2014) Recent advances in the synthesis of pyrroles by multicomponent reactions. Chem Soc Rev 43(13):4633–4657
71. Balu Atar A, Han E, Sohn DH (2019) A solvent and transition metal-free, highly efficient Brønsted acidic ionic liquid promoted one-potthree-component reactions for the synthesis of tetrasubstituted pyrroles. Synth Commun 49(9):1181–1192
72. Hamrahian SA, Rakhtshah J, Mousavi Davijani et al (2018) Copper Schiff base complex immobilized on silica-coated Fe_3O_4 nanoparticles: a recoverable and efficient catalyst for synthesis of polysubstituted pyrroles. Appl Organomet Chem 32 (10)
73. Shalini K, Sharma PK, Kumar N (2010) Imidazole and its biological activities: a review. Der Chemica Sinica 1(3):36–47
74. Feizabad FK, Khandan-Barani K, Hassanabadi A (2017) Glutamic acid as an efficient and green catalyst for the one-pot synthesis of 1,2,4,5-tetrasubstituted imidazoles under thermal, solvent-free conditions. J Chem Res 41(11):673–675
75. Nejatianfar M, Akhlaghinia B Jahanshahi R (2008) Cu(II) immobilized on guanidinated epibromohydrin-functionalized γ-Fe_2O_3@TiO_2 (γ-Fe_2O_3@TiO_2-EG-Cu(II)): a highly efficient magnetically separable heterogeneous nanocatalyst for one-pot synthesis of highly substituted imidazoles. Appl Organomet Chem 32(2):e4095–e4196
76. Cortezano-Arellano O, Hernández-Gasca MA, Ángeles-Beltrán D et al (2018) Diastereoselective synthesis of propargylamines catalyzed by Cu-MCM-41. Tetrahedron Lett 59(25):2403–2406
77. Noreña-Franco L, Hernandez-Perez I, Aguilar-Pliego J et al (2002) Selective hydroxylation of phenol employing Cu–MCM-41 catalysts. Catal Today 75(1):189–195
78. Layek S, Agrahari B, Kumari S et al (2018) [Zn(l-proline)$_2$] catalyzed one-pot synthesis of propargylamines under solvent-free conditions. Catal Lett 148(9):2675–2682
79. Liu X, Tan X, Zhou Y (2019) Cu^0NPs@CMC:an efficient recoverable nanocatalyst for decarboxylative A3 and A3 couplings under neat condition. Res Chem Intermed 45(6):3359–3378

Chapter 6
Solid-State Green Organic Reactions

José Clerigué, María Teresa Ramos, and J. Carlos Menéndez

1 Introduction

Solvents have been traditionally considered essential for energy transfer in organic reactions, besides having additional roles in reaction mechanisms. Although their use is still standard in synthesis, solvents pose a number of safety problems such as fire risks and toxicity; furthermore, discarded solvents are the main source of waste from synthesis, both at laboratory and industrial scales. For this reason, reduction in solvent use and the replacement of conventional volatile organic solvents by alternative reaction media are among the main goals of Green Chemistry [1]. Solid-state chemistry is a part of this effort, and is undergoing a considerable growth in recent years. It should not be confused with solvent-free chemistry, which, besides reactions performed in the solid state, comprises also those performed using liquid reagents that may act as the reaction medium, especially when they are present in large excess. Beyond the above-mentioned environmental aspects, solid-state reactions have potential advantages associated with the very high reagent concentrations that can be achieved, such as higher reaction rates and the possibility to uncover new modes of reactivity.

In this chapter, we will critically summarize the field of solid-state chemistry and its relevance to Green Chemistry, although we will not treat certain aspects that are discussed in other chapters, such as microwave-assisted chemistry in the solid state.

J. Clerigué · M. T. Ramos · J. C. Menéndez (✉)
Unidad de Química Orgánica y Farmacéutica, Departamento de Química en Ciencias Farmacéuticas, Facultad de Farmacia, Universidad Complutense, 28040 Madrid, Spain
e-mail: josecm@farm.ucm.es

J. Clerigué
e-mail: joscleri@ucm.es

M. T. Ramos
e-mail: mtramos@ucm.es

G. Anilkumar and S. Saranya (eds.), *Green Organic Reactions*, Materials Horizons: From Nature to Nanomaterials, https://doi.org/10.1007/978-981-33-6897-2_6

2 Solid-State Photochemical Reactions

In a photochemical reaction, energy in the form of light of the suitable frequency, which corresponds the visible-ultraviolet regions of the spectrum, is absorbed by a molecule, which is promoted to an electronically excited state. The main transitions in organic photochemistry are $\pi \rightarrow \pi^*$ and $n \rightarrow \pi^*$. An excited state of a molecule is a different chemical species from the ground state, since their electronic distributions, geometries, dipole moments and acid or base strengths are different. Thus, the photochemical conditions can render products that are not easily available by ground state chemistry, especially in the area of strained rings [2]. The excited state has a short lifetime and can undergo various physical and chemical processes to release its extra energy. The physical processes are usually summarized in the well-known Jablonski diagrams, and consist of radiative and radiationless processes or energy transfer by collision (photosensitization). On the other hand, several chemical pathways can be taken by the excited molecule, including bond cleavage to render free radicals, photodissociation into smaller molecules, intramolecular rearrangements or isomerizations, and reaction with another molecule to form dimers. Then, secondary processes occur to complete the photochemical reaction.

Photochemical reactions occur more efficiently and selectively in the solid state than in solution, and this is particularly true for the crystalline state due to the ordered molecular structure found in the crystal. Moreover, different products are sometimes obtained in solution and solid state. To carry out solid-state photoreactions, a solid or crystalline sample is irradiated with a xenon or mercury lamp, using filters if a defined wavelength is required, or it is simply exposed to sunlight. No additional energy is normally needed, since photoreactions usually take place at room temperature or below [1]. A suspension of nanocrystals in water running through a continuous flow photoreactor has also been considered to undergo a solid-state photoreaction [3].

2.1 Norrish Type I Reactions

The UV or visible irradiation of aldehydes and ketones results in a $n \rightarrow \pi^*$ transition, *i.e.* the excitation of an electron from the non-bonding *n*-orbital to the antibonding π^*-orbital. Two electronic configurations are possible for the excited state, *singlet*, when the spins are paired, and *triplet*, when the two unpaired electrons have the same spin. In both excited states, α-cleavage can occur to form an alkyl and an acyl radical intermediates, which can react and afford different products depending on the nature of the starting ketone. Loss of a CO molecule from the acyl radical forming a new one, and combination of both radicals to form a C–C bond is known as the Norrish type I reaction (Scheme 1a).

A correlation has been established between the tendency to undergo Norrish type I reactions in the solid state and the radical stabilizing energies (RSE) of the substituents R of the ketones, which have to be above 12–15 kcal/mol [4]. Biradicals

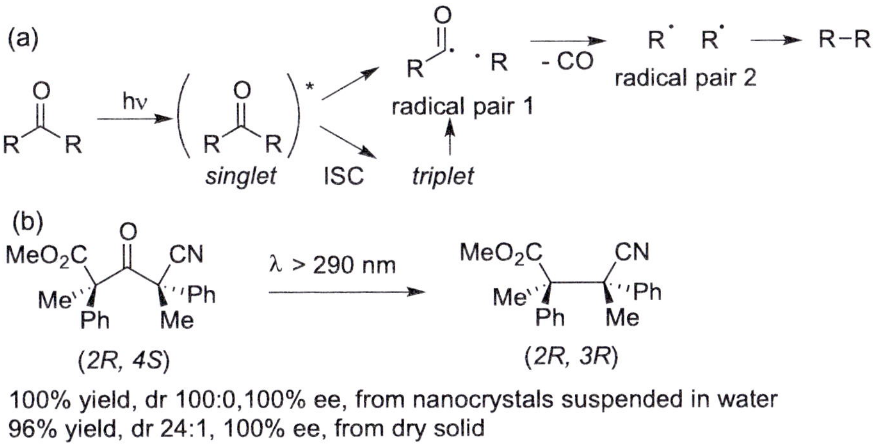

Scheme 1 Norrish type I reaction and an example of its synthetic application

or radical pairs, generated by cleavage of α-bonds and loss of CO, combine with retention of the stereochemistry of the starting ketone, allowing the formation of C–C bonds between two branched substituents, as shown in Scheme 1.b [3].

2.2 *Norrish Type II Reactions*

An alternative path for the excited state resulting from irradiation of a carbonyl compound is known as the Norrish type II reaction (Scheme 2) and involves the intramolecular abstraction of a γ-hydrogen rendering a 1,4-diradical, which undergoes C–C bond cleavage in the singlet state to give an alkene and an enol, which tautomerizes to a ketone. This transformation can also occur in the triplet state, but in

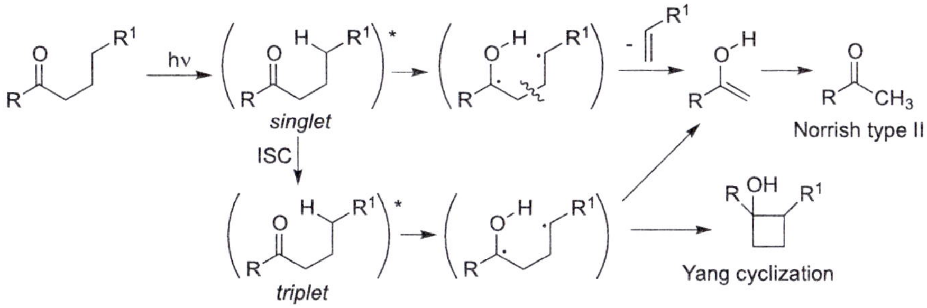

Scheme 2 Norrish type II reaction and the competing Yang cyclization

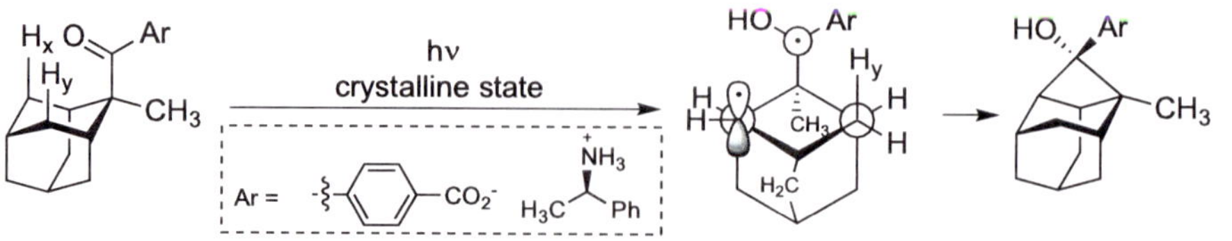

Scheme 3 A photochemical Yang cyclization

such case, it competes with the formation of a cyclobutanol, known as the Yang reaction. Stereoelectronic factors govern the course of the reaction, and ring formation or fragmentation occurs depending on the favoured orbital overlapping.

An interesting example of a Yang cyclization in the solid state involved remote asymmetric induction when starting from a chiral salt substituent (Scheme 3) [5].

2.3 Solid-State Photochemical [2 + 2] Cycloadditions

The solid-state photochemical [2 + 2] cycloadditions of alkenes represent a useful method for producing strained 4-membered rings, difficult to obtain in solution. For the reaction to take place, it is necessary to have double bonds in a parallel arrangement and with a 3.5–4.2 Å distance between them, as determined by an exhaustive study about the photodimerization of three polymorphs of cinnamic acid, which show different behaviour when irradiated (Scheme 4) [6].

Crystal engineering has emerged as a strategy to organize the molecules in an arrangement suitable for the photoreaction to take place. Templating molecules establish non-covalent interactions with the reactants and organize them in the crystal lattice. For instance, the cycloaddition of *trans*-1,2-*bis*(4-pyridyl)ethene is efficiently directed by resorcinol in the solid state via hydrogen bonding, as shown in Scheme 5 [7, 8].

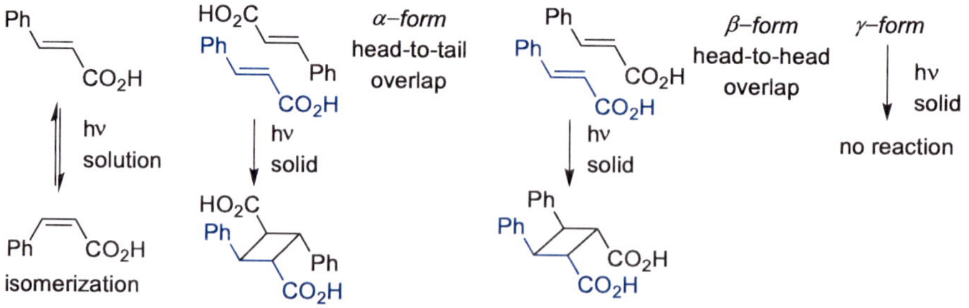

Scheme 4 Solid-state photochemical cyclodimerization of cinnamic acid

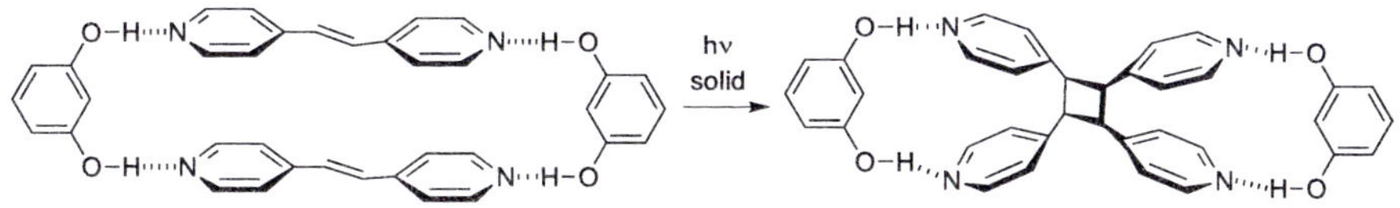

Scheme 5 Resorcinol as a template to control a solid-state photodimerization

3 Mechanochemical Synthesis

3.1 Introduction to Mechanochemistry

The term "mechanochemistry" was first introduced in 1919 by Ostwald to describe the use of mechanical energy to promote chemical transformations. Mortar and pestle are one of the oldest instruments known to humankind, and have been often employed to perform synthetic transformations. However, this approach is not practical unless the reactions are fast, and is not very reproducible because the outcome depends on the milling frequency and hence on the physical strength of the operator. For this reason, modern mechanochemical synthesis is usually performed in dedicated ball mills, which can be planetary or vibratory (Fig. 1a, b). Planetary ball mills receive this name because the milling jar is attached to a rotating disk while it also rotates around its own axis in the opposite direction leading to a composite movement similar to that of a planet in its orbit around a star. In this approach, the energy that drives the reaction comes from the centrifugal force of the balls against the walls of the jar. On the other hand, in vibratory ball mills, the ball is shaken about a central position (hence the alternative name "shaker mills"), and collides with the reactants and the jar walls. This technique is often described as high-speed

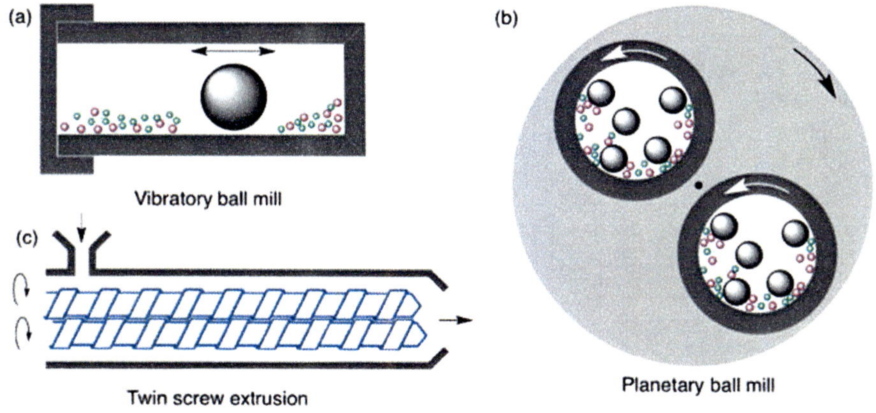

Fig. 1 Movement of the balls in **a** vibratory and **b** planetary ball mills, **c** A schematic twin-screw extrusion setup

ball milling (HSBM) or high-speed vibration milling (HSVM). In order to minimize the shedding of particles into the reacting mixture, milling jars and balls need to be built from hard materials, some of which are stainless steel, tungsten carbide, zirconia, agate, Teflon, nylon, and polyurethane. Besides the nature of these materials, the main factors that influence the outcome of a mechanochemical reaction are the number and mass of balls and the speed of rotation or vibration of the instrument. Additionally, mechanochemical reactions are affected by the use of additives. Inert grinding auxiliaries such as sodium chloride or even sand are sometimes used to increase the efficiency of the mechanochemical process when liquid reagents are employed. On the other hand, Liquid-Assisted Grinding (LAG), which involves the addition of small amounts of a liquid to the reaction mixture, often enables reactions between solid materials that are unproductive under neat milling. Related techniques are Ion- and Liquid-Assisted Grinding (ILAG), involving the addition of a liquid and a small amount of a salt, and Polymer-Assisted Grinding (POLAG).

Another important technique is Twin-Screw Extrusion, in which the starting materials undergo shear and compressive forces due to the rotation of a pair of co-or counter-rotating, intermeshing screws (Fig. 1c). This method allows mechanochemical synthesis to be performed as a continuous, large-scale process that can potentially be adapted to industrial-scale manufacturing [9]. However, since concomitant heating is needed and the reactions often proceed in a melt, it cannot be usually regarded as solid-state chemistry.

Mechanochemical-assisted synthesis is one of the areas of synthetic methodology that has undergone a fastest growth in recent years. We will summarize below its application to some representative areas, and the interested reader is referred to several recent reviews for additional information [10–15].

3.2 Mechanochemical Synthesis of Heterocycles

Heterocycles are the single most important class of bioactive compounds, making up about 60% of drugs and agrochemicals. We summarize below the main aspects of the application of mechanochemistry to their synthesis, and refer the reader to two recent reviews of this subject [16, 17] for more detailed information.

Some well-known heterocyclic syntheses based on name reactions that have been performed under mechanochemical conditions include the Gewald thiophene synthesis [18], the Paal Knorr [19] and Hantzsch [20] pyrrole syntheses, the Phillips benzimidazole synthesis [21], the Hantzsch dihydropyridine synthesis [22], the Biginelli pyrimidine synthesis [23], and the Povarov tetrahydroquinoline synthesis [24]. Beyond these basic reactions, some mechanochemical transformations leading to more complex heterocyclic systems are described below.

Ninhydrin was investigated as a starting material for the quantitative synthesis of fused heterocyclic frameworks via mechanochemical cascade processes [25–27] (Scheme 6). *N*-Arylmethyleneiminium salts were transformed into Tröger's base

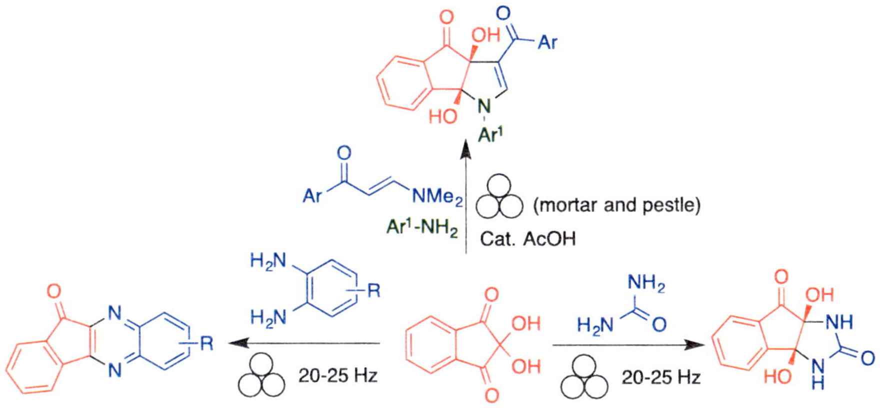

Scheme 6 Some domino mechanochemical processes starting from ninhydrin

derivatives **2** by brief milling (5–10 min) in the presence of water vapour or $MgSO_4 \cdot 7H_2O$ via a three-step domino process shown in Scheme 7 [28].

The mechanochemical Hantzsch pyrrole synthesis has been applied to the preparation of complex heterocyclic systems in the context of diversity-oriented synthesis [29, 30]. This chemistry is exemplified by the double Hantzsch reaction leading to compound **3** via a pseudo-five component reaction (Scheme 8).

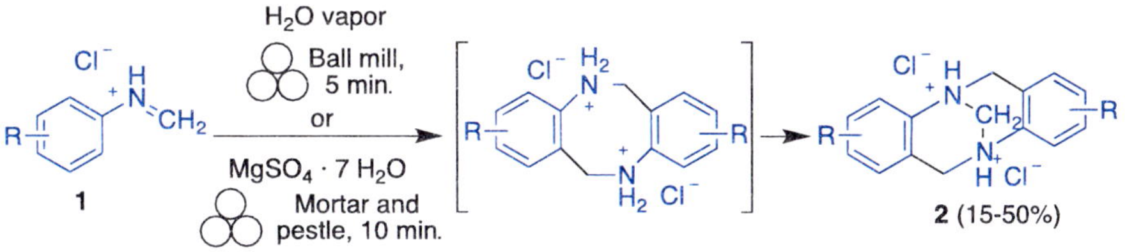

Scheme 7 Solid-state synthesis of Tröger bases

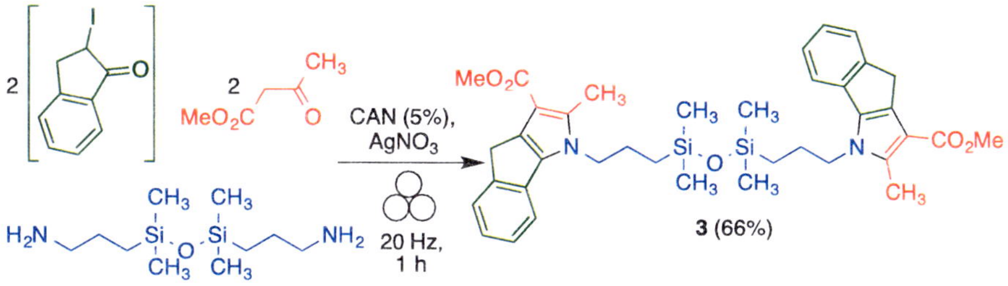

Scheme 8 A mechanochemical double Hantzsch pyrrole synthesis. 2-Iodoindan-1-one was mechanochemically generated in situ from indan-1-one

3.3 Mechanochemical Synthesis of Biological Molecules

Biological molecules often show physicochemical properties that complicate their handling under conventional conditions. Furthermore, achieving selective transformations may not be trivial because of their high degree of functionalization. Mechanochemistry is being explored as a promising way to overcome some of these difficulties.

3.3.1 Carbohydrates

Carbohydrates are the most abundant biomolecules and have key roles in living cells, being involved in their growth, signaling and survival. Natural carbohydrates are not easy to obtain and purify in large quantities, the main hurdles to be overcome being their high level of functionalization with hydroxy and amino groups, which leads to solubility and selectivity issues, and the need to control the anomeric stereochemistry. The intrinsic characteristics of mechanochemical synthesis can help to deal with these obstacles.

Cyclodextrins are cyclic α(1 → 4) glucopyranosides with a wide variety of applications as drug or gene carriers, electrochemical sensors, etc. Thus, the tailored synthesis of these structures is highly appealing, and mechanochemistry may provide a good approach when solution-based methods fail. For instance, as shown in Scheme 9, the synthesis of *per*-6-substituted cyclodextrins has been performed in a planetary ball mill under solvent-free conditions, without the formation of by-products and impurities that plague the same reactions when carried out in solution [31].

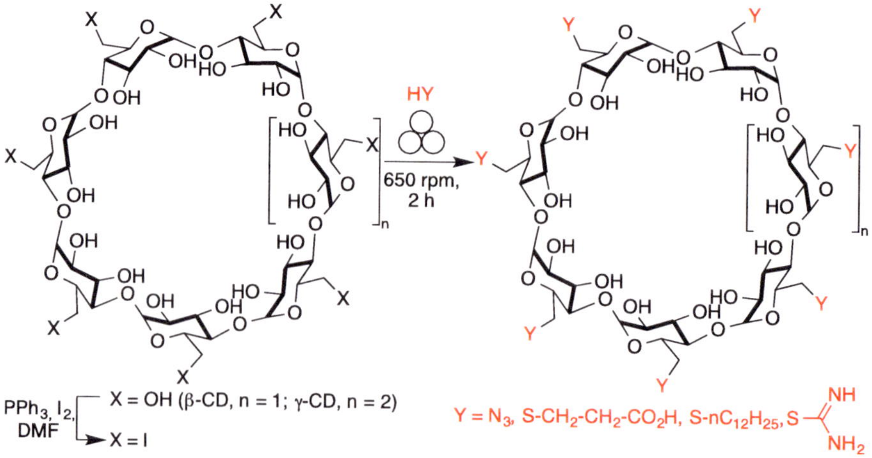

Scheme 9 Mechanochemical functionalization of cyclodextrins

Aryl glycosides are commonly used in oligosaccharide synthesis and to study carbohydrate–lectin interactions. These compounds can be obtained from glycoside halides and phenols in a planetary ball mill in excellent yields, without any purification and avoiding the formation of undesired products associated to the use of phase-transfer catalysts [32].

3.3.2 Amino Acids and Peptides

Besides their enormous biological importance, amino acids and peptides are of high relevance in medicinal chemistry, and as organocatalysts in synthesis. Their poor solubility in organic solvents constitutes the main obstacle when working with these compounds, and therefore mechanochemistry arises as a promising tool to broaden the synthetic scope in this field.

The modification of amino acids in both their amino and carboxyl groups has been explored with ball milling techniques, an example being the deprotection of N-Boc amino acids catalyzed by *p*-toluenesulfonic acid [33].

Amino acids can be connected by amide bonds to form peptides under mechanochemical conditions using several strategies. Those based on activated urethane protected α-amino acid N-carboxyanhydrides (UNCA) have been quite successful and are exemplified by the synthesis of α,α-dipeptides, such as aspartame [34], α,β- and β,β-dipeptides, including L-carnosine [35]. Acid-activating groups can also be applied to the mechanochemical synthesis of peptides. For instance, the synthesis of α,α- and α,β-dipeptides using EDC and HOBt as coupling reagents and Mg–Al hydrotalcite as a recoverable base with good to excellent yields has been described, as summarized in Scheme 10 [36].

Also, the fact that reagents such as HOBt or ethyl cyanohydroxyiminoacetate (Oxyma) are compatible with ball milling conditions is noteworthy, as they can help to suppress racemization. The synthesis of di- and tripeptides by using N-protected-α-aminoacyl benzotriazoles and α-amino acid derivatives with different C-terminal functions is also remarkable, and this protocol also allowed the biotinylation of a pentapeptide [37]. The synthesis of the endogenous analgesic Leu-enkephalin from suitably protected amino acids was performed in 44% overall yield and involved an unprecedented sequence of 9 mechanochemical steps [38].

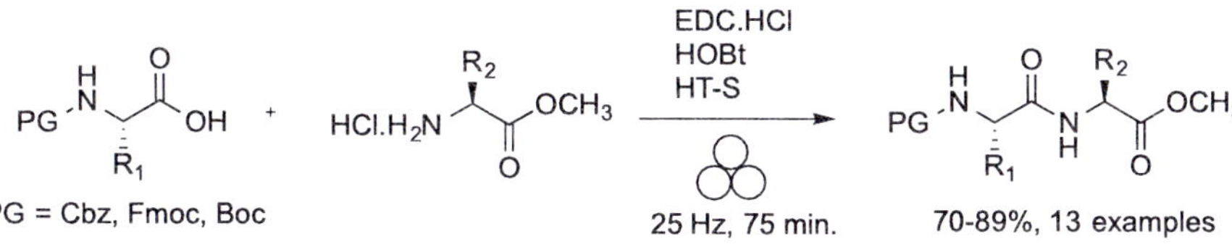

Scheme 10 Mechanochemical synthesis of α,α-dipeptides catalyzed by Mg–Al hydrotalcite

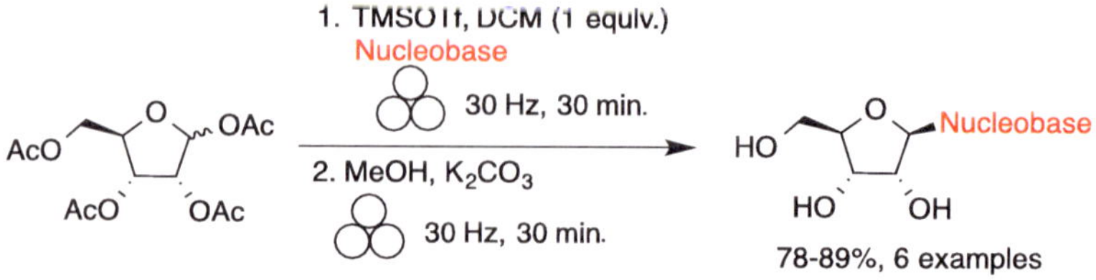

Scheme 11 Mechanochemical Vorbrüggen synthesis of nucleosides

All these developments have led to the conclusion that ball milling has advantages over solid-phase and solution methodologies for the synthesis of small peptides in terms of yield, purity and waste generation [39].

3.3.3 Nucleosides and Nucleotides

Nucleosides, nucleotides and their analogues are crucial in biochemistry and medicinal chemistry. Their synthesis is troublesome because of their poor solubility, the requirement of long reaction times and harsh reaction conditions, and the sensitivity to moisture of some commonly used reagents when working in solution.

The classical Vorbrüggen reaction conditions have been applied successfully to synthesize purine and pyrimidine nucleosides from ribofuranoses and nucleobases in the presence of trimethylsilyl triflate by ball milling with very good yields and regioselectivities (Scheme 11) [40]. Mechanochemistry is also applicable to the modification of nucleosides, as shown with the preparation of 5'-derivatised nucleosides by reacting 5'-halo/tosylnucleosides with suitable nucleophiles [41].

Dinucleoside 5' 5'-polyphosphates can be obtained from commercially available nucleoside-5'-monophosphates (NMP) by vibratory ball milling via their activation with carbonyldiimidazole followed by the addition of nucleoside mono-, di- or triphosphates in a one-pot procedure [42].

3.3.4 Lipids

Lipids are involved in countless biological functions, and, moreover, some drugs, like miltefosine or perifosine, are phospholipids. Nevertheless, examples of lipid synthesis under mechanochemical conditions are scarce.

The esterification of fatty acids (palmitic, estearic and octanoic) with cellulose using acetic anhydride as co-reactant was carried out under ball milling at 80 °C by using a thermostated chamber [43]. Furthermore, an elegant, fully mechanochemical, multi-step synthesis of diacylglycerols from TBDMS-glycidol was recently reported. This starting material was reacted with suitable fatty acids under mechanochemical conditions in the presence of Jacobsen's cobalt(II)-(*S*,*S*) salen complex as catalyst, in an oxidative atmosphere, to afford protected monoacylglycerols in low *ee*, unless the

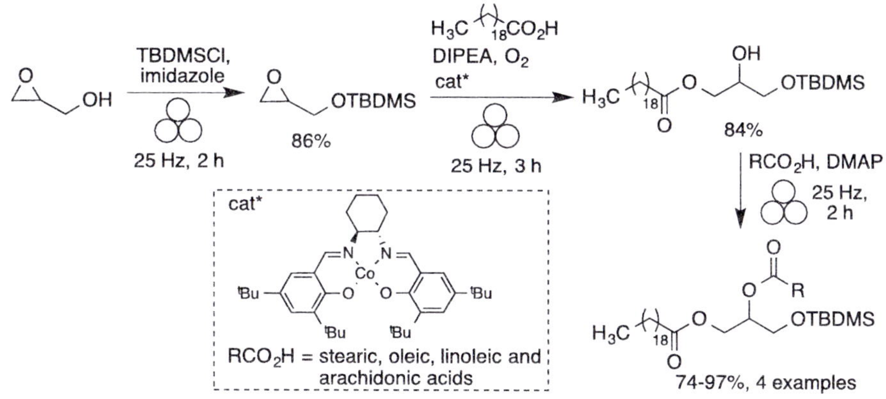

Scheme 12 Fully mechanochemical synthesis of diacylglycerols

sequence started from enantiomerically pure glycidol. Their subsequent esterification, once again under ball milling, with fatty acids using DCC as a coupling reagent, afforded the target diacylglycerols in good to excellent yields (Scheme 12). Since these compounds are difficult to obtain with acceptable purity from natural sources and their conventional synthesis is long and burdensome, this method represents a good example of the relevancy of mechanochemistry to solve difficult synthetic problems [44].

3.4 Mechanochemical Enantioselective Synthesis: Organocatalysis and Biocatalysis

The application of mechanochemical conditions to organocatalytic reactions is a rapidly growing field [45, 46]. Mechanochemistry often gives better results in terms of reactivity and enantiomeric excess than more conventional methodologies, which can be explained by the better activity of many organocatalysts under high concentration.

To give some representative examples, proline itself [47] and proline-containing peptides, such as (*S*)-Pro-(*S*)-PheOMe [48], have been shown to promote the aldol reaction under ball-milling conditions. In these reactions, the pyrrolidine moiety was proposed to activate the ketone reagent via formation of a chiral enamine, while the aldehyde is activated by hydrogen bonding with the amide NH.

Other chiral organocatalysts that have proved to be compatible with mechanochemical activation for aldol, Michael and related reactions include quinuclidine and thiourea derivatives. One example, shown in Scheme 13, is the amination of a β-ketoester with a diazo carboxylate in the presence of the quinidine-type catalyst **4** to give compound **5**, bearing an amino-substituted quaternary stereocenter [49].

The use of enzymes as catalysts in mechanochemical reactions is problematic, as the behaviour of these biomolecules is highly sensitive to temperature and mechanical

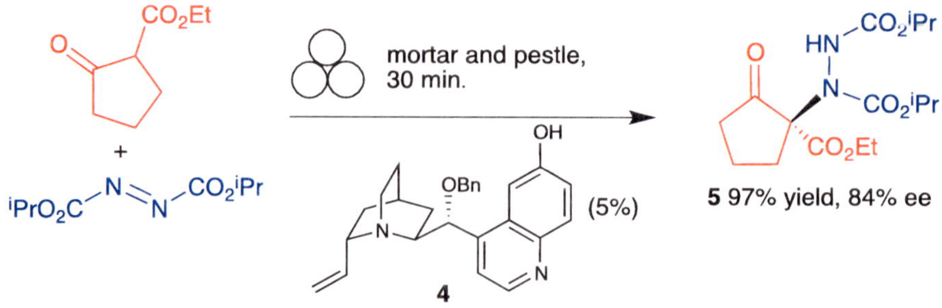

Scheme 13 A mechanochemical, enantioselective organocatalytic amination reaction

stress [50], but these problems can be avoided via enzyme immobilization [51]. One salient example is *Candida antarctica* lipase B (CALB), which has been shown to act as a catalyst for different mechanochemical stereoselective esterifications of primary and secondary alcohols, and also for the selective hydrolysis of primary and secondary esters [52, 53], showing remarkable robustness and recyclability. On the other hand, proteases such as papain have emerged as an alternative to the usual coupling reagents in the mechanochemical peptide synthesis, using either planetary ball milling or twin-screw extrusion. Powdered papain is a suitable catalyst to build oligopeptides, which in solution tend to precipitate while growing, thus paving the way to large-scale, enzyme-catalyzed synthesis of peptides [54]. This method has also been extended to the formation of non-peptidic amide bonds [55].

3.5 Mechanochemical Synthesis of Pharmaceuticals

The role of mechanochemistry in the preparation of pharmaceuticals, which has been described as medicinal mechanochemistry [56], is two-fold. The first aspect is the pursuit of novel synthetic procedures for API fabrication that ideally should be greener and cheaper than those conventionally used, and suitable for large-scale industrial application. On the other hand, mechanochemistry is an interesting approach to modify or control issues related to molecular self-assembly processes in drugs, such as formation of salts, co-crystals, polymorphisms, etc. In this section, we will focus on the first of these aspects.

Bismuth subsalycilate is a drug employed to treat common gastrointestinal discomforts such as heartburn or diarrhoea. An easy procedure for its obtention involves a water-assisted grinding of bismuth oxide with salicylic acid in a 1:2 stoichiometry for one hour, with NH_4NO_3 or KNO_3 as additives. These conditions permit the selective formation of the desired compound instead of other bismuth-subsalycilate complexes with different stoichiometries [57].

Some antidiabetic drugs with a sulphonylurea structure, such as chlorpropamide, tolbutamide and glibenclamide, are easily obtainable in excellent yields by reaction of suitable phenylsulphonamides with alkyl isocyanates under CuCl catalysis, with nitromethane as liquid assistant (Scheme 14) [58].

An elegant mechanochemical synthesis of hydantoins from amino esters was developed, and its generality was proved by its application to the synthesis of the antiepileptic drug phenytoin. Diphenylglycine methyl ester was treated with trimethylsilyl isocyanate in the presence of water under ball milling conditions, and the resulting N-substituted urea intermediate underwent a subsequent cyclization promoted by caesium carbonate under mechanochemical conditions to afford phenytoin in good yield (Scheme 15). This method is more efficient and greener than the one based on the classical Blitz reaction [59, 60]. Other hydantoin-derived APIs, such as nitrofurantoin and dantrolene [61] and etothoin [62], have also been obtained by mechanochemical approaches in the complete absence of organic solvents in all stages of the synthetic process.

Atorvastatin, a cholesterol-lowering drug, is one of the best-selling pharmaceuticals worldwide. Based on a previously reported mechanochemical version of the Hantzsch pyrrole synthesis [20], a simple route to the atorvastatin *tert*-butyl ester (**8**) was achieved by reaction under vibratory ball milling conditions of enaminone (**6**)

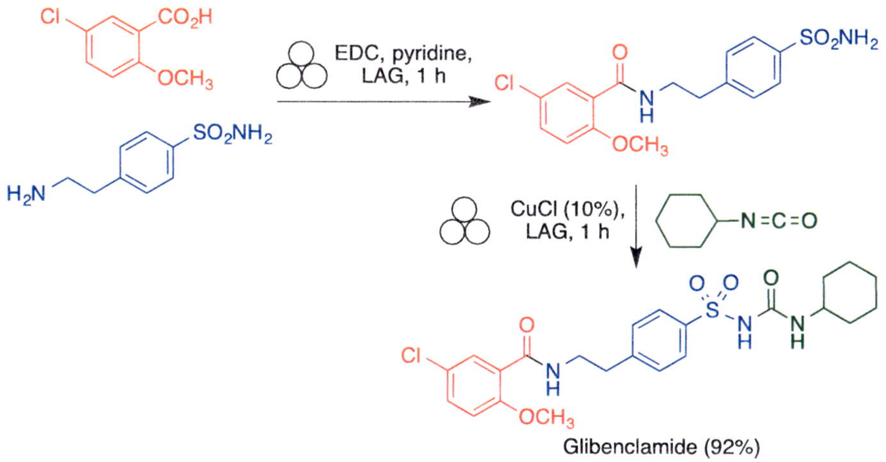

Scheme 14 Synthesis of glibenclamide, an antidiabetic sulphonylurea, in two mechanochemical steps

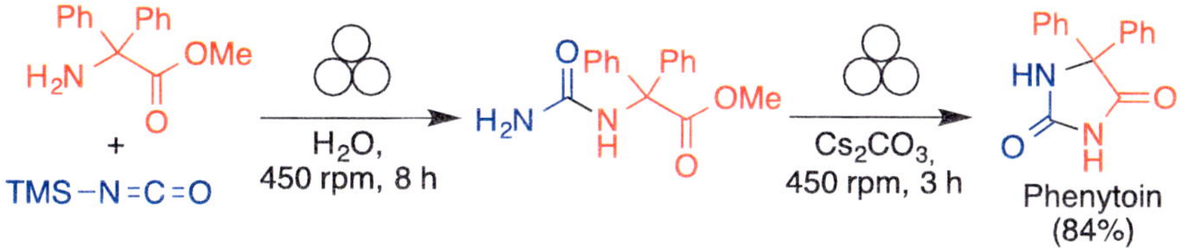

Scheme 15 Mechanochemical synthesis of the anticonvulsant drug phenytoin

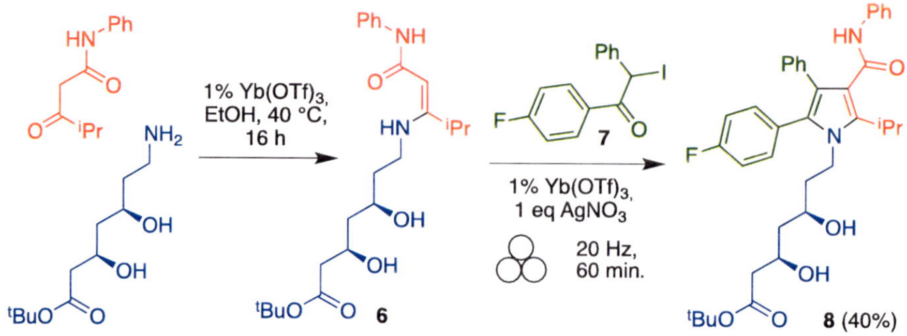

Scheme 16 A mechanochemical reaction under high-speed ball milling as the key step in a very concise synthesis of the cholesterol-lowering drug atorvastatin

with α-iodoketone (**7**) in the presence of ytterbium triflate as a Lewis acid catalyst (Scheme 16). Subsequent acid hydrolysis achieved the side chain deprotection and a concomitant intramolecular esterification affording atorvastatin lactone, which can be transformed into atorvastatin calcium, the therapeutically used form of the drug, in a single, straightforward step [63].

3.6 Prebiotic Mechanochemistry

Some researchers are exploring the prebiotic formation of biological molecules such as monosaccharides and amino acids, promoted by physical forces and catalysed by acidic or basic minerals under dry conditions by simulating these processes in a ball mill. Thus, aldoses and ketoses containing three-to-seven carbon atoms have been obtained through mechanochemical aldol reactions starting from formaldehyde, DL-glyceraldehyde and glycolaldehyde, with calcium hydroxide as catalyst and in the absence of water [64]. Similarly, mechanochemical Strecker reactions have been investigated under conditions believed to simulate those of early Earth from aldehydes, amines and potassium ferrocyanide as a source of HCN, with silicon oxide as catalyst. Treatment of these aminonitriles with paraformaldehyde and sodium hydroxide under ball milling afforded the corresponding α-aminoamides in modest yields [65].

3.7 Mechanochemical Synthesis of Nanostructures

3.7.1 Cage Compounds

There are some interesting examples in the literature of organic molecular cages such as the mechanochemical synthesis of macrobicycles by a [6 + 3 + 2] polycondensation of 4-formylphenylboronic acid (6 equiv.), pentaerythritol (3 equiv.) and (2,4,6-triethylbenzene-1,3,5-triyl)trimethanamine (2 equiv.), with a better yield than the alternative solution procedure [66]. Another multicomponent synthesis of borasiloxane based-macrocycles **9** from 4-formylphenylboronic acid, di(*tert*-butyl)silanediol and diamines was later reported [67], that again proceeded much more efficiently under ball milling than in solution (Scheme 17).

3.7.2 Rotaxanes

A rotaxane can be defined as a mechanically interlocked wheel-and-axle molecular complex, where a linear molecule that bears two bulky stoppers at both ends threads a macrocycle. They are very suitable for the construction of molecular devices, but their synthesis under solution conditions is hampered by the formation of by-products during the stoppering stage. A milestone was reached in this area in 2008 by the one-pot mechanochemical synthesis of [2] and [4]rotaxanes by grinding a macrocycle, a

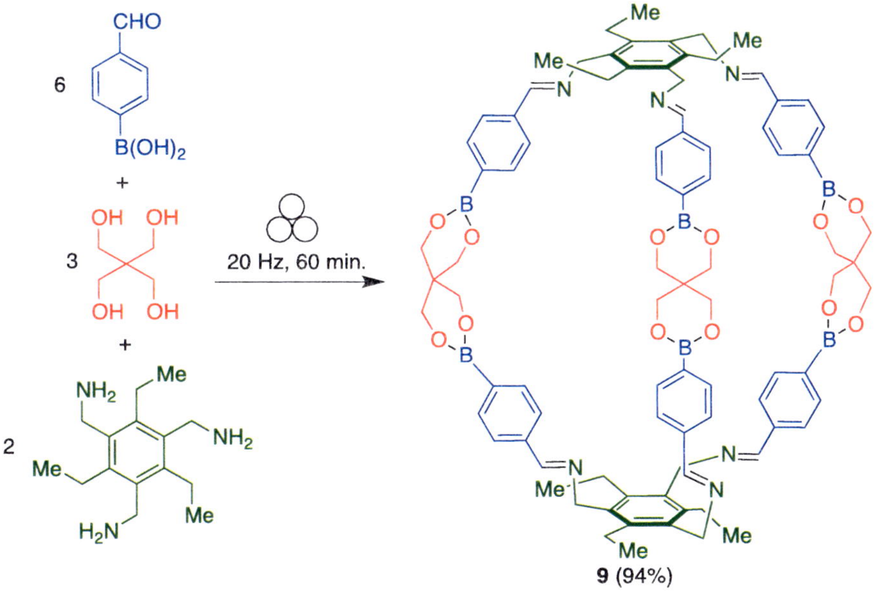

Scheme 17 Mechanochemical synthesis of a macrocyclic cage structure

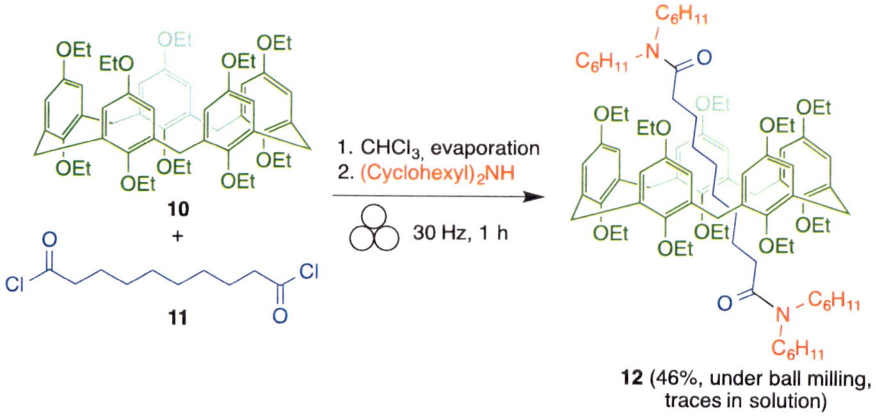

Scheme 18 Mechanochemical synthesis of a rotaxane containing a pillar[5]arene fragment

thread-like component and a stopper, either by a single multicomponent process or sequentially [68].

Remarkably, a library of pillar[5]arene-containing [2]rotaxanes **12** was obtained in good to excellent yields by ball milling of an inclusion complex formed by the pillar[5]arene **10** and dodecanedioyl dichloride **11**, obtained by simple co-evaporation of the starting materials, with primary or secondary amines, giving access to structures whose preparation is almost unfeasible under classical conditions (Scheme 18) [69].

3.7.3 Carbon Nanomaterials

The extraordinary mechanical, physicochemical and electronic properties of all-carbon materials, in special fullerenes, carbon nanotubes and graphene, have led to the prediction of myriad potential applications of these materials. Unfortunately, their synthesis and handling are hampered by their very poor solubility in organic solvents and water. Thus, mechanochemistry can potentially play an important role in this area. Thus, in a demonstration of how can allow the discovery of previously unknown chemistry, milling of fullerene with excess KCN led to its dimerization instead of the hydrocyanation expected from previous solution results [70]. The Prato reaction, a 1,3-dipolar cycloaddition between fullerenes and diazomethine ylides, is the most versatile method to functionalize fullerenes as exemplified by a three-component mechanochemical reaction between aldehydes, *N*-methylglycine and fullerene that afforded fulleropyrrolidines in modest yields, as shown in Scheme 19 [71].

Mechanical stress, when applied to nanotubes, can modify their length, or even transform them into nanoparticles or amorphous materials and therefore solution chemistry is by far the most studied approach to these materials. Nevertheless, some transformations of carbon nanotubes by fine-tuned ball milling have been described in

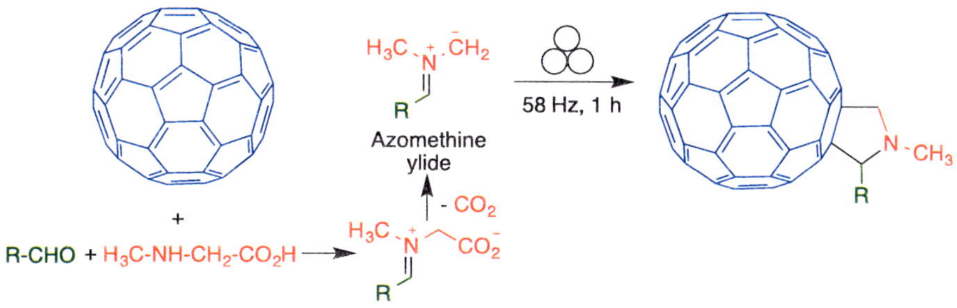

Scheme 19 Mechanochemical functionalization of fullerenes via the Prato reaction

the literature, including, among others, alkylations, arylations and functionalization with diazonium compounds [72]. Interestingly, mechanochemistry has also allowed covalently attaching fullerenes to nanotubes by simply grinding them together in the presence of potassium hydroxide [73].

In the case of graphene, some reported mechanochemical transformations include oxidations, modification of graphene platelets or reactive gas-induced functionalizations. For instance, grinding graphite microflakes with potassium permanganate or ammonium persulfate affords oxidized graphene derivatives bearing carbonyl, epoxy or hydroxy groups [74]. Other procedures to synthesize different graphene nanoplatelets by milling graphite in the presence of different reactive gases or dry ice are possible, affording carboxy-, sulfonic acid- or hydrogen-EFGnPs [75].

3.8 Mechanochemical Polymer Chemistry

Polymers are ubiquitous in all human societies, being widely used in science, technology and industry. The relationship between polymers and mechanochemistry is moving from a "destructive" approach focused on the ability of mechanical forces to degrade polymers in order to generate mechanically responsive materials by inserting mechanophores in their structures, to a "constructive" one aiming at the mechanochemical synthesis of polymers [76]. Ball milling has the inherent advantage of avoiding solubility or gelation problems when constructing large polymeric molecules; for instance, when the synthesis of poly(phenylene vinylenes) was performed by grinding 2-methoxy-5-(2'-ethylhexyloxy)phenylenevinylene monomers in the presence of potassium *tert*-butoxide (Glich method) the problems associated with the classical conditions such as long reaction times and gelation of the reaction mixture were circumvented [77]. Mechanochemical polymerizations based on Suzuki [78] and Friedel-Crafts [79] reactions have also been described. The mechanochemical synthesis of polylactic acid (PLA) polymers is also remarkable and involves the liquid-assisted polymerization of D-lactide, via ring opening with DBU as a base and benzylic alcohol as initiator agent [80].

Mechanical treatments can afford polymers unavailable by conventional methods. Chitosan is produced by deacetylation of chitin, but the poor solubility of the latter and its tendency to depolymerize under harsh deacetylation conditions make it difficult to obtain high-molecular weight chitosans with good deacetylation degrees. A mechanochemical approach has solved this problem and involves an initial amorphization of chitosan by ball milling, followed by its mechanochemical deacetylation using NaOH, and a final ageing under heat and high humidity conditions during several days [81].

Finally, ball milling has also been applied to post-polymerization modifications when there is a need for functional groups that are incompatible with the polymerization procedure. One example is the solid-state functionalization of 4-vinylbenzaldehyde/styrene copolymers and 4-vinylbenzaldehyde homopolymers by treatment with amines, leading to the corresponding Schiff bases [82].

3.9 Mechanochemical Synthesis of Metal-Organic Frameworks

Metal-organic frameworks (MOFs) are organic-inorganic hybrid materials formed by metal ions coordinated to organic ligands. They can be regarded as a class of coordination polymers, characterized by a highly porous crystalline structure and have created high expectations in several fields, including energy storage, gas trapping, drug delivery, catalysis and chemical sensing.

Mechanochemistry has emerged as a promising tool to synthesize MOFs because it avoids the challenging solubility issues that plague the conventional methods. The reactions are often very fast, with one example being finished in just one minute [83]. Metal salts or oxides are the most widely employed metal sources for these reactions, although other options, such as hydrides, have also been explored. Furthermore, twin-screw extrusion has allowed the large-scale synthesis of different metal organic frameworks of Cu, Al and Zn, reaching production rates of several kg/h and with a similar conversion to small-scale synthesis [84].

The role of liquid- and ion-and-liquid-assisted grinding (LAG and ILAG) in MOF synthesis deserves to be highlighted. A small amount of solvent (LAG) can help to template the framework, stabilize the structure by inclusion and achieve a better crystallinity of the product [85]. Sometimes, different MOFs can be obtained from the same starting materials depending on the liquid employed [86].

MOFs, due to their porosity, large surface area and possibility of functionalization, represent promising drug delivery systems. A Cu-based nano-MOF loaded with ibuprofen was synthesized using DABCO and 1,4-benzenedicarboxylic acid as organic moieties (Scheme 20), and showed a very high drug loading capacity and a well-defined drug delivery profile [87].

The encapsulation of enzymes, such as β-glycosidase, invertase, catalase or β-galactosidase, in different Zn- and Zr-based MOFs in a one-pot

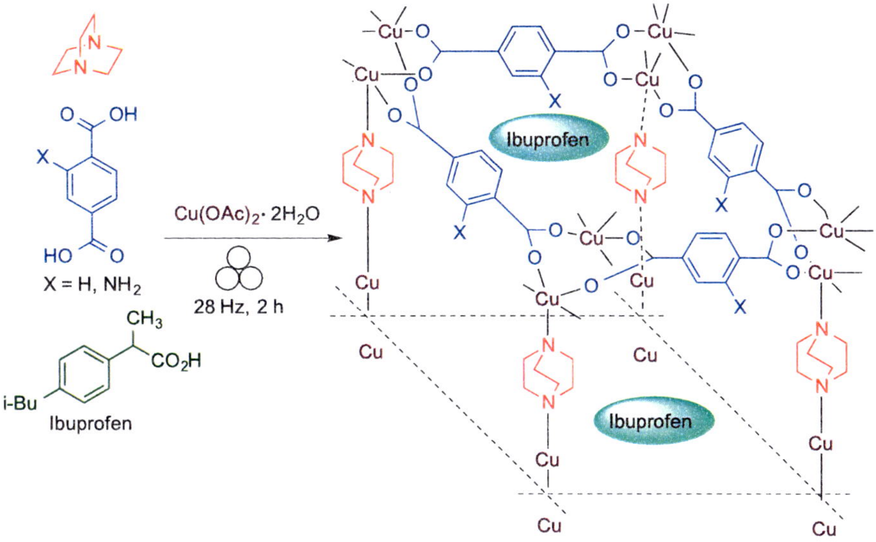

Scheme 20 Mechanochemical synthesis of a MOF-based drug delivery system

synthesis/encapsulation process under mechanochemical conditions, is also noteworthy. These encapsulated enzymes retained their catalytic activity and were protected from inactivation by proteases, even in acidic conditions [88].

4 Reactions Combining Photochemical and Mechanochemical Activation

Mechanochemistry, often by simple manual grinding, has been employed to prepare cocrystals for their subsequent solid-state [2 + 2] photodimerization reactions [89]. Sequential mechanochemical cocrystallization and photodimerization can also be achieved in a single operation by vortex grinding, where a transparent vortexing tube, filled with ball bearings, is exposed to UV light [90]. Photochemical ball milling has been performed by means of transparent jars made up from poly(methylmethacrylate) or glass, surrounded by blue- or green light-emitting diodes, as shown in Fig. 2 [91, 92].

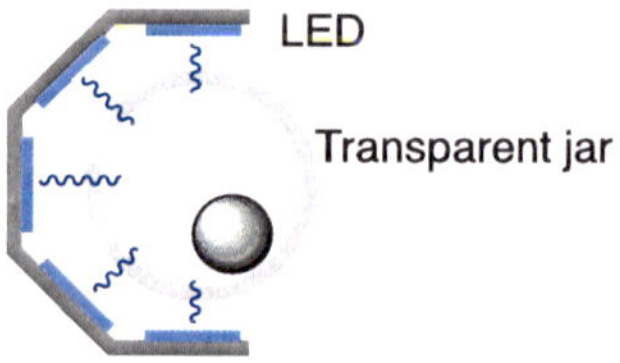

Fig. 2 Schematic experimental setup for ball milling under photochemical conditions

5 Sonochemistry in the Solid State

The first example of solvent-free sonochemistry was published in 2017 and came as a surprise because the most widely accepted mechanism for sonochemical activation requires the energy generated in the collapse of microbubbles formed as a consequence of cavitation phenomena in the liquid phase. Two aldol reactions were performed: the condensation of 1,2-phenylenediamine with *o*-vanillin to afford a salen ligand (**13**), and the formation of a 1,3-indanedione (**14**) from ninhydrin and dimedone (Scheme 21), which took place with full conversion in 60–90 min of sonication, in contrast to refluxes of several hours required conventionally. This approach could provide an alternative to mechanochemistry when reactants, reagents or products are not stable under mechanical stress [93].

Solid-state sonochemistry has also been applied to the generation of some pharmaceutically relevant cocrystals, such as paracetamol-caffeine or aspirin-meloxicam, by neat or liquid-assisted sonication (analogue to liquid-assisted grinding) by green and very simple procedures [94].

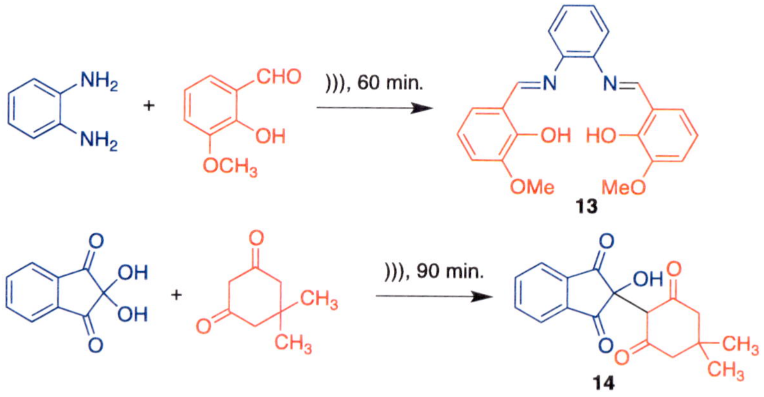

Scheme 21 Two reactions promoted by solid-state sonochemical irradiation

6 Conclusion

Solid-state chemistry, and mechanochemistry in particular, has some potential advantages such as the absence of solubility issues and the potential increase of reactivity or the deviation of the reaction to alternative pathways due to the absence of solvation phenomena. Solid-state synthetic methods also may have an advantage over solution methods in terms of environmental friendliness, although such comparisons are complex, and many factors need to be considered. The absence of solvent in the reactions is a positive feature of mechanochemistry and an advantage over solution chemistry, but it does not automatically make the overall process green since the use of solvents in purification processes needs to be taken into account [1]. Furthermore, energy consumption in reactions carried out by ball milling is lower than that under conventional heating or microwave irradiation, making mechanochemistry preferable from an energetic point of view, although only a few reactions have been studied in this regard [95]. These promising findings place solid-state chemistry in a good starting position in the current endeavour to develop more sustainable synthetic protocols.

References

1. Obst M, König B (2018) Organic synthesis without conventional solvents. Eur J Org Chem 2018:4213–4232
2. Oelgemöller M, Hoffmann N (2016) Studies in organic and physical photochemistry-an interdisciplinary approach. Org Biomol Chem 14:7392–7442
3. Hernández-Linares MG, Guerrero-Luna G, Pérez-Estrada S, Ellison M, Ortín MM, García-Garibay MA (2015) Large-scale green chemical synthesis of adjacent quaternary chiral centers by continuous flow photodecarbonylation of aqueous suspensions of nanocrystalline ketones. J Am Chem Soc 137:1679–1684
4. Campos LM, Dang H, Ng D, Yang Z, Martinez HL, Garcia-Garibay MA (2002) Engineering reactions in crystalline solids: predicting photochemical decarbonylation from calculated thermochemical parameters. J Org Chem 67:3749–3754
5. Ihmels H, Scheffer R (1999) The Norrish type II reaction in the crystalline state: toward a better understanding of the geometric requirements for γ-hydrogen atom abstraction. Tetrahedron 55:885–907
6. Schmidt GMJ (1971) Photodimerization in the solid state. Pure Appl Chem 27:647–678
7. MacGillivray LR, Reid JL, Ripmeester JA (2000) Supramolecular control of reactivity in the solid state using linear molecular templates. J Am Chem Soc 122:7817–7818
8. Grobelny AL, Rathb NP, Groeneman RH (2019) Varying the regiochemistry from a solid-state [2 + 2] cycloaddition reaction within a series of mixed co-crystals based upon isosteric resorcinols. J Photochem Photobiol A: Chem 382:111966–111971
9. Crawford D, Miskimming CKG, Albadarin AB, Walker G, James SL (2017) Organic synthesis by twin screw extrusion (TSE): continuous, scalable and solvent-free. Green Chem 19:1507–1518
10. Hernández JG, Bolm C (2017) Altering product selectivity by mechanochemistry. J Org Chem 82:4007–4019
11. Do JL, Friščić T (2017) Mechanochemistry: a force of synthesis. ACS Cent Sci 3:13–19
12. Howard JL, Cao Q, Browne DL (2018) Mechanochemistry as an emerging tool for molecular synthesis: what can it offer? Chem Sci 9:3080–3094

13. Leonardi M, Villacampa M, Menéndez JC (2018) Multicomponent mechanochemical synthesis. Chem Sci 9:2042–2064
14. Tan D, García F (2019) Main group mechanochemistry: from curiosity to established protocols. Chem Soc Rev 48:2274–2292
15. Friščić T, Cottillo C, Titi HM (2020) Mechanochemistry for synthesis. Angew Chem Int Ed 59:1018–1029
16. El-Sayed TH, Aboelnaga A, El-Atawy MA, Hagar M (2018) Ball milling promoted *N*-heterocycles synthesis. Molecules 23: article nr 1348
17. Leonardi M, Estévez V, Villacampa M, Menéndez JC (2019) Mechanochemical synthesis of biologically relevant heterocycles. In: Ballini R (ed) Green synthetic processes and procedures. Royal Society of Chemistry, London, pp 175–191
18. Shearouse WC, Shumba MZ, Mack J (2014) A solvent-free, one-step, one-pot Gewald reaction for alkyl-aryl ketones via mechanochemistry. Appl Sci 4:171–179
19. Akelis L, Rousseau J, Juskenas R, Dodonova J, Rousseau C, Menuel S, Prevost D, Tumkevicius S, Monflier E, Hapiot F (2016) Greener Paal-Knorr pyrrole synthesis by mechanical activation. Eur J Org Chem 2016:31–35
20. Estévez V, Villacampa M, Menéndez JC (2013) Three-component access to pyrroles promoted by the CAN–silver nitrate system under high-speed vibration milling conditions: a generalization of the Hantzsch pyrrole synthesis. Chem Commun 49:591–593
21. El-Sayeed TH, Aboelnaga A, Hagar M (2016) Ball milling assisted solvent and catalyst free synthesis of benzimidazoles and their derivatives. Molecules 21:1111.
22. Kamur S, Sharma P, Kapoor KK, Hundal S (2008) An efficient, catalyst- and solvent-free, four-component, and one-pot synthesis of polyhydroquinolines on grinding. Tetrahedron 64:536–542
23. Bose AK, Pednekar S, Ganduly SN, Chakraborty G, Manhas MS (2004) A simplified green chemistry approach to the Biginelli reaction using 'Grindstone Chemistry'. Tetrahedron Lett 45:8351–8353
24. Tan YJ, Zhang Z, Wang FJ, Wu HH, Li QH (2014) Mechanochemical milling promoted solvent-free imino Diels-Alder reaction catalyzed by FeCl3: diastereoselective synthesis of cis-2, 4-diphenyl-1, 2, 3, 4-tetrahydroquinolines. RSC Adv 4:35635–35638
25. Kaupp G, Naimi-Jamal MR (2002) Quantitative cascade condensations between o-phenylenediamines and 1,2-dicarbonyl compounds without production of wastes. Eur J Org Chem 2002:1368–1373
26. Kaupp G, Naimi-Jamal MR, Schmeyers J (2002) Quantitative reaction cascades of ninhydrin in the solid state. Chem Eur J 8:594–600
27. Muthusaravanan S, Sasikumar C, Devi Bal B, Perumal S (2014) An eco-friendly three-component regio- and stereoselective synthesis of highly functionalized dihydroindeno[1,2-*b*]pyrroles under grinding. Green Chem 16:1297–1304
28. Kaupp G, Schmeyers J, Boy J (2000) Iminium salts in solid-state syntheses giving 100% yield. J Prakt Chem 342:269–280
29. Estévez V, Sridharan V, Sabaté S, Villacampa M, Menéndez JC (2016) Three-component synthesis of pyrrole-related nitrogen heterocycles by a Hantzsch-type process: comparison between conventional and high-speed vibration milling conditions. Asian J Org Chem 5:652–662
30. Leonardi M, Villacampa M, Menéndez JC (2017) High-speed vibration-milling-promoted synthesis of symmetrical frameworks containing two or three pyrrole units. Beilstein J Org Chem 13:1957–1962
31. Jicsinszky L, Caporaso M, Martina K, Calcio Gaudino E, Cravotto G (2016) Efficient mechanochemical synthesis of regioselective persubstituted cyclodextrins. Beilstein J Org Chem 12:2364–2371
32. Patil PR, Kartha KPR (2008) Solvent-free mechanochemical synthesis of aryl glycosides. J Carbohydr Chem 27:411–419
33. Đud M, Margetić D (2017) Solvent-free mechanochemical deprotection of *N*-Boc group. Int J Org Chem 7:140–144

34. Declerck V, Nun P, Martinez J, Lamaty F (2009) Solvent-free synthesis of peptides. Angew Chem Int Ed 48:9318–9321
35. Hernández JG, Juaristi E (2010) Green synthesis of α, β- and β, β-dipeptides under solvent-free conditions. J Org Chem 75:7107–7111
36. Landeros JM, Juaristi E (2017) Mechanochemical synthesis of dipeptides using Mg-Al hydrotalcite as activating agent under solvent-free reaction conditions. Eur J Org Chem 2017:687–694
37. Gonnet L, Tintillier T, Venturini N, Konnert L, Hernández JF, Lamaty F, Laconde G, Martinez J, Colacino E (2017) *N*-Acyl benzotriazole derivatives for the synthesis of dipeptides and tripeptides and peptide biotinylation by mechanochemistry. ACS Sustainable Chem Eng 5:2936–2941
38. Bonnamour J, Métro TX, Martinez J, Lamaty F (2013) Environmentally benign peptide synthesis using liquid-assisted ball-milling: Application to the synthesis of Leu-enkephalin. Green Chem 15:1116–1120
39. Maurin O, Verdié P, Subra G, Lamaty F, Martinez J, Métro TX (2017) Peptide synthesis: ball-milling, in solution, or on solid support, what is the best strategy? Beilstein J Org Chem 13:2087–2093
40. Crossey K, Cunningham RN, Redpath P, Migaud ME (2015) Atom efficient synthesis of pyrimidine and purine nucleosides by ball milling. RSC Adv 5:58116–58119
41. Eguaogie O, Conlon PF, Ravalico F, Sweet JS, Elder TB, Conway LP, Lennon ME, Hodgson DRW, Vyle JS (2017) Nucleophilic displacement reactions of 5′-derivatised nucleosides in a vibration ball mill. Beilstein J Org Chem 13:87–92
42. Appy L, Depaix A, Bantreil X, Lamaty F, Peyrottes S, Roy B (2019) Straightforward ball-milling access to dinucleoside 5',5'-polyphosphates via phosphorimidazolide intermediates. Chem Eur J 25:2477–2481
43. Hu H, Li H, Zhang Y, Chen Y, Huang Z, Huang A, Zhu Y, Qin X, Lin B (2015) Green mechanical activation-assisted solid phase synthesis of cellulose esters using a co-reactant: effect of chain length of fatty acids on reaction efficiency and structure properties of products. RSC Adv 5:20656–20662
44. Ardila-Fierro KJ, Pich A, Spehr M, Hernández JG, Bolm C (2019) Synthesis of acylglycerol derivatives by mechanochemistry. Beilstein J Org Chem 15:811–817
45. Chauhan P, Chimni SS (2012) Mechanochemistry assisted asymmetric organocatalysis: a sustainable approach. Beilstein J Org Chem 8:2132–2141
46. Ávila-Ortiz CG, Pérez-Venegas M, Vargas-Caporali J, Juaristi E (2019) Recent applications of mechanochemistry in enantioselective synthesis. Tetrahedron Lett 60:1749–1757
47. Rodríguez B, Rantanen T, Bolm C (2006) Solvent-free asymmetric organocatalysis in a ball mill. Angew Chem Int Ed 45:6924–6926
48. Hernández JG, Juaristi E (2011) Asymmetric aldol reaction organocatalyzed by (*S*)-proline-containing dipeptides: improved stereoinduction under solvent-free conditions. J Org Chem 76:1464–1467
49. Chauhan P, Chimni SS (2012) Grinding-assisted asymmetric organocatalysis: a solvent-free approach to the formation of vicinal quaternary and tertiary stereocenters. Asian J Org Chem 1:138–141
50. Valle-Orero J, Rivas-Pardo JA, Tapia-Rojo R, Popa I, Echelman DJ, Haldar S, Fernández JM (2017) Mechanical deformation accelerates protein ageing. Angew Chem Int Ed 129:9873–9878
51. Hanefeld U, Cao L, Magner E (2013) Enzyme immobilisation: fundamentals and application. Chem Soc Rev 42:6211–6212
52. Pérez-Venegas M, Reyes-Rangel G, Neri A, Escalante J, Juaristi E (2017) Mechanochemical enzymatic resolution of *N*-benzylated-β^3-amino esters. Beilstein J Org Chem 13:1728–1734
53. Weißbach U, Dabral S, Konnert L, Bolm C, Hernández JG (2017) Selective enzymatic esterification of lignin model compounds in the ball mill. Beilstein J Org Chem 13:1788–1795
54. Hernández JG, Ardila-Fierro KJ, Crawford D, James SL, Bolm C (2017) Mechanoenzymatic peptide and amide bond formation. Green Chem 19:2620–2625

55. Ardila-Fierro KJ, Crawford D, Körner A, James SL, Bolm C, Hernández JG (2018) Papain-catalysed mechanochemical synthesis of oligopeptides by milling and twin-screw extrusion: application in the Julia-Colonna enantioselective epoxidation. Green Chem 20:1262–1269
56. Tan D, Loots L, Friščić T (2016) Towards medicinal mechanochemistry: evolution of milling from pharmaceutical solid form screening to the synthesis of active pharmaceutical ingredients (APIs). Chem Commun 52:760–7781
57. André V, Hardeman A, Halasz I, Stein RS, Jackson GJ, Reid DG, Duer MJ, Curfs C, Duarte MT, Friščić T (2011) Mechanosynthesis of the metallodrug bismuth subsalicylate from Bi_2O_3 and structure of bismuth salicylate without auxiliary organic ligands. Angew Chem Int Ed 50:7858–7861
58. Tan D, Štrukil V, Mottillo C, Friščić T (2014) Mechanosynthesis of pharmaceutically relevant sulfonyl-(thio)ureas. Chem Commun 50:5248–5250
59. Konnert L, Reneaud B, de Figueiredo RM, Campagne JM, Lamaty F, Martinez J, Colacino E (2014) Mechanochemical preparation of hydantoins from amino esters: application to the synthesis of the antiepileptic drug phenytoin. J Org Chem 79:10132–10142
60. Colacino E, Porcheddu A, Charnay C, Delogu F (2019) From enabling technologies to medicinal mechanochemistry: an eco-friendly access to hydantoin-based active pharmaceutical ingredients. React Chem Eng 4:1179–1188
61. Colacino E, Porcheddu A, Halasz I, Charnay C, Delogu F, Guerra R, Fullenwarth J (2018) Mechanochemistry for "no solvent, no base" preparation of hydantoin-based active pharmaceutical ingredients: nitrofurantoin and dantrolene. Green Chem 20:2973–2977
62. Porcheddu A, Delogu F, De Luca L, Colacino E (2019) From Lossen transposition to solventless "medicinal mechanochemistry" ACS Sustain Chem Eng 7:12044–12051
63. Estévez V, Villacampa M, Menéndez JC (2014) Concise synthesis of atorvastatin lactone under high-speed vibration milling conditions. Org Chem Front 1:458–463
64. Lamour S, Pallmann S, Haas M, Trapp O (2019) Prebiotic sugar formation under nonaqueous conditions and mechanochemical acceleration. Life 9:52
65. Bolm C, Mocci R, Schumacher C, Turberg M, Puccetti F, Hernández JG (2018) Mechanochemical activation of iron cyano complexes: a prebiotic impact scenario for the synthesis of α-amino acid derivatives. Angew Chem Int Ed 57:2423–2446
66. Içli B, Christinat N, Tönnemann J, Schüttler C, Scopelliti R, Severin K (2009) Synthesis of molecular nanostructures by multicomponent condensation reactions in a ball mill. J Am Chem Soc 131:3154–3155
67. Pascu M, Ruggi A, Scopelliti R, Severin K (2013) Synthesis of borasiloxane-based macrocycles by multicomponent condensation reactions in solution or in a ball mill. Chem Comm 49:45–47
68. Hsueh SY, Cheng KW, Lai CC, Chiu SH (2008) Efficient solvent-free syntheses of [2]- and [4]rotaxanes. Angew Chem Int Ed 47:4436–4439
69. Holler M, Stoerkler T, Louis A, Fischer F, Nierengarten JF (2019) Mechanochemical solvent-free conditions for the synthesis of pillar[5]arene-containing [2]rotaxanes. Eur J Org Chem 21:3401–3405
70. Wang GW, Komatsu K, Murata Y, Shiro M (1997) Synthesis and X-ray structure of dumb-bell-shaped C_{120}. Nature 387:583–586
71. Wang GW, Zhang TH, Hao EH, Jiao LJ, Murata Y, Komatsu K (2003) Solvent-free reactions of fullerenes and *N*-alkylglycines with and without aldehydes under high-speed vibration milling. Tetrahedron 59:55–60
72. Zhu SE, Li F, Wang GW (2013) Mechanochemistry of fullerenes and related materials. Chem Soc Rev 42:7535–7570
73. Li X, Liu L, Qin Y, Wu W, Guo ZX, Dai L, Zhu D (2003) C_{60} modified single-walled carbon nanotubes. Chem Phys Lett 377:32–36
74. Posudievsky OY, Khazieieva OA, Koshechko VG, Pokhodenko VD (2012) Preparation of graphene oxide by solvent-free mechanochemical oxidation of graphite. J Mater Chem 22:12465–12467
75. Jeon IY, Choi HJ, Jung SM, Seo JM, Kim MJ, Dai L, Baek JB (2012) Large-scale production of edge-selectively functionalized graphene nanoplatelets via ball milling and their use as metal-free electrocatalysts for oxygen reduction reaction. J Am Chem Soc 135:1386–1393

76. Li J, Nagamani C, Moore JS (2015) Polymer mechanochemistry: from destructive to productive. Acc Chem Res 48:2181–2190
77. Ravnsbæk JB, Swager TM (2014) Mechanochemical synthesis of poly(phenylene vinylenes). ACS Macro Lett 3:305–309
78. Grätz S, Wolfrum B, Borchardt L (2017) Mechanochemical Suzuki polycondensation—from linear to hyperbranched polyphenylenes. Green Chem 19:2973–2979
79. Grätz S, Zink S, Kraffczyk H, Rose M, Borchardt L (2019) Mechanochemical synthesis of hyper-crosslinked polymers: influences on their pore structure and adsorption behaviour for organic vapors. Beilstein J Org Chem 15:1154–1161
80. Ohn N, Shin J, Kim SS, Kim JG (2017) Mechanochemical ring-opening polymerization of lactide: liquid-assisted grinding for the green synthesis of poly(lactic acid) with high molecular weight. Chemsuschem 10:3529–3533
81. Di Nardo T, Hadad C, Van Nhien AN, Moores A (2019) Synthesis of high molecular weight chitosan from chitin by mechanochemistry and aging. Green Chem 21:3276–3285
82. Ohn N, Kim JG (2018) Mechanochemical post-polymerization modification: solvent-free solid-state synthesis of functional polymers. ACS Macro Lett 7:561–565
83. Zhang R, Tao CA, Chen R, Wu L, Zou X, Wang J (2018) Ultrafast synthesis of Ni-MOF in one minute by ball milling. Nanomaterials 8:E1067
84. Crawford D, Casaban J, Haydon R, Giri N, McNally T, James SL (2015) Synthesis by extrusion: continuous, large-scale preparation of MOFs using little or no solvent. Chem Sci 6:1645–1649
85. Friščić T (2012) Supramolecular concepts and new techniques in mechanochemistry: cocrystals, cages, rotaxanes, open metal–organic frameworks. Chem Soc Rev 41:3493–3510
86. Friščić T, Fábián L (2009) Mechanochemical conversion of a metal oxide into coordination polymers and porous frameworks using liquid-assisted grinding (LAG). CrystEngComm 11:743–745
87. Nadizadeh Z, Naimi-Jamal MR, Panahi L (2018) mechanochemical solvent-free in situ synthesis of drug-loaded $\{Cu_2(1,4\text{-bdc})_2(\text{dabco})\}_n$ MOFs for controlled drug delivery. J Solid State Chem 259:35–42
88. Wei TH, Wu SH, Huang YD, Lo WS, Williams BP, Chen SY, Yang HC, Hsu YS, Lin ZY, Chen XH, Kuo, PE, Choy LY, Tsung CK, Shieh FK (2019) Rapid mechanochemical encapsulation of biocatalysts into robust metal-organic frameworks. Nat Commun 10: article nr. 5002
89. Papaefstathiou GS, MacGillivray LR (2001) Discrete versus infinite molecular self-assembly: control in crystalline hydrogen-bonded assemblies based on resorcinol. Org Lett 3:3835–3838.
90. Friščić T, MacGillivray LR (2003) 'Template-switching': a supramolecular strategy for the quantitative, gram-scale construction of a molecular target in the solid state. Chem Commun 2003:1306–1307
91. Hernández JG (2017) Mechanochemical borylation of aryldiazonium salts; merging light and ball milling. Beilstein J Org Chem 13:1463–1469
92. Štrukil VS, Sajko I (2017) Mechanochemically-assisted solid-state photocatalysis (MASSPC). Chem Commun 53:9101–9104
93. Crawford D (2017) Solvent-free sonochemistry: sonochemical organic synthesis in the absence of a liquid medium. Beilstein J Org Chem 13:1850–1856
94. Roy D, James SL, Crawford D (2019) Solvent-free sonochemistry as a route to pharmaceutical co-crystals. Chem Commun 55:5463–5466
95. Do JL, Friščić T (2017) Chemistry 2.0: developing a new, solvent-free system of chemical synthesis based on mechanochemistry. Synlett 28:2066–2092

Chapter 7
Green in Analytical Chemistry

Mihkel Koel

Nature does not care what we call it, she just keeps on doing it.
Richard P. Feynman, The Feynman Lecture on Physics

1 Introduction

1.1 The Purpose of Analytical Chemistry

Analytical chemistry performs the following functions:

- deals with qualitative and quantitative information about the nature, amount and identity of elements and molecules in our environment;
- provides sufficient information with the appropriate analytical sensitivity, selectivity and accuracy to make decisions and solve problems;
- provides specific information according to the requirements of the end-users of the chemical information.

The roots of contemporary chemistry lie in chemical analysis for the purpose of obtaining and accumulating knowledge and formulating theories in chemistry and related areas, while providing the motivation for developing new technologies and logistical solutions. Analytical chemistry forms the basis of standards and specifications justifying the need for new laws and administrative prescriptions, and also provides the means for following the rules related to chemicals and materials. From another perspective, analytical chemistry with its comprehensive systems is an important tool for demonstrating the environmental benignancy of new methods, processes and products, and for supporting product life cycle analysis (LCA) (Fig.1).

Analytical chemistry laboratories and the analysts who staff them are in an ideal position to solve particular problems. The demand for this type of laboratory is

M. Koel (✉)
Tallinn University of Technology, Ehitajate 5, Tallinn 19086, Estonia
e-mail: mihkel.koel@ttu.ee; Mihkel.koel@taltech.ee

G. Anilkumar and S. Saranya (eds.), *Green Organic Reactions*, Materials Horizons: From Nature to Nanomaterials, https://doi.org/10.1007/978-981-33-6897-2_7

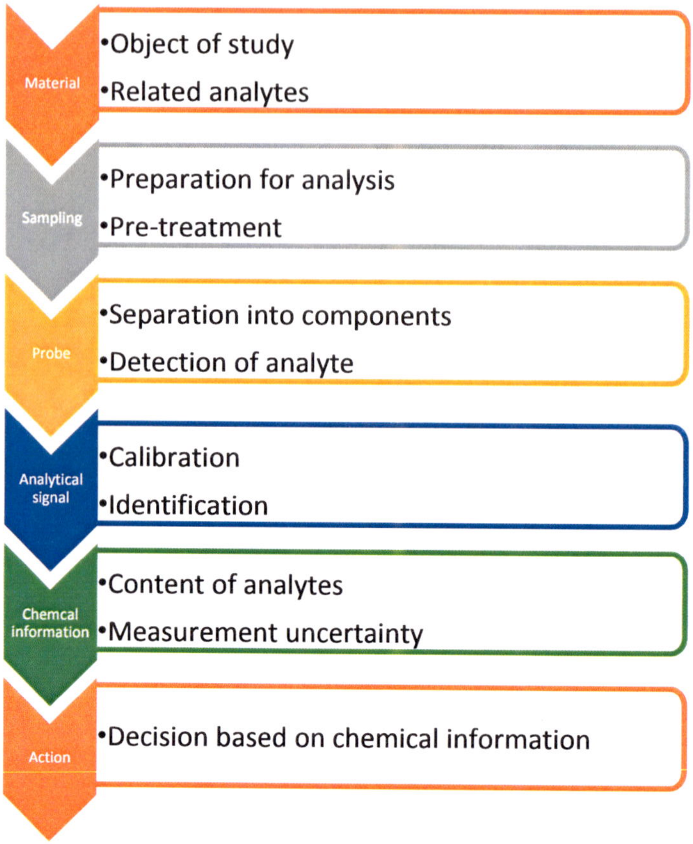

Fig. 1 Analytical system

growing, as is the number of analyses in various areas of activity. There are analytical chemistry laboratories in academia, and for public services such as environmental monitoring and control (air, water, soil), and monitoring the toxicity of food and feed, etc. In addition, there are industrial analytical chemistry laboratories related to process and production quality control (Fig. 2).

Monitoring and quality-control laboratories require a significant amount of chemicals and energy for a large number of runs. Their total waste production and energy consumption is substantial. Usually, laboratories are considered small-scale activities compared with factories in the chemical industry, but because these laboratories are ubiquitous, with very different analytical processes and a substantial number of samples to analyse, they are comparable with the fine chemicals industry, which has a similar E-factor (ratio of by-products to the end product) of 25–100 [1].

International regulations are a good example of the power and societal influence of analytical chemistry. Developing the analytical performance necessary for the detection of toxic compounds has provided a firm basis for international agreements:

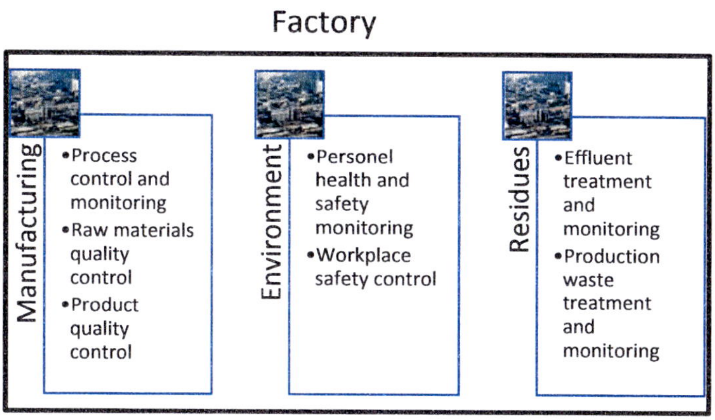

Fig. 2 Steps related to industrial production

- Regulation of the emission of volatile organic compounds (VOC): 1985—Vienna convention; 1987—Montreal Protocol; 1990—US Clean Air Act
- Ban on persistent organic pollutants (POP) (the 'dirty dozen') 2001—Stockholm Protocol
- Kyoto Protocol to reduce greenhouse gas emissions (human-made CO_2 emissions): 1997; enacted 16 February 2005.

The next step is to improve the analytical procedures required to monitor these and future agreements in accordance with the principles of green analytical chemistry.

Optimising analytical procedures involves providing the required information:

- with appropriate analytical parameters (sensitivity, selectivity, accuracy and precision);
- in a manner that is inherently safe, nontoxic and environmentally benign;
- with the least possible consumption of material and energy;
- with minimum generation of waste.

In the light of trends in analytical chemistry, in which intensive research is being undertaken in the areas of automation, systems integration, miniaturisation and sensors, the use of new materials, especially those whose physical or chemical properties can be controllably altered in response to an external stimulus, is very attractive. In terms of social responsibility, it has been proposed that analytical chemistry be defined as 'research that develops and optimises analytical processes with respect to the consumption of material and energy and the generation of waste, inherent safety, non-toxicity and environmental friendliness' [2].

Analytical chemists must be guided by the following considerations: What kind of information is actually required—the precise number of molecules, or basic chemical knowledge about the object? Must every molecule be counted in every analysis? How should the quality of the analytical data be measured? What will happen to the waste? What is included in the cost of the analysis? The choice of the method of analysis and related procedures depends on the answers to these questions as well as the origin of the sample. There are myriad procedures for sample pre-treatment and separation prior to analysis, and a wide choice of methods for detection and identification. It is important to observe the principle of fitness for purpose—the property of data produced by a measurement process that enables the user to make technically correct decisions for a stated purpose [3].

It is useful when developing an analytical approach to follow the guidelines from the International Conference on Harmonisation (ICH) Q8 (R2) on Quality by Design (QbD), which describe this step as 'a systematic approach to development that begins with predefined objectives and emphasises product and process understanding and process control, based on sound science and quality risk management' [4]. The guidelines stress that the method and performance characteristics of the process of analysis need to be designed to fulfil the specific objectives of the study rather than being based on the performance of tests or quality control of previously released batches. At this stage, it is important to assess the methods and analytical procedures in terms of safety and risk, and the economy and efficiency of the instrumentation. The use of mathematical methods for planning analytical procedures and data processing would be advantageous. Simple methods can often provide enough information to make a decision and solve the problem. Analytical quality by design allows for the development of a robust and cost-effective analytical method that is applicable throughout the life cycle of the product and facilitates regulatory flexibility [5].

1.2 Developments in Analytical Chemistry

Trends in analytical chemistry feature the development of automated procedures and equipment for real-time and point-of-interest applications using new materials (Table 1):

Automation and miniaturisation using sensors and systems integration with high throughput for multiple tasks;
Real-time, portable and highly sensitive analytical instruments capable of detecting and quantifying biological or chemical hazards;
Nanomaterials, photonics and improved bioassay techniques in electro-optics and new materials;
Smart materials with physical or chemical properties that can be controllably altered in response to an external stimulus;
Computing and communication technology in analysis, especially in sensor-related research and information/data handling.

In general, the development of instrumental methods leads to efficient use of energy, especially when the method is highly automated and uses a minimal amount of sample. The hyphenation of several methods for sample treatment and the separation of components, or the integration of separation with complex methods of detection, also enable efficient use of energy. The miniaturisation and related development of

Table 1 Current challenges in analytical chemistry

Microsystems	Miniaturisation: lab-on-a-chip and total analysis systems
	Microfluidics, chip-to-world interfacing
	Automated sampling strategies
	Remote (wireless) control and data acquisition, and analysis
Sensors	Wide selection of sensing principles
	Highly selective, sensitive sensors; biosensors
	Stable, no need for calibration
	Real time: instantaneous response, no kinetic effects
New materials	Composites, alloys, biomaterials
	Polymers, molecularly imprinted materials
	Stimuli-responsive materials
	Nanostructured materials
Information processing	Multiple, localised simultaneous measurements to provide • A spectrum of information • Information on different species/processes • Analysis in the frequency domain
	Chemometrics and multivariate analysis techniques

sensors and the introduction of information-processing technologies provide a basis for efficient and environmentally benign analytical chemistry. New materials lend extra flexibility to the development of analytical procedures that reduce the amount of waste. Smart materials are also used in miniaturisation, the development of sensors and the automation of analytical procedures, which are important factors in greening these processes. They also have a major influence on the time and cost of analysis.

In analytical chemistry, the main sources of environmental problems and waste generation are solvents and harmful chemicals; therefore, greening of the analysis involves reducing or eliminating the use of solvents and chemicals. This is highly relevant to a discussion of methods of separation, sample preparation and treatment, which are the biggest consumers of solvents and other reagents. This discussion gave rise to the 3R movement—reduction, replacement and recycling [6].

R1—Reduction–In chromatography, both gas and liquid using narrow-bore columns, which entail smaller particles, higher pressures and shorter columns, substantially reduces solvent consumption and waste generation [7]. Additional economy is gained from microcolumns, which require only a small fraction of the chromatographic stationary phase.

Elevated temperatures also alter the properties of separation media and solid supports, increasing efficiency and reducing the use of solvents [8]. However, the approach involves special temperature-resistant packing materials and also carries a potential risk of on-column degradation of thermally labile compounds, in addition to solubility problems with hydrophobic compounds.

R2—Replacement—There are 250–300 solvents available to chemists working in academia and industry (plus an infinite number of solvent mixtures), and this number is increasing. The use of solvents is always related to occupational safety and health, which are important components of environmental sustainability [9]. Intensive research is being conducted in order to find a replacement for environmentally harmful solvents.

The greenest solvent in separation science is carbon dioxide in a supercritical state ($scCO_2$) used as an eluent or extracting agent [10]. This approach is being utilised in large-scale industrial extraction applications and also in preparative chromatography [11]. Mixed-phase SFC solvents are not as environmentally benign as single-phase CO_2, but they are significantly easier to dispose of or recycle than mixed organic-aqueous LC solvents. Because $scCO_2$ is inert, the materials as well as most of the instrumentation used in liquid chromatography can also be employed in supercritical fluid chromatography.

Selection of the most appropriate solvent should always include consideration of alternative solvents that takes into account toxicity, cost, safety, workability, as well as chromatographic selectivity and elution strength [12] (Table 2).

Procedures have been developed for assessing the environmental risks related to solvent emissions whereby specific hazards—toxicological, environmental persistence or photochemical ozone creation—have been identified for each component. The results support informed solvent selection [13].

Amongst these alternatives to common organic solvents is a group of so-called neoteric solvents (neoteric denoting 'modern' or 'recent in origin', derived from the Greek *neoterikos* meaning 'younger'), which includes perfluorinated (fluorous) solvents, and room-temperature ionic liquids. Most of these solvents have tunable properties that enable them to replace specific solvents with one that will accommodate different processes.

Fluorous (perfluorinated) solvents such as perfluoroalkenes, perfluorolkyl ethers and perfluoro-alkylamines are generally chemically inert, non-toxic, non-inflammable and thermally stable. Their weak solubility in water is due to their low surface tension, low intermolecular interaction, high density and low dielectric constant. They usually possess a limited, temperature-dependent miscibility with conventional organic solvents, forming biphasic systems with such solvents at ambient temperatures. These biphasic organic/fluorous solvents with their different solubilities for reactants, catalysts and products can be considered smart systems, the thermomorphic effect of which can be used to change a reaction from heterogeneous to homogeneous with concomitant mass transfer advantages [14].

Low-temperature molten salts or ionic liquids (IL) are very popular solvents at the moment and the search for advantageous applications is continuing in both analytical chemistry and separation science [15, 16]. IL are mainly used as additives in liquid chromatography and capillary electrophoresis, in which they are able to greatly increase separation efficiency by modifying the mobile (buffer) or stationary (capillary surface) phases [17]. However, they are not a likely to be an appropriate medium in chromatographic methods [18]. One must be wary of declaring these new solvents green. Ionic liquids are inert and not volatile, which are good properties for developing environmentally benign processes, but some of them are potentially toxic, and their environmental footprint is considerable due to complex and multistep synthesis. For that reason, attention is being directed towards bio-based ionic liquids synthesised from amino acids, carbohydrates, lignin and other renewable sources [19].

Deep eutectic solvents (DES) are an emerging alternative to IL. DES possess physical and chemical properties similar to those of IL, but DES outperform IL in terms of biodegradability, toxicity profiles and solubility properties for both hydrophilic and lipophilic compounds [20]. DES were developed by mixing hydrogen-bond acceptors (choline, tetramethyl ammonium and tetrabutyl ammonium chloride) with hydrogen-bond donors (urea, glycerol, ethylene glycol) [21]. DES have recently been created by combining primary metabolites with bio-renewable starting materials, e.g. sugar alcohols, sugars, and amino and organic acids, which makes them very attractive in the development of green technologies [22] and green additives in separation processes.

Table 2 Classification of solvents. [Reproduced with permission from: Prat D., Hayler J., Wells A. (2014) A survey of solvent selection guides. Green Chem 16:4546–4551]

Recommended	Water, Ethanol, *iso*-Propanol, n-Butanol, Ethyl acetate, *i*-Propyl acetate, *n*-Butyl acetate, anisole, sulfolane
Recommended or problematic	Methanol, t-Butanol, benzyl alcohol, ethylene glycol, acetone, Methyl ethyl ketone, Methyl isobutyl ketone, cyclohexanone, Methyl acetate
Problematic	Methyl tetrahydrofurane, heptane, methyl cyclohexane, toluene, xylenes, chlorobenzene, acetonitrile, dimethyl sulfoxide
Problematic or hazardous	Methyl tertbutyl ether, tetrahydrofurane, cyclohexane, Dichloromethane, formic acid, pyridine
Hazardous	Diisopropyl ether, 1,4-dioxane, dimethoxyethane, pentane, hexane, *N,N'*-dimethyl formamide, *N,N'*-dimethyl acetamide, *N*-methyl pyrrolidone, methoxy-ethanol, triethylamine
Highly hazardous	Diethyl ether, benzene, chloroform, carbon tetrachloride 1,2-dichloroethane, nitromethane

R3—Recycling—Whenever possible, proper equipment and solvent recycling should be used in the laboratory in order to reduce the amount of waste generated. However, this is expensive, and the energy requirements unfortunately prevent it from being a truly green approach.

Solvent recycling and recovery requires well organised and economically viable waste-solvent management that is aimed at minimising hazardous waste, reducing energy input and decreasing the emission of toxic substances. A variety of treatment methods are available. Solvent recovery by distillation and solvent incineration are common industrial-scale technologies. Waste-solvent treatment can be assessed by life cycle analysis whereby all the human and environmental impacts during the entire life cycle of the solvent (including raw material extraction, solvent production, energy and ancillary consumption, as well as waste-solvent treatment) are considered. Recovery and reuse of material and energy must also be included in the assessment. Life cycle analysis provides the justification for choosing the right treatment option: recycling to produce fresh solvent, or incineration to obtain energy.

The principles of green chemistry advocate reducing energy use and employing instruments that are suited to the purpose [23]. Reducing energy use is closely related to laboratory operations, because chemistry and analytical laboratories tend to be energy-intensive facilities, consuming many times more energy than other academic buildings. There are large numbers of heated and cooled areas, intensive ventilation and fume hoods. There is an enormous variety of specialised laboratory equipment that requires electricity and water—both room-temperature and chilled. Laboratory systems based on good laboratory practice (GLP) make economical use of equipment time and optimise analytical procedures.

Instrumental analytical chemistry strives to apply the principles of green engineering [24] to equipment and method design, focussing on optimising the consumption of materials and minimising their diversity, and avoiding excessive complexity. This has a substantial economic impact in control laboratories with a substantial number of analyses.

However, the need for applications and the trends in academic research are moving in opposite directions. Inaccurate assumptions about chemical analysis are partly to blame: instrumental methods are thought to be the only solution, and it is deemed necessary to acquire a new model of instrument every year. It is commonly believed that simple and robust solutions to analytical procedures produce low-quality data, despite the fact that every method has an optimal area of application, and discarding serviceable instruments and producing new ones are the most wasteful aspects of analytical chemistry (Fig. 3). A similar and unproven approach is related to the greenness of new solvents (such as ionic liquids) and new materials (nanoparticles), which must be thoroughly tested for environmental benignity.

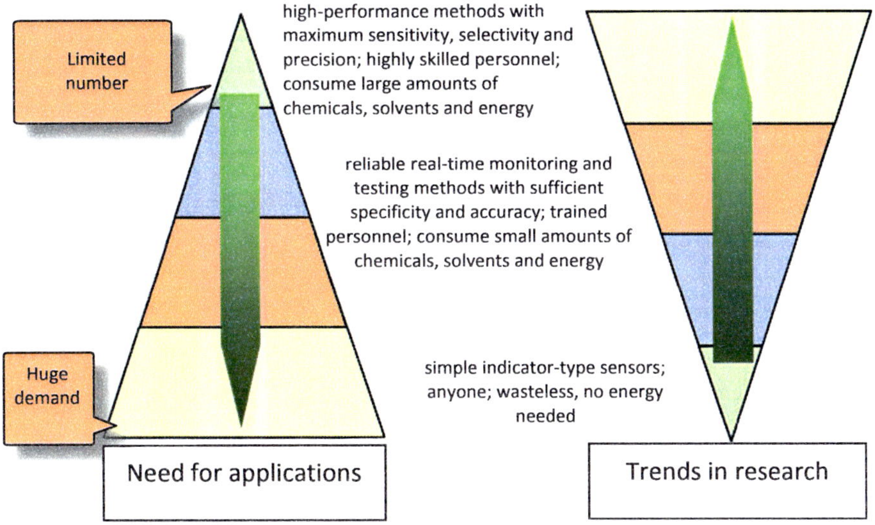

Fig. 3 Trends and needs in analytical chemistry research. Adapted from [Reproduced with permission from: Koel (2016), Do we need green chemistry? Green Chem 18:923–930]

2 The Influence of Green Chemistry on Analytical Chemistry

Green chemistry has been defined as the utilisation of techniques and methodologies that reduce or eliminate the use or generation of feedstocks, products, by-products, solvents, reagents, etc. that are hazardous to human health or the environment [25]. This implies the prevention of pollution before it happens rather than decontamination after the fact. It also points the way to changing the image of chemistry that involves designing chemical products and processes that reduce or eliminate the use and/or generation of hazardous substances. From an economic perspective, this can help industry save money by using less energy and fewer/safer chemicals, thus reducing the costs of pollution control and waste disposal. The following statement was made nearly twenty-five years ago: '.... the efforts taken to reduce the undesirable side effects of the tools of the trade. Indeed, it is time to enhance the role of analytical chemistry to take a lead towards the preserving of our environment rather than to measure its deterioration' [26]. An important development is the use of renewable raw material or feedstocks wherever technically and economically practicable.

The guidelines for making the discipline environmentally benign by design are derived from two sets of principles—those of green chemistry [24] and green engineering [27]—that are directly related to analytical chemistry.

The principles of green chemistry for the design of new methods (the number indicates the principle [24]) are the following:

prevent waste (1);
use safer solvents and auxiliaries (5);
design for energy efficiency (6);
avoid chemical derivatisation (8)
use safer chemistry to minimise the potential for chemical accidents (12).

The principles of green engineering for the design of new instruments (the number indicates the principle [29]) are the following:

ensure that all inputs and outputs are inherently non-hazardous (1);
maximise mass, energy, space and time efficiency (4);
limit underutilised and unnecessary materials and energy (8);
minimise material diversity (9);
design for a commercial ‘afterlife’ (11).

The principles related to engineering emphasise limiting the expenditure of underutilised and unnecessary materials and energy; avoiding overly complicated systems, and constructing instruments to withstand projected operating conditions over their expected lifetime, based on life cycle analysis. It would also be useful and economical to consider end-of-life recycling of an instrument.

One can also find principles of green analytical chemistry in the literature; however, these largely coincide with those listed above [28].

Greening analytical chemistry not only means reducing the use of solvents and harmful chemicals, but also utilising instruments that fit their purpose, optimising analytical processes with respect to the consumption of material and energy and the generation of waste, ensuring inherent safety and environmental friendliness, and providing the requisite analytical information with the necessary sensitivity, accuracy and precision.

Life cycle analysis (LCA) is currently the most commonly used tool for defining and quantifying problems related to the manufacture, use and disposal, including recycling, of a chemical product, and is already included in the standards [29]. LCA is by nature a general approach that greatly assists in estimating the greenness of a product. The need for LCA in the design of materials, methods and instruments is implicit in the principles green chemistry [30] (Fig. 4).

2.1 4S Approach

M. Koel and M. Kaljurand have proposed a ‘4S’ approach for green analytical chemistry [31] whose keywords reflect the challenges presented in Table 1:

- Specificity—using specific methods that are inherently environmentally benign;
- Smaller dimensions—performing analyses using microsystems (spots, channels, columns);
- Simplicity—obtaining the necessary information with simpler methods that do not require external energy and consume a minimal amount of material;

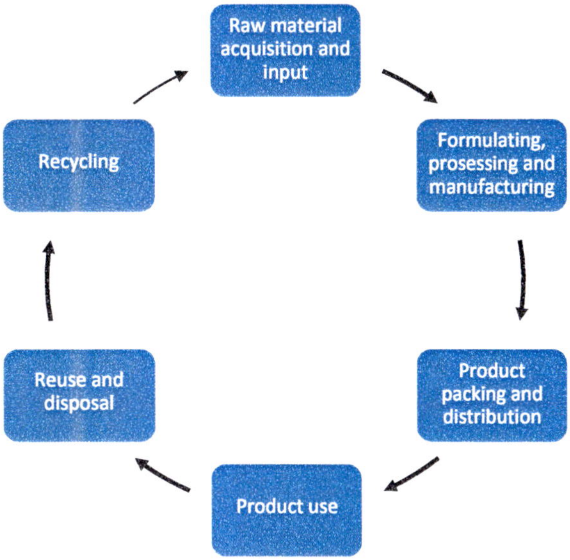

Fig. 4 Life cycle analysis considerations

- Statistics—applying statistical and mathematical methods in data processing that permit the use of simpler measurement processes, thereby avoiding sample pretreatment and shortening analysis time.

S1—Specific methods—The best example of employing different methods to solve similar problems is using capillary electrophoresis (CE) in place of HPLC.

The advantages of capillary electrophoretic techniques are their flexibility with a wide selection of separation media (buffers); high-efficiency separations that can be combined with chromatographic separation mechanisms; low solvent consumption; and shorter analysis times relative to other techniques. Different detectors can also be used in CE, which makes it more effective than HPLC for many applications.

On the other hand, it is possible to select an appropriate analytical procedure in chromatography, especially when supercritical fluids are used as solvents. This enables a link between gas and liquid, providing a continuum of mobile phase (solvent) properties. There are official analytical methods prescribed by the US EPA using supercritical CO_2 as a reference method for extracting PAHs (EPA Method 3561) and PCBs (EPA Method 3562) from solid environmental matrices, and also for extracting recoverable petroleum hydrocarbons (EPA Method 3560) [32].

Similarly, ionic liquids can provide additional flexibility for modifying existing analytical methods and increase the specificity of analytical procedures because of their unique properties. This attribute might best be described as chemical tunability—ionic liquids are a class of solvents possessing similar physical properties, but different chemical behaviour. IL are also comprised of distinct functional groups, which has given rise to the development of task-specific ionic liquids (TSIL).

IL have been used as additives or even as extraction solvents in various modes of liquid-phase microextraction including dispersive liquid–liquid, single drop and hollow fibre liquid-phase microextraction [33], as well as in aqueous biphasic systems [34].

IL are used as solvents in optical spectroscopy to study the properties and behaviour of a wide range of solutes under conditions that are not amenable to the use of organic solvents. A variety of transition metal complexes, which are unstable in other media, can be studied in room-temperature ionic liquids without the risk of solvation and solvolysis, allowing reliable solution spectra to be recorded for these species [35, 36].

Although the applicability of ionic liquids to mass spectrometry is dubious, IL have advanced in one area of that field: matrix-assisted laser-desorption/ionisation mass spectrometry (MALDI-MS) [37]. Because of their ability to operate in a high vacuum, IL can be used as matrices, solvents, probes and mobile phases for LDI-MS [38].

As a result, IL for gas–liquid chromatography are now listed in commercial catalogues [39]. The difference in selectivity between ionic liquids and methylphenyl polysiloxane, the conventional GC stationary phase, is due to the unique solvation characteristics of IL, which allow them to serve as useful dual-nature stationary phases. Mixed ionic-liquid stationary phases also provide more flexibility to control and optimise selectivity for complicated mixtures of analytes.

Multifunctional materials, the physical or chemical properties of which can be controlled by an external stimulus, and are thus called stimuli-responsive or **smart materials** [40], are receiving a great deal of attention. These materials react to environmental changes and adjust their physical properties in a predictable manner, and are therefore good candidates for designing intelligent systems and adaptive structures in analytical chemistry.

Multifunctional materials with physical or chemical properties that sense changes in the environment and respond to external stimuli provide flexibility in the development of new analytical procedures beside with improved analytical parameters and properties that will reduce the amount of waste. The use of different external stimuli (light, magnetic and electric fields, temperature, pH of the environment, etc.) has been tested in several applications, mainly analytical and bioanalytical, resulting in increased flexibility and economy as well as higher selectivity and sensitivity of the analysis. Smart materials are being successfully used in miniaturisation, the development of sensors, and the automation of analytical procedures, which are important factors in greening these processes. They also have a major influence on the time and cost of analysis.

The selection of materials on the basis of which smart applications can be developed is surprisingly wide: inorganic materials including different allotropes of carbon with chemically modified surfaces; polymeric materials in a variety of forms; compounds made up of different materials, as well as those that are between a solid and liquid state.

Smart materials in the form of new types of sorbents are front-runners in analytical chemistry. Carbon black is currently being replaced by carbon nanotubes and modifications of graphene. Solid-phase microextraction fibres can be modified by means of sol-gel chemistry, which facilitates the formation of selective coatings with improved adsorptive capacity. In bioanalysis, molecularly imprinted polymers are being successfully developed as highly selective tools for techniques such as solid-phase extraction, binding assays and biosensors. On the basis of molecularly imprinted polymers, new hybrid materials are being devised with different geometric structures of nanomaterial composites that provide either online isolation or pre-concentration of target small molecules from complex matrices. These materials facilitate greener processes by simplifying analyses and providing a substantial reduction in the amount of solvents and chemicals required, or by making use of environmentally benign aqueous media. However, little progress has been made in extending these discoveries to the development of materials and methods suitable for industrial-scale processing.

Many smart materials require new green commercial production techniques. This gives rise to a need for basic as well as engineering research, and coordination of the two between the industrial and research communities. Toxicology and analysis protocols have to be developed and constantly updated to reflect advances in the science [41].

S2—Smaller dimensions—There are several approaches to reducing the dimensions of analytical procedures. First, microflow and capillary columns can be used. Electrophoretic separation of analytes can easily be incorporated into these miniature systems. This has been applied in lab-on-chip technology wherein fluid transport and separation is affected in micrometer—and even nanometre-scale channels. Performing measurements on a microchip platform in which functionalised microsystems integrate multiple sample-handling processes (pre-concentration/extraction, chemical/biochemical derivatisation) with the measurement (detection) step substantially reduces the amount of solvents and chemicals [42]. However, there is still a major hurdle: the 'world-to-chip' interface. The introduction of nanolitres of sample is required for processing and/or separation, and it is difficult to obtain commercially available samplers.

The second approach is to perform chemistry and the analysis in small droplets using a technique called 'digital droplet microfluidics', whereby it is possible to combine sample preparation and analysis.

The amount of sample available for analysis is often very small, as in bioscience and nanoscience, in which individual cells or nanoparticles and sometimes even single molecules are analysed. This alone provides a justification for miniaturisation.

Another rationale for miniaturisation is the need to access the sampling site with portable highly sensitive analytical instruments, and to perform measurements in real time. Small sensors are easier to integrate into flexible automated measuring systems that can provide high throughput for multiple tasks.

Automation and hyphenation of instrumental methods is an important factor in conserving energy and materials. The new materials also enable a significant reduction in the size of the sample required for analysis as well as in the amount of waste generated. Miniaturisation of analytical devices with sophisticated sensors provides point-of-interest instruments with minimal consumption of energy and chemicals. Since separation methods and sample treatment are the biggest consumers of solvents and other reagents, these areas should be the targets of new materials and technologies.

S3—Simpler methods—High-performance instrumental methods are not always the only solution. The amount of information obtained with these complicated instruments often exceeds what is required and is not fully utilised. Every method has its optimal application and it is useful to have a variety of instruments available. Sufficient chemical information can be successfully obtained by methods that do not require sample pre-treatment, or in which the whole process can be conducted with equipment that does not have any moving parts. Simple methods with sufficient specificity and accuracy can work very well. They generate little or no waste and consume minimal amounts of energy or none at all.

Many examples of applications based on spot-test analysis can be cited [43]. In the past, these were the main analytical tests, whereby reactions produced various colours of spots on paper, and the necessary information was obtained by visual inspection. The work of Whitesides [44] has recently sparked a resurgence of interest in paper-based analytical assays. Paper can be used as an inexpensive, biodegradable, renewable, flexible substrate for prototype 'labs-on-a-chip'. It is a material that is universally available and compatible with many biological and chemical assays. The porous nature of the material eliminates the need for external pumps or energy. After use, this 'device' can be easily disposed of [45, 46]. An increasing array of lateral flow devices is now available as well—porous material or membranes are combined with immunoassay testing—as in the common pregnancy test.

Ordinary communications equipment (mobile phones, cameras, scanners, etc.) can be put to highly advantageous use with respect to cost, simplicity and portability, and offers many opportunities for point-of-care medical diagnosis, and on-site monitoring of environmental, agricultural and nutritional parameters [47]. The term 'smartphone-based colorimetry' has been applied to the process of using a smartphone to take a photo of a paper-based array of spots, identify their colour and calculate the concentration of the analytes [48, 49]. The process can be accelerated through automation by using a simple robotic system for spotting. This system has been successfully used for quantitative colorimetric determination of phytochemicals in plant extract samples [50], which led to the conclusion that the paper-based colorimetric test 'is faster, yields better reproducibility and higher throughput'. It has been demonstrated that red wine polyphenols, flavonoids and anthocyanins can be quantified on paper microzones via a digital camera and remote computer with dedicated free software [51]. There are instances in which colorimetric spectroscopic analysis of extracts is not possible because of precipitates, and in those cases, colorimetric paper microzone (PμZ) assays can produce reliable results. This was successfully demonstrated by an analysis of phenolic compounds involving ionic liquid extraction of plant biomass. In this case, spectroscopic analysis of extracts with imidazolium-based ionic liquids was not possible due to complex formation with the reagent and deposits from the solution. A PμZ assay produced reliable results free from interference and well within acceptable detection parameters [52].

There is a great need for simple indicator-type sensors with sufficient specificity and accuracy, which do not require trained personnel, generate little or no waste, and consume minimal amounts of energy or none at all. These simple devices and sensors are used to obtain chemical information from the point of care, including on-site process analysis for the widest use. The most important and exacting cases may be analysed in a laboratory. Laboratory environments in which analyses are performed have an influence on the greenness of analytical chemistry. Conserving energy and making the most economical use of equipment time must be considered in laboratory operations, and these optimised procedures are closely connected with good laboratory practices [53].

S4—Statistics—Information relevant to the analyte signal can be generated from data obtained by mathematical manipulation of the measurement results. This kind of mathematical extraction of chemical information from output data is called 'chemometrics'. A definition of chemometrics is provided by Otto: 'chemometrics is the chemical discipline that uses mathematical and statistical methods to design or select optimal measurement procedures and experiments, and to provide maximum chemical information by analysing chemical data' [54].

Although data processing is not included amongst the principles of green chemistry, it can be considered part of an environmentally benign laboratory system by virtue of the fact that it is possible to extract useful information from raw measurement data using mathematical procedures rather than chemical processing, which helps to prevent human error and saves time.

Contemporary analytical instruments provide large sets of data—usually a complex stream of serial or parallel data—spectra or images. Data rates in electronic interfacing can range from hundreds to gigabits of information per second. The difficulty of interpreting these huge data sets in chemistry was recognised and multivariate statistics for the extraction of useful information were proposed as early as the 1970s [55]. Smart data-treatment algorithms can replace or supplement the work of experienced scientists in high-throughput screening procedures. Various automated algorithms for deconvoluting spectra, and assigning and classifying spectral lines, have already been developed, enabling large sample sets to be analysed with little expert involvement. Large chromatography-mass-spectrometry datasets can be evaluated without the supervision of analysts to ascertain optimal instrument performance, and identify the sources of variation over years of data collection.

Chemometrics, which is generally performed using multivariate statistical procedures, could be considered as an alternative to chemical data processing, enabling the use of much simpler measurement processes that usually avoid sample pre-treatment and entail a shorter analysis time. It is a promising route to the ultimate greening of analytical chemistry since it enables results to be obtained solely from calculations, thus reducing the need for complicated measurements involving chemicals and solvents [56]. The approach does not eliminate the need for measurement completely;

however, it allows the results to be obtained with much simpler measurement procedures. For example, the use of transmittance measurements in NIR spectroscopy and partial least squares calibration has been proven to be a powerful tool for the determination of the peroxide index in edible oil without using solvents or derivatisation reagents [57]. Another example is recording individual concentrations and spectra of mixtures using excitation–emission spectroscopy by means of a chemometric technique known as PARAFAC [58], which can determine the concentration and spectra of individual compounds in a mixture even when the component spectra/concentration profiles overlap.

Similarly, another widely used chemometric method—principal component analysis (PCA)—for examining the relationships between samples (patients, food samples, organisms, chromatographic columns, spectra) and variables (compound concentrations, spectral peaks, chromatographic peak areas, elemental compositions) is to decompose the initial large amounts of data into simple components that can be represented graphically, thus allowing the relationships between objects to be easily visualised. Chemometric data treatment is widely used for direct vibrational spectrometric data [59]—near infrared (NIR), mid-infrared (mid-IR) or Raman spectrometry, fluorescence, UV-Vis spectroscopy and nuclear magnetic resonance (NMR)—as it shortens the length of analysis time and improves the performance of these methods [60].

Design of experiments (DOE) is another chemometric method that can contribute significantly to the greening of analytical methods by enabling rapid optimisation of analytical processes. DOE drastically reduces the number of experiments required to determine the optimal value of each factor by varying these values in parallel. It allows for determining not only the main effects of each factor, but also the interactions between factors. Various ways of combining factors include the Box-Behnken, Doehler and central composite design methods [61].

Laboratories are under pressure to increase throughput and shorten response times to produce data for decision making, while maintaining or improving the quality of the data. The solution to increasing the productivity of a laboratory involves sensitive, highly efficient instruments, automated operations and fast communications. Automation in particular is closely related to the elimination of defects and human error in laboratory processes. The resulting real-time data is important for process control and product quality, and for minimising the environmental impact. Combining automation, sensors and networking on one platform reflects the definition of an internet of things (IoT): 'a dynamic global network infrastructure with self-configuring capabilities based on standard and interoperable communication protocols where physical and virtual things have identities, physical attributes, and virtual personalities and use intelligent interfaces, and are seamlessly integrated into the information network [62]. The main features of an IoT, in addition to substantial cost savings from automated processes, are communications—a smart device collects and communicates information—and the ability to be remotely or automatically controlled. Another driver of the internet of things is the wider availability of devices/sensors with advanced application protocols, which can be networked in real time to provide reliable information about the surrounding world (Fig. 5).

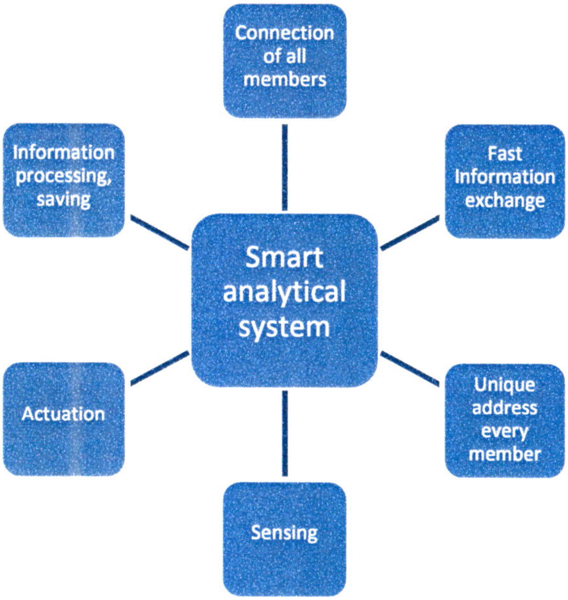

Fig. 5 Characteristics of smart analytical systems

3 Conclusion

A green approach is becoming increasingly important not only for developers in research laboratories but also for providers of analytical instruments and techniques. Assessing progress in adopting the principles of green chemistry has been always problematic, and some benchmarks and multilevel indicators have been proposed [63]. Using new materials and improving the metrological quality of analysis furthers the movement towards green chemistry on the molecular level by assessing the environmental benignity of the materials and compounds, and on the product level by determining the safety, efficiency and economy of the methods.

Conducting research in which analytical processes are optimised in order to provide the required information in a manner that is inherently safe, non-toxic and environmentally benign, and with the least possible consumption of materials and energy and generation of waste leads to the development of more environmentally safe and economical methods of chemical analysis. These new principles are guiding analytical chemistry towards becoming more environmentally benign. The future seems to be bright: representatives of companies are saying: 'Green chemistry can deliver for people, planet and profit' [64]; academic chemistry societies support green chemistry research programs and networks; new curricula for kindergartens through post-graduate schools have been developed, and the educated professionals who emerge from them are able to influence thinking with regard to sustainable development.

The new functional materials are good candidates for making analytical chemistry greener because of their flexibility and much wider range of application. These materials are not yet widespread and are still being studied; however, as mentioned previously, there are already many promising examples of the ways in which the analytical parameters of procedures are being improved, and the use of energy and chemicals/solvents reduced. The goal of analytical chemistry is to obtain accurate and unambiguous signals directly from the point of interest—whether it is in the interior or on the surface, solid, liquid or gas—and to present this knowledge in a precise, selective and economical manner.

The quality of measurements targeting the specific needs and demands of end-users depends on the use of instrumentation and procedures according to their optimal area of application. Reliable and fit-for-purpose analytical methodologies also result in financial savings by reducing the use of solvents and reagents (especially toxic ones) that are potentially released into the environment. The overall economy obtained from the selection of instruments and methods that suit the purpose in a given situation not only conserves materials and energy but also reduces the use of chemicals and solvents, and reduces the health and safety hazards for the analyst as well as for the environment.

References

1. Sheldon RA (1992) Organic synthesis—past, present and future. Chem Ind 23:903–906
2. Koel M (2016) Do we need green analytical chemistry? Green Chem 18:923–931
3. Thompson M, Ramsey MH (1995) Quality concepts and practices applied to sampling—an exploratory study. Analyst 120:261–270
4. International Conference on Harmonization (ICH) (2009) Tripartite guidelines,' ICH Q8 (R2): Pharmaceutical Development. London
5. Peraman R, Bhadraya K, Reddy PY (2015) Analytical quality by design: a tool for regulatory flexibility and robust analytics. Int J Anal Chem Article ID 868727
6. Welch CJ, Wu N, Biba M et al (2010) Greening analytical chromatography. TrAC—Trends Anal Chem 29(7):667–680
7. Cielecka-Piontek J, Zalewski P, Jelinska A, Garbacki P (2013) UHPLC: the greening face of liquid chromatography. Chromatographia 76(21–22):1429–1437
8. Shaaban H, Górecki T (2015) Current trends in green liquid chromatography for the analysis of pharmaceutically active compounds in the environmental water compartments. Talanta 132(15):739–752
9. Schulte PA, McKernan LT, Heidel DS et al (2013) Occupational safety and health, green chemistry, and sustainability: a review of areas of convergence. Environ Health A Global Access Sci Sour 12(1):31
10. Brunner G, Steinkopff (2013) Gas extraction: an introduction to fundamentals of supercritical fluids and the application to separation processes (Topics in Physical Chemistry)
11. Poole CF (ed) (2017) Supercritical fluid chromatography (Handbooks in Separation Science), Elsevier
12. Shen Y, Chen B, van Beek TA (2015) Alternative solvents can make preparative liquid chromatography greener. Green Chem 17:4073–4081
13. Tobiszewski M, Namieśnik J, Pena-Pereira F (2017) Environmental risk-based ranking of solvents using the combination of a multimedia model and multi-criteria decision analysis. Green Chem 19(4):1034–1042

14. *Handbook of Fluorous Chemistry* (2004) Gladysz JA, Curran DP, Horváth IT (Eds), Wiley-VCH
15. Koel M (ed) (2016) Analytical applications of ionic liquids. World Scientific
16. De Los Rios AP, Fernandez FJH (ed) (2014) Ionic liquids in separation technology. Elsevier
17. García-Álvarez-Coque MC, Ruiz-Angel MJ, Berthod A, Carda-Broch S (2015) On the use of ionic liquids as mobile phase additives in high-performance liquid chromatography: a review. Anal Chim Acta 883:1–21
18. Escudero LB, Grijalba AC, Martinis EM, Wuilloud RG (2013) Bioanalytical separation and preconcentration using ionic liquids. Anal Bioanal Chem 405(24):7597–7613
19. Hulsbosch J, De Vos DE, Binnemans K, Ameloot R (2016) Biobased ionic liquids: solvents for a green processing industry? ACS Sustain Chem Eng 4(6):2917–2931
20. Dai Y, van Spronsen J, Witkamp G-J et al (2013) Ionic liquids and deep eutectic solvents in natural products research: mixtures of solids as extraction solvents. J Nat Prod 76:2162–2173
21. Tan T, Zhang M, Wan Y, Qiu H (2016) Utilization of deep eutectic solvents as novel mobile phase additives for improving the separation of bioactive quaternary alkaloids. Talanta 149:85–90
22. Dai Y, van Spronsen J, Witkamp G-J et al (2013) Natural deep eutectic solvents as new potential media for green technology. Anal Chim Acta 766:61–68
23. Warner PT, Anastas JC (1998) Green chemistry: theory and practice. Oxford University Press, New York
24. Anastas PT, Zimmerman JB (2003) Design through the twelve principles of green engineering. Env Sci Tech 37(5):94A–101A
25. Anastas PT (1999) Green chemistry and the role of analytical methodology development. Crit Rev Anal Chem 29(3):167–175
26. de la Guardia M, Ruzicka J (1995) Towards environmentally conscientious analytical chemistry through miniaturization, containment and reagent replacement. Analyst 120(2):17N–17N
27. Anastas PT, Zimmerman JB (2003) Design through the 12 principles of green engineering. Environ Sci Technol 37(5):94A–101A
28. Gałuszka A, Migaszewski Z, Namieśnik J (2013) The 12 principles of green analytical chemistry and the significance mnemonic of green analytical practices. TrAC Trends Anal Chem 50:78–84
29. ISO 14044 (2006) Environmental management—life cycle assessment—requirements and guidelines. International Organization of Standardization: Geneva, Switzerland
30. Anastas PT, Lankey RL (2000) Life cycle assessment and green chemistry: the yin and yang of industrial ecology. Green Chem 2:289–295
31. Koel M, Kaljurand M (2012) Benign design in analytical chemistry. Crit Rev Anal Chem 42:192–195
32. http://www.epa.gov/Hazardous Waste Test Methods/SW-846
33. Martinis EM, Berton P, Wuilloud RG (2014) Ionic liquid-based microextraction techniques for trace-element analysis. TRAC-Trends Anal Chem 60:54–70
34. An K, Trujillo-Rodriguez MJ, Pino V, Anderson JL (2017) Non-conventional solvents in liquid phase microextration and aqueous biphasic systems. J Chromatogr A 1500:1–23
35. Appleby D, Hussey CL, Seddon KR, Turp JE (1986) Room-temperature ionic liquids as solvents for electronic absorption-spectroscopy of halide-complexes. Nature 323:614–616
36. Paschoal VH, Faria LFO, Ribeiro MCC (2017) Vibrational spectroscopy of ionic liquids. Chem Rev 117:7053–7112
37. Armstrong DW, Zhang L-K, He L, Gross ML (2001) Ionic liquids as matrixes for matrix-assisted laser desorption/ionization mass spectrometry. Anal Chem 73:3679–3686
38. Abdelhamid HN (2016) Ionic liquids for mass spectrometry: matrices, separation and microextraction. TrAC -Trends Anal Chem 77:122–138
39. Sigma-Aldrich: ionic liquid capillary GC columns. http://www.sigmaaldrich.com/analyticalchromatography/analytical-products.html?TablePage=101691909. Accessed 16 Feb 2016
40. Wei M, Gao Y, Li X, Serpe MJ (2017) Stimuli-responsive polymers and their applications. Polym Chem 8:127–143

41. Nath D, Banerjee P (2013) Green nanotechnology—A new hope for medical biology. Environ Toxicol Pharmac 36(3):997–1014
42. Shen Y, van Beek TA, Zuilhof H, Chen B (2013) Hyphenation of optimized microfluidic sample preparation with nano liquid chromatography for faster and greener alkaloid analysis. Anal Chimica Acta 797:50–56
43. Feigl F (1946)Qualitative analysis by spot tests (3rd edn). Elsevier, New York, USA
44. Martinez AW, Phillips ST, Carrilho E et al (2008) Simple telemedicine for developing regions: camera phones and paper-based microfluidic devices for real-time, off-site diagnosis. Anal Chem 80:3699–3707
45. da Silva AS, Bedatty F, Flavio C et al (2011) A simple and green analytical method for determination of glyphosate in commercial formulations and water by diffuse reflectance spectroscopy. Spectrochim Acta Part A Mol Biomol Spec 79(5):1881–1885
46. Noor MO, Hrovat D, Moazami-Goudarzi M et al (2015) Ratiometric fluorescence transduction by hybridization after isothermal amplification for determination of zeptomole quantities of oligonucleotide biomarkers with a paper-based platform and camera-based detection. Anal Chimica Acta 885:156–165
47. Grudpan K, Kolev SD, Lapanantnopakhun S et al (2015) Applications of everyday IT and communications devices in modern analytical chemistry: a review. Talanta 136:84–94
48. Hong JI, Chang BY (2014) Development of the smartphone-based colorimetry for multi-analyte sensing arrays. Lab Chip 14:1725–1730
49. Campos AR, Knutson CM, Knutson TR et al (2016) Quantifying gold nanoparticle concentration in a dietary supplement using smartphone colorimetry and google applications. J Chem Educ 93:318–321
50. Vaher M, Borissova M, Seiman A et al (2014) Automatic spot preparation and image processing of paper microzone-based assays for analysis of bioactive compounds in plant extracts. Food Chem 143:465–471
51. Vaher M, Kaljurand M (2012) The development of paper microzone-based green analytical chemistry methods for determining the quality of wines. Anal Bioanal Chem 404:627–633
52. Aid T, Kaljurand M, Vaher M (2015) Colorimetric determination of total phenolic contents in ionic liquid extracts by paper microzones and digital camera. Anal Methods 7:3193–3199
53. Koel M (2012) Energy savings in analytical chemistry. In: De la Guardia SM, Garrigues A (eds) Handbook of green analytical chemistry. Wiley, Hoboken, pp 291–319
54. Otto M Chemometrics. Wiley-VCH Verlag GmbH & Co, KGaA, Weinheim, p 328
55. Kowalski BR (ed) (1977) Chemometrics: theory and application. ACS symposium series 52. American Chemical Society, Washington, DC
56. Armenta S, Garrigues S, de la Guardia M (2008) Green analytical chemistry. TrAC - Trends Anal Chem 27:497–511
57. Moh MH, Man YBC, van de Voort FR, Abdullah WJW (1999) Determination of peroxide value in thermally oxidized crude palm oil by near infrared spectroscopy. J Am Oil Chem Soc 76:19
58. Bro R (1997) PARAFAC. Tutorial and applications. Chemom intel lab systems 38(2):149–171
59. Mark H, Workman J, (2018) Chemometrics in spectroscopy (2nd edn). Academic Press
60. Gredilla A, de Vallejuelo SF, Elejoste N et al (2016) Non-destructive Spectroscopy combined with chemometrics as a tool for green chemical analysis of environmental samples: a review. TrAC Trends Anal Chem 76:30–39
61. Max Morris (2017) Design of experiments. Texts Statist Sci. Routledge
62. Sundmaeker H, Guillemin P, Friess P, Woelfflé S (eds) (2010) CERP-IoT. Vision and challenges for realising the internet of things. European Commission: Brussels, Belgium Luxembourg
63. Tickner JA, Becker M (2016) Mainstreaming green chemistry: the need for metrics. Curr Opin Green Sustain Chem 1:1–4
64. Tucker JL, Faul MM (2016) Drug companies must adopt green chemistry. Nature 534:27–29

Chapter 8
Green in Pharmaceutical Chemistry

Sankaran Radhika, Mohan Neetha, and Gopinathan Anilkumar

1 Introduction

The green synthetic reactions and routes are based on the 12 principles of green chemistry and green engineering. Many pharmaceutical manufacturers use enormous quantities of large varieties of organic solvents and after the process, release the solvents as waste which is generally more than 85% of the total ingredients of a reaction [1, 2]. Usually pharmaceutical industry disposes these hazardous and toxic waste into the environment as a part of their production. In this context, the implementation of green chemistry to pharmaceutical industry projects itself as a boon to the nature. In the recent years, the utilization of greener protocols in pharmaceutical industry has received appreciable importance. This is because of safety, reliability in products, efficiency, keeping of environmental sustainability in manufacturing procedures, cost-effectiveness [3] and minimum production of waste. Application of green principles in pharmaceutical chemistry has been proved through microwave chemistry, sonochemistry, nano-chemistry, biocatalysis, green solvents such as ionic liquids and atom economic procedures [4].

Evaluation of greenness is not limited to lowering of chemical waste alone, but also concentrates on different metrics. The major metrics which are followed in pharmaceutical industry include resources, life cycle assessment, cleaning with maintenance, measurement of chemistry and process efficiency, measuring process parameters and

S. Radhika · M. Neetha · G. Anilkumar (✉)
School of Chemical Sciences, Mahatma Gandhi University, Priyadarsini Hills, Kottayam, Kerala 686560, India
e-mail: anilgi1@yahoo.com

S. Radhika
e-mail: radhikasreekrishna@gmail.com

M. Neetha
e-mail: neethamohan131144@gmail.com

G. Anilkumar and S. Saranya (eds.), *Green Organic Reactions*, Materials Horizons: From Nature to Nanomaterials, https://doi.org/10.1007/978-981-33-6897-2_8

emissions, energy measurement, measurement of substance toxicity, degradation potential measurement and inherent safety measurement [5].

Applications of enzymes in chemical production have become a very hopeful and a growing toolbox in medicinal chemistry. Biocatalysis exhibits enantio, regio and chemo-selectivity in chemical reactions. Nanosystems are yet another class of compounds exhibiting remarkable pharmaceutical significance. Nanoparticles being environmentally innocuous and economical are widely employed in therapeutics [6]. All these greener pharmaceutical merits inspired researchers to well explore them in the field of novel drug delivery and synthesis.

1.1 Nanochemistry in Pharmaceutical Industry

Zhu and co-workers developed gelatin-based graphene nanosheets (GNS), by a one-pot green procedure. They proved the efficiency of the synthesized gelatin—GNS as a potential intramolecular drug carrier taking doxorubicin (DOX), a usual anticancer medicine as example in human breast adenocarcinoma cells (MCF-7). Gelatin—GNS is non-toxic to MCF-7 cells, and therefore, DOX can be released to the cells from DOX@gelatin–GNS without any toxicity, which enhanced the therapeutic capacity [7]. Doxorubicin hydrochloride was also encapsulated by chitosan-low density lipoprotein (LDL) nanoparticles (CS-LDL-DOX NPs) obtained by simple mixing of solution of chitosan and suspension of LDL [8]. The NPs displayed lower cytotoxicity as a result of extended release. Compared to free doxorubicin, CS-LDL-DOX NPs showed higher cellular uptake owing to endocytosis of cancer cells. An approach of drug delivery mode using porphyran capped gold nanoparticles (AuNps) was explored by Pokharkar and his group. Depending on pH variation, the greener AuNps were recognized as good drug vehicles for the release of DOX in human glioma cell line, i.e. basic pH lowers the release of the anticancer drug [9]. Ma and co-workers have established a green protocol towards the synthesis of reduced graphene oxide (rGO), and the same was analysed for its efficiency to act as nano-carrier for pH-sensitive drug transport [10]. The prototype drug chosen was doxorubicin hydrochloride which was efficiently fastened to the nano-rGO surface through π-π stacking interaction. The nanohybrid displayed excellent stability, good drug loading, pH-sensitivity and sustainable delivery of drugs. The hybrid was cytotoxic towards MCF-7 [11] and A549 [12] cancer cells. Another approach towards the use of reduced graphene oxide as nanocarrier was presented by Palai et al. [13]. Here also, the prototype drug was doxorubicin. The rGO was decorated with PEG-supported nanoceria forming the nanocomposite. The nanocomposite also exhibited good drug loading and pH-sensitive drug release similar to the work of Ma et al. It displayed promising anti-cancer functions and lower side effects compared to pure graphene oxide.

Jahangarian and co-workers as well as Lam et al. reported a study of various nano-based systems for efficient drug delivery [14, 15]. The various drug delivery agents used were, nano-metal compounds (γ-Fe_2O_3 NPs [16], Mg NPs [17], silica

NPs [18], MoS_2 nanosheet [19]), polymer nanocomposites (polyethylene-grafted chitosan [20], dextrin-poly(lactide) nanogel [21], core–shell micelle carriers [22]), quantum dots (graphene oxide [27], CdTe [23]) and carbon nano-tubes [24–26]. Polyelectrolyte thin films were used as drug delivery agents by Sripriya et al. [27].

Poly allylamine hydrochloride and dextran sulphate were electrostatically adsorbed to form the assembly of the multilayers. Silver nanoparticles were then introduced into the film. Silver nanoparticles were exhibiting anti-microbial activities. Similarly, moxifloxacin hydrochloride, an anti-bacterial drug, could also be incorporated into the film along with Ag NPs to provide enhanced anti-bacterial functions. Upon laser irradiation, the silver NPs would absorb energy and heats up the film and leads to its rupture releasing the nanoparticles and the drug. Hydrofluoric acid-independent protocol for the synthesis of iron (III) trimesate MIL (Material of Institute Lavoisier)-100 nanoparticles is reported [28]. This method utilized greener steps avoiding the usage of toxic HF. Higher payloads of azidothymidine triphosphate (AZT-TP), an anti-viral drug, could be encapsulated by these HF-free solids and also released the drug in a progressive style. The AZT-TP loaded MIL-100 NPs could restrain the in vitro viral replication by successfully penetrating into the HIV-affected cells. Kang and co-workers reported the green synthetic procedure of silver nanoparticles in water utilizing chitosan oligomer as reducing and stabilizing agent [29]. A CHI-Ag ointment was prepared using this Ag nanoparticles, which is studied for burn wound healing in-vitro. Silica nanoparticles are yet another class of drug delivery agents [30]. Green synthesis of which could avoid vigorous labour and the use of harmful chemicals. Drug encapsulated here was a protein which could exhibit anti-cancer properties. The drug was incorporated during the time of formation of biosilica which upon hydrolysis, delivered the protein to the target site, and thus, a potent drug delivery system (DDS) was developed. Biocompatibility of AuNps, prepared from *Ganoderma spp. mycelia,* was tested using MDA-MB-231 breast cancer cells and found that these are biocompatible and non-cytotoxic [31]. Thiruvengadam et al. have put forward a green strategy towards the synthesis of copper nanoparticles using the extract of *Millettia pinnata* as the reducing agent [32]. The synthesized nanoparticles displayed anti-inflammatory properties utilizing albumin denaturation and membrane stabilization.

Ginsenoside is a bioactive constituent of herb Panax ginseng. Ginsenoside Rb1 could self-assemble along with anti-cancer drugs via a green protocol forming nanoparticles exhibiting superlative anti-cancer properties both in vivo and in vitro [33]. Lower side effects and higher potency of gensenoside were revealed during the anti-tumour analysis, making them remarkable nanocarriers for the treatment of cancer.

Hydrophilic, aromatic and low molecular-weight drugs (HALMD) were non-covalently attached to aquaphilic aromatic polymers via a novel green methodology to develop nanocarriers [34]. The aqueous solution of aromatic polymer was mixed with aqueous solution of HALMD to generate nanocarriers. This system displayed high drug encapsulation values and efficient drug association.

Nanofibers based on polycaprolactone/gelatin, supported with halloysite nanotubes (HNTs), were made by Pavliňáková and co-workers, following the principles of green chemistry. Cytotoxic studies of this nanofibers in NIH-3T3 mouse fibroblasts disclosed that these are convenient for biomedical applications like drug delivery, since they are non-toxic to cells [35]. Bhat et al. have utilized nanocellulose as a medium for efficient drug delivery [36]. Cellulose being a natural fibre is renewable and provides a greener approach towards the drug delivery. Nanocellulose has immense hydroxyl moieties on its surface and thus can be easily tuned to interact with different drug molecules helping in efficient loading and release of drugs. Chu et al. have described a biomineralization strategy based on metal–organic framework (MOF) NPs for the treatment of cancer [37]. Biomineralization is a process of hardening or stiffening of tissues via minerals produced by living organisms [38]. The biomineralization through a self-assembly procedure is applied to render biomacromolecular exoskeletal protection to preserve bioactivity and immune shielding. MOFs are efficient supporting materials due to their high encapsulation capacity and low toxicity. Hence, incorporation of both was presented as a novel and efficient protocol towards cancer treatment.

1.2 Biocatalysis in Pharmaceutical Industry

Among various catalysis, biocatalysis and chemoenzymatic catalysis gained a special attention in pharmaceutical chemistry due to the absence of chemicals used externally. Some of the enzymes used in pharmaceutical synthesis are racemases [39], oxidases [40], dehydrogenases [41], lipases [42], etc. Various novel enzymes were found to be effective in ketone reduction, which imparted several pharmaceutically precious alcohols, including chiral secondary alcohols [43]. Lipases that are derived from *Candida antarctica* exhibited good enantioselectivity in tertiary alcohol synthesis [44]. In 2008, Woodley published a review on greener processes in pharmaceutical chemistry using biocatalysis as the theme, which describes various reaction categories and its corresponding enzymes [45].

Zhang and co-workers exhibited total synthesis of tambromycin, an antibiotic, via biocatalytic C-H functionalization utilizing lysine hydroxylase, KDO1 (Scheme 1) [46].

Atorvastatin and Rosuvastatin are synthetic fat and "bad" cholesterol-lowering statins having a 3,5-dihydroxyacidic side moiety, precursor of statin side chain [47]. The side chain preparation involves the utilization of two enzymes—ketoreductase and halohydrin dehydrogenase. Deoxyribose 5-phosphate aldolases mediated aldol reaction is also needed for the synthesis of statin side chain. Another approach towards atorvastatin was a three-enzyme green protocol involving ketoreductase, NADP-dependent glucose dehydrogenase and halohydrin dehalogenase [48]. The use of enzymes for the manufacture of the drug made the reaction environment friendly at the same time it also improved the feasibility of the reaction.

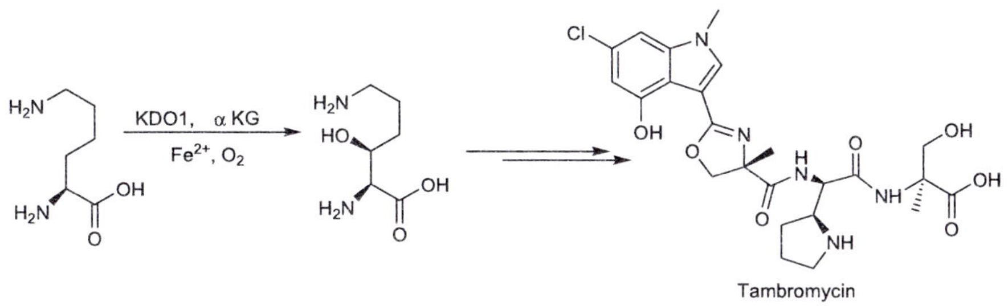

Scheme 1 Total synthesis of tambromycin

Biosynthesis of Penicillins and Cephalosporins, two important β-lactam antibiotics, needed the aid of nonribosomal peptide synthetase (NRPS) [49] and DAOC synthetase, respectively. Penicillin G is formed from isopenicillin, by the action of isopenicillin-N transacylase. Isopenicillin is obtained through isopenicillin synthetase and ACV NRPS synthetase. Biosynthesis of Cephalosporin C is mediated by DAOC acyltransferase, DAOC hydroxylase, DAOC synthetase and isopenicillin epimerase [50]. Likewise, glycosyltransferases [51], thioesterase (Tyc A TE) [52], lipases [53], amycolatopsis orientalis SC 15,847 and lipolases [54] are very essential in the synthesis of glycopeptide antibiotics (such as vancomycin and teicoplanin), tyrocidine, paclitaxel, epothilone F and pregabalin, respectively. Pérez noted that several putative enzymes contained in the zoantharian *Protopalythoa variabilis* could efficiently be transformed into beneficial biomedicals and pharmaceutical products [55].

The significance of enzymatic processes in the production of drugs was again discussed by W. Cabri. The impact of green approaches in the modern synthesis of ST1535, idarubicin, rubitecan and 7-aminocephalosporanic acid (7-ACA) was considered. The use of mild reaction conditions in the enzymatic transformation of 7-ACA with more than 99.5% purity makes this synthetic protocol more suitable than chemical method [56]. An asymmetric aldol condensation between cyclic ketones with isatin derivatives, catalysed via nuclease p1 derived from *Penicillium citrinum* in solvent-free condition was established by Lin et al. for the first time. This reaction provides a chance to construct biologically active molecules in pharmaceutical industry [57].

1.3 Reactions of "in-Water" and "on-Water" Chemistry

The traditional method in the synthesis of rabeprazole, an anti-ulcer drug, follows an eco-unfriendly strategy involving the oxidation of sulphide derived from 1*H*-benzo[*d*]imidazole by *m*-CPBA. A more environmentally favoured method (Scheme 2) suggested the use of sodium hypochlorite in water in the oxidation step, affording the product with better yield percentage [58].

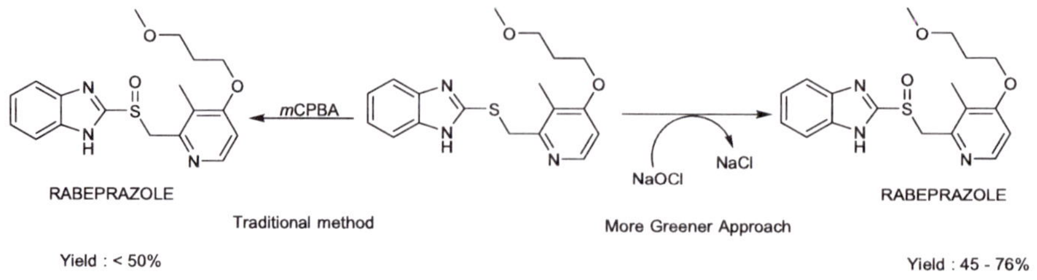

Scheme 2 Synthesis of rabeprazole

Substituted tetrahydroquinolines are important class of compounds used in medicinal chemistry such as in anti-cancer therapy and Alzheimer's diseases. Vivancos et al. discussed the formation of tetrahydroquinolines in water. Tetrahydroquinolines were synthesized by the reduction of quinoline derivatives using H_2, triazoylidene-containing iridium complex as ligand and water as solvent at 90 °C (Scheme 3) [59].

An "on-water" chemistry of anti-proliferative 3-benzoylquinoxalines synthesis through activated carbon (AC)-assisted aerobic benzylic oxidation was reported (Scheme 4) [60]. This provides an interest in medicinal chemistry, since 3-benzoylquinoxalinone moiety is very much alike with tubulin inhibitors, which are important in cancer treatment. Quinoline ring having electron-donating group at the second position yielded more product than that having electron-withdrawing group at the same position.

Ceborska disclosed the synthesis of several folic acid complexes using water as the medium. Small α-cyclodextrin unit resulted in the formation of an exclusion compound, while β- and γ-cyclodextrins gave 2-rotaxane, both are widely used in drug delivery and prodrug synthesis [61].

Scheme 3 Reduction of quinolone in water

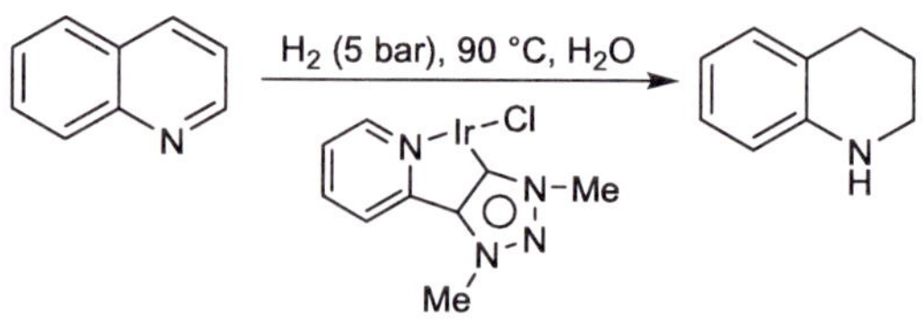

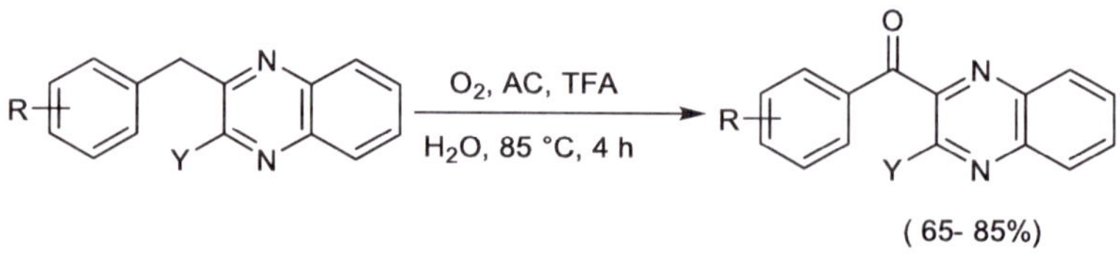

Scheme 4 Activated carbon mediated synthesis of 3-benzoylquinoxalines

Endothelins are vasorestricting peptides and are responsible for various vascular diseases of lungs, heart and brain [62]. Endothelin-A antagonist ABT-546 was obtained in efficient yields when water was used as a co-solvent in the final steps of the drug synthesis [63]. Garcia et al. established a novel green protocol towards succinyl-β-cyclodextrin in water media [64]. The synthesized molecules were found to have guest–host affinity which have been improved by the introduction of acidic groups. Albendazole was selected as the model drug. This non-toxic derivative could function as excipient to develop oral dosages, especially in parasitic diseases treatment.

1.4 Reactions Mediated by Ionic Liquid (IL)

Ionic liquids are green alternatives to organic solvents today. Shaabani et al. discussed the usage of 1-butyl-3-methylimidazolium bromide ([bmim][Br]) IL in the synthesis of anti-viral drug, 3-amino-imidazo[1,2-α]pyridines (Scheme 5) [65]. In 2004, another work was reported about the preparation of L-4-boronophenylalanine, a potent drug in boron neutron capture therapy, using [bmim][BF_4] and [bmim][PF_6] [66]. Synthesis of several caffeic acid phenethyl ester derivatives, which are effective anti-proliferative compounds, was also reported with ([bmim][NTf_2]) IL as solvent [67]. The investigation for new non-steroidal anti-inflammatory medicines revealed the use of ionic liquids, especially ionic liquids having imidazolium-base in Friedel–Crafts reactions [68].

Nucleoside analogues are potential drugs in cancer treatment and chemotherapy. The utilization of [bmim][PF_6] IL in the synthesis of pyrimidine nucleoside-thiazolini-4-one hybrids, which are good anti-parasitic drugs, was reported by Zhang et al. [69]. Fan et al. used [bmim][BF_4] IL in hybrid compound synthesis [70]. Pyrimidine nucleosides with pyrano[4,3-*c*]pyranes and pyrano[3,2-*c*]pyridines show excellent anti-leishmanial and anti-viral property, respectively. Reported papers suggested that ionic liquids gained good attraction in nucleoside chemistry as a solvent, since both deoxyribo and ribonucleosides have good solubility in them which provided milder reaction environment and better selectivity. Methanesulfonate-based ILs were realized to solubilize deoxynucleosides well in all reactions. For example, MOEPy.OMs and MOEMIM.OMs embellished the solubility of thymidine [71]. ILs proved its efficiency as good reaction solvents through various nucleoside modification reactions like protection, acylation, alteration in nucleoside base and through enzyme catalysed reactions. Malhotra et al. replaced DMA with ILs, keeping all other

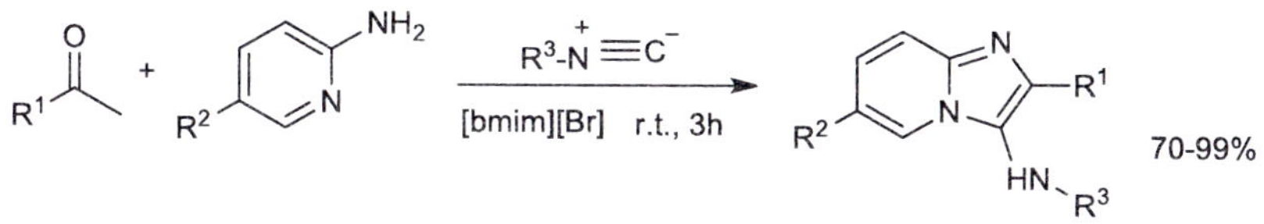

Scheme 5 [bmim][Br] IL in 3-amino-imidazo[1,2-α]pyridine synthesis

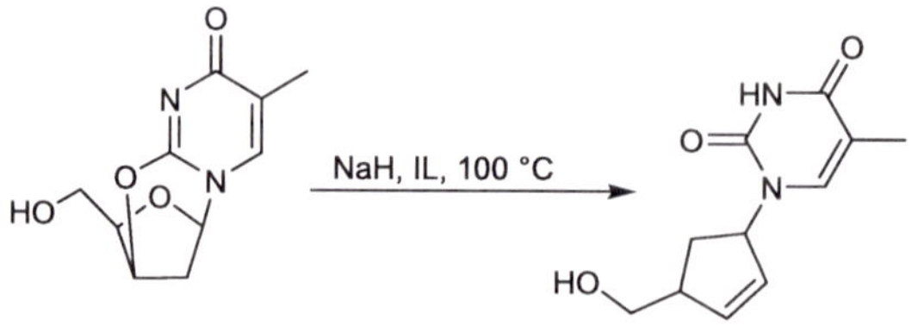

solvents	time (min.)	yield (%)
DMA	30	81
[MoeMIm][TFA]	5	93
[BMIm][TFA]	10	91
[MoeMIm][Ms]	10	89

Scheme 6 Influence of IL in d4T synthesis

conditions as previously reported, in the synthesis of d4T, an important anti-HIV drug (Scheme 6). They noticed that completion of reaction took place within 5–10 min. followed by an easy workup procedure. Similarly, an anti-HSV antiviral medicine, Trifluridine (TFT), which is manufactured from 2′-deoxyuridine, was achieved in ILs with above 90% yield [72].

1.5 Green Synthesis of Pharmaceutically Relevant Compounds

Compounds bearing benzimidazole nucleus are known for their excellent analgesic, anti-viral, anti-microbial and anti-inflammatory properties and showed significant application in medicine praxis. Gioia et al. put forward a green and sustainable protocol for the synthesis of 2-substituted or 1,2-disubstituted benzimidazoles from *o*-phenylenediamine with several aldehydes using deep eutectic solvents (DES) (Scheme 7) [73]. This strategy afforded benzimidazoles with high yield in very short reaction time through easy work up procedures.

Dithiocarabamates are compounds showcasing various pharmacological and biological functions including anti-glaucoma [74], anti-fungal [75], anti-proliferative [76], anti-tumour [77], etc. A catalyst-free novel green strategy towards dithiocarbamate synthesis is reported by Asadipour et al. [78].

Coumarins are important therapeutic molecules having anti-inflammatory, anti-allergic, anti-fungal and anti-tumour features. Even though several enzymatic methods for the production of natural coumarins were generated, novel report on the characteristics, industrial utility, features and classification of natural coumarin derived from ascomycetes and basidiomycetes fungi was done by Oliveira et al. and

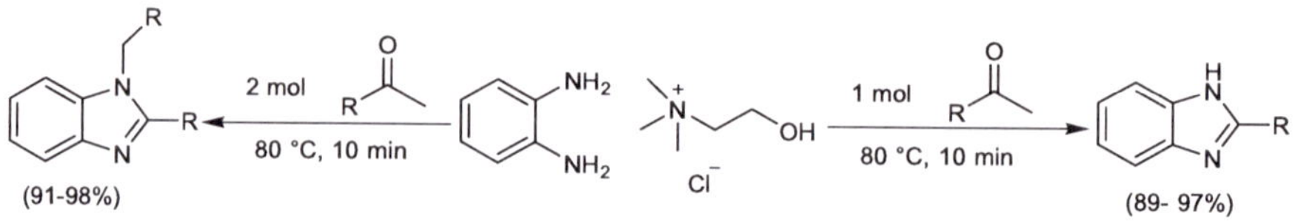

Scheme 7 Synthesis of benzimidazoles using DES

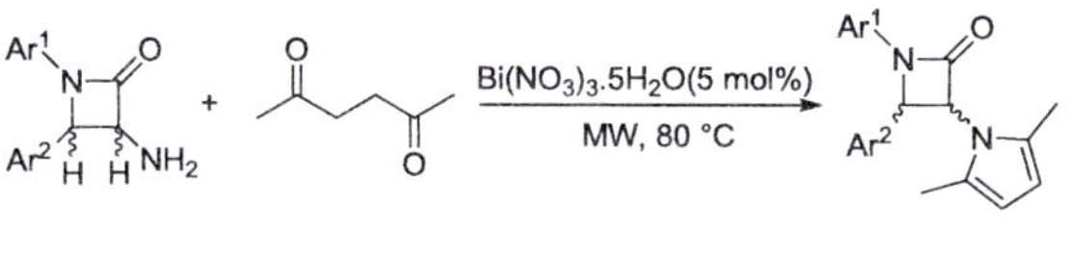

Ar^1	Ar^2
phenyl	phenyl
p-methoxyphenyl	*p*-methoxyphenyl
6- chrysenyl	2-thiophenyl
	ferrocenyl
	2-furyl

Scheme 8 Microwave synthesis of N-(2-azetidinonyl) 2,5-disubstituted pyrroles

compared fungal coumarin to coumarin, which are synthesized by chemical means. Natural synthesis of coumarin eliminated the formation of unwanted side products that are toxic [79].

2-Azetidinones and pyrroles have marked their existence in the pharmacological world. Many natural and synthetic drugs exhibiting anti-diabetic [80], anti-parkinsonian [81] and anti-cancer [82] activities have these moieties as their major contents. Bandyopadhyay and co-workers have developed a green protocol involving microwave synthesis of *N*-(2-azetidinonyl) 2,5-disubstituted pyrroles having both pharmaceutically active pyrrole and 2-azetidinones (Scheme 8) [83].

1.6 Case Studies

Sertraline an anti-depressant drug could be effectively synthesized by substituting the conventional tetrahydrofuran by ethanol [84]. The use of ethanol has also improved the reaction by eliminating the difficult waste separation processes and their corresponding disposal.

Simvastatin drug lowers the cholesterol in the body and thus finds effective in the treatment and prevention of dyslipidemia and cardiovascular diseases. Semisynthetic routes towards its synthesis require large labour, and the process is very expensive than the starting material lovastatin. Thus, a greener approach towards its synthesis was established by Tang and co-workers [85]. The strategy involved a whole-cell biocatalytic procedure and could synthesize simvastatin very efficiently even in industrial scale.

Celebrex® a COX-2 anti-inflammatory agent contains the active ingredient celecoxib. The conventional method for the synthesis of celecoxib required the elimination of hydrazine and regioisomers obtained during cyclization. Greenness of the reaction lies in the usage of innocuous solvents (isopropanol and methanol) and avoiding the use of hydrazine. The product could be now isolated by water dilution and cooling [86]. Type-two diabetes could be treated by making DPP-IV enzyme inactive. This could be achieved by Sitagliptin, a drug that increases the level of incretin that helps to control blood sugar by improving the release of insulin. The green synthesis of which utilizes asymmetric catalytic hydrogenation and resulted in high yields, optical purity and reduced the number of reaction steps considerably.

Fortunak et al. have put forth a green methodology towards piperaquine tetraphosphate [87]. Piperaquine belongs to class of 4-aminoquinoline known for anti-malarial activities. Conventional synthesis proceeds through unwanted side reactions. The present green strategy results in 92–93% yield in two synthetic steps.

Pregabalin a γ-aminobutyric acid was synthesized for curing different central nervous system abnormalities like anxiety, epilepsy, social phobia and neuropathic pain. The various processes in the traditional synthesis were Knoevenagel condensation, cyanation, hydrolysis, reduction and decarboxylation; all of which reduced the yields of the product considerably. The process was not economically viable and recycling of the unwanted γ-amino acid enantiomer also could not be carried out. A biocatalytic approach utilizing lipolase could tackle the issues of cost and environment viability increasing the yields at the same time [88].

Diterpenoid alkaloid paclitaxel isolated from *Taxus brevifolia* is found to exhibit pronounced anti-cancer nature and is prescribed mainly against ovarian, breast and lung cancer. Isolation of which is required killing of these trees for their barks. Thus, a semisynthetic methodology was designed from 10-deacetylbaccatin III as the starting material. But the methodology was complex and environmentally noxious. Hence, a greener approach towards the synthesis of paclitaxel proceeded through *T. brevifolia* cell fermentation, followed by extraction and recrystallization [89]. The steps involved were a range of enzymatic conversions like hydroxylation, oxidation, acylation and oxetene ring generation.

Neurodegenerative disorders like Huntington's and Alzheimer's disease can be effectively treated by a potent pharmaceutical agent LY300164 which utilizes air oxidation and biocatalytic reduction both of which are environmental-friendly strategies [90]. Quinapril hydrochloride is used for hypertension and congestive heart failure treatment. Methylene chloride, hydroxybenzotriazole, DCC and toluene were used to remove the acetic acid in the conventional synthesis which promoted the generation of diketopiperazine that could minimize the yield of quinapril. The redesign of reaction led to the utilization of N-carboxyanhydride as the precursor which would eliminate the acetic acid usage and allowing the removal of DCC, at the same time avoiding the employment of chlorinated solvent [91]. All of which improved the yield of quinapril to 90%.

Conventional synthesis of ibuprofen, an anti-inflammatory drug was laborious and generated good amounts of chemical waste. A nickel-catalysed three step strategy was employed by Cann and co-workers which improved the synthetic efficiency to a good extent [92]. The methodology was effective in reducing the waste production to about 1%.

Anti-epileptics are drugs that prevent the brisk, repetitive stimulation in the brain that results in epilepsy, a seizure disorder [93]. Phenytoin an anti-epileptic drug could be efficiently synthesized by a green protocol involving the synthesis of 5,5-disubstituted hydantoins from potassium cyanate and methyl esters of amino acids via a mechanochemical approach [94]. Another approach towards anti-epileptic drug was put forward by Sahoo and co-workers [95]. They carried out the synthesis of a range of 3-(aryl)-2-thioxo-2,3-dihydroquinazolin-4(1*H*)-ones via microwave irradiation. Microwave condition provided a simple protocol, at the same time increased the

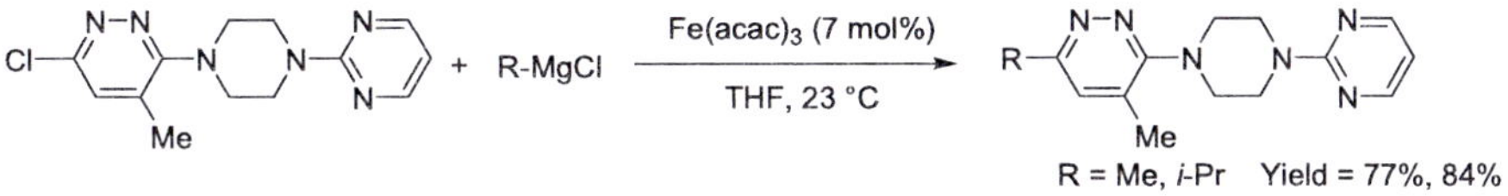

Scheme 9 Cross-coupling of 2-chloropyridazine with alkyl magnesium chloride

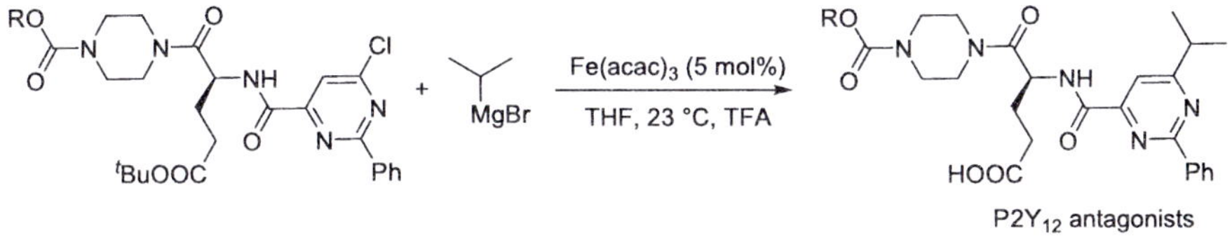

Scheme 10 Synthesis of $P2Y_{12}$ antagonists

pace and the feasibility of the reaction towards the synthesis of these anti-convulsant agents.

A green perspective towards the synthesis of acetaminophen an anti-pyretic drug was developed by Korto et al. [96]. This iron-catalysed methodology involved electrophilic aromatic substitution, hydrogenation and acylation. The reaction proceeded only in three steps with minimum wastage. $Fe(acac)_3$ catalysed cross-coupling of Grignard reagent with 2-chloropyridazine resulted in the formation of an anti-neuroinflammatory drug for the healing of Alzheimer's disease [97]. The use of $Fe(acac)_3$ could attach the alkyl part of Grignard reagent to the polynitrogenated precursor efficiently (Scheme 9).

Anti-platelet agents help to lower the formation of blood clots by reducing their capacity to stick together. $P2Y_{12}$ antagonists an anti-platelet agent was synthesized by a coupling of isopropylmagnesium bromide and choro-pyrimidine via iron catalysis by Caroff et al. (Scheme 10) [98].

1.7 Miscellaneous Reactions

An inclusion complex MOL/PAc-β-CD was prepared by Lim from peracetylated-*β*-cyclodextrin (PAc-*β*-CD) and molsidomine (MOL), a nitrovasodilating drug in supercritical carbon dioxide [99]. In vitro drug release studies of the drug from MOL/PAc-*β*–CD complex was studied. Upon complexation, the MOL release was found to be retarded because of the aquaphobic properties of the PAc-*β*–CD.

A much greener pathway in the solid-phase production of anti-sense oligonucleotide drugs through the modification of the present phosphoramidite-based method was disclosed by Sanghvi et al. [100]. The greener modification involves the eradication of halogenated solvents, reduction of waste materials, removal of toxic reagents, usage of green solvents and reuse of protecting groups and amidites.

Song et al. could answer the problem of low encapsulation of drug for the treatment of bone diseases [101]. Enlightened by the lotus leaf phenomenon, they put forward a lucid and green approach to overcome this difficulty. A slurry of calcium sulphate dihydrate bone cement particles, water and drug (ibuprofen and acetaminophen) was dripped on a superhydrophobic surface forming a microreactor. The drug gets encapsulated via a self-setting mechanism in the microreactor. Thus, improved drug loading could be achieved to facilitate bone healing functions.

2 Conclusion

Green chemistry lays the foundation towards sustainable development, as it persuades the researches to focus on products and procedures that represent a good process development. From the foregoing discussion, it is evident that application of green chemistry in pharmaceutical industry is an emerging area. The methodologies derived from green chemistry allow to achieve robustness in the synthesis of desired products in cheaper and profitable rates. Principles of green chemistry can be applied to the manufacture of drugs very efficiently. Drug synthesis and delivery could now be achieved via various green protocols including nanomaterials, polymers, enzymes, ionic liquids, etc. The field of green chemistry in pharmaceutical chemistry is thus a prominent area that opens up wide opportunities to aspiring researchers.

References

1. Constable DJC, Jinenez-Gonzalez C, Henderson RK (2007) Perspective on solvent use in the pharmaceutical industry. Org Proc Res Devel 11(1):133–137
2. Slater CS, Savefski M (2007) A method to characterize the greeness of solvents used in the pharmaceutical manufacturing. Environ Sci Heath a Environ Sci Eng Toxic Hazard Subst Control 42(11):1595–1605
3. Tucker JL (2006) Green Chemistry, a Pharmaceutical Perspective. Organ Process Res Develop 10(2):315–319
4. Tauro SJ, Gawad JB (2013) Green Chemistry: A Boon to Pharmaceutical Synthesis. Journal of Scientific Research 2(7):67–69
5. Jiménez-González C, Constable DJC, Ponder CS (2012) Evaluating the "Greenness" of chemical processes and products in the pharmaceutical industry—a green metrics primer. Chem Soc Rev 41:1485–1498
6. Kumar S, Pandita LVD (2015) Green synthesis of therapeutic nanoparticles: an expanding horizon. Nanomedicine 10(15):2451–2471
7. Zhu J-J, Wang C, Liu K et al (2011) Green and facile synthesis of highly biocompatible graphene nanosheets and its application for cellular imaging and drug delivery. J Mater Chem 21:12034–12040
8. Tian J, Xu S, Deng H et al (2017) Fabrication of Nanaki SG, Pantopoulos K, Bikiaris DN (2011) Synthesis of biocompatible poly(ε-caprolactone)-block-poly(propylene adipate) copolymers appropriate for drug nanoencapsulation in the form of core-shell nanoparticles. Int J Nanomedicine 6:2981–2995

9. Venkatpurwar V, Shiras A, Pokharkar V (2011) Porphyran capped gold nanoparticles as a novel carrier for delivery of anticancer drug: In vitro cytotoxicity study. Int J Pharm 409(1–2):314–320
10. Ma N, Zhang B, Liu J et al (2015) Green fabricated reduced graphene oxide: evaluation of its application as nano-carrier for pH-sensitive drug delivery. Int J Pharm. 496(2):984–992
11. Lee AV, *Oesterreich S, Davidson NE* (2015) MCF-7 Cells—Changing the Course of Breast Cancer Research and Care for 45 Years. J Natl Cancer Inst. 107(7):djv073.
12. Foster KA et al (1998) Characterization of the A549 cell line as a type II pulmonary epithelial cell model for drug metabolism. Exp Cell Res. 243(2):359–366
13. Palai PK, Mondal A, Chakraborti CK et al (2019) Doxorubicin Loaded Green Synthesized Nanoceria Decorated Functionalized Graphene Nanocomposite for Cancer-Specific Drug Release. J Clust Sci. 30(6):1565–1582
14. Jahangirian H, Lemraski EG, Webster TJ et al (2017) A review of drug delivery systems based on nanotechnology and green chemistry: green nanomedicine. Int J Nanomedicine 12:2957–2978
15. Lam PL, Wong WY, Bian Z et al (2017) Recent advances in green nanoparticulate systems for drug delivery: efficient delivery and safety concern. Nanomedicine 12(4):357–385
16. White MA, Johnson JA, Koberstein JT, Turro NJ (2006) Toward the syntheses of universal ligands for metal oxide surfaces: controlling surface functionality through click chemistry. J Am Chem Soc 128(35):11356–11357
17. Kumar CSSR, Mohammad F (2011) Magnetic nanomaterials for hyperthermia-based therapy and controlled drug delivery. Adv Drug Deliv Rev 63(9):789–808
18. Li J, Wu S, Wu C et al (2016) Versatile surface engineering of porous nanomaterials with bioinspired polyphenol coatings for targeted and controlled drug delivery. Nanoscale 8(16):8600–8606
19. Li X, Gong Y, Zhou X et al (2016) Facile synthesis of soybean phospholipid-encapsulated MoS2 nanosheets for efficient in vitro and in vivo photothermal regression of breast tumor. Int J Nanomedicine 11:1819–1833
20. Zhao MD, Cheng JL, Yan JJ et al (2016) Hyaluronic acid reagent functional chitosan-PEI conjugate with AQP2-siRNA suppressed endometriotic lesion formation. Int J Nanomedicine 11:1323–1336
21. Das D, Patra P, Ghosh P et al (2016) Dextrin and poly(lactide)-based biocompatible and biodegradable nanogel for cancer targeted delivery of doxorubicin hydrochloride. Polym Chem 7(17):2965–2975
22. Nanaki SG, Pantopoulos K, Bikiaris DN (2011) Synthesis of biocompatible poly(ε-caprolactone)-block-poly(propylene adipate) copolymers appropriate for drug nanoencapsulation in the form of core-shell nanoparticles. Int J Nanomedicine 6:2981–2995
23. Zhu HY, Zhang SY, Ling Y et al (2015) pH-responsive hybrid quantum dots for targeting hypoxic tumor siRNA delivery. J Control Release 220:529–544
24. Huang N, Wang H, Zhao J et al (2010) Single-wall CNTs assisted photothermal cancer therapy: animal study with a murine model of squamous cell carcinoma. Laser Surg Med 42(9):638–648
25. Liu XW, Tao HQ, Yang K et al (2011) Optimization of surface chemistry on single-walled CNTs for in vivo photothermal ablation of tumors. Biomaterials 32(1):144–151
26. Jeyamohan P, Hasumura T, Nagaoka Y et al (2013) Accelerated killing of cancer cells using a multifunctional single-walled CNT-based system for targeted drug delivery in combination with photothermal therapy. Int J Nanomedicine 8(1):2653–2667
27. Sripriya J, Anandhakumar S, Achiraman S (2013) Laser receptive polyelectrolyte thin films doped with biosynthesized silver nanoparticles for antibacterial coatings and drug delivery applications. Int J Pharm 457(1):206–213
28. Agostoni V, Horcajada P, Couvreur P et al (2013) "Green" fluorine-free mesoporous iron(III) trimesate nanoparticles for drug delivery. Green Materials 1(4):209–217
29. Kang Y, Park W. (2012) Study on Synthesis Chitosan Oligomer Stabilized Silver Nanoparticles Using Green Chemistry and Their Burn Wound Healing Effects. MRS Proceedings, 1453, Mrss12–1453-aaa05–08. doi:https://doi.org/10.1557/opl.2012.1341.

30. Steven CR, Busby GA, Grant MH (2014) Bioinspired silica as drug delivery systems and their biocompatibility. J Mater Chem B 2(31):5028–5042
31. Gurunathan S, Han J, Park JH et al (2014) A green chemistry approach for synthesizing biocompatible gold nanoparticles. Nanoscale Res Lett 9(1):248
32. Thiruvengadam M, Chung I, Gomathi T et al (2019) Synthesis, characterization and pharmacological potential of green synthesized copper nanoparticles. Bioproc Biosyst Eng 42(11):1769–1777
33. Dai L, Liu K, Si C et al (2016) Ginsenoside nanoparticle: A new green drug delivery system. J Mater Chem B 4(3):529–528
34. Villamizar-Sarmiento MG, Molina-Soto EF, Guerrero J et al (2019) A New Methodology to Create Polymeric Nanocarriers Containing Hydrophilic Low Molecular-Weight Drugs: A Green Strategy Providing a Very High Drug Loading. Mol Pharm 16(7):2892–2901
35. Pavliňáková V, Fohlerová Z, Pavliňák D et al (2018) Effect of halloysite nanotube structure on physical, chemical, structural and biological properties of elastic polycaprolactone/gelatin nanofibers for wound healing applications. Mater Sci Eng, C 91:94–102
36. Bhat AH, Khan I, Usmani MA et al (2019) Cellulose an ageless renewable green nanomaterial for medical applications: An overview of ionic liquids in extraction, separation and dissolution of cellulose. Int J Biol Macromol. 129:750–777
37. Chu C, Su M, Zhu J et al (2019) Metal-Organic Framework Nanoparticle-Based Biomineralization: A New Strategy toward Cancer Treatment. Theranostics 9(11):3134–3149
38. Sigel A, Sigel H, Sigel RKO, (eds) (2008). Biomineralization: From Nature to Application. Metal Ions in Life Sciences. 4. Wiley VCH Weinheim Germany.
39. Schnell B, Faber K, Kroutil W (2003) Enzymatic racemisation and its application to synthetic biotransformations. Adv Synth Catal 345:653–666
40. Dembitsky V (2003) Oxidation, epoxidation and sulfoxidation reactions catalysed by haloperoxidases. Tetrahedron 59(26):4701–4720
41. Nakamura K, Yamanaka R, Matsuda T, Harada T (2003) Recent developments in asymmetric reduction of ketones with biocatalysts. Tetrahedron Asymmetry 14:2659–2681
42. Thiel F (2000) Enhancement of selectivity and reactivity of lipases by additives. Tetrahedron 56(19):S2905-2919
43. Ni Y, Xu J-H (2012) Biocatalytic ketone reduction: A green and efficient access to enantiopure alcohols. Biotechnol Adv 30(6):1279–1288
44. Kourist R, Bornscheuer UT (2011) Biocatalytic synthesis of optically active tertiary alcohols. Appl Microbiol Biotechnol 91(3):505–517
45. Woodley JM (2008) New opportunities for biocatalysis: making pharmaceutical processes greener. Trends Biotechnol 26(6):321–327
46. Zhang X, King-Smith E, Renata H (2018) Total Synthesis of Tambromycin by Combining Chemocatalytic and Biocatalytic C—H Functionalization. Angew Chem Int Ed 57(18):5037–5041
47. Müller M (2005) Chemoenzymatic synthesis of building blocks for statin side chains. Angew Chem Int Ed 44(3):362–365
48. Ma SK, Gruber J, Davis C et al (2010) A green-by-design biocatalytic process for atorvastatin intermediate. Green Chem. 12:81–86
49. Byford MF, Baldwin JE, Shiau CY, Schofield CJ (1997) The mechanism of ACV synthetase. Chem Rev 97(7):2631–2639
50. Yeh WK, Queener SW (1990) Potential industrial use of cephalosporin biosynthetic enzymes and genes. Ann New York Acad Sci 613:128–141
51. Ran N, Rui E, Liu J et al (2009) Chemoenzymatic Synthesis of Small Molecule Human Therapeutics. Curr Pharm Des 15:134–152
52. Trauger JW, Kohli RM, Mootz HD, Marahiel MA, Walsh CT (2000) Peptide cyclization catalysed by the thioesterase domain of tyrocidine synthetase. Nature 407:215–218
53. Patel RN, Banerjee A, Ko RY, Howell JM, Li WS, Comezoglu FT et al (1994) Enzymic preparation of (3R-cis)-3-(acetyloxy)-4-phenyl-2-azetidinone: a taxol side-chain synthon. Biotechnol Appl Biochem 20:23–33

54. Martinez CA, Hu S, Dumond Y, Tao J, Kelleher P, Tully L (2008) Development of a chemoenzymatic manufacturing process for pregabalin. Org Proc Res Dev 12:392–398
55. Huang C, Pérez CD, Rádis-Baptista G et al (2018) The Holo-Transcriptome of the Zoantharian Protopalythoa variabilis (Cnidaria: Anthozoa): A Plentiful Source of Enzymes for Potential Application in Green Chemistry. Industrial and Pharmaceutical Biotechnology. Mar Drugs 16(6):207
56. Cabri W (2009) Catalysis: The pharmaceutical perspective. Catal Today 140(1–2):2–10
57. Wu Q, Lin X-F, Shen Z et al (2014) Enzymatic enantioselective aldol reactions of isatin derivatives with cyclic ketones under solvent-free conditions. Biochimie 101:156–160
58. Huang J, - Y, Li S – J, Wang Y – G, (2006) TEMPO-linked metalloporphyrins as efficient catalysts for selective oxidation of alcohols and sulphide. Tetrahedron Lett 47:5637–5640
59. Vivancos Á, Beller M, Albrecht M (2018) NHC-based Iridium Catalysts for Hydrogenation and Dehydrogenation of N-Heteroarenes in Water under Mild Conditions. ACS Catal 8:17–21
60. Zhang W, Wu L, Zhou K, Guan Q et al (2020) Activated carbon/Brønsted acid-promoted aerobic benzylic oxidation under "on-water" condition: Green and efficient synthesis of 3-benzoylquinoxalinones as potent tubulin inhibitors. Eur J Med Chem 186:111894
61. Ceborska M, Zimnicka M, Pietrzak M et al (2012) Structural diversity in native cyclodextrins/folic acid complexes – from [2]-rotaxane to exclusion compound. Org Biomol Chem 10(27):5186–5188
62. Davenport AP, Hyndman KA, Dhaun N et al (2016) Endothelin. Pharmacol Rev 68(2):357–418
63. Barnes DM, Ji J, Fickes MG et al (2002) Development of a Catalytic Enantioselective Conjugate Addition of 1,3-Dicarbonyl Compounds to Nitroalkenes for the Synthesis of Endothelin-A Antagonist ABT-546. Scope, Mechanism, and Further Application to the Synthesis of the Antidepressant Rolipram. J Am Chem Soc 124(44):13097–13105.
64. Garcia A, Leonardi D, Lamas MC (2016) Promising applications in drug delivery systems of a novel β-cyclodextrin derivative obtained by green synthesis. Bioorg Med Chem Lett 26(2):602–608
65. Shaabani A, Soleimani E, Maleki A (2006) Ionic liquid promoted one-pot synthesis of 3-aminoimidazo[1,2-*a*]pyridines. Tetrahedron Lett 47:3031–3034
66. Zaidlewicz M, Cytarska J, Dzielendziak A et al (2004) Synthesis of boronated phenylalanine analogues with a quaternary center for boron neutron capture therapy. Arkivoc 3:11–21
67. Kurata A, Kitamura Y, Irie S et al (2010) Enzymatic synthesis of caffeic acid phenethyl ester analogues in ionic liquid. J Biotechnol 148(2–3):133–138
68. Si6dmiak T, Marszall MP, Proszowska A (2012) Ionic Liquids: A New Strategy in Pharmaceutical Synthesis. Mini-Rev Org Chem 9(6):203–208
69. Zhang X, Li X, Li D et al (2009) Ionic liquid mediated and promoted eco-friendly preparation of thiazalidinone and pyrimidine nucleoside-thiazolidinone hybrids and their antiparasitic activities. Bioorg Med Chem Lett 19(22):6280–6283
70. Fan X, Feng D, Qu Y et al (2010) Practical and efficient synthesis of pyrano[3,2-*c*]pyridine, pyrano[4,3-*b*]pyran and their hybrids with nucleoside as potential antiviral and antileishmanial agents. Bioorg Med Chem Lett 20(3):809–813
71. Kumar V, Parmar VS, Malhotra SV (2011) Ionic Liquids: Neoteric Solvents for Nucleoside Chemistry. Curr Org Synth 8:777–786
72. Kumar V, Malhotra SV (2008) Synthesis of nucleoside-based antiviral drugs in ionic liquids. Bioorg Med Chem Lett 18:5640–5642
73. Gioia MLD, Cassano R, Costanzo P et al (2019) Green synthesis of privileged benzimidazole scaffolds using active deep eutectic solvent. Molecules 24(16):2885
74. Carta F, Aggarwal M, Maresca A et al (2012) Dithiocarbamates strongly inhibit carbonic anhydrases and show antiglaucoma action in vivo. J Med Chem 55:1721–1730
75. Turan-Zitouni G, Ozdemir A, Guven K (2005) Synthesis of some 1[(*N*, N-disubstitutedthio carbamoylthio)acetyl]-3-(2-thienyl)-5-aryl-2-prazoline derivatives and investigation of their antibacterial and antifungal activities. Arch Pharm Chem Life Sci 338:96–104

76. Cao SL, Han Y, Yuan CZ, Y et al (2013) Synthesis and antiproliferative activity of 4-substituted-piperazine-1-carbodithioate derivatives of 2,4-diaminoquinazoline. Eur J Med Chem 64:401–409
77. Cao SL, Feng YP, Jiang YY et al (2005) Synthesis and in vitro antitumor activity of 4(3H)-quinazolinone derivatives with dithiocarbamate side chains. Bioorg Med Chem Lett 15(7):1915–1917
78. Asadipour a, Shams Z, Eskandari K, et al (2017) Efficient, straightforward, catalyst-free synthesis of medicinally important S-alkyl/benzyl dithiocarbamates under green conditions. Res Chem Intermed 44(2):1295–1304
79. Costa TM, Tavares LBB, de Oliveira D (2016) Fungi as a source of natural coumarins production. Appl Microbiol Biotechnol 100(15):6571–6584
80. Banik I, Banik BK (2013) Synthesis of β-Lactams and Their Chemical Manipulations Via Microwave-Induced reactions. Top Heterocycl Chem 30:183–222
81. Mehta PD, Sengar NPS, Pathak AK (2010) 2-Azetidinone – A new profile of various pharmacological activities. Eur J Med Chem 45(12):5541–5560
82. Banik BK, Banik I, Becker FF (2010) Novel Anticancer β-Lactams. Top Heterocycl Chem 22:349–373
83. Bandyopadhyay D, Rhodes E, Banik B (2013) A green, chemoselective, and practical approach toward N- (2-azetidinonyl) 2,5-disubstituted pyrroles. RSC Adv 3:16756–16764
84. Taber GP, Pfisterer DM, Colberg JC (2004) A New and Simplified Process for Preparing N-[4-(3,4-Dichlorophenyl)-3,4-dihydro-1(2H)-naphthalenylidene]methanamine and a Telescoped Process for the Synthesis of (1S-cis)-4-(3,4-Dichlorophenol)-1,2,3,4-tetrahydro-N-methyl-1-naphthalenamine Mandelate: Key Intermediates in the Synthesisof Sertrailine Hydrochloride. Org Proc Res Dev 8(3):385–388
85. Xie X, Tang Y (2007) Efficient Synthesis of Simvastatin by Use of Whole-Cell Biocatalysis. Appl Environ Microbiol 73(7):2054–2060
86. Talley JJ, Penning TD, Collins PW et al (1997) Synthesis and Biological Evaluation of the 1,5-Diarylpyrazole Class of Cyclooxygenase-2 Inhibitors: Identification of 4-[5-(4-Methylphenyl)- 3- (trifluoromethyl)-1*H*-pyrazol-1-yl]benzenesulfonamide (SC-58635, Celecoxib). J Med Chem 40(9):1347–1365
87. Fortunak JMD, Byrn SR, Dyson B et al (2013) An Efficient, Green Chemical Synthesis of the Malaria Drug. Piperaquine. Trop J Pharm Res 12(5):791–798
88. Jennings RA, Kissel WS, Le TV et al (2003) An enantioselective synthesis of (S)-(+)-3-aminomethyl-5-methylhexanoic acid via asymmetric hydrogenation. J Org Chem 68(14):5731–5734
89. Patel RN (1998) Tour de Paclitaxel: Biocatalysis for semisynthesis. Ann Rev Microbiol 98:361–395
90. Anderson BA, Hansen MM, Harkness AR et al (1995) Application of a Practical Biocatalytic Reduction to an Enantioselective Synthesis of the 5H–2,3-Benzodiazepine LY300164. J Am Chem Soc 117(49):12358–12359
91. Jennings S. (Pfizer) (2005). A Green Process for the Synthesis of Quinapril Hydrochloride: Summary for the Presidential Green Chemistry Challenge Awards Program.
92. Cann MC, Connelly MC (2000) The BHC company synthesis of Ibuprofen, real world case, green chemistry. ACS, Washington DC
93. Abou-Khalil BW (2016) Antiepileptic Drugs. Continuum:Lifelong Learning in Neurology. 22(1):132–156.
94. Konnert L, Colacino E, Reneaud B et al (2014) Mechanochemical preparation of Hydantoins from Amino Esters: Application to the Synthesis of the Antiepileptic Drug Phenytoin. J Org Chem. 79(21):10132–10142
95. Sahoo BM, Dinda SC, Kumar BVVR, Panda JR (2013) Green Synthesis and Evaluation of 3-(aryl)-2-thioxo-2,3-dihydroquinazolin-4(1H)-ones. Int J Pharm Sci Nanotech 6(2):2046–2052
96. Korto GK et al (2013) The steps synthesis of Acetaminophen. Sr seraphim Gibbons undergraduate symposium 26.

97. Zhou W, Zhong G, Rao X (2012) Identification of Aminopyridazine-Derived Antineuroinflammatory Agents Effective in an Alzheimer's Mouse Model. ACS Med Chem Lett 3(11):903–907
98. Caroff E, Meyer E, Treiber A (2014) Optimization of 2-phenyl-pyrimidine-4-carboxamides towards potent, orally bioavailable and selective P2Y(12) antagonists for inhibition of platelet aggregation. Bioorg Med Chem Lett. 24(17):4323–4331
99. Lee MY, Ganapathy HS, Lim KT (2010) Controlled drug release applications of the inclusion complex of peracetylated-β-cyclodextrin and water-soluble drugs formed in supercritical carbon dioxide. J Phys Chem Solids 71(4):630–633
100. Sanghvi YS, Ravikumar VT, Scozzari AN et al (2001) Applications of green chemistry in the manufacture of oligonucleotide drugs. Pure Appl Chem 73(1):175–180
101. Song B, Hu G (2018) Lotus Leaf-Inspired Bone Cement Particles with Ultrahigh Drug Encapsulation Capacity. ACS Appl Bio Mater 1(1):175–184

Chapter 9
Green Organocatalysis

Puja Basak and Pranab Ghosh

1 Introduction

Green chemistry has been designed as a methodology for achieving sustainable development during the last decades [1]. It is the area of chemistry that is focused to innovate such technologies that eliminate or reduce the use and generation of hazardous substances. In, 1988, Paul Anastas and John C. Warner described green chemistry as a set of 12 principles [2]. Catalysis, one of the principles of green chemistry that includes heterogeneous catalysis, enzyme catalysis and organocatalysis which has been introduced to promotean eco-friendly and minimum waste production process in the field of sustainable chemistry [3]. This branch of chemistry is found to minimize the environmental impact of chemical processes [4].

The term green organocatalysis refers to the catalysis, where the reaction rate is governed by the substoichiometric amount of the small organic molecule. The organocatalysts consist of carbon, nitrogen, hydrogen, sulfur, phosphorus, and they may be chiral or achiral [5, 6]. The metal-free nature of organocatalyst makes an undeniable contribution to the green chemistry. Nowadays, many organic chemists have set their goal to employ green organic catalysts which are an alternative to the heavy and toxic transition metals. Green organic catalysts generally provide mild reaction conditions, insensitivity to moisture, a wide range of substrate applicability and diverse reaction mechanisms. The one pot tandem organic transformation governed by green organic catalyst is particularly important owing to low cost, potential application in large-scale industrial production, immobilization of the catalyst and use of mild reaction conditions. Green organocatalysts can be Lewis acids, Lewis bases,

P. Basak · P. Ghosh (✉)
Department of Chemistry, University of North Bengal, Siliguri, West Bengal, India
e-mail: pizy12@yahoo.com

P. Basak
e-mail: pujabasakchem@gmail.com

G. Anilkumar and S. Saranya (eds.), *Green Organic Reactions*, Materials Horizons: From Nature to Nanomaterials, https://doi.org/10.1007/978-981-33-6897-2_9

Bronsted acids, and Bronsted bases. Due to the absence of metal, organocatalysts are found to be in the heart of green chemistry.

2 Background of Green Organocatalysts

Organocatalysts have opened up a new era toward the synthesis of fine chemicals and bioactive, drug molecules. The first organocatalytic reaction was Liebig's synthesis of oxamide from dicyan and water [7]. Amino acids also act as organocatalyst, and the amine-catalyzed Knoevenagel-type condensation of aldehydes and carboxylic acids or esters was demonstrated by Dakin in 1909 [5]. The application of small oligopeptides and simple amino acids in the enamine-type reaction mechanism was first discovered by Langenbeck [6, 7]. Cinchona alkaloid, natural products like strychnine, brucine, and different amino acids were the first organocatalysts, tested in several organic reactions [8, 9]. Although organocatalysts are small organic molecules, these have a profound effect on the synthesis of building blocks of life. In this regard, *L*-isovaline and *L*-alanine are particularly useful for the C–C bond formation reaction [10]. Asymmetric catalysis has been found to be the most powerful area to synthesize optically active molecules. Bredig and Fiske first published the asymmetric synthesis with a small organic molecule in 1912 [11]. The use of cinchona as bifunctional organocatalyst was first shown by various 1,2 and 1,4 additions in organic transformation [12, 13]. List et al. reported proline-catalyzed asymmetric aldol synthesis in 2000 which amplified the previous research endeavors on organocatalysts [14, 15].

3 Classification of Organocatalysis

Type-I: activation of the organic reaction based on the nucleophilic/electrophilic properties of the catalyst (Scheme 1). This type of mechanism is generally governed by biomolecules like proline, amines, cinchona alkaloids [16, 17].

Type-II: Reactive intermediates are formed by organic molecules, and the chiral catalyst is involved in this type of mechanism which is consumed during the reaction and regenerated in parallel catalytic mechanism as it is reported by Shu and Shi in

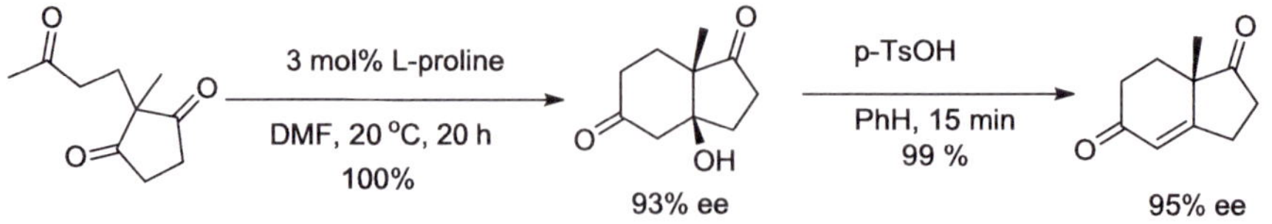

Scheme 1 Asymmetric synthesis of optically active bicyclic diketone

Scheme 2. The Shi epoxidation is a chemical reaction that describes the asymmetric epoxidation of alkenes with ozone and fructose derived catalyst [18].

Type-III: Phase transfer catalysts are one type of organocatalysts. This organocatalyst forms host–guest complex with the substrate and then phase transfer reactions occur (Scheme 3) [19].

Type-IV: Asymmetric transformation from substrate to desired product accelerated by a molecular cavity in which the catalyst may select from competing substrates, depending on the structure and size of the substrate (Scheme 4) [20].

Scheme 2 An efficient ketone-catalyzed epoxidation using hydrogen peroxide as oxidant

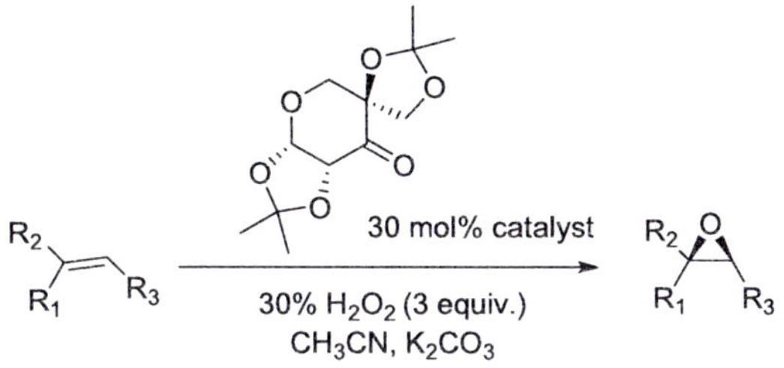

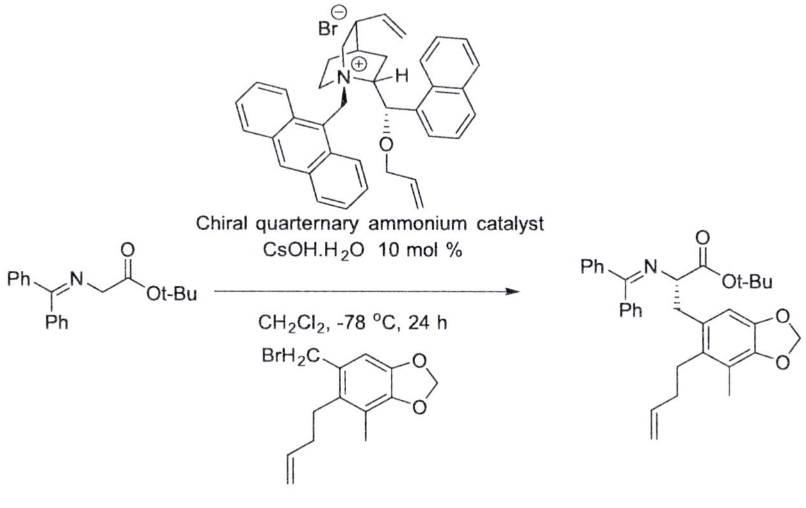

Scheme 3 Enantioselective enolate alkylation by phase transfer catalyst

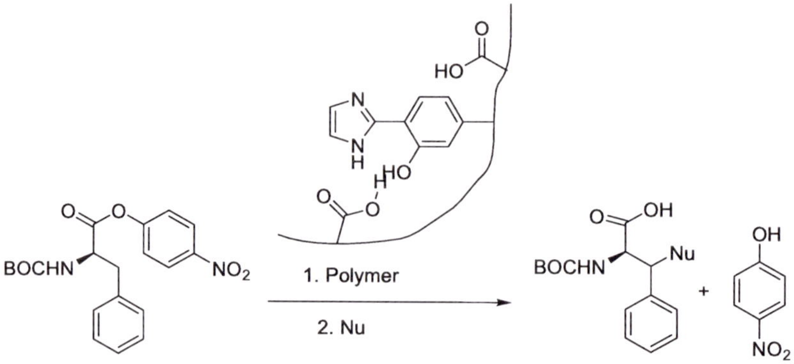

Scheme 4 Enantioselective ester hydrolysis catalyzed by imprinted polymers

4 Some Examples of Green Organocatalysts

4.1 Proline and Its Derivatives

A unique amino acid proline is extensively used as an organocatalyst in a variety of organic synthesis. Proline is optically active pyrrolidine derivative, and it has a rigid ring structure. The reactions catalyzed by proline generally involve activation by the nucleophilic/electrophilic property of the catalyst. It is inexpensive, readily available in both enantiomeric forms and nontoxic catalyst which is emerging as a complement to the metal catalyst. This amino acid contains both a carboxylic acid moiety functioning as a Brønsted acid and a nucleophilic secondary amino group which implies its bifunctional character [21]. Considering the above properties, proline has been employed as a green catalyst for organic reactions. Proline-catalyzed Robinson annulations were one of the earliest examples of green reactions catalyzed by organocatalyst [22, 23] although it is argued that the first proline-catalyzed asymmetric aldol reaction was reported by Hajos and Parrish (Scheme 5) [24–26].

Another group Eder, Sauer, and Wiechert and the group of Hajos and Parrish reported another enantioselective Aldol reaction catalyzed by (*S*)-proline (Scheme 6). Because of its wide application in synthetic organic chemistry, it is further explored to asymmetric aldol reaction [24], α-alkylation [27], Mannich reaction [28], Michael addition [29], and α-amination [30] of carbonyl compounds. Asymmetric Robinson annulation catalyzed by (*S*)-proline is known as the Hajos–Wiechert aldol reaction, Hajos–Parrish reaction or Hajos–Parrish Robinson annulations. The experimental and the theoretical results reveal that nucleophilic addition of the neutral enamine to the carbonyl group occurs simultaneously with the hydrogen transfer from the proline carboxylic acid moiety to the developing alkoxide.

Eder, Sauer, and Wiechert reported (*S*)-proline-catalyzed Michael addition followed by aldol condensation. The vinyl ketone reacts with the 1,3-diketone

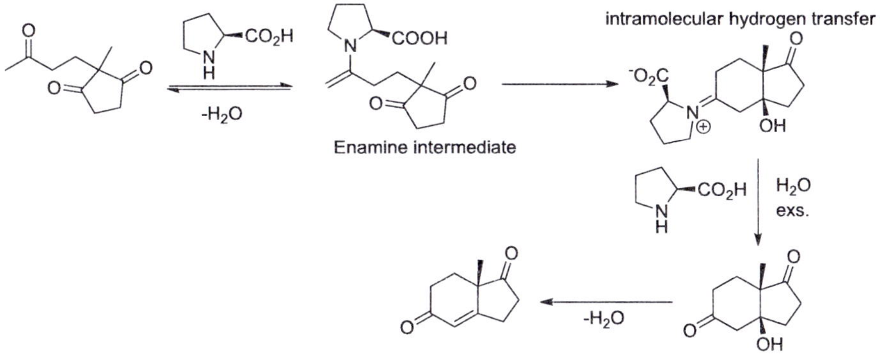

Scheme 5 Mechanism of proline-catalyzed asymmetric aldol condensation

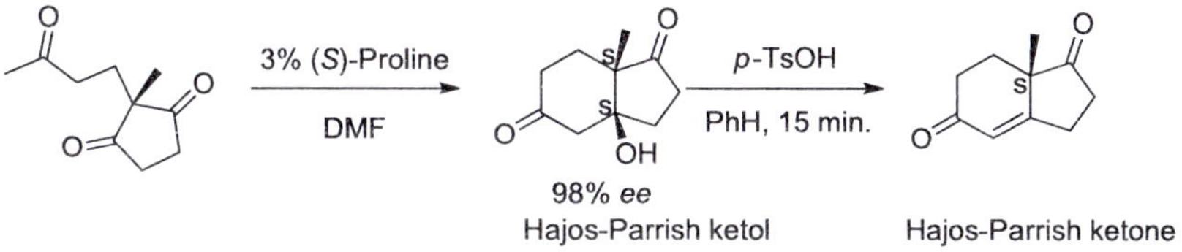

Scheme 6 Asymmetric Robinson annulation catalyzed by (*S*)-proline

compound through Michael's addition, and the final optically active bicyclic compound was obtained by successive aldol condensation (Scheme 7) [31].

Vignola and List had extended the catalytic activity of proline in case of asymmetric intramolecular α-alkylation of aldehydes which is a central C–C bond-forming reaction in organic chemistry (Scheme 8). [32] This process provides chiral substituted pyrrolidines, cyclopropanes, and cyclopentanes, in high yields and *ee*'s. However, the possibility of side reaction, e.g., racemization, aldolization, or catalyst alkylation, did not occur in the following alkylation process which illustrates the mildness and selectivity of enamine catalysis.

The Mannich reaction is a useful reaction used to synthesize *β*-amino-ketone from ketone, aldehyde, and amine. It may be direct with unmodified ketone and indirect with enolate equivalents. List et al. in the year 2000 reported a three-component direct proline-catalyzed highly enantioselective Mannich reaction [28]. The reaction

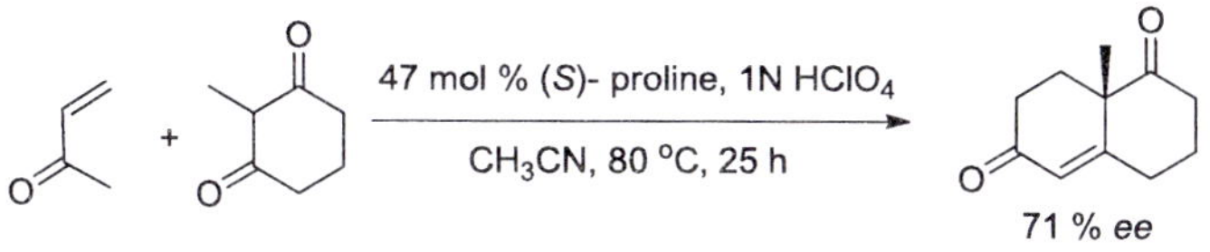

Scheme 7 Proline-catalyzed Michael addition followed by aldol condensation

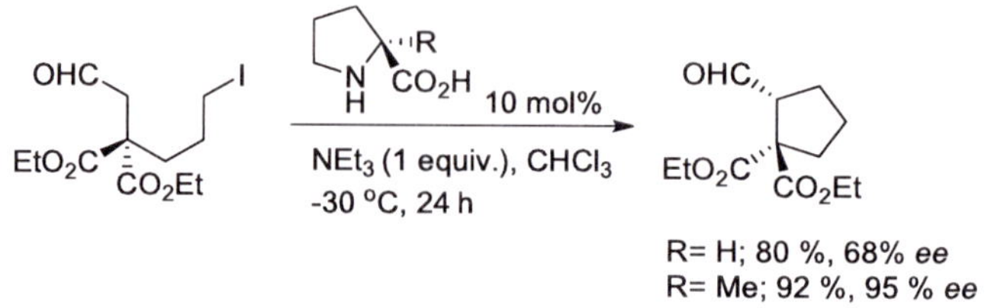

Scheme 8 Asymmetric intramolecular α-alkylation of aldehydes

proceeds through the catalytic activation of carbonyl compounds by proline and its derivatives via nucleophilic enamine intermediate (Scheme 9).

In the above three-component organocatalytic Knoevenagel condensation involving the reaction of 4-nitrobenzaldehyde and 1,3-indandione in methanol at ambient temperature in the presence of *L*-proline or pyrrolidine catalysis furnished the expected 2-(4-nitro-benzylidene)-indan-1,3-dione (Scheme 10).

Proline and its derivatives act as bifunctional catalysts in various organic reactions. Scheme 11 describes proline-catalyzed α-oxyamination of aldehydes with nitrosobenzene. Highly versatile aldehyde products are directly formed without

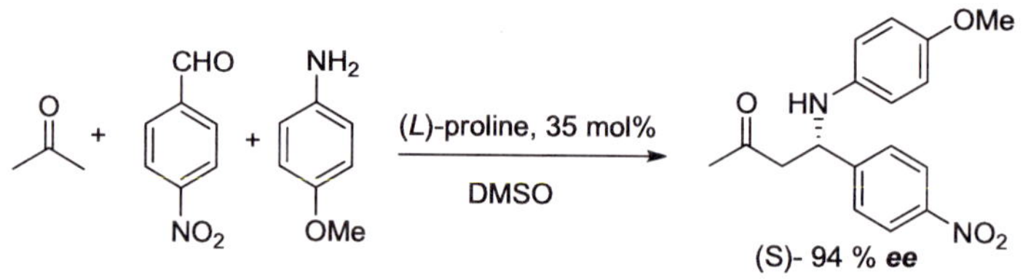

Scheme 9 Three-component asymmetric Mannich reaction catalyzed by proline

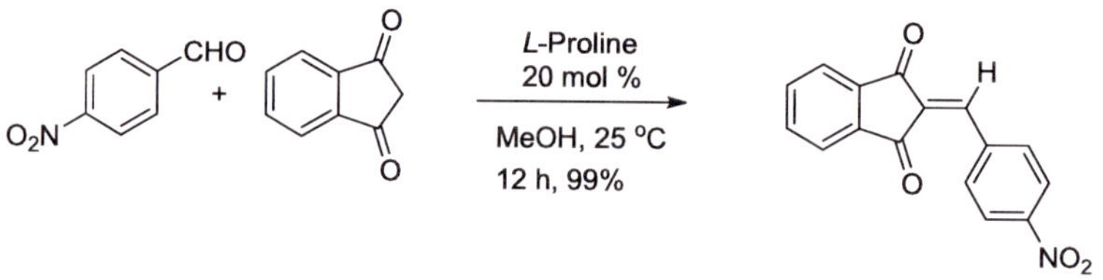

Scheme 10 Proline-catalyzed Knoevenagel condensation

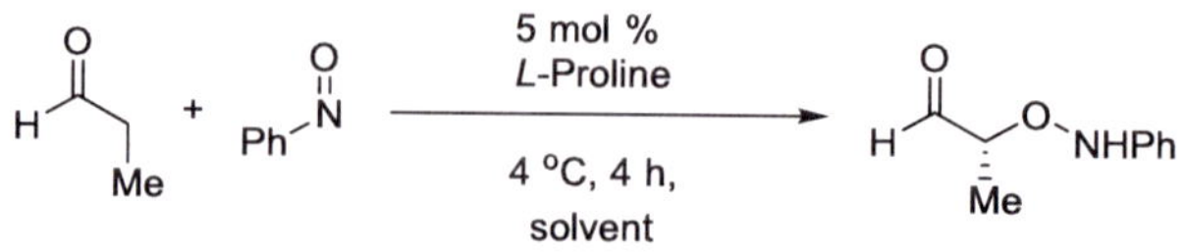

Scheme 11 Proline-catalyzed oxyamination

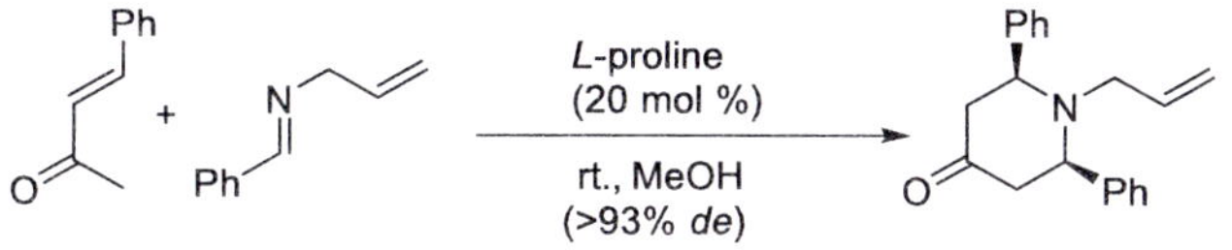

Scheme 12 Imino-Diels–Alder reaction catalyzed by proline

any pre-activation via an enol derivative is attractive from operational, atom, and step-economy standpoints.

Amine-catalyzed imino-Diels–Alder reaction between acyclic α,β-unsaturated ketones and aldimines, providing an efficient single-step route to the synthesis of *meso*-2,6-disubstituted-4-piperidones is reported (Scheme 12) [33]. Here, *trans*-4-phenyl-3-buten-2-one and *N*-allyl benzaldimine were reacted in the presence of *L*-proline organocatalyst at room temperature to furnish the desired imino-Diels–Alder product, (2*R*,6*S*)-*N*-allyl-2,6-diphenyl-4-piperidone, in 65% yield with >93% diastereoselectivity. The research contribution from prof. Ahrendt et al. to the field of proline-catalyzed organic reaction is acknowledged by many chemists that may play a vital role to know more about the proline organocatalyst [33–36].

4.2 N-Heterocyclic Carbene (NHC) as Organocatalysts

N-Heterocyclic carbenes (NHCs) have received considerable attention in the past several years as catalyst for organic transformations. Chiral NHC catalyzed organic transformation has found to be promising in the field of green chemistry. Although many examples are known, the most common NHCs are thiazolylidene, imidazolinylidene, imidazolylidene, and triazolylidene (Fig. 1 I–IV). These neutral molecules exist in singlet and triplet forms and contain divalent C atom with 6 electrons in its valence shell [37]. The rapid success of NHC as organocatalyst has resulted in the development of a variety of benzoin-type reactions.

The benzoin condensation reaction represents one of the earliest known C–C bond-forming reactions catalyzed by N-heterocyclic carbenes (NHCs) [38, 39]. In 2005, Xu and Xia used N-alkyl substituted imidazolium carbene-catalyzed homo-benzoin condensation at mild reaction conditions (Scheme 13). Although the electron-rich aromatic aldehydes exert the benzoin product with good yield whereas, the electron-deficient and aliphatic aldehydes react slowly in this proposed reaction [38, 39].

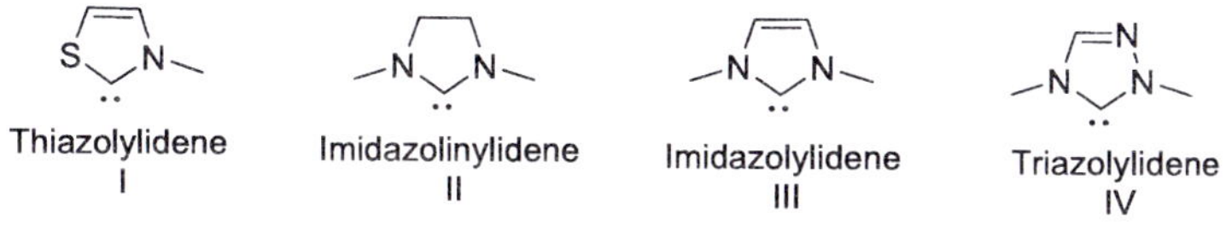

Fig. 1 Most common NHCs

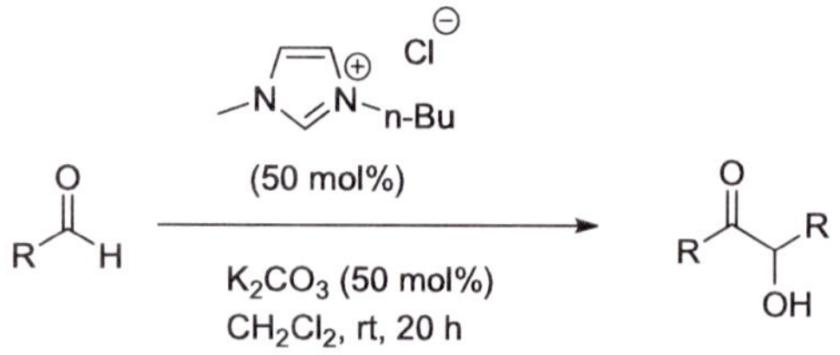

Scheme 13 Imidazolium carbene-catalyzed homo-benzoin condensation

The chemoselective intermolecular cross-benzoin condensation catalyzed by NHC for aromatic and aliphatic aldehydes was reported by Yang and co-workers. The chemoselectivity was only achieved with a large excess of the aliphatic aldehyde (molar ratio of 1:15) [40]. Interestingly, the identical yield and chemoselectivity are achieved upto five times of the reaction (Scheme 14).

Another example of a chemoselective cross-benzoin condensation reaction between aldehydes and α-ketoesters catalyzed by electron-deficient triazolium-derived NHC was reported so far [41]. Surprisingly, the main competing reaction hydroacylation did not occur under the condition of the reaction. A variety of acyl donors aromatic and aliphatic aldehydes have been reacted with the electrophilic partner α-ketoesters and exerted the desired acyloin products with high yields. (Scheme 15).

A facile NHC-mediated transformation of aromatic aldehydes to carboxylic acids in the presence of carbon dioxide is described in Scheme 16. Among all of the

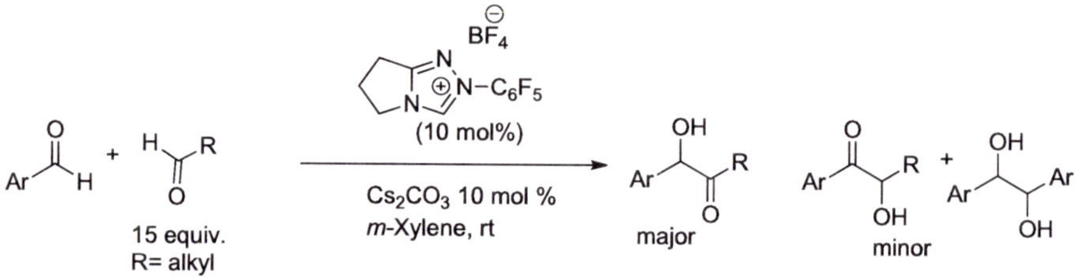

Scheme 14 Selective intermolecular cross-benzoin condensation reactions of aromatic and aliphatic aldehydes

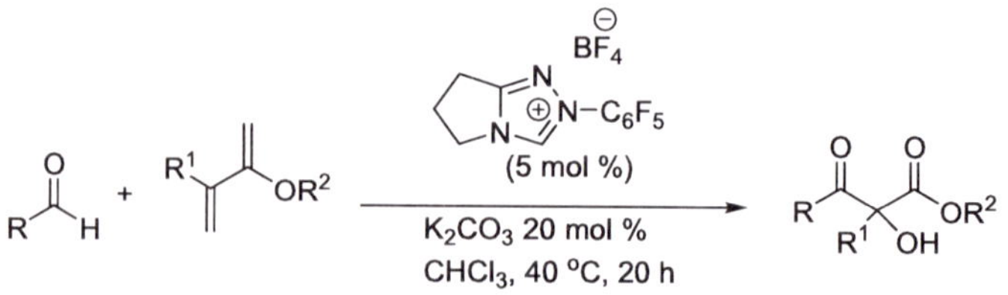

Scheme 15 Cross-benzoin reactions of aldehydes and α-ketoesters

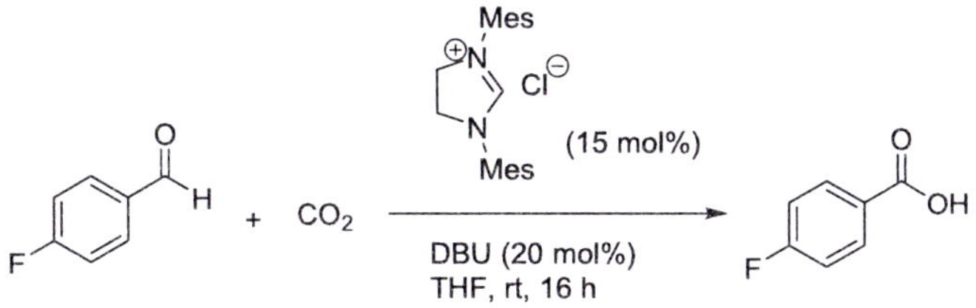

Scheme 16 NHC-mediated transformation of aromatic aldehydes to carboxylic acids

organocatalysts, NHCs have been found to be the most promising green catalyst for the conversion of aldehydes to carboxylic acid [42].

4.3 DMAP

4-(*N,N*-Dimethylamino)pyridine (DMAP) is a derivative of pyridine and is an effective nucleophilic base catalyst for the esterification of alcohols with acid anhydrides, hydrosilylation, the Baylis–Hillman reaction, tritylation, the Steglich rearrangement, etc. [43–51]. The term organocatalysis implies the use of small organic molecules as catalyst, that is, metal-free catalysts. In the case of Steglich esterification, DMAP acts as an acyl hydrogen transfer agent [48]. The reaction leads to the formation of amides with amines because of the nucleophilic nature of amine. If the side reaction occurs, then the rate of esterification is slow which lowers the yield of the reaction. The side reaction involves 1,3-rearrangement of the *O*-acyl intermediate, and the rearrange product is unable to react with alcohol; thus, the reaction is stopped. DMAP suppresses this side reaction, in the following manner (Scheme 17):Scheme 19Selective N-acylation of phenylhydrazine catalyzed by DMAP

Interestingly, DMAP has been employed as the most frequently used catalyst in recent years for the organic transformative reactions. In this reaction, selective *N*-acylation of 4-bromophenylhydrazine has occurred in presence of a small amount of DMAP in dry pyridine [52].

4.4 MacMillan's Imidazolidinone Organocatalysts

MacMillan's catalyst was developed by Professor David MacMillan at Caltech, generally imidazolidinone-based organocatalysts are named MacMillan's catalyst and are used in a variety of asymmetric transformations. The enantioselective Diels–Alder reaction involving organocatalyst (5*S*)-2,2,3-trimethyl-5-phenylmethyl-4-imidazolidinone monohydrochloride was first reported by MacMillan in his pioneering work in 2000 (Scheme 20).

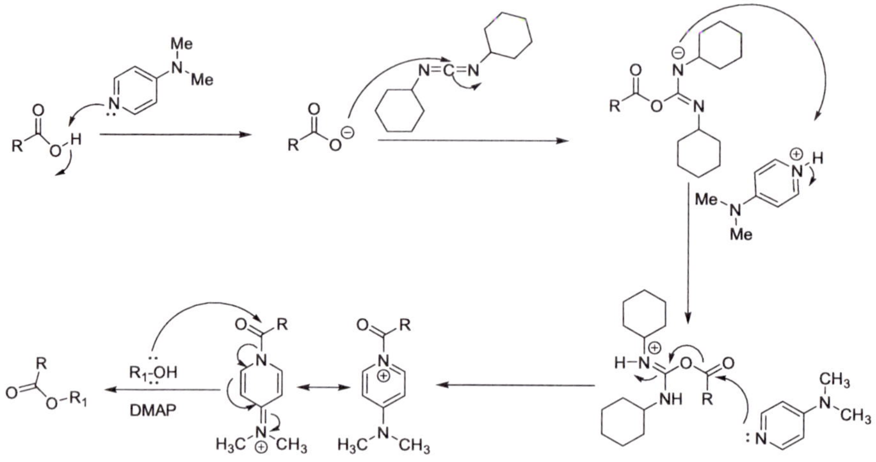

Scheme 17 Mechanism of Steglich esterification catalyzed by DMAP

Scheme 18 Selective *O*-acylation by DMAP

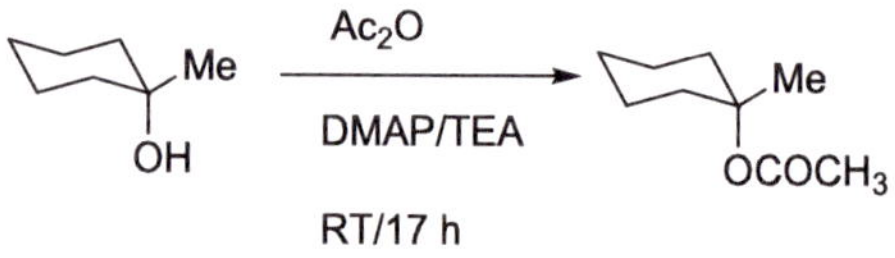

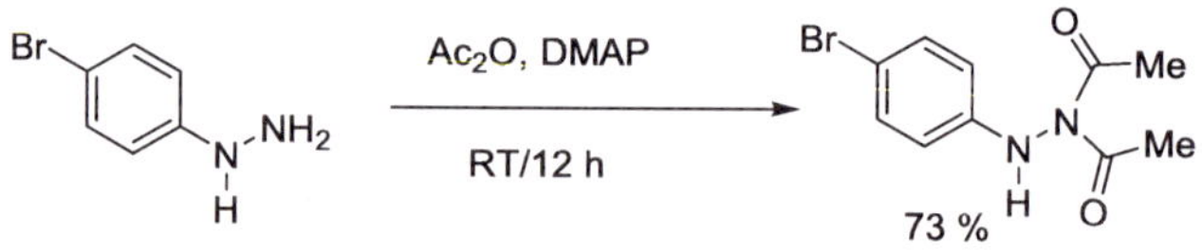

Scheme 19 Selective *N*-acylation of phenylhydrazine catalyzed by DMAP

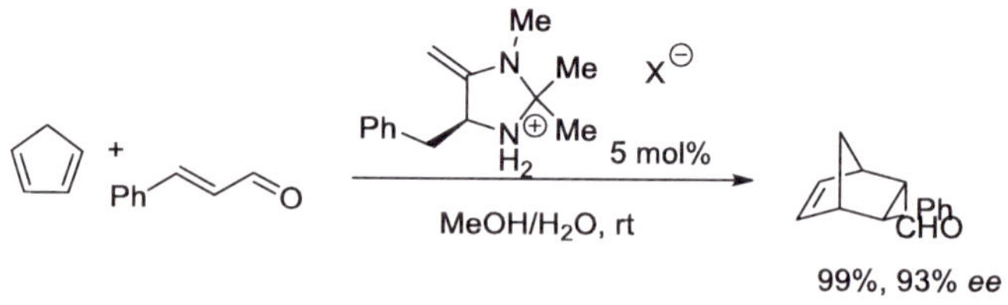

Scheme 20 Diels–Alder reaction using MacMillan's catalyst

Other organic transformative reactions, e.g., Friedel–Crafts alkylations [53], α-chlorinations [54] using MacMillan's catalyst, were also reported with high enantioselectivity. While observing the Friedel–Crafts alkylations of indoles, MacMillan

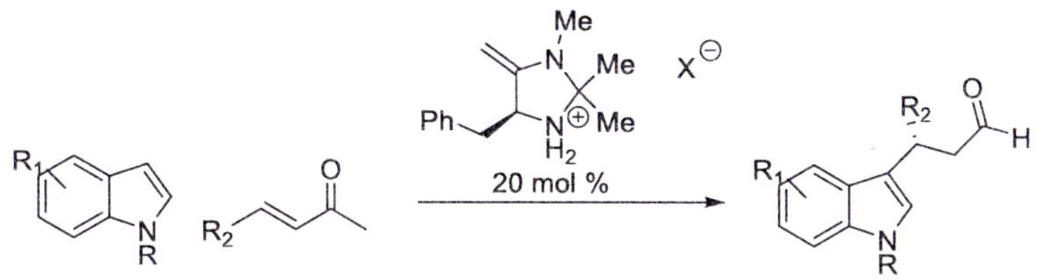

Scheme 21 Friedel–Crafts alkylation of indole using MacMillan's catalyst

found an optimized structure in (2*S*,5*S*)-(−)-2-*tert*-butyl-3-methyl-5-benzyl-4-imidazolidinone that can be shown in Scheme 21.

4.5 Thiourea

Among the organocatalysts, urea and thiourea are particularly important to accelerate organic transformation in past years. Thiourea is a green organic catalyst due to its nontoxic and metal-free, water-tolerant property. The interaction between the substrate and thiourea is purely hydrogen-bonding interaction, and unlike other organocatalyst, it participates in the reaction through non-covalent interaction (hydrogen-bonding). These small organic molecule catalyze a wide range of non-stereoselective and stereoselective reaction [55–65]. The hydrogen-bonding between the carbonyl substrate and thiourea occurs in two places because of the coplanar amino substituents in thiourea.

The above scheme represents the catalytic enantioselective Friedel–Crafts alkylation of indoles with nitroalkenes for the first time [62]. The formation of a double hydrogen bond by the organocatalyst plays a vital role in activating carbonyl groups and related compounds as shown in Fig. 2. The Friedel–Crafts alkylation reaction involving aromatic and heteroaromatic compounds with nitroalkenes and the activated double hydrogen-bonding motif present in the *bis*[3,5-*bis*(trifluoromethyl)phenyl]thiourea was first developed by Schreiner for Diels–Alder

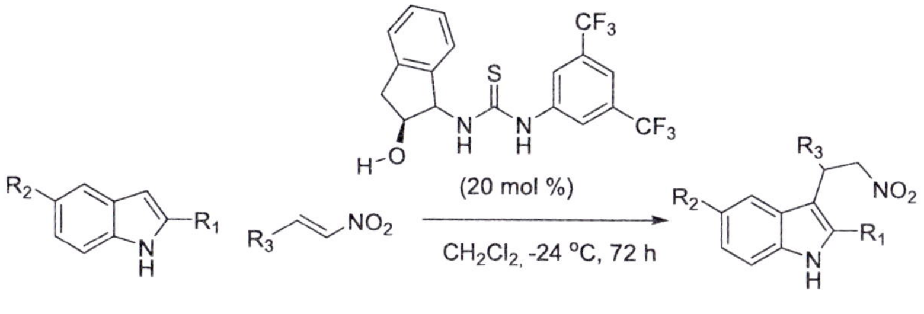

Scheme 22 Addition of indole to *trans*-*β*-nitrostyrene to give optically active 2-indolyl-1-nitro derivatives

Fig. 2 Possible bifunctional mode of action of the catalyst

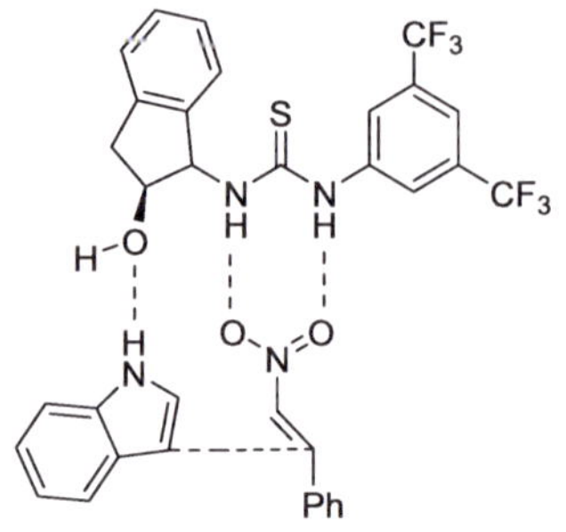

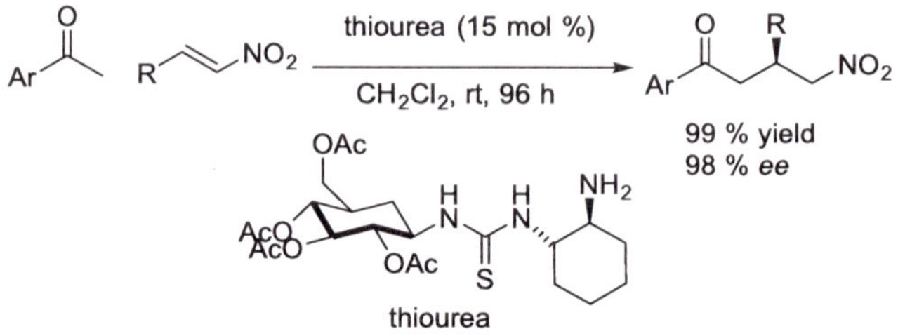

Scheme 23 Fructose derived thiourea catalyzed Michael addition reaction

reaction [60 (a,b)]. As the extension of this research work, thiourea catalyzed Michael addition was also tried by the chemists to find out the superiority of this organocatalyst (Scheme 23).

5 Conclusion

In conclusion, unlike metal catalysis organocatalysis provides a simple, nontoxic, mild, environmentally friendly, cost-effective methodology for drug designing and fine chemical synthesis. In view of the increasing interest in green chemistry, eco-friendly technologies for organic transformative reactions have attracted much consideration. The organic catalysts have versatile applications as a catalyst as well as they find commercial importance in industries also. Organocatalysts are going to be reconstructed or remodified in the future, and this will make an accessible catalytic activity in the industry for environmentally benign and sustainable technologies. In this book, chapter 36 schemes on organocatalytic reaction is represented which will be helpful for readers to make a clear conception about green organic catalysts.

References

1. Lancaster M (2002) Green Chemistry an introductory text. The Royal Society of Chemistry, Cambridge, UK
2. Anastas PT, Warner JC (1998) Green Chemistry: theory and practice. Oxford University Press, NewYork, USA
3. Clark JH (2001) Catalysis for green chemistry. Pure Appl Chem 73:103–111
4. Clark JH, Rhodes CN (2000) Clean synthesis using porous inorganic solid catalysts. RSC Clean Technology Monographs, Cambridge, UK
5. Dakin HD (1909) The catalytic action of amino-acids, peptide and proteins in effecting certain syntheses. J Biol Chem 7:4
6. Enders D, Grondal C, Hüttl MRM (2007) Asymmetric organocatalytic domino reactions. Angew Chem Int Ed 46:1570–1581
7. Jv L (1860) Ueber die Bildung des Oxamids aus Cyan. Annalen-Der-Chemie Und Pharmacie 113:246–247
8. Vavon MM, Peignier P (1929) L'application des alcalo¨ıdes dans la synth'ese organique. Bulletin de la Soci´et´e Chimique de France 45:293
9. Wegler R (1932) Uber die mitverschiedenerReaktionsgeschwindigkeiterfolgende Veresterung der optischen Antipoden eines Racemates durch opt. akt Katalysatoren. Justus Liebigs Annalen Der Chemie 498:62–73
10. Pizzarello S, Weber AL (2004) Prebiotic amino acids as asymmetric catalysts. Science 303:1151
11. Bredig G, Fiske PS (1912) BiochemischeZeitschrift 46:7
12. Wynberg H (1986) Asymmetric catalysis by alkaloids. Top Stereochem 16:87–12
13. Wynberg H (1985) Catalytic asymmetric synthesis of chiral 4substituted 2-oxetanones. J Org Chem 50:1977–2197
14. List B, Lerner RA, Barbas CF (2000) Proline-catalyzed direct asymmetric aldol reactions. J Am ChemSoc 122:2395–2396
15. Bui T, Barbas CF (2000) A proline-catalyzed asymmetric Robinsonannulation reaction. Tetrahedron Lett 41:6951–6954
16. Hajos ZG, Parrish DR (1974) Synthesis and conversion of 2-methyl-2-(3-oxobutyl)-1,3-cyclopentanedione to the isomeric racemic ketols of the [3.2.1]bicyclooctane and of the perhydroindane series. J Org Chem 39:1615–1621
17. Bredig G, Fiske PS (1912) Durch Katalysatoren bewirkte asymmetrische Synthese. Biochem Z 46:7–23
18. Shu L, Shi Y (2000) An efficient ketone-catalyzed epoxidation using hydrogen peroxide as oxidant. J Org Chem 65:8807–8810
19. Corey EJ, Xu F, Noe MC (1997) A rational approach to catalytic enantioselective enolate alkylation using a structurally rigidified and defined chiral quaternary ammoniumsalt under phase transfer conditions. J Am Chem Soc 119:12414–12415
20. Sellergren B, Karmalkar RN, Shea KJ (2000) Enantioselective ester hydrolysis catalyzed by imprinted polymers. J Org Chem 65:4009–4027
21. Ogstone AG (1958) Nature 181:1462–1465
22. Yamada S-I, Otani G (1969) Asymmetric synthesis with amino acid II asymmetric synthesis of optically active 4,4-disubstituted-cyclohexenone. Tetrahedron Lett 10:4237–4240
23. Hajos ZG, Parrish DR (1971) United States Patent US, 3975442
24. Hajos ZG, Parrish DR (1974) Asymmetric synthesis of bicyclic intermediates of natural product chemistry. J Org Chem 39:1615–1621
25. Hajos ZG, Parrish DR (1971) German Patent DE, 2102623
26. Hajos ZG, Parrish DR (1971) United States Patent, US, 3975442
27. Vignola N, List B (2004) Catalytic asymmetric intramolecular α-alkylation of aldehydes. J Am Chem Soc 126:450–545
28. List B (2000) The direct catalytic asymmetric three-component Mannich reaction. J Am Chem Soc 122:9336–9337

29. Alexakis A, Andrey O (2002) Diamine-catalyzed asymmetric Michael additions of aldehydes and ketones to nitrostyrene. Org Lett 4:3611–3614
30. Ramachary B, Anebouselvy K, Chowdari Naidu S et al (2004) Direct organocatalytic asymmetric heterodomino reactions: the knoevenagel/diels alder/epimerization sequence for the highly diastereoselective synthesis of symmetrical and nonsymmetrical synthons of benzoannelated centropolyquinanes. J Org Chem 69:5839–6584
31. Eder U, Sauer G, Wiechert R (1971) Hajos-Parrish-Eder-Sauer Wiechert reaction. Angew Chem Int Ed 10:496–497
32. Alexakis A, Andrey O (2002) Diamine-catalyzed asymmetric Michael additions of aldehydes and ketones to nitrostyren. Org Lett 4:3611–3614
33. List B, Lerner RA, Barbas CF (2000) Proline-catalyzed direct asymmetric aldol reactions. J Am Chem Soc 122:2395–2396
34. Aznar F, Garca A-B, Cabal M-P (2006) Proline-catalyzed imino-Diels–Alder reactions: synthesis of meso-2,6-diaryl-4-piperidones. Adv Synth Catal 348:2443–2448
35. Calderon F, Fernandez R, Sanchez F et al (2005) Asymmetric aldol reaction using immobilized proline on mesoporous support. Adv Synth Catal 347:1395–1403
36. Bortolini O, Cavazzini A, Giovannini PP et al (2013) A combined kinetic and thermodynamic approach for the interpretation of continuous-flow heterogeneous catalytic processes. Chem Eur J 19:7802–7808
37. Moore JL, Rovis T (2010) Lewis base catalysts 6: carbene catalysts. Top Curr Chem 291:1–51
38. Xu LW et al (2005) Efficient and mild benzoin condensation reaction catalyzed by simple 1-N-alkyl-3-methylimidazolium salts. Tetrahedron Lett 46:5317–5320
39. Iwamoto K et al (2006) Benzoin reaction in water as an aqueous medium catalyzed by benzimidazolium salt. Tetrahedron Lett 47:7175–7177
40. Jin MY, Kim SM, Mao H et al (2014) Chemoselective and repetitive intermolecular cross-acyloin condensation reactions between a variety of aromatic and aliphatic aldehydes using a robust N-heterocyclic carbene catalyst. Org Biomol Chem 12:1547–2155
41. Thai K, Langdon SM, Bilodeau F et al (2013) Highly chemo- and enantioselective cross-benzoin reaction of aliphatic aldehydes and α-ketoesters. Org Lett 15:2214–2217
42. Nair V, Varghese V, Paul RR et al (2010) NHC catalyzed transformation of aromatic aldehydes to acids by carbon dioxide: an unexpected reaction. Org. Lett 12:2653–2655
43. Sakakura A, Kawajiri K, Ohkubo T et al (2007) Widely useful DMAP-catalyzed esterification under auxiliary base- and solvent-free conditions. J Am Chem Soc 129:14775–21477
44. Ranarsson U, Grehn L (1998) Novel amine Chemistry based on DMAP-catalyzed acylation. Acc Chem Res 31:494–501
45. Xu S, Held I, Kempf B, Mayr H et al (2005) The DMAP-catalyzed acetylation of alcohols-a mechanistic study (DMAP = 4-(dimethylamino)-pyridine). Chem Eur J 11:4751–4757
46. Scriven EFV (1983) 4-Dialkylaminopyridines: super acylation and alkylation catalysts. Chem Soc Rev 12:129–161
47. Berry DJ, Charles VD, Stephanie SM et al (2001) Catalysis by 4-dialkylaminopyridines. Arkivoc 5:201–226
48. Steglich W, Hoefle (1969) G *N,* N-dimethyl-4-pyridinamine, a very effective acylation catalyst. Angew Chem Int Ed 8:981–981
49. Held I, Von den Hoff P, Stephenson DS et al (2008) Domino catalysis in the direct conversion of carboxylic acids to esters. Adv Synth Cat 11:1891–1900
50. Hoefle G, Steglich W et al (1972) 4-Dialkylaminopyridines as acylation catalysts; III[1] Acylation of Sterically Hindered Alcohols. Synthesis 11:619–621
51. Banwell MG et al (1995) Trifluoromethanesulfonic anhydride–4-(*N*, N-dimethylamino)pyridine as a reagent combination for effecting Bischler-Napieraiski cyclisation under mild conditions: application to total syntheses of the*Amaryllidaceae* alkaloids N-methylcrinasiadine, anhydrolycorinone, hippadine and oxoassoanine. J Chem Soc Chem Commun 24:2551–2553
52. Nicolaou KC, Snyder SA, Montagnon T et al (2002) The Diels-Alder reaction in total synthesis. Angew Chem Int Ed 41:1668–1698

53. Gaunt MJ, Johansson CCC, McNally A et al (2007) Enantioselective organocatalysis. Drug Discovery Today 12:8–27
54. Wende RC, Schreiner PR (2012) Evolution of asymmetric organocatalysis: multi- and retrocatalysis. Green Chem 14:1821–2184
55. Mike K, Schreiner PR (2009) (Thio)ureaOrganocatalysts. In: Pihko PM (ed) Hydrogen bonding in organic synthesis. Wiely, New York.
56. Wittkopp A, Schreiner PR (2000) The Chemistry of Dienes and Polyenes. In: Rappoport Z (ed) Diels-Alder reactions in water and in hydrogen-bonding environments. Wiley Inc, New York
57. Schreiner PR (2003) Metal-free organocatalysis through explicit hydrogen bonding interactions. Chem Soc Rev 32:289–296
58. Taylor MS, Jacobsen EN (2006) Asymmetric catalysis by chiral hydrogen-bond donors. Angew Chem Intd Ed 45:1520–1543
59. Curran DP, Kuo LH (1995) Acceleration of a dipolar Claisen rearrangement by Hydrogen bonding to a soluble diaryl^urea. Tetrahedron Lett 36:6647–6650
60. (a) Liu K, Cui HF, Nie J et al (2007) Highly enantioselective michael addition of aromatic ketones to nitroolefins promoted by chiral bifunctional primary amine-thiourea catalysts based on saccharide. Org Lett 9:923–925 (b) Okino T, Hoashi Y, Takemoto Y (2003) Enantioselective Michael reaction of malonates to nitroolefins catalyzed by bifunctional organocatalysts. J Am Chem Soc 125:12672–12673
61. Lee Y, Klausen RS, Jacobsen EN (2011) Thiourea-catalyzed enantioselective iso-pictet–spengler reactions. Org Lett 13:5564–5567
62. Herrera RP, Sgarzani V, Bernardi L et al (2005) Catalytic enantioselective Friedel-crafts alkylation of indoles with nitroalkenes by using a simple thiourea organocatalyst. Angew Chem Int Ed 44:6576–6657
63. Sohtome Y, Takemura N, Takagi R et al (2008) Thiourea-catalyzed Morita–Baylis–Hillman reaction. Tetrahedron 64:9423–9942
64. Wittkopp A, Schreiner PR (2003) Metal-free, noncovalent catalysis of Diels-Alder reactions by neutral hydrogen bond donors in organic solvents and in water. Chem Eur J 9:407–414
65. Yoon TP, Jacobsen EN (2005) Highly enantioselective thiourea-catalyzed nitro-Mannich reaction. Angew Chem Int Ed 44:466–468

Chapter 10
Environmentally Benign Organic Solvent: A Green Approach

Bijeta Mitra, Gyan Chandra Pariyar, and Pranab Ghosh

1 Introduction

Most of the solvents are originated from non-renewable sources like petroleum, which is incompatible to the very basic principle of "Green Chemistry" [1]. It is now widely renowned that for the development of sustainable and green synthetic methodologies there is a very basic requirement of scientific and potential alternation of volatile organic compounds (VOCs) as solvents and reaction media [2]. The use of greener solvent which should be environmentally benign, non-toxic, non-volatile, inflammable, inexpensive, bio-degradable, and recyclable is enormously inspiring in modern synthetic organic chemistry as it involves in designing greener synthetic protocols [3, 4]. With these points in mind, there is the exploration of alternative green solvents which would be superior for human health and environment.

Use of non-conventional [5] reaction media in organic synthesis is a constraint for making the mother earth green. Besides water, the most common of these solvent systems include as ionic liquids (ILs) and fluorous solvents are generally considered as green, but they do not convince these criteria and often acquire problems with high cost, unknown toxicity, and poor or no biodegradability. In the past decades, ionic liquids have gained a focus of rising interest for many organic transformations.

Herein, we focus on the organic synthesis using unconventional biodegradable solvents, such as polyethylene glycols (PEGs) and some solvents derived from biomass viz glycerol, ethyl lactate, gluconic acid aqueous solution (GAAS), and

B. Mitra · P. Ghosh (✉)
Department of Chemistry, University of North Bengal, Siliguri, West Bengal, India
e-mail: pizy12@yahoo.com

B. Mitra
e-mail: bijetamitra50@gmail.com

G. C. Pariyar
Department of Food Technology, University of North Bengal, Siliguri, West Bengal, India
e-mail: pariyarg@gmail.com

G. Anilkumar and S. Saranya (eds.), *Green Organic Reactions*, Materials Horizons: From Nature to Nanomaterials, https://doi.org/10.1007/978-981-33-6897-2_10

ethanol. Using bio-based derivatives is a current paradigm of green chemistry. [6, 7] Solvents derived from biomass have enormous potential for use in medicinal chemistry and pharmaceutical industry with many advantages over conventional organic solvent [8]. Application of green solvents in organic chemistry often helps with recovery of valuable transition metal catalysts [9].

2 Organic Synthesis in Polyethylene Glycol

Application of PEGs (molecular wt. 200–6000) as alternative medium for organic synthesis is exemplified below.

A facile protocol for the synthesis of pyrrolo[1,2-*a*]quinoxalines **1** using benzylamine derivatives and 1-(2-aminoaryl) pyrrole has been described by Patil et al. [10] (Scheme 1). This synthetic protocol involves oxidation of benzylamine derivatives into the respective aldehydes followed by condensation with 1-(2-aminoaryl)pyrrole in PEG-400 as a green and reusable solvent using $K_2S_2O_8$ as an oxidant.

Another green protocol was designed by Akbarzadeh et al. for the synthesis of densely functionalized 2-aminothiophene derivatives **2** through one-pot three-component reaction from enolizable carbonyl compounds, malononitrile or ethyl cyanoacetate and elemental sulfur in PEG-600, as an eco-friendly reaction medium, without using any basic catalyst under ultrasonic with the recovery and reusability of the solvent at least five times without significant loss of its activity (Scheme 2) [11].

A novel Co(III)-catalyzed oxidative annulations of aromatic aldehydes with internal alkynes for accessing isocoumarins **3** is described by Tao et al. [12]

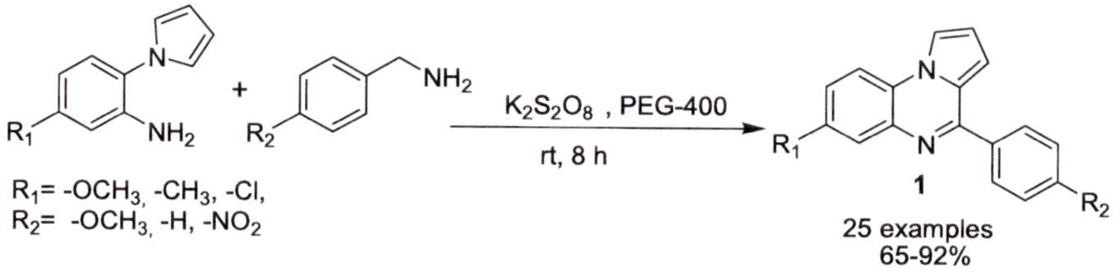

Scheme 1 PEG-400 mediated synthesis of pyrrolo[1,2-*a*] quinoxalines

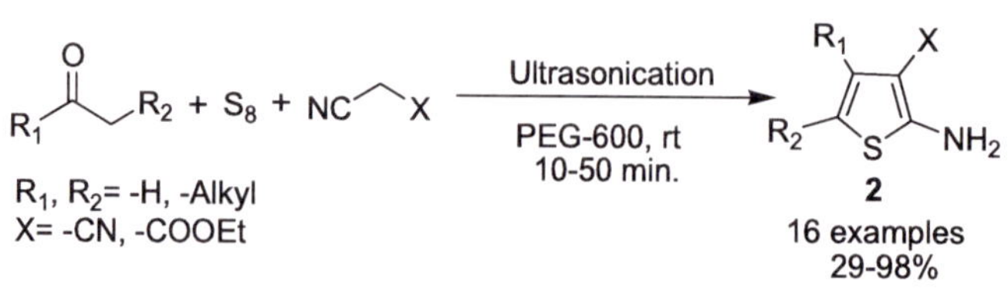

Scheme 2 PEG-600 mediated method for the synthesis of 2-aminothiophene derivatives

(Scheme 3) which is achieved by oxidation, weak chelation-assisted C-H bond functionalization, and annulation cascades with excellent functional group compatibility, high atom economy, and step efficiency by using environmentally benign solvent PEG-400. This is the first recyclable Co(III)/CuO/PEG-400 system for oxidative annulation of aromatic aldehydes with internal alkynes involving C-H functionalization.

A ligand, base, metal oxidant free Pd-catalyzed C(sp^2)–H activation, followed by isocyanide insertion to synthesize indolizine **5** and imidazoline-fused heterocyclics **4** in PEG-400 as a recyclable solvent was designed by Kumar et al. for the first time in the organic synthetic literature (Scheme 4) [13]. The present transformation is considered as a greener strategy using oxygen as external oxidant.

A facile strategy for the synthesis of pyrazolopyrano pyrimidines **6** via one-pot, four-component reaction of barbituric acid, hydrazine hydrate, aromatic aldehydes, and ethyl acetoacetate in PEG-400 as a green solvent in absence of any additional catalyst at ambient temperature has been described by Kardooni et al. [14] (Scheme 5).

A robust one-pot synthesis of amide **7** derivatives was achieved by Liang et al. in good yields via the direct oxidative amidation of aldehydes with amines under PEG-400 as solvent and NaOCl/Bu_4NHSO_4 as an oxidant system (Scheme 6) [15].

Bi et al. have synthesized a series of α-mono-fluorinated acetoacetamides **8** via selective α-electrophilic mono-fluorination under mild condition with industrialized selectfluor as the F^+ source in PEG-400 (Scheme 7) [16]. This protocol avoided the use of base or metal catalyst and in most cases proceeded in nearly quantitative

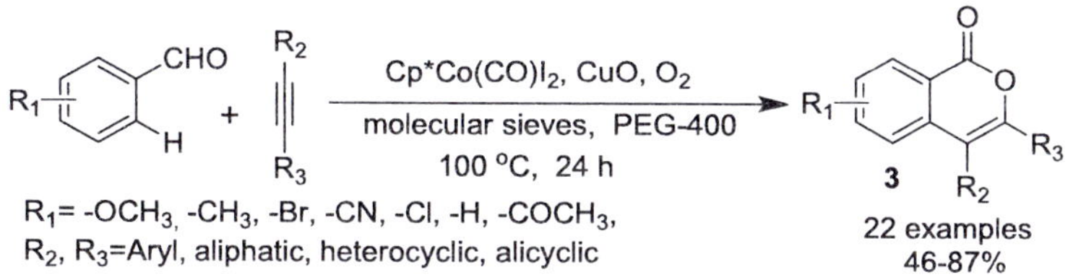

Scheme 3 Cobalt(III)-catalyzed oxidative annulation via C-H functionalization in PEG-400

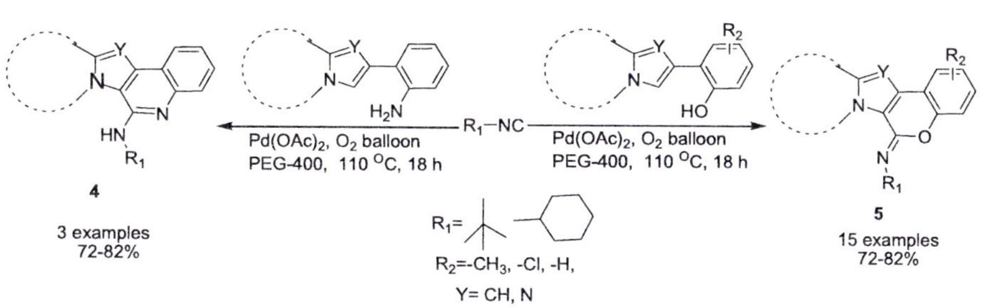

Scheme 4 C(sp^2)-H activation and isocyanide insertion in PEG-400

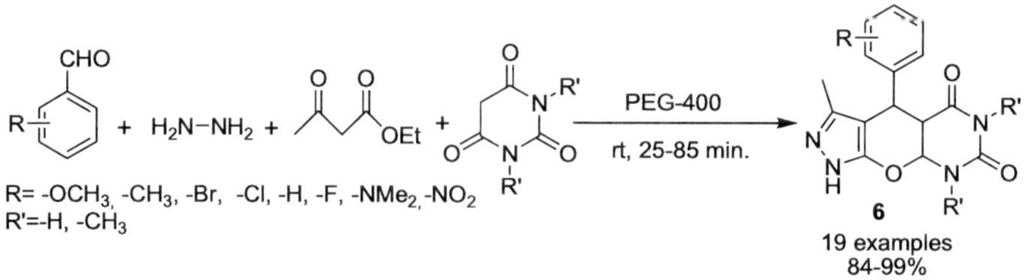

Scheme 5 Synthesis of pyrazolopyranopyrimidines in PEG-400

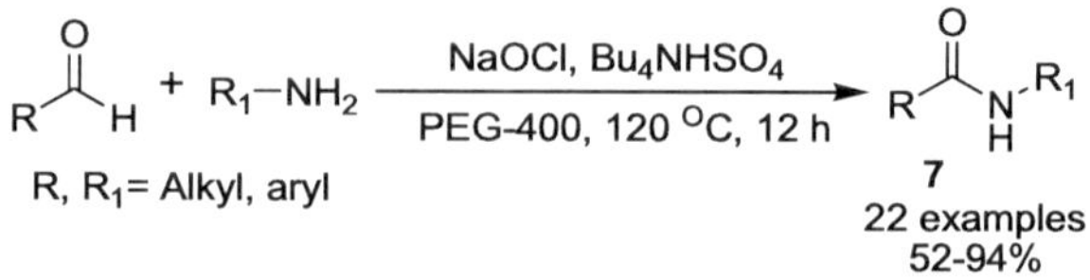

Scheme 6 Metal-free synthesis of amides by oxidative amidation in PEG-400

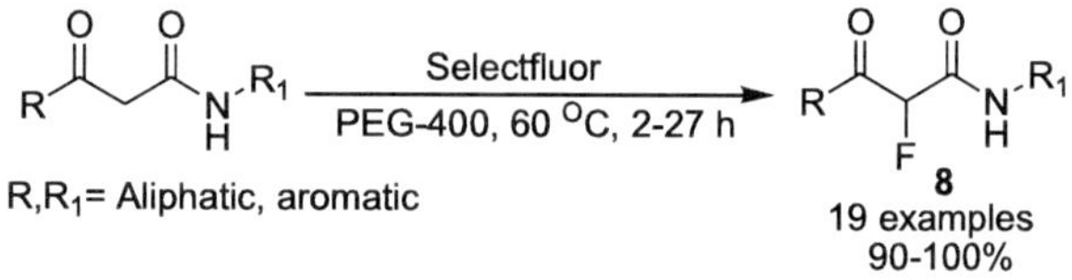

Scheme 7 Electrophilic mono-fluorination in PEG-400

conversions regardless of the electronic nature of the diversity substituent. The ratio of mono- and di-fluorinated products was more than 30:1.

Khan et al. demonstrated an environmentally benign one-pot three-component protocol for the synthesis of a series of 2-aminoselenopyridine **9** derivatives from easily accessible materials aldehydes, malononitrile, and benzeneselenol in PEG-400 using ultrasound (Scheme 8) [17]. In this process, a total of four new bonds such as one C–N, one C–Se, and two C–C were formed in one pot.

Ghom et al. first time designed a metal-free and greener approach for the one-pot direct iodination and dehydrogenation of dihydrobenzo[*a*]carbazoles using periodic acid in PEG-400 with high regioselectivity (Scheme 9) [18]. The solvent has been successfully recycled up to five times without any loss of activity in an aqueous

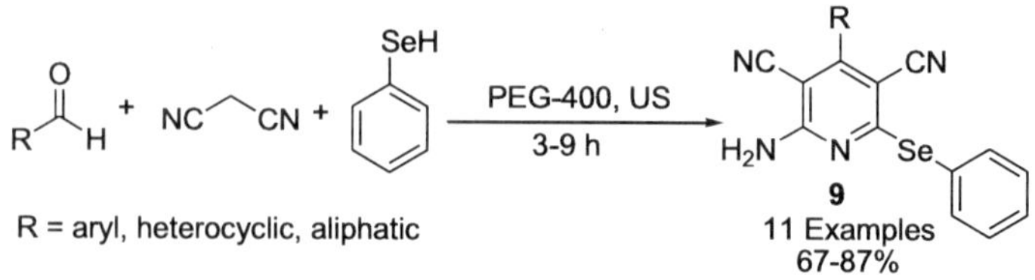

Scheme 8 Synthesis of highly functionalized selenopyridines in PEG-400

Scheme 9 A one-pot direct regioselective iodination using periodic acid in PEG-400

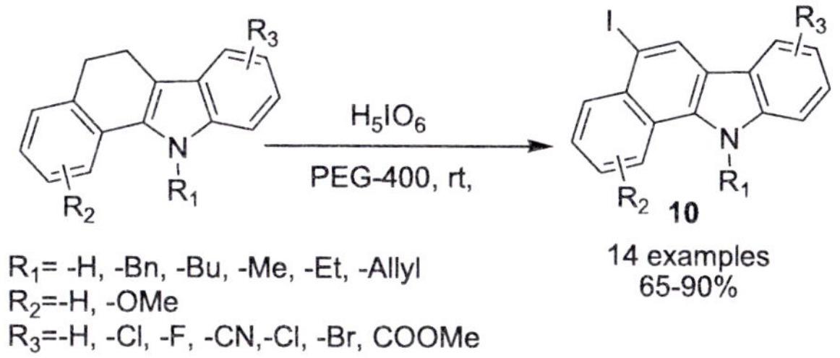

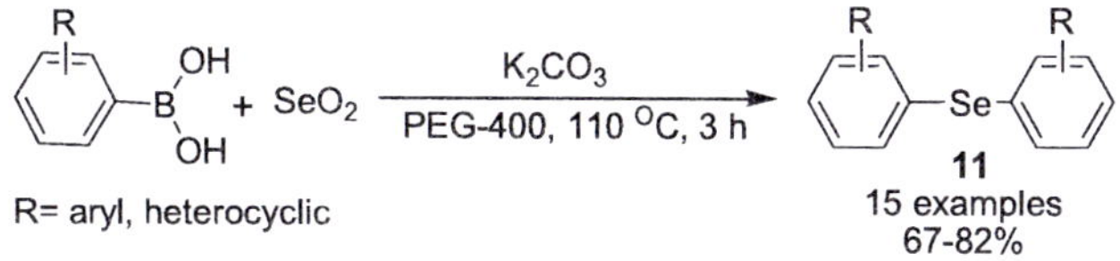

Scheme 10 Metal-free synthesis of diaryl selenides using SeO_2 as a selenium source

medium. In addition, the reaction discovers a new path for green synthesis of iodo-derived benzocarbazoles.

Another eco-friendly synthetic protocol has been developed for designing diaryl selenium compounds **11** by Kumar et al. (Scheme 10) [19]. Compared with other selenium sources the new source SeO_2 gives a superior low-cost synthetic route, which may be useful for large-scale synthesis.

3 Organic Synthesis in Glycerol

Glycerol has been used as a medium in organic synthesis often being more efficient than conventional organic solvents, and its applications are summarized below.

Singh et al. demonstrated one-pot three-component tandem approach for the preparation of a bioactive hybrid scaffold pyrido[2,3-*d*]pyrimidines **12** in glycerol, a biodegradable, and reusable promoting medium derived by the reaction of 6-amino-1-methyluracil, aldehydes, and malononitrile (Scheme 11) [20]. The advantages of this methodology include 100% atom economy, broad substrate scope, high yields, and eco-friendliness.

Mitra et al. promoted one-pot multi-component strategy which contribute to sustainability by exploring the green solvent glycerol to design the diverse biologically important scaffold thioamide **13** (Scheme 12) and 4*H*-thiopyran **14** (Scheme 13) [21] in good yields without any supplementary catalyst, the reaction time for these approaches was also shorter, and they have reported a wide range of compounds. Thioamide was designed by the one-pot three-component reaction of aldehydes, amines, and molecular sulfur while thiopyran has been developed by

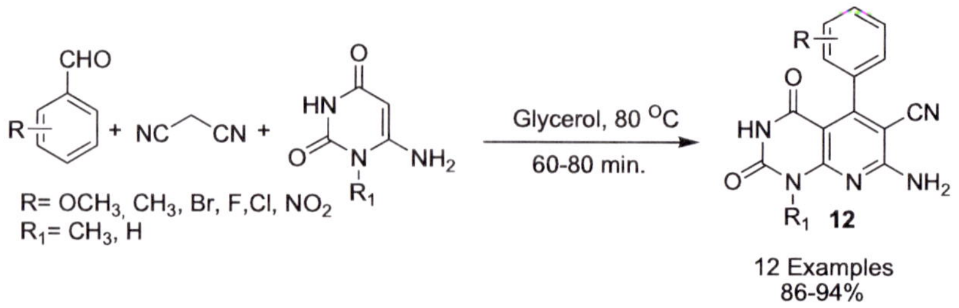

Scheme 11 General strategy for the synthesis of pyrido[2,3-*d*]pyrimidines in glycerol

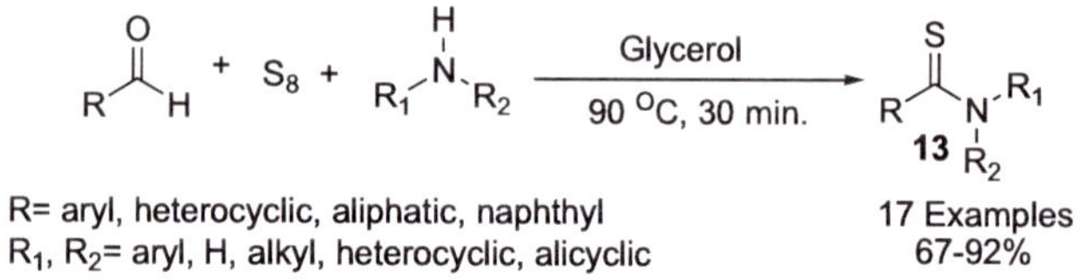

Scheme 12 Catalyst-free synthesis of thioamide from aldehydes and amines in glycerol

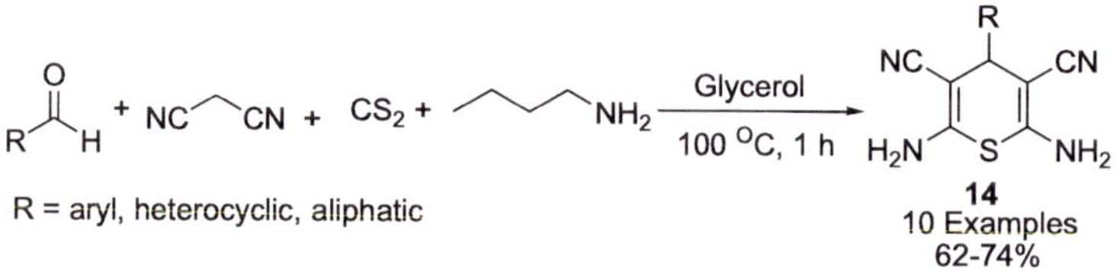

Scheme 13 Catalyst-free one-pot four-component synthesis of 4*H*-thiopyran in glycerol

one-pot four-component synthesis of aldehyde, malononitrile, carbon disulfide, and butylamine.

Nagasundaram et al. described glycerol mediated eco-friendly protocol for the expedient access of diverse array of 4,4′-(arylmethylene)bis(3-methyl-1*H*-pyrazol-5-ol) **15** and 2-aryl-2,3-dihydroquinazolin-4(1*H*)-one **16** motifs under catalyst-free conditions. For these approaches, a mixture of ethyl acetoacetate, hydrazine hydrate, and aromatic aldehyde afford the corresponding 4,4′-(arylmethylene)bis(3-methyl-1*H*-pyrazol-5-ols) (Scheme 14a) [22] and a mixture of aromatic aldehyde or ketone and anthranilamide in glycerol produce dihydroquinazolin-4(1*H*)-one (Scheme 14b) [22].

One-pot, three-component condensation reaction between aromatic aldehydes, *α*-naphthol and (phenylsulfonyl)acetonitrile for the synthesis of 4-(aryl)-3-(phenylsulfonyl)-4*H*-benzo[*h*]chromen-2-amine **17** derivatives has been developed in glycerol, which is a greener inexpensive and non-toxic compound, by Morshedi et al. (Scheme 15) [23]. The method involves domino Knoevenagel condensation/Michael addition and cyclization cascade.

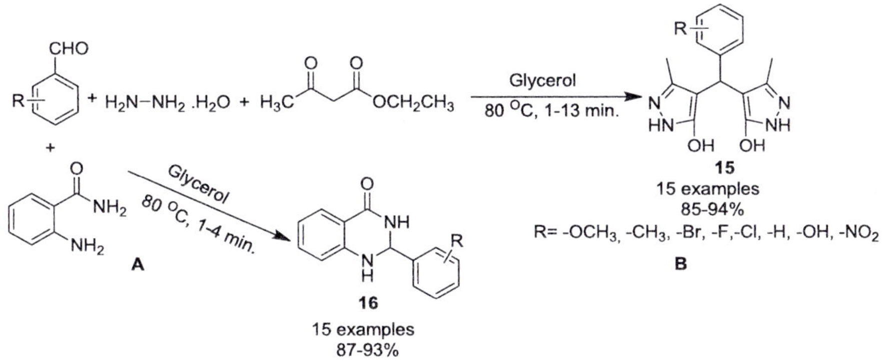

Scheme 14 Glycerol-assisted synthesis

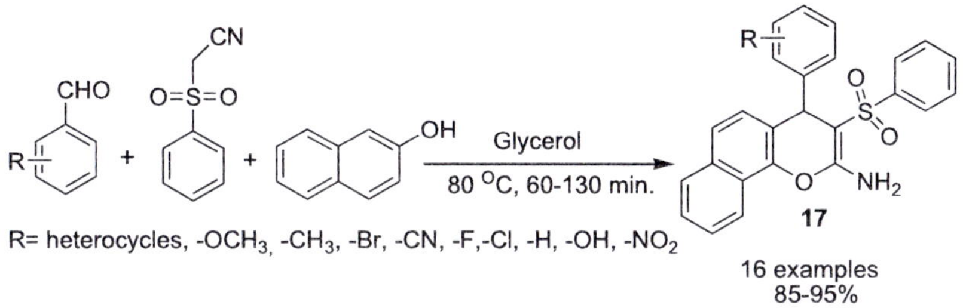

Scheme 15 Synthesis of 4-(aryl)-3-(phenylsulfonyl)-4*H*-benzo[*h*]chromen-2-amine derivatives

Nickel-catalyzed Suzuki coupling reactions of phenyl boronic acids with aryl diazonium salts in glycerol as a reaction medium are discussed by Bhojane et al. [24] (Scheme 16). Various aryl diazonium salts were efficiently reacted with aryl boronic acids under optimized conditions to give the respective diaryl compounds **18** in good yields.

A copper-catalyzed ligand-free C-S cross-coupling reaction of aryl halides with thiols has been performed in the presence of glycerol to construct diversified functionalized thiols **19** with good chemoselectivity by Dubey et al. [25] (Scheme 17). The catalytic system is found to be active for trickier and less reactive aryl bromides.

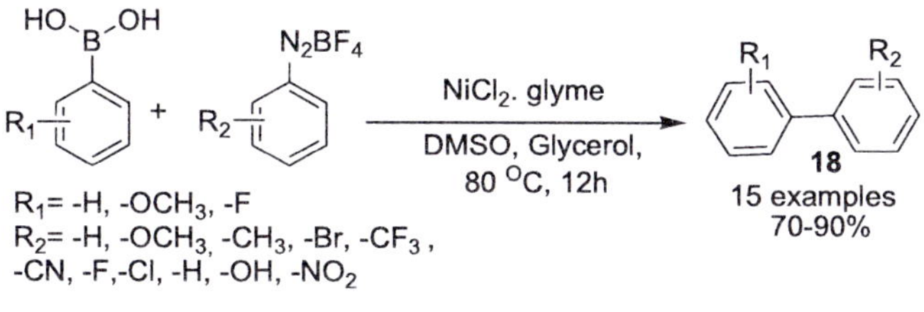

Scheme 16 Nickel-catalyzed Suzuki coupling reactions using aryl diazonium salts in glycerol

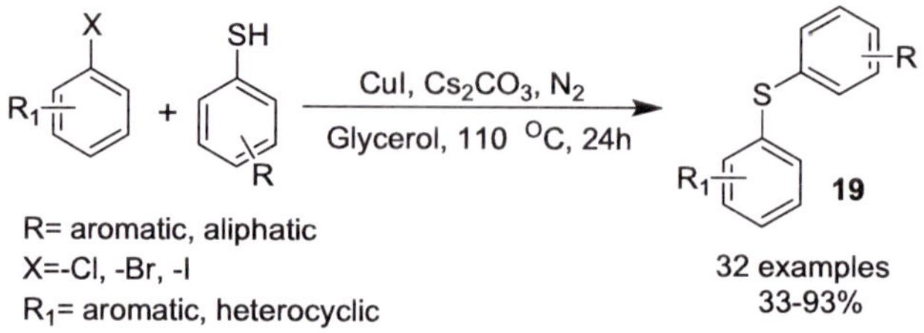

Scheme 17 Copper-mediated ligand-free C-S cross-coupling reaction in glycerol

Scheme 18 Synthesis of 1,4-disubstituted 1,2,3-triazoles

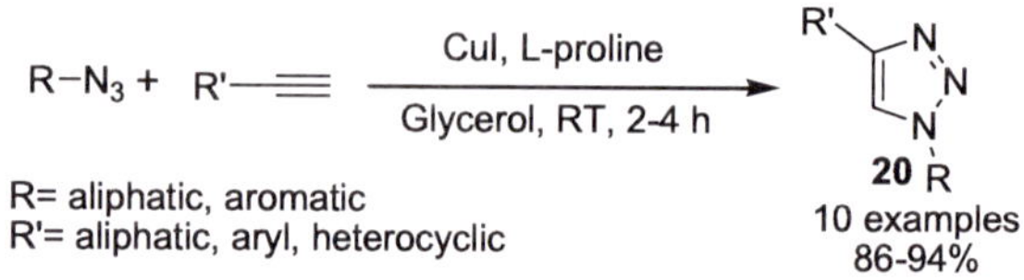

Scheme 19 Synthesis of nitriles from aldehydes using glycerol

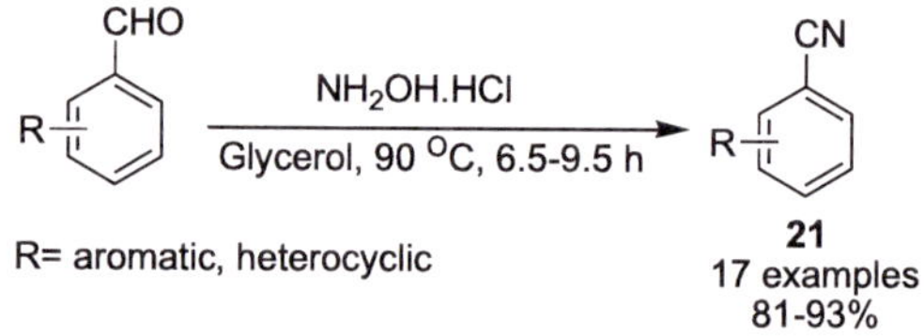

Synthesis of 1,4-disubstituted 1,2,3-triazoles **20** through CuAAC reaction is known as click reaction. Pasupuleti et al. [26] (Scheme 18) reported the first CuI catalyzed click reaction without an inert atmosphere by employing the CuI/*L*-proline system in glycerol. This approach may help both academic as well as industries for production of triazole moiety.

Ingale et al. demonstrated one-pot catalyst-free protocol for the synthesis of nitriles **21** from aldehydes by using hydroxyl amine hydrochlorides in glycerol solvent (Scheme 19) [27]. This protocol was efficiently used for the transformation of aromatic aldehydes bearing electron-withdrawing and electron-donating groups into aryl nitriles in good yields.

4 Organic Reactions in Ethanol

Ethanol is also considered as a major solvent in the synthesis of organic compounds, and hence, it is regarded as a vital component in a number of manufacturing industries. Some of the synthesis in organic chemistry using ethanol as a solvent can be summarized as follows.

Babu et al. used ethanol as an environmental friendly solvent for the synthesis of 1,2-disubstituted acetylenes **22** from organoboron derivatives and alkynyl bromides with microwave irradiation in presence of CuI as a catalyst and K_3PO_4 as base (Scheme 20) [28]. This strategy does not need any ligand or sealed tube conditions and produce 1,2-diarylacetylenes in good yields.

One-pot four-component protocol was reported by Etivand et al. for the synthesis of imidazo[1,2-*a*]pyridines **23** by the reaction of glyoxal monohydrates, ethyl acetoacetate, hydrazine hydrate, and 2-aminopyridine (Scheme 21) [29]. This regioselective reaction was performed in presence of tetrapropyl ammonium bromide as a catalyst using ethanol as a solvent under reflux condition and high yield of the desired product was reported.

Ethanol has been reported to be used as an environmentally friendly solvent by Depa et al. in one-pot three-component coupling of *N*-alkyl anilines, indoles, and aldehydes with catalytic quantity of molecular iodine for the synthesis of 3-aminoalkyl indoles **24** at room temperature (Scheme 22) [30].

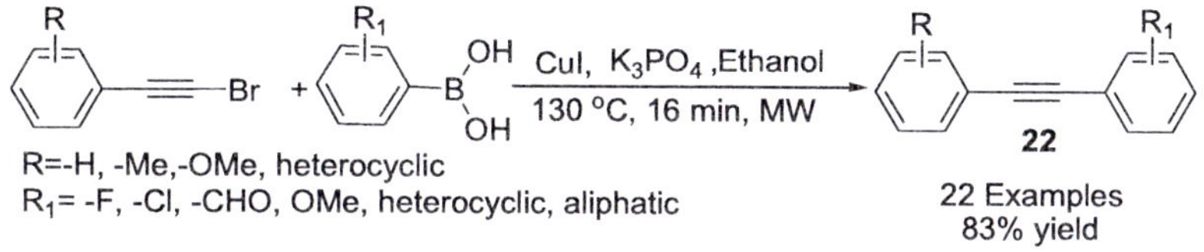

Scheme 20 Cu-catalyzed Suzuki coupling of alkynylbromides with boronic acids in ethanol

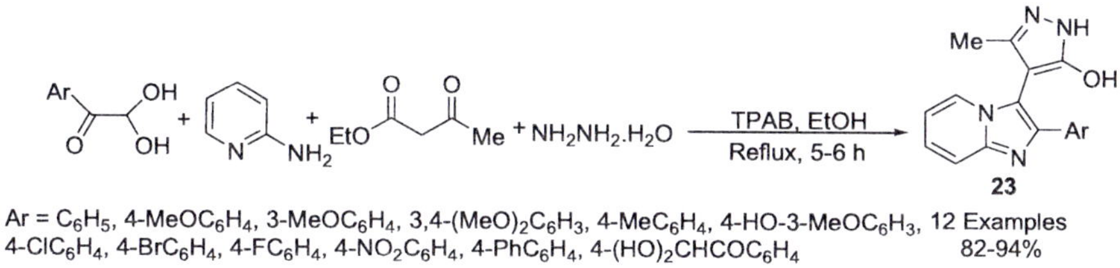

Scheme 21 One-pot, four-component synthesis of imidazo[1,2-*a*]pyridines in ethanol

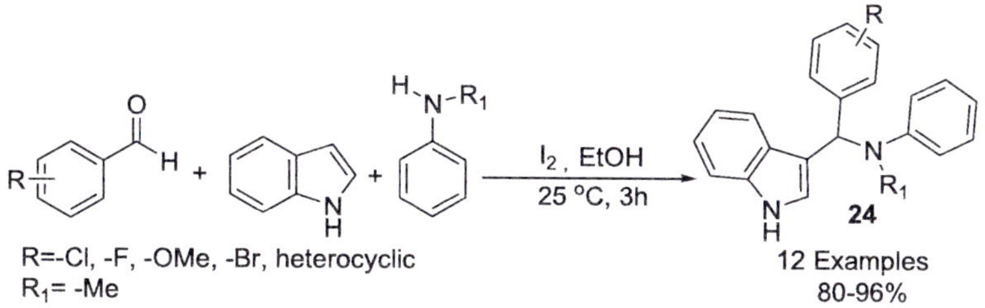

Scheme 22 Synthesis of 3-aminoalkyl indole in ethanol

Ghafuri et al. were successful to synthesize highly ordered sulfonated mesoporous graphite carbon nitride (ompg-C_3N_4/SO_3H) organocatalyst. The organocatalyst was then employed in one-pot synthesis of 2,3-dihydroquinazolin-4(1*H*)-ones **25** by condensation of anthranilamide with aldehydes or ketones. Ethanol played an important role of solvent during the course of the reaction (Scheme 23) [31].

Catalyst-free synthesis of 12-substituted-3,3-dimethyl-3,4,5,12-tetrahydrobenzo[*b*]acridine-1,6,11(2*H*)-triones **26** by cyclocondensation of 2-amino-1,4-naphthoquinone, aldehydes, and dimedone was achieved by Kamalifar et al. where ethanol was taken as a reaction medium considering it being an environment friendly solvent (Scheme 24) [32].

Alifu et al. introduced the synthesis of (*E*)-*N*-benzylidene-2-(hydroxymethyl)benzohydrazide derivatives **27** by reaction between phthalide, aldehydes, and hydrazine hydrate (Scheme 25) [33]. This one-pot synthesis was performed using environment friendly ethanol as solvent at 78 °C with wide range of compounds furnishing moderate to excellent yields.

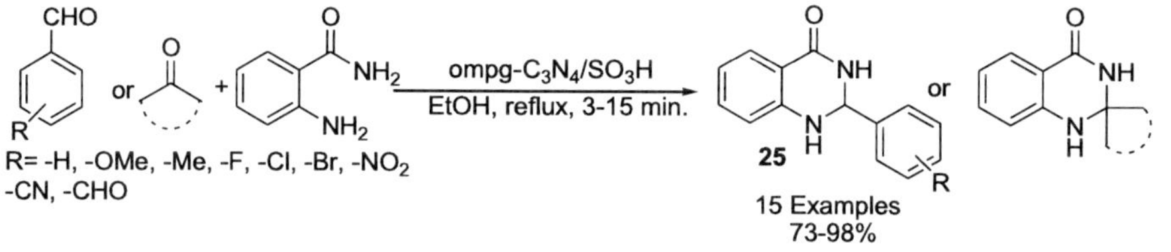

Scheme 23 Synthesis of 2,3-dihydroquinazolin-4(1*H*)-one derivatives in ethanol

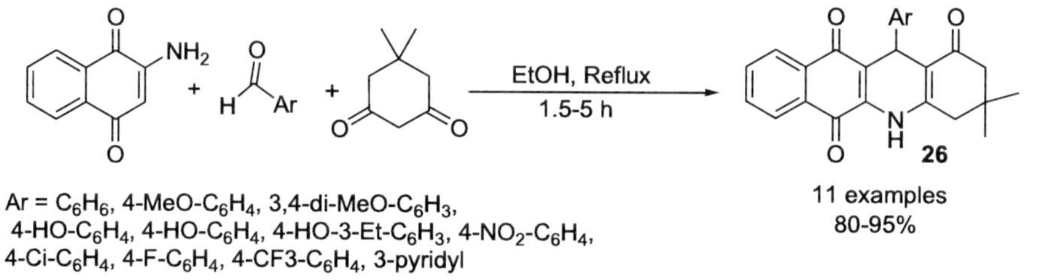

Scheme 24 Three-component synthesis of acridine-1,6,11(2*H*)-triones in ethanol

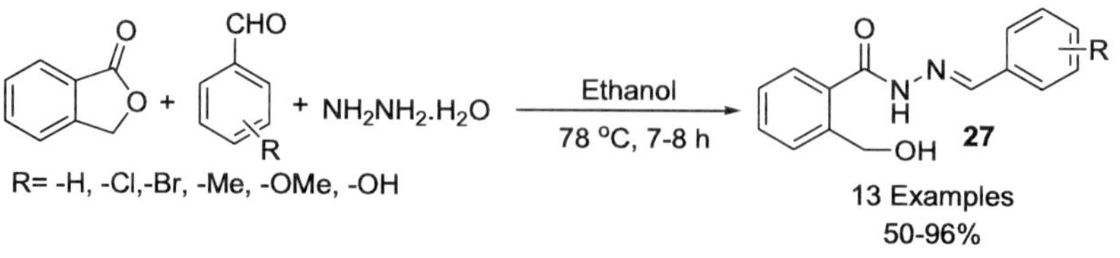

Scheme 25 Synthesis of *N*′-benzylidene-2-hydroxymethylbenzohydrazides in ethanol

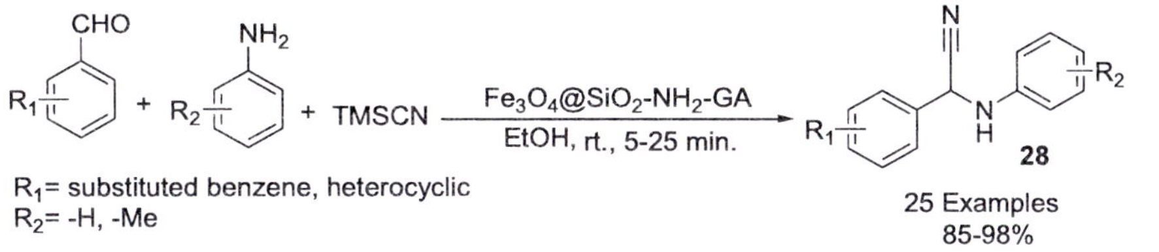

Scheme 26 Synthesis of α-aminonitriles in ethanol

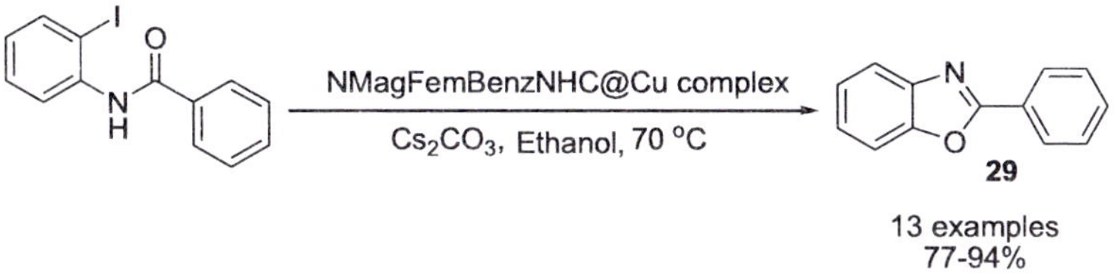

Scheme 27 Synthesis of benzoxazole via intramolecular *O*-arylation of *o*-iodoanilide

Maleki et al. prepared nano Fe_3O_4@SiO_2-NH_2-GA and explored this as an efficient catalyst for the preparation of α-aminonitriles **28** using one-pot Strecker reaction between aniline, aromatic amides and trimethylsilyl cyanide (Scheme 26) [34]. The catalyst was characterized using SEM–EDX, FT-IR, thermogravimetric analysis, and vibrating-sample magnetometer curve.

Naikwade et al. established the structure of NMagFemBenzNHC@Cu complex. This complex prepared was characterized by them using various analytical technologies such as FT-Raman, FT-IR, X-ray photoelectron spectroscopy, TEM, XRD, and vibrating sample magnetometer analysis. The complex was then exploited as a catalyst in intramolecular *O*-arylation of *o*-iodoanilides using heterogeneous conditions using ethanol as a green solvent (Scheme 27) [35] to design benzoxazole derivative **29**.

The protocol describes the method for synthesis of nano-cellulose-OSO_3H by the preparation of nano-cellulose as a support which was followed by treatment with chlorosulfonic acid. The nano-cellulose-OSO_3H prepared was then utilized for synthesis of 6-amino-2-(hydroxymethyl)-8-aryl-7-(phenylsulfonyl)pyrano[3,2-*b*]pyran-4(8*H*)-one derivatives from kojic acid, phenylsulfonylacetonitrile and aldehyde through one-pot multi-component reaction to get **30** in presence of ethanol under reflux conditions as reported by Sadeghi et al. [36] (Scheme 28).

5 Organic Reactions in Ethyl Lactate

Owing to the superior nature of ethyl lactate, the contribution of it as a medium for organic synthesis is exemplified below.

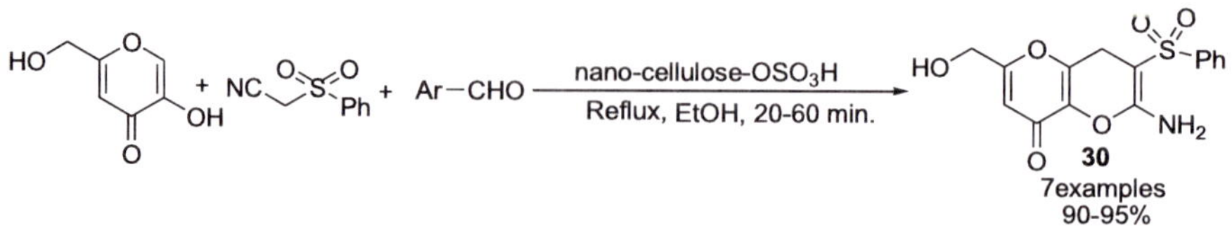

Scheme 28 Synthesis of pyran-4(8*H*)-one derivatives in ethanol

Gao et al. reported one-pot three-component condensation of various aldehydes, amines, and triethyl phosphate to produce α-aminophosphonates **31** in water-ethyl lactate under ultrasonic irradiation conditions at room temperature (Scheme 29) [37] without any additional catalyst.

A greener approach was developed by Shen et al. for the synthesis of 4(3*H*)-quinazolinones **32** using camphorsulfonic acid as a catalyst in an aqueous ethyl lactate solution (Scheme 30) [38]. Various types of quinazolinone derivatives were obtained by cyclization of 2-aminobenzamides with a wide range of 1,3-diketones via C–C bond cleavage in moderate to excellent yields. Ethyl lactate gave yields of a few times higher than that in conventional solvents.

Spirooxindole-pyran derivatives **33** were synthesized by Zhang et al. via one-pot, three-component reaction of enolizable C-H activated compounds, isatins, and malononitrile under visible-light irradiation in water–ethyl lactate at room temperature (Scheme 31) [39]. Advantages of this protocol involve absence of catalyst, no chromatographic separation, and applicability for large-scale synthesis.

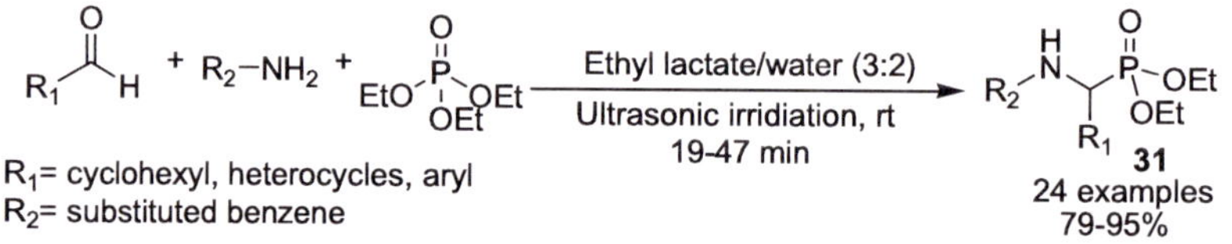

Scheme 29 One-pot synthesis of α-aminophosphonates in aqueous ethyl lactate

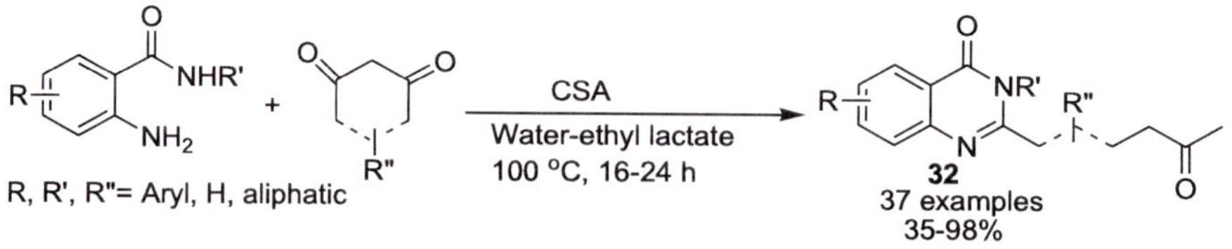

Scheme 30 Synthesis of 4(3*H*)-quinazolinones in aqueous ethyl lactate

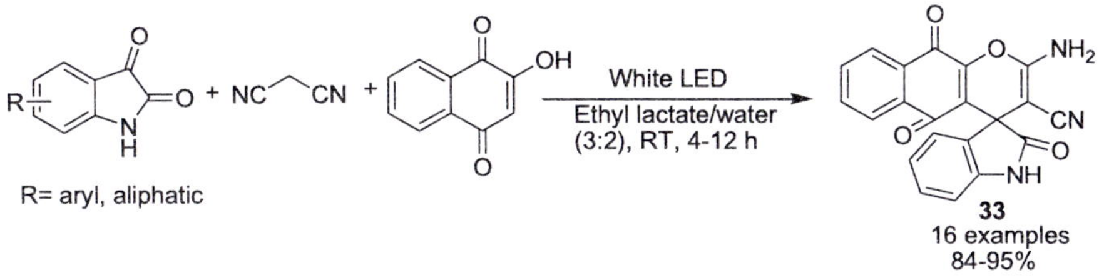

33

Scheme 31 Catalyst-free synthesis of spirooxindole-pyran derivatives in aqueous EL

6 Organic Synthesis in Gluconic Acid Aqueous Solution

Gluconic acid aqueous solution (GAAS) should be considered as a green solvent, and its contribution as a medium for organic synthesis is discussed below.

Guo et al. demonstrated a three-component reaction of barbituric acids, isatins, 1,3-dicarbonyl compounds in 50% GAAS without any additional catalyst affording functionalized spirooxindoles **34** (Scheme 32) [40]. Their structures were established by IR, ^{1}H NMR and ^{13}C NMR spectra, and elemental analysis. The structure was further confirmed by XRD (R^1 = Me, R^2 = 4-MeOC$_6$H$_4$, Ar = Ph).

The four-component reaction of amines, aldehydes, 1,3-dicarbonyl compounds, and nitromethane was carried out in 50% GAAS to synthesize functionalized poly-substituted pyrroles **35** by Li et al.(Scheme 33) [41]. Other solvents showed less efficiency. Unexpectedly, when gluconic acid aqueous solution was used as solvent,

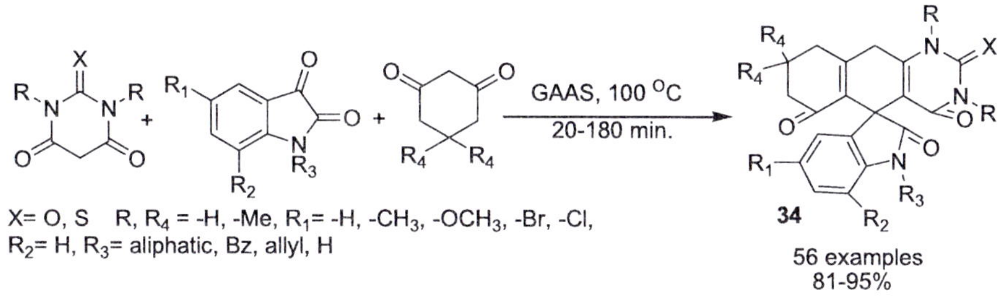

34

Scheme 32 One-pot three-component synthesis of spirooxindoles in GAAS

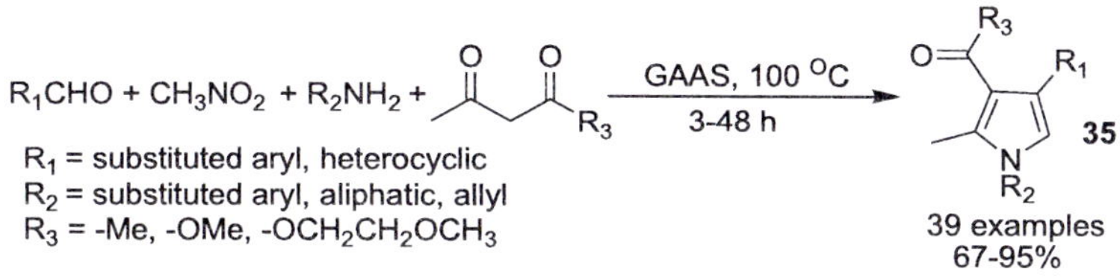

35

Scheme 33 Synthesis of functionalized pyrroles in gluconic acid aqueous solution the reaction yield was drastically enhanced without any added catalyst and could be recycled and reused several times without significant loss of its efficiency. The structure of compound (R^1 = -Ph, R^2 = 4-ClC_6H_5-, R^3 = -Me) was also established by single-crystal X-ray crystallography.

7 Conclusion

It is very difficult to find a perfect green solvent for all types of organic synthesis. Each type of green solvent offers a unique combination of properties and associated techniques for various applications. This chapter helps to find and gather some contributing information about the green solvent like bio-based solvent, PEG, glycerol, and renewable solvents like ethanol, ethyl lactate, and gluconic acid. All these solvents have enormous advantages but a few drawbacks also, as we know that nothing is perfect but they share excellent health and safety profile. These solvents perform as a reaction medium as well as sometimes the role of the catalyst, ligand, and additive in organic synthesis with the aim to reduce the environmental hazard making the mother earth pollution-free. This type of nontoxic, easily accessible, and completely biodegradable solvents are good for human health and environment.

Acknowledgements One of the authors B.M. is thankful to Council of Science Industrial Research (CSIR), New Delhi, for financial support.

References

1. Adam D (2000) Clean and green, but are they mean? Nature 407:938–940
2. Andrade CKZ, Alves LM (2005) Environmentally benign solvents in organic synthesis: current topics. Curr Org Chem 9:195–218
3. Capello C, Fischer U, Hungerbuhler K (2007) What is a green solvent? A comprehensive framework for the environmental assessment of solvents. Green Chem 9:927–934
4. Alfonsi K, Colberg J, Dunn PJ et al (2008) Green chemistry tools to influence a medicinal chemistry and research chemistry based organization. Green Chem 10:31–36
5. Henderson RK, Jimenez-Gonzalez C, Constable DJC et al (2011) Expanding GSK's solvent selection guide- embedding sustainability into solvent selection starting at medicinal chemistry. Green Chem 13:854–862
6. Sheldon RA (2014) Green and sustainable manufacture of chemicals from biomass: state of the art. Green Chem 16:950–963
7. Corma A, Iborra S, Velty A (2007) Chemical routes for the transformation of biomass into chemicals. Chem Rev 107:2411–2502
8. Gu Y, Jérôme F (2013) Bio-based solvents: an emerging generation of fluids for the design of eco-efficient processes in catalysis and organic chemistry. Chem Soc Rev 42:9550–9570
9. Yan N, Xiao C, Kou Y (2010) Transition metal nanoparticle catalysis in green solvents. Coord Chem Rev 254:1179–1218
10. Patil BN, Lade JJ, Pardeshi SD et al (2019) Polyethylene-glycol-(PEG-400) mediated environmentally benign protocol for the synthesis of pyrrolo[1,2-*a*] quinoxalines from benzyl amines at room temperature. Chem Select 4:11362–11366

11. Akbarzadeh A, Dekamin MG (2017) A facile and environmentally benign polyethylene glycol 600-mediated method for the synthesis of densely functionalized 2-aminothiophene derivatives under ultrasonication. Green Chem Lett Rev 10:315–323
12. Tao L-M, Li C-H, Chen J et al (2019) Cobalt(III)-catalyzed oxidative annulation of benzaldehydes with internal alkynes via C-H functionalization in poly(ethylene glycol). J Org Chem 84:6807–6812
13. Kumar KS, Rajesham B, Kumar NP et al (2018) A ligand/additive/base-free $C(sp^2)$-H activation and isocyanide insertion in PEG-400: synthesis of indolizine/imidazoline-fused heterocycles. Chem Select 3:4581–4584
14. Kardooni R, Kiasat AR (2019) A green, catalyst-free synthesis of pyrazolopy rano pyrimidines in polyethylene glycol as a biodegradable medium at ambient temperature. Mol Divers 23:639–649
15. Liang J, Lv J, Shang Z (2011) Metal-free synthesis of amides by oxidative amidation of aldehydes with amines in PEG/oxidant system. Tetrahedron 67:8532–8535
16. Bi J, Zhang Z, Liu Q et al (2012) Catalyst-free and highly selective electrophilic monofluorination of aceto acetamides: facile and efficient preparation of 2-fluoroacetoacetamides in PEG-400.Green Chem 14:1159–1162
17. Khan MN, Karamthulla S, Choudhury LH et al (2015) Ultrasound assisted multicomponent reactions: a green method for the synthesis of highly functionalized selenopyridines using reusable polyethylene glycol as reaction medium. RSC Adv 5:22168–22172
18. Ghom MH, Naykode MS, Humne VT et al (2019) A one-pot direct regioselective iodination of Fischer-Borsche product using periodic acid in PEG-400.Tetrahedron Lett 60:1029–1031
19. Kumar RU, Reddy KHV, Satish G et al (2016) Metal free synthesis of diaryl selenides using SeO_2 as a selenium source. Tetrahedron Lett 57:4138–4141
20. Singh S, Saquib M, Singh M et al (2016) A catalyst free, multicomponent-tandem, facilesynthesis of pyrido[2,3-*d*]pyrimidines usingglycerol as a recyclable promoting medium. New J Chem 40:63–67
21. Mitra B, Pariyar GC, Ghosh P (2019) Glycerol: A Benign Solvent-Assisted Metal-Free One-Pot Multi-Component Synthesis of 4H-Thiopyran and Thioamides from Easily Accessible Precursors. ChemistrySelect 4:5476–5483
22. Nagasundaram RRN, Meignanasundar D, Vadivel P et al (2017) Glycerol assisted eco-friendly strategy for the facile synthesis of 4,40-(arylmethylene)bis(3-methyl-1*H*-pyrazol-5-ols) and 2-aryl-2,3-dihydroquinazolin-4(1*H*)-ones under catalyst-free conditions. Res Chem Intermed 43:1767–1782
23. Morshedi A, Shaterian HR (2018) Green approaches to synthesis of novel and broad-range diversity of 4-(aryl)-3-(phenylsulfonyl)-4*H*benzo[*h*]chromen-2-amine derivatives. Res Chem Intermed 44:7219–7230
24. Bhojane JM, Sarode SA, Nagarkar JM (2016) Nickel–glycerol: an efficient, recyclable catalysis system for Suzuki cross coupling reactions using aryl diazonium salts. New J Chem 40:1564–1570
25. Dubey AV, Gharat SB, Kumar AV (2017) Glycerol as a recyclable solvent for copper-mediated ligand-free C-S cross-coupling reaction: application to synthesis of gemmacin precursor. Chem Select 2:4852–4856
26. Pasupuleti BG, Bez G (2019) CuI/L-proline catalyzed click reaction in glycerol for the synthesis of 1,2,3-triazoles.Tetrahedron Lett 60:142–146
27. Ingale AP, Patil SM, Shinde SV (2017) Catalyst-free, efficient and one pot protocol for synthesis of nitriles from aldehydes using glycerol as green solvent. Tetrahedron Lett 58:4845–4848
28. Babu SA, Saranya S, Rohit KR et al (2019) Ligand-free Cu-catalyzed suzuki coupling of alkynyl bromides with boronic acids in ethanol under microwave irradiation. Chem Select 4:1019–1022
29. Etivand N, Khalafy J, Marjani AP (2019) Facile, one-pot, four-component synthesis of a new series of imidazo[1,2-*a*]pyridines in presence of TPAB in EtOH under reflux conditions. Res Chem Intermed 45:3379–3394

30. Depa N, Erothu H (2019) One-pot three-component synthesis of 3-aminoalkyl indoles catalyzed by molecular iodine. Chem Select 4:9722–9725
31. Ghafuri H, Goodarzi N, Rashidizadeh A et al (2019) ompg-C3N4/SO3H: an efficient and recyclable organocatalyst for the facile synthesis of 2,3-dihydroquinazolin-4(1*H*)-ones. Res Chem Intermed 45:5027–5043
32. Kamalifar S, Kiyani H (2019) An expeditious and green one-pot synthesis of 12-substituted-3,3-dimethyl-3,4,5,12-tetrahydrobenzo[*b*] acridine-1,6,11(2*H*)-triones. Res Chem Intermed 45:5975–5987
33. Alifu Z, Nizhamu M, Ablajan K (2019) Efficient synthesis of *N*′-benzylidene-2-hydroxymethylbenzohydrazides from the one-pot reaction of phthalide, hydrazine and aldehydes. Res Chem Intermed 45:4779–4788
34. Maleki A, Azadegan S, Rahimi J (2019) Gallic acid grafted to amine-functionalized magnetic nanoparticles as a proficient catalyst for environmentally friendly synthesis of α-aminonitriles. Appl Organometal Chem 33:e48
35. Naikwade A, Jagadale M, Kale D et al (2019) Intramolecular *O*-arylation using nano-magnetite supported N-heterocyclic carbene-copper complex with wingtip Ferrocene. Appl Organometal Chem 33:e5066
36. Sadeghi B (2019) Synthesis of novel 6-amino-2-(hydroxymethyl)-8-aryl-7-(phenylsulfonyl) pyrano[3,2-*b*]pyran-4(8*H*)-one derivatives catalyzed by nano-cellulose-OSO_3H. Res Chem Intermed 45:4897–4906
37. Gao G, Chen MN, Mo LP et al (2019) Catalyst free one-pot synthesis of α-aminophosphonates in aqueous ethyl lactate. Phosphorus Sulfur Silicon Relat Elem 194:528–532
38. Shen G, Zhou H, Du P et al (2015) Bronsted acid-catalyzed selective C-C bond cleavage of 1,3-diketones: a facile synthesis of 4(3*H*)-quinazolinones in aqueous ethyl lactate. RSC Adv 5:85646–85651
39. Zhang M, Fu QY, Gao G et al (2017) Catalyst-free, visible-light promoted one-pot synthesis of spirooxindole-pyran derivatives in aqueous ethyl lactate. ACS Sustain Chem Eng 5:6175–6182
40. Guo RY, Wang P, Wang GD et al (2013) One-pot three-component synthesis of functionalized spirooxindoles in gluconic acid aqueous solution. Tetrahedron 69:2056–2061
41. Li BL, Li PH, Fang XN et al (2013) One-pot four-component synthesis of highly substituted pyrrolesin gluconic acid aqueous solution. Tetrahedron 69:7011–7018

Chapter 11
Green Chemistry on C–H Activation

Vahid Khakyzadeh and Sahra Sheikhaleslami

1 Introduction

1.1 Green Chemistry and C–H Bond Activation

The origin of the most majority of organic molecules and compounds is natural gas and petroleum-based unrenewable feedstocks; therefore, one of the most challenging issues in chemistry is how to make these feedstocks useful by breaking and forming the new C–C bonds and modifying the C–H bonds into other functional groups [1]. On the other hand, synthetic chemistry continuously encounters challenges to produce selectively and efficiently organic molecules, whether the synthesis of small or complex structures [2] and also many strategies of organic synthesis are being developed in order to extend the chemical toolbox [3].

The traditional procedures, known as the functional group interconversion strategies [4], have achieved this aim by initial C–H bond functionalizations which are then followed by a modifiable sequence of steps to introduce desired functional groups or C–C bonds building the desired skeletons [5].

Notwithstanding its merited place in organic chemistry that the introductory chapters of organic chemistry textbooks intensely concentrate on radical C–H bond functionalizations (halogenations) which are followed by substitution and elimination reactions, it suffers from a perceptual disadvantage. For example, it leads to profoundly futile processes by requiring several reaction steps from unfunctionalized feedstocks to functionalized products, during which undesired by-products are unfortunately generated that leads to high E factors (kg waste/kg product) [6].

V. Khakyzadeh (✉) · S. Sheikhaleslami
Faculty of Science, Department of Chemistry, K. N. Toosi University of Technology, P.O. Box 15875-4416, Tehran, Iran
e-mail: v.khakyzadeh@kntu.ac.ir

S. Sheikhaleslami
e-mail: sahra.sh73@gmail.com

G. Anilkumar and S. Saranya (eds.), *Green Organic Reactions*, Materials Horizons: From Nature to Nanomaterials, https://doi.org/10.1007/978-981-33-6897-2_11

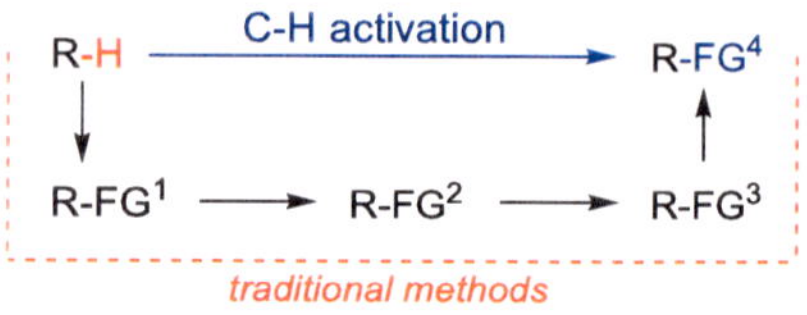

Scheme 1 C–H bond activation and functionalization vs traditional methods

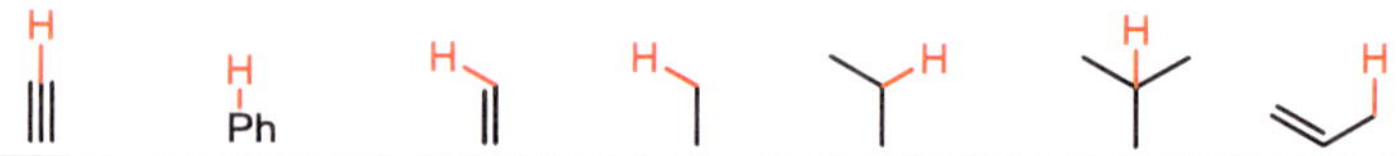

BDE (kJ/mol)	552.2	473.0	460.2	410.8	397.9	389.5	361.1

Scheme 2 Bond dissociation energies and pKa values of selected hydrocarbon C–H bonds

In contrast, a significantly greener and sustainable approach is represented by the direct use of otherwise inert C–H bonds as latent functional groups. C–H bond activations contribute a new path to introducing functional groups by preventing lengthy synthetic operations and reducing the by-products and thereby blocking waste production and facilitating direct access to desired target structures even at the late stages of synthesis (Scheme 1) [7].

Synthetic C–H activation catalysts struggle to achieve similar selectivities as remarkable chemoselectivity of enzymatic C–H oxidations. Some critical challenges in catalyst development are considering the significant energies that are needed to directly cleavage of C–H bond [8]. The bond dissociation energies (BDEs) and acidities of common C–H bonds in hydrocarbons are shown in Scheme 2. The quantities of BDEs are very close to each other and are between 361.1 and 552.2 kJ/mol, which are one of the most difficult bonds to cleave among other bonds. Furthermore, an isolated C–H bond in a molecule has a very low reactivity owing to the large kinetic barrier associated with the C–H bond cleavage and a polar nature of this bond which do not possess suitable lone pairs to coordinate with a catalyst (Scheme 2) [9].

Considering all these critical challenges, the importance of site- and product-selective transformation of unactivated C–H bonds into other functional groups and especially catalytic ones is largely accepted by modern chemists, and it is under active study for several years and still regarded as the Holy Grail in chemistry for its step efficiency, atom economy, and potential as a method for late-stage functionalization of complex organic molecules [10].

2 Proposed Mechanisms for C–H Bond Activation

The promotion of technical facilities and laboratory proficiency has increased the scientific abilities to explore the mechanistic foresight of a chemical reaction. During the last decade, due to the prospering interest on catalytic C–H functionalization and owing to the deep understanding of the elementary steps in homogeneous catalysis systems, many researchers have developed significant improvements in the activity and performances in catalytic systems for C–H activation [11]. There are some well-established mechanisms for C–H bond activation that mainly fall under these categories: (1) oxidative addition (OA); (2) electrophilic aromatic substitution (S_E Ar); (3) σ-bond metathesis (σBM); (4) single-electron transfer (SET); concerted metalation deprotonation (CMD); and (5) base-assisted intramolecular electrophilic-type substitution (BIES). Herein, we give a brief explanation for each one of them [12]:

Oxidative addition (OA): This is the most common mechanism by which a R-H bond is cleaved and a M-R bond and a M-H bond are formed. It frequently occurs by having an electron-rich and low-oxidation late transition metal centers (Re, Fe, Ru, Os, Ir, Pt) interacting strongly with the C–H bond. The σ-C–H bond coordinates to the metal and a dπ-back donates to the σ*-C–H orbital, lowering its bond order and resulting in the bond cleavage and oxidizing the reaction center in two units. Oxidative addition reaction leads to the creation of a reactive organometallic species containing a hydride and alkyl/aryl ligands at the oxidized metal center (Scheme 3).

Electrophilic aromatic substitution (SEAr): This classification of electrophilic substitution has emerged from the mechanistic pathway by which the hydrogen atom of the substrate is replaced by a metal and thus it acts as a Lewis acid. This reaction is based on the electronic interaction between the π-electronic cloud of the substrate and the electrophilic metal center which acts as Lewis acid forming a new C(aryl)–M bond. In contrast with the oxidative addition, metal oxidation state stays without any changes (Scheme 4).

The vicinal C(aryl)–H bond could be easily lost as a proton by re-aromatization or by the action of a base, as a result of the acidity enhancement on this bond. Under circumstances in which the base is in the coordination area of the metal center, the mechanism is acknowledged as a **base-assisted intramolecular electrophilic-type substitution (BIES)** which has more recently been put forward to account for the often-preferred reactivity of electron-rich arenes (Scheme 5).

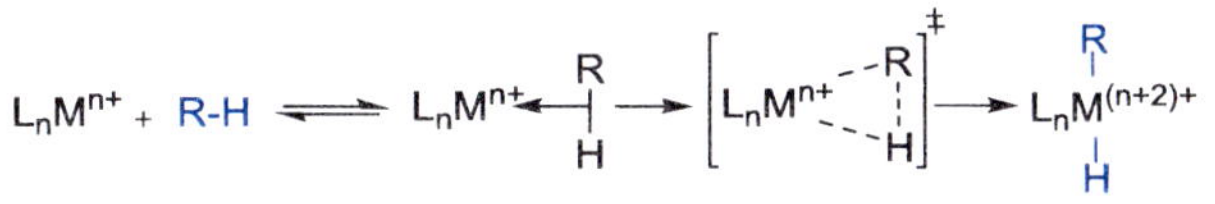

Scheme 3 Oxidative addition (OA)

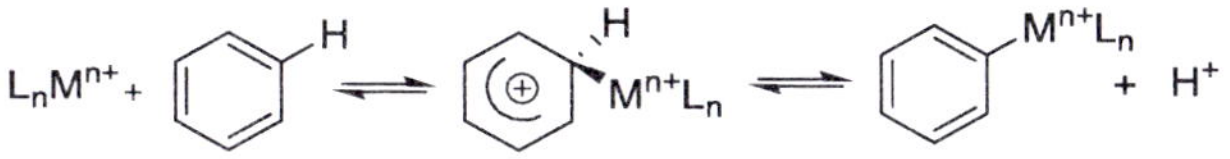

Scheme 4 Electrophilic aromatic substitution (SEAr)

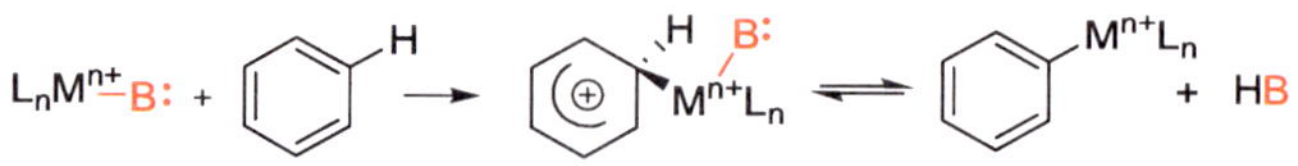

Scheme 5 Base-assisted intramolecular electrophilic substitution (BIES)

σ-Bond metathesis (σBM): This mechanism is favored for electron-poor metal centers with a high oxidation state. Oxidative addition is not possible with transition metals having d^0 electronic configuration (groups 3 and 4, lanthanides and actinides) and thus preferred mechanism is σ-bond metathesis. Cleavage and formation of bonds go through a four-membered square transition state without changing the oxidation state at the metal center (Scheme 6). This is usually common for late or post-transition metals (Pd^{2+}, Pt^{2+} or Pt^{4+}, Hg^{2+}) and the new C-M and C–H bonds are made without containing any metal hydride species.

Single-electron transfer (SET): It contains two steps, that each step involves one electron. Homolytic cleavage of the C–H bond conduce formation of the metal hydride species and a carbon-centered radical (Scheme 7) and construction of the alkyl/aryl-hydride metal oxidized species occurs after recombination reaction between the radical and the metal center.

Concerted metalation deprotonation (CMD): Close contiguity of this bond to the metal center that is usually promoted by a directing donor group is the key point in this mechanism. At the same time, the metal center possesses a coordinated base that promotes the deprotonation of the C–H bond in a concerted fashion while the formation of the C–M bond is occurring (Scheme 8).

Scheme 6 σ-bond metathesis (σ-BM)

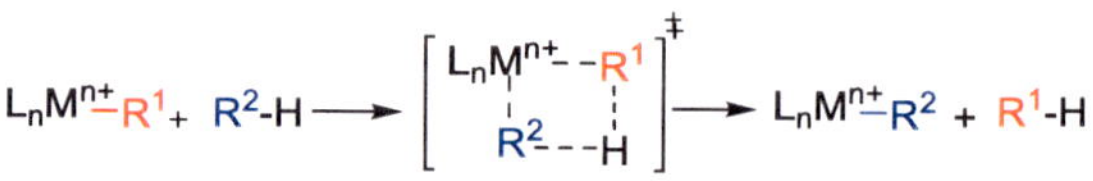

Scheme 7 Single-electron transfer (SET)

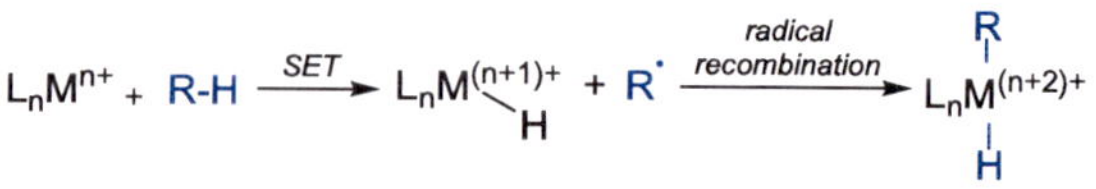

Scheme 8 Concerted metalation deprotonation (CMD)

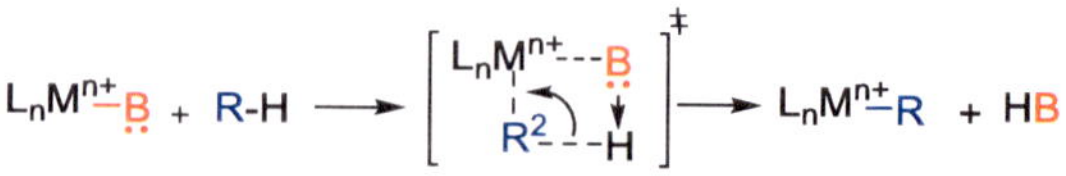

3 Classification of Green C–H Activations

There are some circumstances that affect green principles of C–H activation such as solvent, oxidant, metal catalysts, and so on. For many years, chemists are working on these circumstances to make it greener and safer methodology. From this point of view, we can classify green C–H activation from the traditional green methods to the developed new methods:

- Green protocols for C–H bond activation:
  a. Transition metal-catalyzed C–H activation
  b. Transition metal-free C–H activation
  c. Green solvent/solvent-free C–H activation
  d. Green oxidant/oxidant-free C–H activation
  e. Direct C–H functionalization
- Electrochemical C–H bond activation
- C–H bond activation under flow condition
- Electrochemical C–H bond activation under flow condition
- Photocatalytic C–H bond activation
- C–H bond activation using ball milling and transition metal catalysts

4 Green Protocols for C–H Bond Activation

Past decades have witnessed the emergence of C–H functionalizations as a particularly powerful tool for molecular syntheses, [13] with enabling applications to material sciences, late-stage diversification, natural product synthesis, and pharmaceutical industries, among others [14]. However, most C–H functionalization protocols suffer from stoichiometric amounts of costly and/or toxic transition metal oxidants that make undesired metal-containing by-products [15]. Therefore, applying green methods in C–H activation is highly desirable. In the next parts, we will introduce some green protocols and methods for the aforesaid purpose.

4.1 Applicable Metals in C–H Activation

Metal-catalyzed C–H functionalization chemistry provides the step economical and original construction of C–C, and C-X (X=N, O, etc.) bonds, commencing from hydrocarbon fragments without the necessity of prior non-catalytic oxidation steps and pre-functionalization of substrates [16]. Thus far, the vast majority of C–H functionalization advances continue to heavily rely on precious 4d or 5d transition metal catalysts such as Pd, Ir, Rh, and Ru [17]. Unfortunately, these 4d and 5d transition metals are not only cost-intensive but are generally comparatively toxic. Given the cost-effective and sustainable nature of earth-abundant first row transition metal and also less toxicity of these metals, the evolution of 3d metal catalysts such as Sc, V, Mn, Fe, Ti, Cr, Co, Ni, Cu, Zn, for C–H activation has attained remarkable recent momentum as green alternatives [18].

The C–H activation by 3d-based metals continues to largely undergo single-electron transfer manifolds, setting the platform for more reactivities and selectivities [19]. More widespread applications of these 3d metal catalysts are moderately disturbed by their substantial oxophilicity that leads to reducing chemo-selectivities and functional group tolerance. Manganese-, cobalt-, and iron-catalyzed C– H activations emerged as potent systems for various C–H alkylations, alkenylations, and arylations [20]. Besides, according to the d6-electron configuration of manganese (I) and cobalt (III) complexes, these metals accomplish the C–H transformations, in which the oxidation state of metal remains without changing during the entire catalytic cycle [16]. In terms of versatility, Ni [21] and Cu [22] catalysis were utilized in alkylation and arylation reactions under mild conditions. Furthermore, the Ni and Cu catalysis regimes were not restricted to redox-neutral C–H transformations. Indeed, these manifolds proved particularly powerful for oxidative C–H functionalizations.

The combinations of C–H activation and cross-coupling reactions give countless opportunities to synthesize complex molecules via Mizoroki–Heck and Suzuki-type cross-coupling C–H functionalization reactions [23].

There are many practical methods of metal-catalyzed C–H activation operated in the synthesis of medicinally valuable molecules like lithospermic acid, piperaborenine B, losartan, valsartan, anacetrapib, and oxazolidinone antibacterial using transition metals [24].

4.2 *Transition Metal-Free C–H Activation*

Most of the transition metal catalysts are normally very expensive, and the supporting ligands are generally even more costly and sometimes there are obstacles in their preparation. Also, as aforesaid, several transition metals are toxic. On the other hand, various transition metal catalysts are normally sensitive to oxygen (O_2) and moisture. Moreover, in cases which high efficiency and selectivity of transformations are challenging, it is essential to use special additives and co-catalysts. Consequently, transition metal-free conditions became more attractive than classic transition metal-catalyzed reactions [25]. The usage of hypervalent iodine reagents [26], diazonium salts [27], or employing electrochemistry methods [28] are some of the interesting examples of transition metal-free processes.

The mixture of electron-rich arenes and hypervalent iodine(III) reagents results in radical cation as a selective and efficient SET oxidizing agent that provides a series of direct C–H functionalization products under mild conditions [29]. In these reactions, the application of polar and low-nucleophilic protic solvent such as multi-fluorosubstituted alcohol is crucial because it affects stabilizing the reactive cationic intermediates and preventing other probable side-reactions (Scheme 9) [30].

Kita and co-workers made significant progress in hypervalent iodine(III)-promoted metal-free C–H functionalization reactions on a variety of heteroaromatic compounds such as thiophenes and pyrroles (Scheme 10) [31].

Direct fluorination to get the mono- and difluoromethylated arenes in the presence of selectfluor and potassium persulfate was disclosed by Yi and co-workers [32] (Scheme 11).

The ligand- and transition metal-free direct C–H functionalization of quinones and naphthoquinones with diaryliodonium salts through the radical pathway which led to the synthesis of aryl naphthoquinones as β-secretase inhibitors in moderate to good yields was introduced by Wang et al. [33] (Scheme 12).

Scheme 9 C–H functionalization with azides, acetate, and 1,3-dicarbonyl compounds

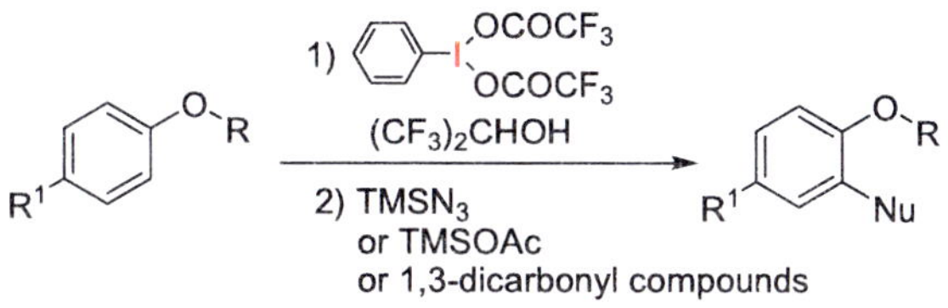

Scheme 10 C–H functionalization with thiophenes and pyrroles

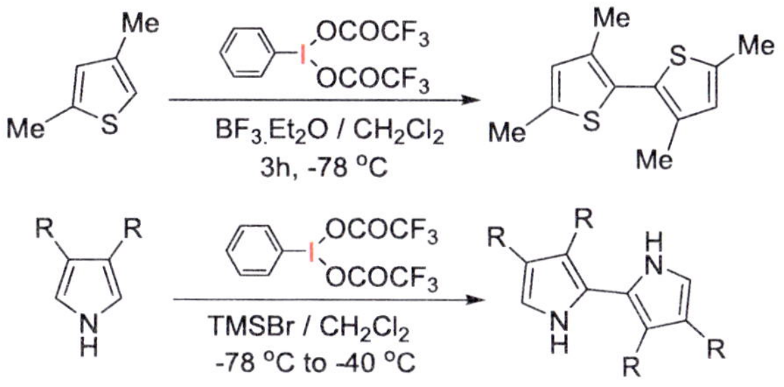

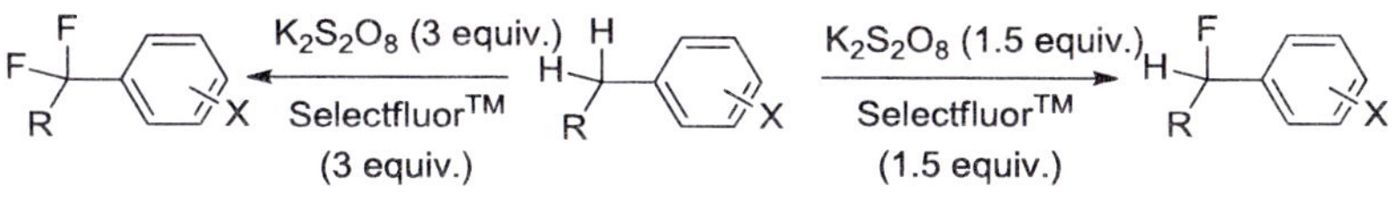

Scheme 11 Transition metal-free direct benzylic C–H fluorination

Scheme 12 Transition metal-free direct C–H functionalization of Quinones

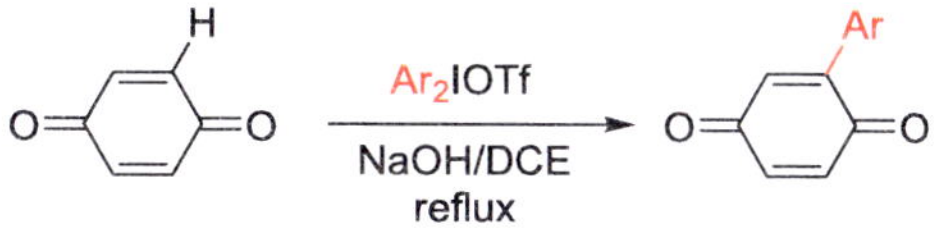

Scheme 13 Transition metal-free C3 arylation of indoles with aryl halides

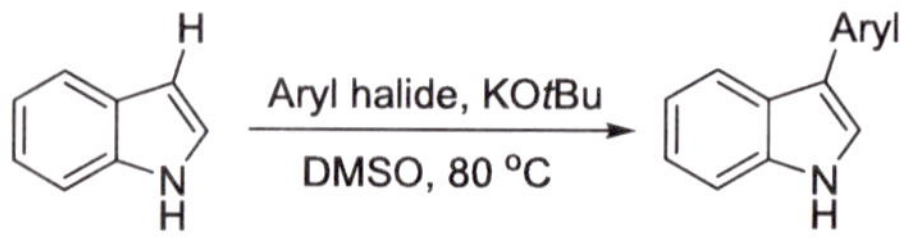

In another example, a transition metal-free regio-selective coupling reaction of indoles and aryl halides using KO^{t}Bu and degassed solvent was reported (Scheme 13) [34].

Recently, many chemists are working on expanding modern metal-free aerobic C–H bond functionalization reactions which have utilized aldehydes, ethers, benzylamines, and glycine derivatives [35].

4.3 Green Solvent/solvent-Free C–H Activation

It has been determined that most of the waste produced in the chemical synthesis originates from the usage of solvents which are frequently used in large quantities in comparison with the other reactants. Accordingly, it is not unexpected that significant attention has been dedicated to the definition of "green solvents" [36]. In general, bio-based solvents coming from the biomass feedstock can be recognized as "greener" alternatives [37]. But none of the alternative solvents are as green as water. However, there is still a gap between the use of water as a reaction medium and practical green chemistry. Another alternative procedure is conducting reactions under solvent-free conditions [38].

During the past five years, employing greener solvents such as water, dialkyl carbonates, and PEGs in the ruthenium and palladium-catalyzed direct arylation of aromatic heterocycles have increased significantly [39]. Using dialkyl carbonates resulted in the facilitated work-up procedure and decreased the production of wastes and PEGs as a solvent provided the recycling of the catalyst and using water led to higher rates and cleaner reactions in several cases particularly with ruthenium. So far, there are no examples of these reactions that have been described in other alternative media such as ionic liquids or supercritical CO_2. Also, carbonates have been determined as a suitable solvent for palladium-catalyzed direct arylation reactions and the reaction was found to be more selective than in DMF, dioxane, and other previously used toxic solvents since using this solvent resulted in fewer traces of side products [40].

Wang and Wu developed a green step-economic and sustainable approach to construct a C–C bond without any organic solvents or additives. They reported the example of water-mediated C–H activation of arenes using sulfoxonium ylides (Scheme 14) [41].

In the same year, Yao and co-workers developed a method for C–C bond formation via a terminal alkyne C–H bond activation and synthesizing quinoline derivatives

Scheme 14 Water-mediated C–H activation

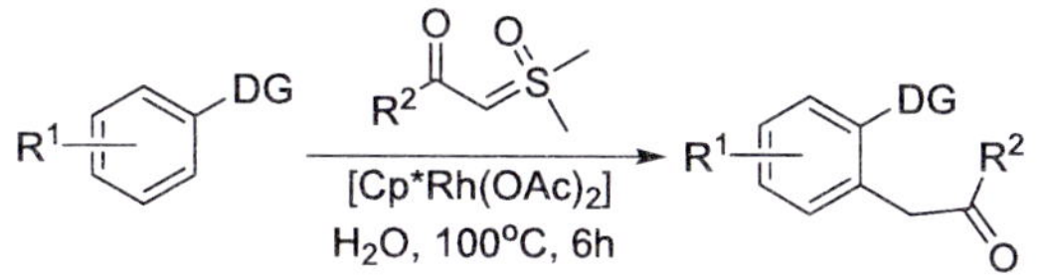

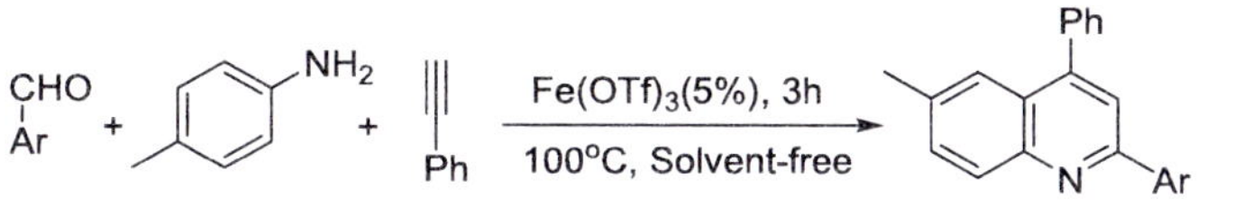

Scheme 15 Green synthesis of quinoline derivatives

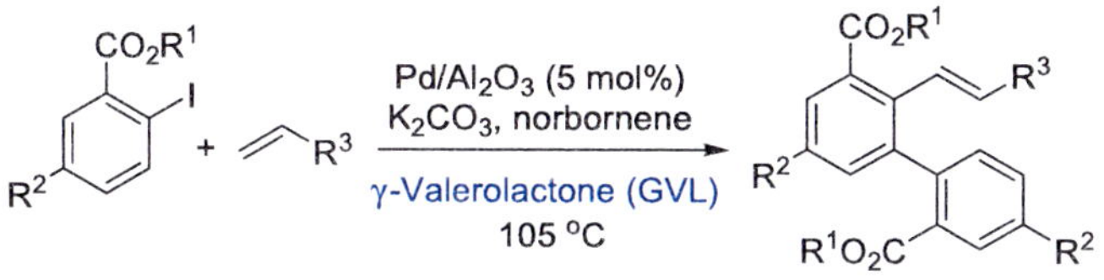

Scheme 16 Catellani reaction catalyzed by palladium/Al_2O_3 in GVL

using stable and inexpensive $Fe(OTf)_3$ as a catalyst under solvent-free condition (Scheme 15) [42].

Recently, major advances have been represented in using biomass-derived solvents such as; glycerol, 2-methyl tetrahydrofuran, ethyl lactate, and γ-valerolactone (GVL), in transition metal-catalyzed couplings, including Suzuki–Miyaura, Mizoroki–Heck, Sonogashira–Hagihara reactions, and C–H functionalizations [37]. Ackermann, Vaccaro, and co-workers reported the first example of a palladium-catalyzed Catellani reaction using GVL as a solvent instead of the frequently employed DMF, DMA, or acetonitrile (Scheme 16) [43].

4.4 Green Oxidant/Oxidant-Free C–H Activation

One of the recognized and extremely attractive conversions from environmental and economical viewpoints is the construction of C–C bonds from two C–H bonds under oxidative conditions that reduced waste and reaction steps. Despite excellent improvements being made, key difficulties remain [44]. Therefore, the development of green and user-friendly C–H bond activation procedures in the lack of chemical oxidants would be instantly desirable. Oxygen gas is well known for an ideal and readily available green oxidant and its solo by-product is desired for cross-coupling C–H activation but suffers from weak reactivity.

The coupling reaction of 2,6-lutidine and internal alkynes which begun with C(sp^3)-H bond activation via σ-bond metathesis and then mediated by a non-metallocene cationic alkylhafnium complex to give five-membered carbocyclic compounds was firstly developed by Mashima and co-workers (Scheme 17) [45].

Four years later Hu and co-workers reported a palladium-catalyzed oxidative Suzuki coupling reaction of 3-unsubstituted indolizines at the 3-position with aryl boronic acids to produce 3-aryl-indolizine. The distinguished advantage of this method was using O_2 as a green oxidant (Scheme 18) [46].

Gong and Meggers have also introduced the first example of an asymmetric photoredox dehydrogenative cross-coupling of two Csp^3-H groups catalyzed by a chiral rhodium complex and with molecular oxygen as the oxidant in 2015 (Scheme 19) [47].

The rhodium-catalyzed oxidative dehydrogenative cross-coupling of arylamines with electron-rich arenes under mild aerobic conditions resulting in the synthesis of non-symmetrical biaryl amines, in excellent yields and high selectivities, is another example of using green oxidant (Scheme 20) [48].

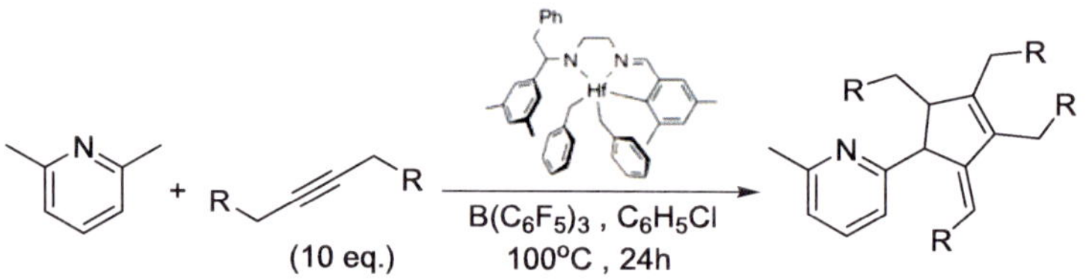

Scheme 17 Coupling reaction of 2,6-lutidine and internal alkynes

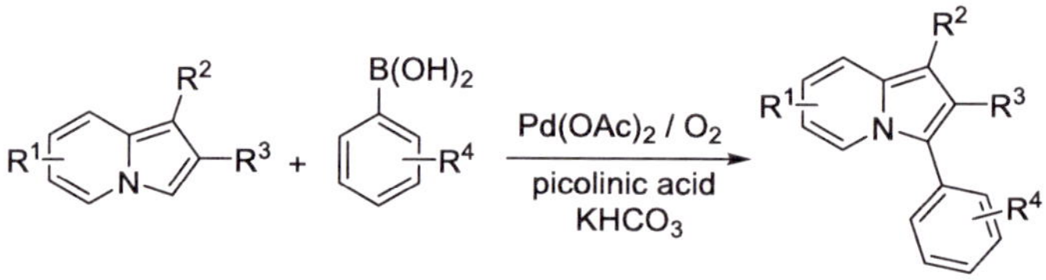

Scheme 18 Palladium-catalyzed oxidative Suzuki coupling reaction

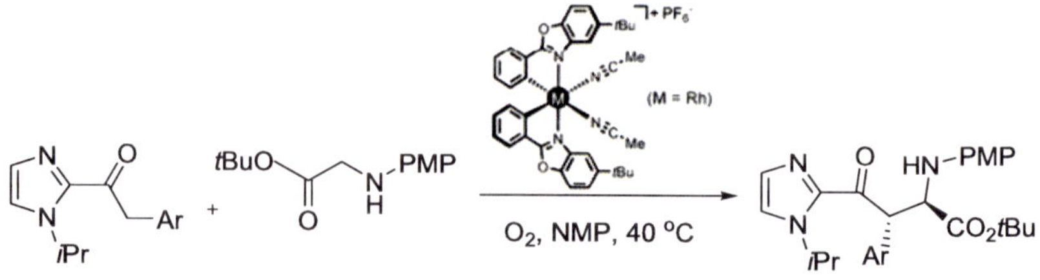

Scheme 19 Asymmetric green cross-coupling of two Csp^3-H groups

Scheme 20 Rhodium-catalyzed synthesis of non-symmetrical biaryl amines

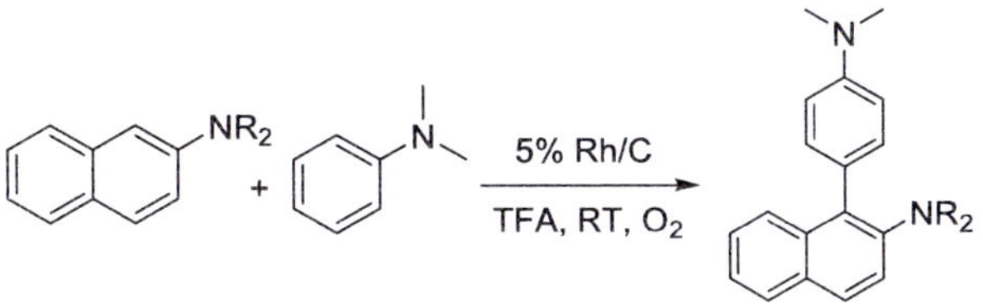

Scheme 21 Synthesis of *p*-hydroxybenzoates

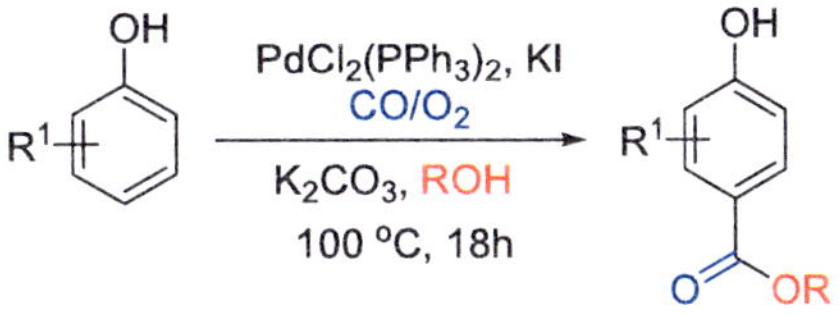

In 2018, Bhanage reported an effective protocol for the synthesis of valuable *p*-hydroxybenzoates directly from phenols by palladium-catalyzed aerobic oxidative carbonylation of phenolic C–H bond, proceeding through oxidative iodination. Using O_2, high selectivity, no co-catalyst, co-solvent, and external ligand are some advantages of this method (Scheme 21) [49].

4.5 Direct C–H Functionalization

Owing to the complexity of organic substrates, several types of C–H bonds can be found in their chemical skeletons. However, in terms of selectivity, controlling reactivity on one single bond is very challenging. For this reason, the use of a donor group (DG) as a directing group is a very broadly applied strategy to selectively activate C–H bonds [50]. Direct C–H functionalization procedures are based on the use of directing Lewis bases covalently linked to the substrate. Heteroatom-based groups are the most used directing groups, although alkenes can also be effective. This field has been much studied, especially for arene C–H activation [51]. For the past fifty years, *ortho*-selectivity via cyclometalation is the most utilized way of C–H activation; more recently devised special directing groups allow also *meta*-selectivity [52]. Prominently, directed C–H activation allows to activate unactivated C(sp^3)-H bonds, too [53].

5 Electrochemical C–H Bond Activation

Combining metal-catalyzed C–H activation with electrocatalysis, have resurfaced as a viable platform for sustainable transformations due to its inherent advantages and unique characters such as replacement of dangerous and toxic chemicals by electric current, less waste production, applying few amounts of chemicals, and affording fewer reaction steps than traditional methods [54]. Moreover, a majority of the preparative electrochemical reactions even with high activation energies, can be carried out at ambient temperature since the energy of the electrochemical system is controlled by the applied electrode potential [55]. There are two kinds of cell setups in electrochemistry systems: (1) divided cell, the anode and cathode are separated by a porous and ion-exchange membrane; (2) undivided cell setup the anode and cathode are being placed in the same cell in a significantly more user-friendly manner [56]. Also, organic electrochemistry can be classified into two categories; direct electrolysis without redox mediator and indirect electrolysis [57] with redox mediators.

Advantages of redox-mediated electrolysis are avoiding obstacles related to heterogeneous electron transfer such as overpotentials and conducting electrolysis at lower potentials which lead to accelerating the reaction rate and can feature beneficial effects in terms of chemoselectivity and robustness with bypassing probable side reactions [58]. The low atom economy for this strategy was the most important motivation for applying direct metallaelectrocatalytic C–H activations. Recently the scope of useful direct electrochemical ruthenium (II) catalysis was extended by Ackermann, Qiu, Mei and Xu [59]. Also, there is plenty of reported rhodium-catalyzed oxidative C–H activation reaction with stoichiometric metal oxidants during the last decades [60]. More recently Ackermann reported the novel rhodaelectro-catalyzed C–H activation in which there was no need for stoichiometric chemical oxidants [61]. Mei and co-workers recently reported a notable palladaelectro-catalyzed C(sp^3)-H oxygenation of substituted oxime ethers [62]. Inspired from this approach, further palladium-catalyzed electrochemical C(sp^2)-H activations, including direct acetoxylations [63], acylations [64], methylations, and alkylations [65] have been reported. In 2017, the Ackermann group [66] made a significant success in reaching full resource economy through direct metallaelectrocatalytic C–H oxygenation using inexpensive alcohols as a coupling partner with 3d transition metals and in undivided cell setups. Significant progress was achieved by Ackermann [67], Lei [68], and Ye [69] in this field (Scheme 22). Lately, there are a few reports of copper-catalyzed electrocatalytic C–H amination [70] and also a new electrooxidative nickel catalysis without redox mediators with a full resource economy [71].

6 C–H Bond Activation Under Flow Condition

Flow technologies are one of the most assuring tools for providing more extended applicability of C–H bond functionalization reactions [72]. Some features of flow condition are listed as follows: (a) more reliable and safer [73]; (b) easy temperature controlling [74]; (c) efficient mixing of two phases and contacting between reagents in different phases (mainly gases and liquids or solids and liquids) [75]; (d) the high surface area-to-volume ratio of flow reactors [76]; (e) easier scale-up of the reaction; and (f) simple separation of the catalyst from the reaction products [77].

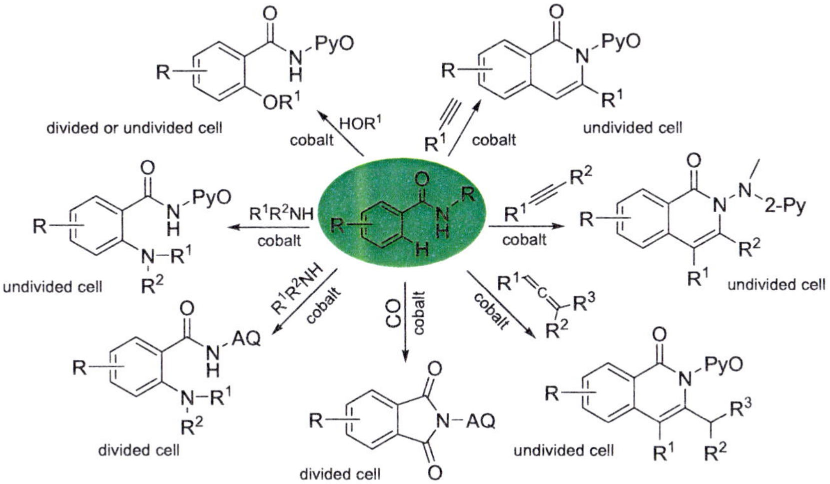

Scheme 22 Co-catalyzed electrocatalytic C–H activations

Scheme 23
Palladium-catalyzed trifluoromethylthiolation of amides in flow

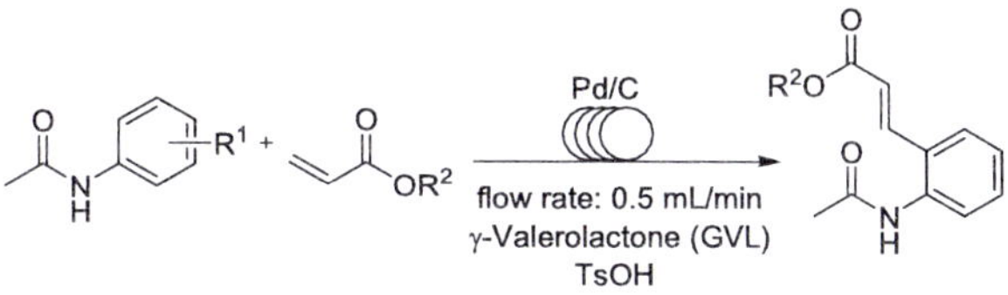

As an example, the *ortho*-selective C–H alkenylation of acetanilides in heterogeneous continuous flow conditions was discovered by Ackermann, Vaccaro, and co-workers. The reaction was catalyzed by Pd/C, required benzoquinone (BQ) as the terminal oxidant and a strong Brønsted acid as an additive in biomass-derived γ-valerolactone (GVL). The advantages of performing this reaction in flow were the shorter reaction times and improvement in the stability of the catalyst compared to batch conditions (Scheme 23) [78].

7 Electrochemical C–H Bond Activation Under Flow Condition

Very recently, there are some examples of merging organic electrochemistry with flow technologies. For instance, Xu and co-workers reported an electrochemical protocol to synthesize 4H-1,3-benzoxazines from the cyclization of readily available N-benzylamides. Most of the researches were conducted in batch, with only a single reaction being conducted in an electrochemical microreactor. The reactor was containing a platinum foil as the cathode and a graphite layer as the anode,

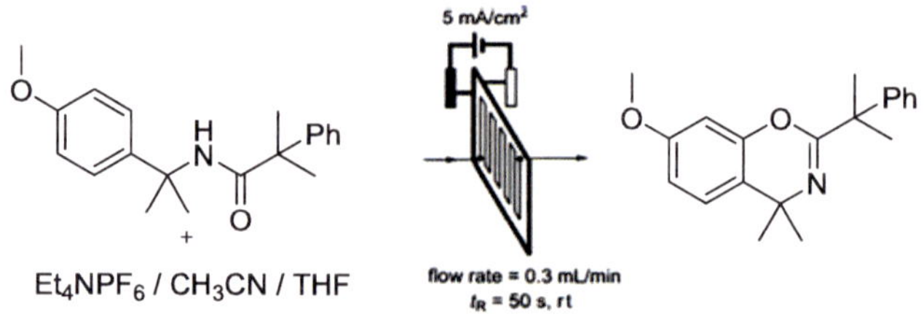

Scheme 24 Electrochemical synthesis of a 4H-1,3-benzoxazine in flow

separated by a fluorinated ethylene propylene (FEP) membrane. Notably, the use of flow conditions made it possible to reduce the amount of the supporting electrolyte, to perform the reaction at ambient temperature and to scale-up the process in high yields (Scheme 24) [79].

8 Photocatalytic C–H Bond Activation

Visible light photoredox catalysis has rebirthed due to the improvement of radical chemistry in organic synthesis. The simplicity of reaction setups, as well as mild reaction conditions, low price, green and clean energy sources, and the broad applicability of them led to resolve some of the contemporary challenges and scarcities in several synthetic methodologies using transition metals or strong oxidants. One of the most influential accomplishments in photocatalysis was the activation of molecular oxygen [80]. Despite its poor reactivity as a diradical, it can be transformed into the superoxide radical anion upon single-electron transfer (SET), which can undergo H-radical abstraction or lead to the in situ formation of hydrogen peroxide (Scheme 25, top) [81].

Additionally, transition metal chromophores as photocatalysts [82] have sufficient potential in their excited states and can induce C–H functionalization reactions by the direct oxidation of the substrates to the corresponding radical or radical cation without the necessity of a stoichiometric amount of an oxidant or a pre-functionalization of

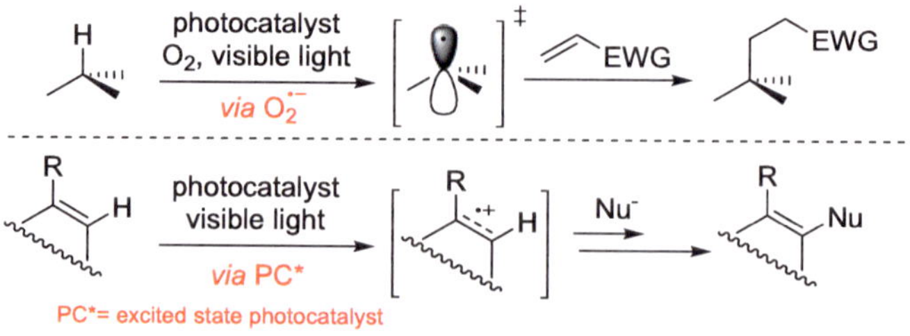

Scheme 25 Photocatalytic C–H bond activation pathways

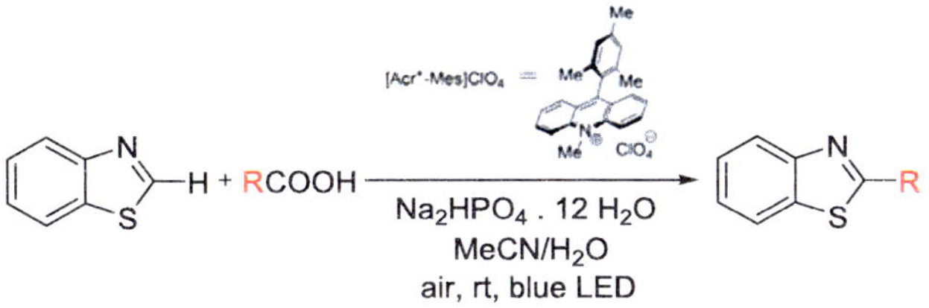

Scheme 26 Mes-Acr$^+$ catalyzed C–H alkylation reaction

Scheme 27 Mechanochemical C–H bond alkylation of indoles

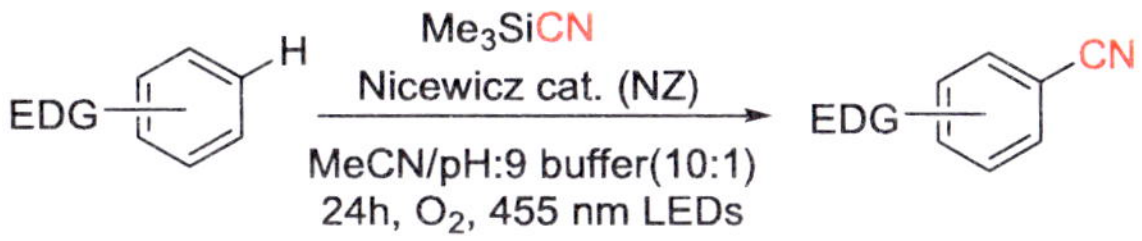

the substrate (Scheme 25, bottom). Generally used photoorganocatalysts are mostly heterocycles and organic dyes such as Eosin Y, methylene blue, and riboflavin [83].

Although most of the reported visible light-mediated catalytic C–H functionalization methods have been performed by expensive metal-based Ru or Ir photoredox complexes but their mild reaction conditions and the use of stable and cheap organic dyes could justify its green manner [84]. In 2019, an efficient photoredox-induced decarboxylative C2-alkylation of benzothiazoles was produced by Wang and co-workers [85]. The reaction was carried out by using a catalytic amount of 9-mesityl-10-methyl acridinium perchlorate as photocatalyst and O_2 as an oxidant with a blue LED under transition metal free conditions at 25 °C. Broad scope, high yields, mild reaction conditions, and easy work-up were its highlighted advantages (Scheme 26).

In 2017, the group of Nicewicz reported the direct C–H cyanation of electron-rich arenes via visible light photoredox catalysis [86]. The authors applied an acridinium salt derivative (NZ) as the photocatalyst for the single electron oxidation of aromatic compounds into the radical cationic species, which were subsequently trapped by TMSCN to produce nitrile. The employed oxygen atmosphere made the addition of external oxidants unnecessary. Also, one year later, the same group, applied a similar photocatalytic methodology for the visible light-mediated C–H bond azidation of aliphatic compounds with acridinium (Scheme 27) [87].

9 C–H Bond Activation Using Ball Milling and Transition Metal Catalysts

Mechanochemistry has attracted increasing attention from chemists. Mechanochemistry methods are applicable to distribute energy for chemical processes as efficient as possible via grinding, ball milling, shearing, and kneading [88].

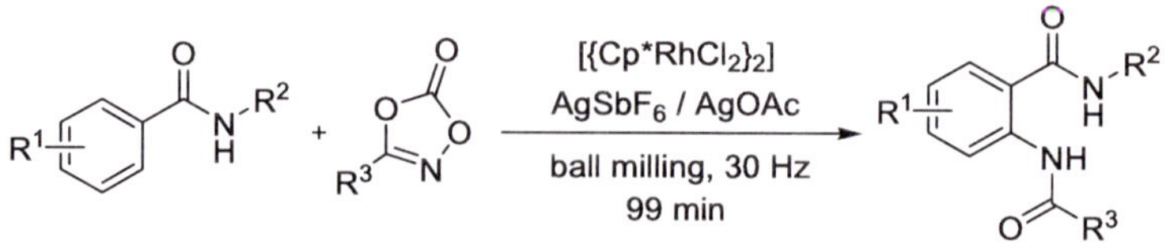

Scheme 28 Mechanochemical Rh(III)-catalyzed C–H bond amidation of benzamides

Ball-milling-induced reactions are cost-effective, efficient, and green methods which are typically operated in a stainless-steel jar with plenty of balls rotating at high speed (60–800 rpm). These reactions are performed under mild conditions, without any organic solvents (or minimum amounts of solvent), at approximately ambient temperature and in relatively short reaction times. The first mechanochemical C–H functionalization was reported by Bolm and co-workers [89]. Although the discovery was remarkable, the solventless process was limited in terms of catalyst efficiency. Since then chemists have been trying to expand the scope of mechanochemical C–H functionalizations in the presence of transition metals such as Pd, Rh, Ru, Co, and Ir and apply it for a variety of important functionalizations, including halogenation, amidation, alkynylation, and dehydrogenative coupling and also utilize it for other important organic reactions [90]. As a ball-milling reaction example, Bolm and co-workers reported a procedure for the direct mechanochemical rhodium-(III)-catalyzed C(sp^2)-H bond amidation of the arenes using a 1,4,2-dioxazol-5-one as the nitrogen source and amidating agent and using ball milling in a 25 mL ZrO_2 milling jar with one ZrO_2 ball of 15 mm diameter at 30 Hz. The reaction proceeds in the presence of $[Cp^*RhCl_2]_2$, $AgSbF_6$, and AgOAc under solvent-free conditions without additional heating. The *ortho* amidated products were formed in great yields and in shorter reactions times (99 min.) in comparison with the solution and exhibited benefits of mechanistic techniques to the standard solvent-mediated protocols (Scheme 28) [91].

10 Conclusion

During the last decade, considerable advances in C–H functionalization reactions have witnessed the importance of these transformations. These protocols give us powerful tools to create organic building blocks of complex structures in molecular science. On the other hand, their step- and atom-economic nature, are completely in agreement with green chemistry protocols. The applicability of C–H bond activation reactions is currently under remarkable investigation in related chemistries which will be reported in due course.

References

1. Hess J, Bednarz D et al (2011) Petroleum and health care: evaluating and managing health care's vulnerability to petroleum supply shifts. Am J Publ Health 101:1568–1579
2. (1) Chen K, Baran PS (2009) Total synthesis of eudesmane terpenes by site-selective C–H oxidations. Nature 459:824–828; (2) Jørgensen L, McKerrall SJ et al (2013) 14-step synthesis of (+)-ingenol from (+)-3-carene. Science 341:878–882
3. Hesp KD, Bergman RG, Ellman JA (2011) Expedient synthesis of N-acyl anthranilamides and β-enamine amides by the Rh (III)-catalyzed amidation of aryl and vinyl C–H bonds with isocyanates. J Am Chem Soc 133:11430–11433
4. Das P, Dutta A et al (2014) Heterogeneous ditopic ZnFe 2 O 4 catalyzed synthesis of 4 H-pyrans: further conversion to 1, 4-DHPs and report of functional group interconversion from amide to ester. Green Chem 16:1426–1435
5. Walling C, Jacknow BB (1960) Positive halogen compounds. I. The radical chain halogenation of hydrocarbons by t-butyl hypochlorite[1]. J Am Chem Soc 82:6108–6112
6. Davies HML, Du Bois J et al (2011) C–H Functionalization in organic synthesis. Chem Soc Rev 40:1855–1856
7. Potavathri S, Pereira KC et al (2010) Regioselective oxidative arylation of indoles bearing N-alkyl protecting groups: dual C—H functionalization via a concerted metalation—deprotonation mechanism. Am Chem Soc 132:14676–14681
8. Blanksby SJ, Ellison GB (2003) Bond dissociation energies of organic molecules. Acc Chem Res 36:255–263
9. Roudesly F, Oble J, Poli G (2017) Metal-catalyzed CH activation/functionalization: the fundamentals. J Mol Catal a: Chem 426:275–296
10. Shang R, Ilies L, Nakamura E (2017) Iron-catalyzed C–H bond activation. Chem Rev 117:9086–9139
11. Li JJ (2015) CH bond activation in organic synthesis. CRC press
12. (1) Goldman AS, Goldberg KI (2004) Organometallic C–H bond activation: an introduction, ACS; (2) Lapointe D, Fagnou K (2010) Overview of the mechanistic work on the concerted metallation–deprotonation pathway. Chem Lett 39:1118–1126; (3) Ackermann L (2011) Carboxylate-assisted transition-metal-catalyzed C—H bond functionalizations: mechanism and scope. Chem Rev 111:1315–1345; (4) Balcells D, Clot E, Eisenstein O, (2010). C–H bond activation in transition metal species from a computational perspective. Chem Rev 110:749–823; (5) Gallego D, Baquero EA, (2018) Recent advances on mechanistic studies on C–H activation catalyzed by base metals. Open Chem 16: 1001–1058
13. He J, Wasa M, Chan KS et al (2017) Palladium-catalyzed alkyl C–H bond activation. Chem Rev 117:8754–8786
14. Park Y, Kim Y et al (2017) Transition metal-catalyzed C–H lamination: scope, mechanism, and applications. Chem Rev 117:9247–9301
15. Hickman AJ, Sanford MS (2012) High-valent organometallic copper and palladium in catalysis. Nature 484:177–185
16. Gandeepan P, Müller T et al (2019) 3d transition metals for C–H activation. Chem Rev 119(4):2192–2452
17. Choy PY, Wong SM et al (2018) Recent developments in palladium-catalysed non-directed coupling of (hetero) arene C–H bonds with C–Z (Z= B, Si, Sn, S, N, C, H) bonds in bi (hetero) aryl synthesis. Org Chem Front 5:288–321
18. Egorova KS, Ananikov VP (2017) Toxicity of metal compounds: knowledge and myths. Organometallics 36:4071–4090
19. Gallego D, Baquero EA (2018) Recent advances on mechanistic studies on C–H activation catalyzed by base metals. Open Chem 16:1001–1058
20. (a) Liu W, Ackermann L (2016) Manganese-catalyzed C–H activation. ACS Catal 6:3743–3752; (b) Shang R, Ilies L, Nakamura E, (2017) Iron-catalyzed C–H bond activation. Chem Rev 117:9086—9139; (c) Wang H, Moselage M et al (2016) Selective synthesis of indoles by cobalt (III)-catalyzed C–H/N–O functionalization with nitrones. ACS Catal 6:2705–2709

21. Castro LCM, Chatani N (2015) Nickel catalysts/N, N′-bidentate directing groups: an excellent partnership in directed C–H activation reactions. Chem Lett 44:410–421
22. Daugulis O, Do HQ et al (2009) Palladium-and copper-catalyzed arylation of carbon—hydrogen bonds. Acc Chem Res 42:1074–1086
23. Shi S, Nawaz KS et al (2018) Advances in enantioselective C–H activation/mizoroki-heck reaction and Suzuki reaction. Catalysts 8:90
24. Basu D, Kumar S et al (2018) Transition metal catalyzed CH activation for the synthesis of medicinally relevant molecules: a review. J Chem Sci 130:71
25. Sun CL, Shi ZJ (2014) Transition-metal-free coupling reactions. Chem Rev 114:9219–9280
26. Gu Y, Wang D (2010) Direct C-3 arylation of N-acetylindoles with anisoles using phenyliodine bis (trifluoroacetate)(PIFA). Tetrahedron Lett 51:2004–2006
27. Zhang YP, Feng XL (2016) Metal-free, C–H arylation of indole and its derivatives with aryl diazonium salts by visible-light photoredox catalysis. Tetrahedron Lett 57:2298–2302
28. Morofuji T, Shimizu A et al (2012) Metal-and chemical-oxidant-free C–H/C–H cross-coupling of aromatic compounds: the use of radical-cation pools. Angew Chem Int Ed 51:7259–7262
29. Kita Y, Tohma H et al (1994) Hypervalent iodine-induced nucleophilic substitution of para-substituted phenol ethers. Generation of cation radicals as reactive intermediates. J Am Chem Soc 116:3684–3691
30. Eberson L, Hartshorn MP et al (1996) Making radical cations live longer. Chem Commun 18:2105–2112
31. Kita Y, Takada, et al (1996) Hypervalent iodine reagents in organic synthesis: nucleophilic substitution of p-substituted phenol ethers. Pure Appl Chem 68:627
32. Ma JJ, Yi WB et al (2015) Transition-metal-free C–H oxidative activation: persulfate-promoted selective benzylic mono-and difluorination. Org Biomol Chem 13:2890–2894
33. Wang D, Ge B et al (2014) Transition metal-free direct C–H functionalization of quinones and naphthoquinones with diaryliodonium salts: synthesis of aryl naphthoquinones as β-secretase inhibitors. J Org Chem 79:8607–8613
34. Chen J, Wu J (2017) Transition-metal-free C3 arylation of indoles with aryl halides. Angew Chem Int Ed 56:3951
35. Shamsabadi A, Chudasama V (2019) Recent advances in metal-free aerobic C–H activation. Org Biomol Chem 17:2865–2872
36. Jimenez-Gonzalez C, Ponder CS et al (2011) Using the right green yardstick: why process mass intensity is used in the pharmaceutical industry to drive more sustainable processes. Org Process Res Dev 15:912–917
37. Santoro S, Ferlin F et al (2017) Biomass-derived solvents as effective media for cross-coupling reactions and C–H functionalization processes. Green Chem 19:1601–1612
38. Fu XP, Liu L et al (2011) “On water”-promoted direct alkynylation of isatins catalyzed by NHC–silver complexes for the efficient synthesis of 3-hydroxy-3-ethynylindolin-2-ones. Green Chem 13:549–553
39. Fischmeister C, Doucet H (2011) Greener solvents for ruthenium and palladium-catalysed aromatic C–H bond functionalisation. Green Chem 13:741–753
40. Schäffner B, Schäffner F et al (2010) Organic carbonates as solvents in synthesis and catalysis. Chem Rev 110:4554–4581
41. Nie R, Lai R et al (2019) Water-mediated C–H activation of arenes with secure carbene precursors: the reaction and its application. Chem Commun 55:11418–11421
42. Yao C, Qin B et al (2012) One-pot solvent-free synthesis of quinolines by C–H activation/C–C bond formation catalyzed by recyclable iron (III) triflate. RSC Adv 2:3759–3764
43. Rasina D, Kahler-Quesada A et al (2016) Heterogeneous palladium-catalysed Catellani reaction in biomass-derived γ-valerolactone. Green Chem 18:5025–5030
44. Sambiagio C, Schönbauer D et al (2018) A comprehensive overview of directing groups applied in metal-catalysed C–H functionalisation chemistry. Chem Soc Rev 47:6603–6743
45. Tsurugi H, Yamamoto K et al (2010) Oxidant-free direct coupling of internal alkynes and 2-alkylpyridine via double C—H activations by alkylhafnium complexes. J Am Chem Soc 133:732–735

46. Hu H, Liu Y et al (2014) Palladium catalyzed oxidative Suzuki coupling reaction of indolizine at the 3-position using oxygen gas as the only oxidant. RSC Adv 4:24389–24393
47. Tan Y, Yuan W et al (2015) Aerobic Asymmetric dehydrogenative cross-coupling between two C—H groups catalyzed by a chiral-at-metal rhodium complex. Angew Chem Int Ed 54:13045–13048
48. Matsumoto K, Yoshida M et al (2016) Heterogeneous rhodium-catalyzed aerobic oxidative dehydrogenative cross-coupling: nonsymmetrical biaryl amines. Angew Chem 128:5358
49. Gaikwad VV, Bhanage BM (2018) Palladium-catalyzed aerobic oxidative carbonylation of C–H bonds in phenols for the synthesis of p-hydroxybenzoates. Eur J Org Chem 22:2877–2881
50. Rostami A, Khakyzadeh V et al (2018) Co (II)-catalyzed regioselective clean and smooth synthesis of 2-(aryl/alkyl-thio) phenols via sp2 CH bond activation. Molecular Catalysis 452:260–263
51. Albrecht M (2010) Cyclometalation using d-block transition metals: fundamental aspects and recent trends. Chem Rev 110:576–623
52. Patra T, Watile R et al (2016) Sequential meta-C–H olefination of synthetically versatile benzyl silanes: effective synthesis of meta-olefinated toluene, benzaldehyde and benzyl alcohols. Chem Commun 52:2027–2203
53. Herrmann P, Bach T (2011) Diastereotopos-differentiating C–H activation reactions at methylene groups. Chem Soc Rev 40:2022–2038
54. Yoshida JI, Kataoka K et al (2008) Modern strategies in electroorganic synthesis. Chem Rev 108:2265–2299
55. Frontana-Uribe BA, Little RD et al (2010) Organic electrosynthesis: a promising green methodology in organic chemistry. Green Chem 12:2099–2119
56. Cardoso DS, Šljukić B et al (2017) Organic electrosynthesis: from laboratorial practice to industrial applications. Org Process Res Dev 21:1213–1226
57. Kärkäs MD (2018) Electrochemical strategies for C–H functionalization and C–N bond formation. Chem Soc Rev 47:5786–5865
58. Meyer TH, Finger LH et al (2019) Trends in Chemistry 1:63–76
59. Qiu Y, Tian C et al (2018) Electrooxidative ruthenium-catalyzed C—H/O—H annulation by weak O-coordination. Angew Chem Int Ed 57:5818–5822
60. Song G, Wang F et al (2012) C-C, C–O and C–N bond formation via rhodium (iii)-catalyzed oxidative C–H activation. Chem Soc Rev 41:3651–3678
61. Qiu Y, Kong WJ et al (2018) Electrooxidative rhodium-catalyzed C—H/C—H activation: electricity as oxidant for cross-dehydrogenative alkenylation. Angew Chem Int Ed 57:5828
62. Yang QL, Li YQ et al (2017) Palladium-catalyzed C (sp3)—H oxygenation via electrochemical oxidation. J Am Chem Soc 139:3293–3298
63. Li YQ, Yang QL et al (2017) Palladium-catalyzed C (sp2)–H acetoxylation via electrochemical oxidation. Org Lett 19:2905–2908
64. Ma C, Zhao CQ et al (2017) Palladium-catalyzed C–H activation/C–C cross-coupling reactions via electrochemistry. Chem Commun 53:12189–12192
65. Yang QL, Li CZ et al (2018) Palladium-catalyzed electrochemical C–H alkylation of arenes. Organometallics 38:1208–1212
66. Sauermann N, Meyer TH et al (2017) Electrochemical cobalt-catalyzed C–H oxygenation at room temperature. J Am Chem Soc 139:18452–18455
67. Tian C, Massignan L et al (2018) Electrochemical C—H/N—H activation by water-tolerant cobalt catalysis at room temperature. Angew Chem Int Ed 57:2383
68. Tang S, Wang D et al (2018) Cobalt-catalyzed electrooxidative CH/NH [4+2] annulation with ethylene or ethyne. Nat Commun 9:798
69. Yu Y, Zheng P et al (2018) Electrochemical cobalt-catalyzed C–H or N–H oxidation: a facile route to synthesis of substituted oxindoles. Org Biomol Chem 16:8917–8921
70. Yang QL, Wang XY et al (2018) Copper-catalyzed electrochemical C–H amination of arenes with secondary amines. J Am Chem Soc 140:11487–11494
71. Zhang SK, Samanta RC et al (2018) Nickel-catalyzed electrooxidative C—H amination: support for nickel (IV). Chem Eur J 24:19166

72. Santoro S, Ferlin F et al (2019) C–H functionalization reactions under flow conditions. Chem Soc Rev 48:2767–2782
73. Gutmann B, Cantillo D et al (2015) Continuous-flow technology—a tool for the safe manufacturing of active pharmaceutical ingredients. Angew Chem Int Ed 54:6688–6728
74. Mandrelli F, Buco A et al (2017) The scale-up of continuous biphasic liquid/liquid reactions under super-heating conditions: methodology and reactor design. Green Chem 19:1425–1430
75. Lévesque F, Seeberger PH (2012) Continuous-flow synthesis of the anti-malaria drug artemisinin. Angew Chem Int Ed 51:1706–1709
76. Su Y, Straathof NJ et al (2014) Photochemical transformations accelerated in continuous-flow reactors: basic concepts and applications. Chem Eur J 20:10562–10589
77. Vaccaro L, Curini M et al (2018) Definition of green synthetic tools based on safer reaction media, heterogeneous catalysis, and flow technology. Pure Appl Chem 90:21–33
78. Ferlin F, Santoro S et al (2017) Heterogeneous C–H alkenylations in continuous-flow: oxidative palladium-catalysis in a biomass-derived reaction medium. Green Chem 19:2510–2514
79. Xu F, Qian XY et al (2017) Synthesis of 4 H-1, 3-benzoxazines via metal-and oxidizing reagent-free aromatic C–H oxygenation. Org Lett 19:6332–6335
80. Fabry DC, Rueping M (2016) Merging visible light photoredox catalysis with metal catalyzed C–H activations: on the role of oxygen and superoxide ions as oxidants. Acc Chem Res 49:1969–1979
81. Zeitler K (2009) Photoredoxkatalyse mit sichtbarem Licht. Angew Chem 121:9969–9974
82. Karkas MD, Porco JA Jr et al (2016) Photochemical approaches to complex chemotypes: applications in natural product synthesis. Chem Rev 116:9683–9747
83. Ravelli D, Fagnoni M et al (2013) Photoorganocatalysis. What for? Chem Soc Rev 42:97–113
84. Romero NA, Nicewicz DA (2016) Organic photoredox catalysis. Chem Rev 116:10075–10166
85. Wang B, Li P et al (2019) Visible-light induced decarboxylative C2-alkylation of benzothiazoles with carboxylic acids under metal-free conditions. Org Biomol Chem 17:115–121
86. McManus JB, Nicewicz DA (2019) Direct C–H cyanation of arenes via organic photoredox catalysis. J Am Chem Soc 139:2880–2883
87. Margrey KA, Czaplyski WL et al (2018) A general strategy for aliphatic C–H functionalization enabled by organic photoredox catalysis. J Am Chem Soc 140:4213–4217
88. Wang GW (2013) Mechanochemical organic synthesis. Chem Soc Rev 42:7668–7700
89. Cheng H, Hernández JG et al (2017) Mechanochemical ruthenium-catalyzed hydroarylations of alkynes under ball-milling conditions. Org Lett 19:6284–6287
90. Howard JL, Cao Q et al (2018) Mechanochemistry as an emerging tool for molecular synthesis: what can it offer? Chem Sci 9:3080–3094
91. Hermann GN, Bolm C (2017) Mechanochemical rhodium (III)-catalyzed C–H bond amidation of arenes with dioxazolones under solventless conditions in a ball mill. ACS Catal 7:4592–4596

Chapter 12
Atom Economic Green Organic Reactions

Mohan Neetha, Sankaran Radhika, and Gopinathan Anilkumar

1 Introduction

The construction of organic molecules with giant complex structures was a big challenge to synthetic organic chemists from past decade itself. Even though considerable methods provide such complex compounds, the problem associated with these were severe starting material consumption and waste production. Synthetic chemists were in search for reactions which reduced waste generation and consumed minimum raw materials, and afforded good efficiency and selectivity. This point of question can only be solved, when the starting materials are completely consumed to synthesize the desired product alone. This context arouses the opinion of atom economy as the second principle of green chemistry.

Among the twelve principles of green chemistry, the second principle emphasizes the conversion efficiency of reactions. Simply, atom economy is the most generally used metric to calculate the efficiency [1, 2] of greener reactions. Trost has popularized the concept of atom economy [3], which gives the idea that each atom of every starting material being employed or utilized is incorporated in the main product of the chemical reaction, i.e., for a protocol to be 100% atom economic, whole atoms in the reactant should be converted to the desired product and result in minimal waste production. It is measured as follows [4]:

$$\text{Atom economy} = (\text{molecular weight of desired product}/\text{molecular weight of all products}) \times 100\%$$

M. Neetha · S. Radhika · G. Anilkumar (✉)
School of Chemical Sciences, Mahatma Gandhi University, Priyadarsini Hills, Kottayam, Kerala 686560, India
e-mail: anilgi1@yahoo.com

M. Neetha
e-mail: neethamohan131144@gmail.com

S. Radhika
e-mail: radhikasreekrishna@gmail.com

G. Anilkumar and S. Saranya (eds.), *Green Organic Reactions*, Materials Horizons: From Nature to Nanomaterials, https://doi.org/10.1007/978-981-33-6897-2_12

From the equation, it is evident that hundred percentage atom economy of a reaction could be achieved, only if the desired product alone is formed.

The combination of one pot, atom economic and step economic (PASE) organic synthesis is also found to be significantly greener. Such reactions are important path in the synthesis of various crowded tetrahydropyran-4-ones [5].

From a much simpler and transparent angle, the various atom-economic green reactions are segregated into catalyst-free, water-assisted, pericyclic, solvent-free, ionic liquid-mediated, metal-free dehydrogenative coupling and miscellaneous reactions.

1.1 Catalyst-Free Reactions

Bai and co-workers established an innocuous aerobic oxidation of *N*-alkyl iminium salts employing potassium *tert*-butoxide [6]. This protocol is a green approach towards the formation of various lactams from iminium salts of phenanthroline, quinoline, phthalazine, isoquinoline and phenanthridine. The reaction proceeded well under room temperature in the presence of base and solvent without the help of any form of catalysts.

A catalyst-free one-pot green synthetic methodology was used by Mohamadpour for the preparation of dihydropyrano[2,3-*c*]pyrazole derivatives utilizing glycerol as the reaction media through four-component Knoevenagel–Michael cyclocondensation (Scheme 1) [7]. This high atom economic protocol yielded the product from direct workup procedure without any column separation.

Another catalyst-free protocol for the formation of 3,3′-spirooxindole derivatives by the reaction of a carbonyl compound, malanonitrile and isatins in water was developed by Li et al. [8]. This eco-friendly reaction afforded 18 new 3,3′-spirooxindole derivatives in 75–95% yield. A thermal catalyst-free reaction for the synthesis of 3,3′-spirooxindoles was also reported [9].

A novel one-pot green reaction was reported with isocyanides, alcohols or thiols and elemental sulphur affording *O*-thiocarbamates and dithiocarbamates (Scheme 2)

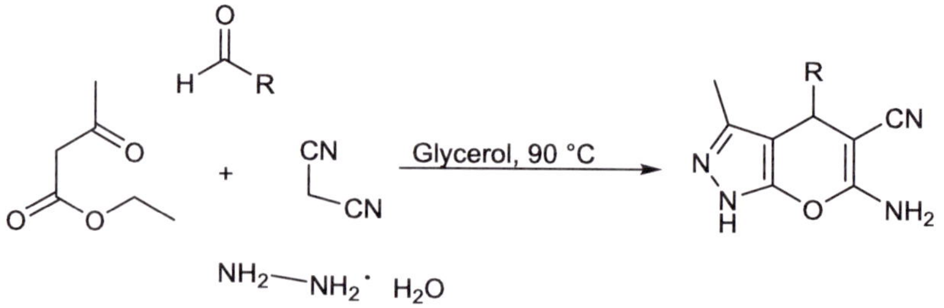

Scheme 1 Synthesis of dihydropyrano[2,3-*c*]pyrazoles

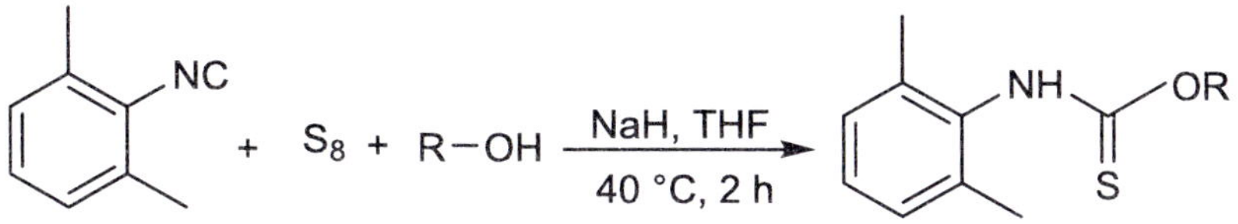

Scheme 2 Synthesis of *O*-thiocarbamates

[10]. This catalyst-free method includes the production of isothiocyanate intermediate, an important scaffold in synthetic chemistry. Good functional group tolerance, moderate to good yields and high atom economy are the major features of this reaction.

1.2 Water-Assisted Reactions

$B(C_6F_5)_3$-catalysed synthesis of indole-substituted 1,4-naphthoquinones in water was developed by Dong et al. [11]. This high atom economical strategy proceeded via the coupling of 1,4-naphthoquinones with the C-3 position of derivatives of indole (Scheme 3). The reaction could render the desired products in moderate to efficient yields without the aid of any organic solvent and base.

Chate and co-workers developed a synthetic procedure towards benzylpyrazolyl coumarins as well as pyrano[2,3-*c*]pyrazoles integrated by isonicotinic acid hydrazide [12]. 4-Hydroxycoumarin, aldehydes, hydrazine hydrate/phenyl hydrazine hydrate and ethyl acetoacetate reacted together to yield benzylpyrazolyl coumarins whereas pyrano[2,3-*c*]pyrazoles were obtained by the reaction of ethyl acetoacetate, aldehyde, dicyanomethane with isoniazid. Both the reactions were catalysed by 2-aminoethanesulphonic acid in water affording high atom economy.

A novel scheme for quinoline-4-carboxylic acid synthesis via the green Doebner reaction with V_2O_5/Fe_3O_4 as catalyst in water was reported by Khillare and co-workers (Scheme 4) [13]. Catalyst recyclability by simple filtration, short period of reaction, high atom economy, use of water as solvent and less percentage of catalyst are the major attraction of this green Doebner reaction.

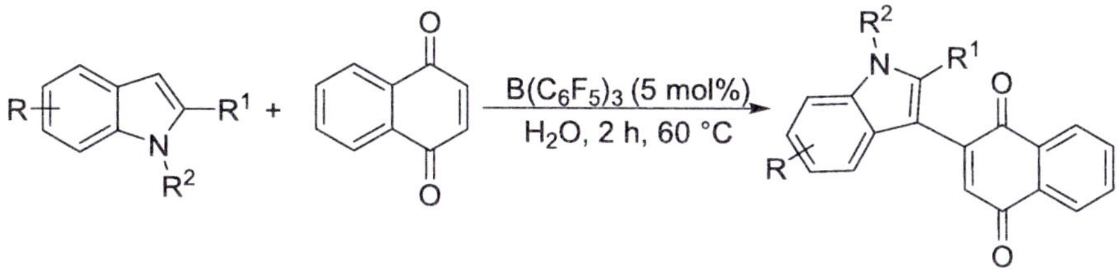

Scheme 3 Synthesis of indole-substituted 1,4-naphthaquinones

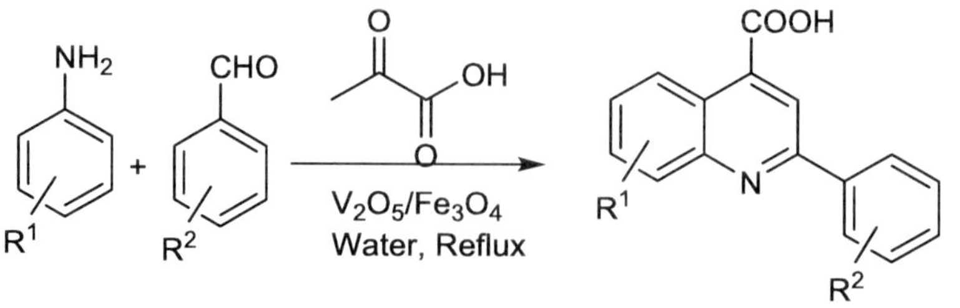

Scheme 4 Synthesis of quinoline-4-carboxylic acids using V_2O_5/Fe_3O_4

Choudhury achieved an effective regioselective base-catalysed synthesis of varieties of new spiro-pyrrolidine-oxindoles using water as the solvent. Isatins, malonitrile and hydantoin or thiohydantoin sequentially reacted in the presence of triethylamine at 70 °C and delivered spirooxindoles tethered with pyrrolizidine unit [14]. A facile lipase-catalysed milder synthetic method to 3,3′-spirooxindoles was reported by Zhang [15]. This novel reaction in water afforded products through Knoevengel–Michael–cyclization, in an eco-favourable manner.

Graphene oxide-catalysed thiolation of indoles was projected as an atom economic strategy by Chen and co-workers [16]. This protocol employed thiols in water for carrying out thiolation, forming 3-sulphenylindoles (Scheme 5). The reaction was performed under organic solvent-free conditions and exhibited wide substrate scope with efficient tolerance towards various functional groups.

Synthesis of various 1,4-disubstituted 1,2,3-triazoles through a one-pot three-component reaction was developed by Nasr-Esfahani and co-workers [17]. The reaction between sodium azide and alkynes with α-bromo ketones or organic halides was catalysed by Cu(II)-TD@nSiO_2 (copper immobilized on nanosilica triazine dendrimer) in (2:1) water:ethanol medium and afforded high atom economy

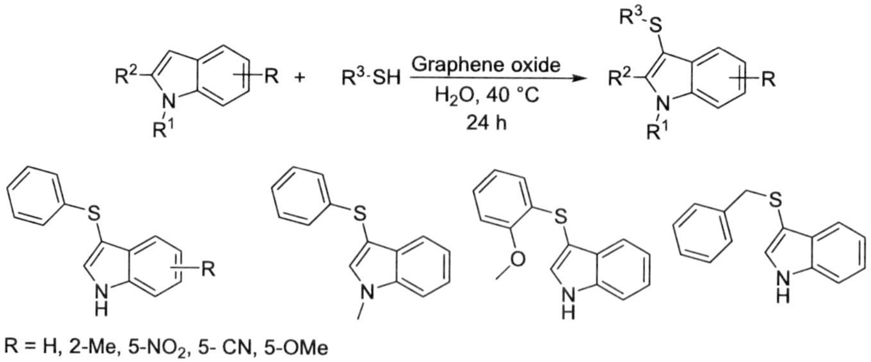

Scheme 5 Substrate scope analysis towards various 3-sulphenylindoles

(Scheme 6). Bis and tris-1,4-substituted 1,2,3-triazoles were also synthesized effectively using the same catalyst.

Nataraj et al. [18] developed a green and highly atom economic synthetic procedure towards N-substituted carbazoles from 2-aminobiaryls. The organic solvent-free and transition metal-free protocol was carried out using peroxodisulphate in water (Scheme 7). The mechanism involved an intramolecular oxidative radical cyclization of the 2-aminobiaryls, resulting in a radical moiety which then underwent an in situ reoxidation.

Brahmachari et al. [19] achieved a green one pot strategy for the synthesis of 3,3′-(arylmethylene)bis(4-hydroxy-2*H*-chromen-2-one), a biologically important biscoumarin derivative, by the reaction of 4-hydroxycoumarin with aromatic aldehydes, utilizing ceric ammonium nitrate (CAN) in aqueous medium at room temperature (Scheme 8). This protocol is cost-effective, environmentally suitable and highly atom economic. Both electron withdrawing and releasing functional groups afforded

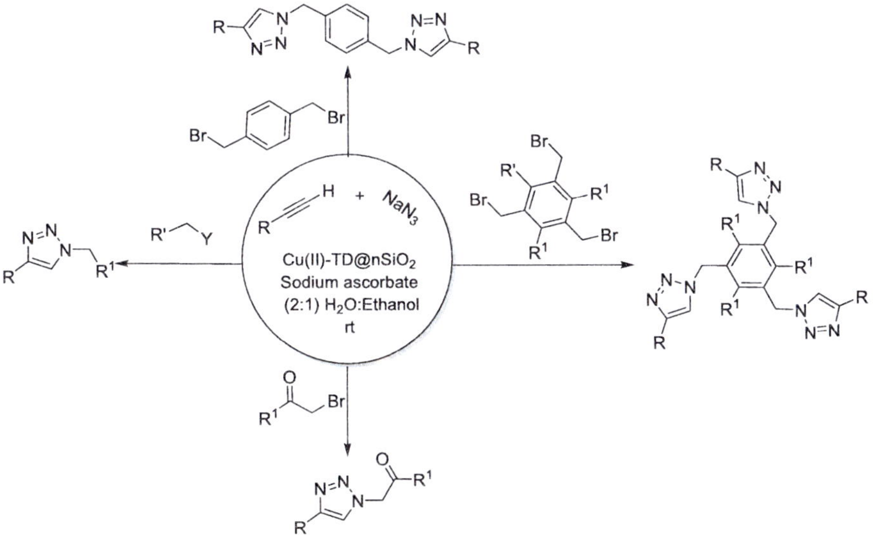

Scheme 6 Cu(II)-TD@nSiO$_2$ catalysed synthesis of 1,4-disubstituted 1,2,3-triazoles, bis and tris-1,4-substituted 1,2,3-triazoles

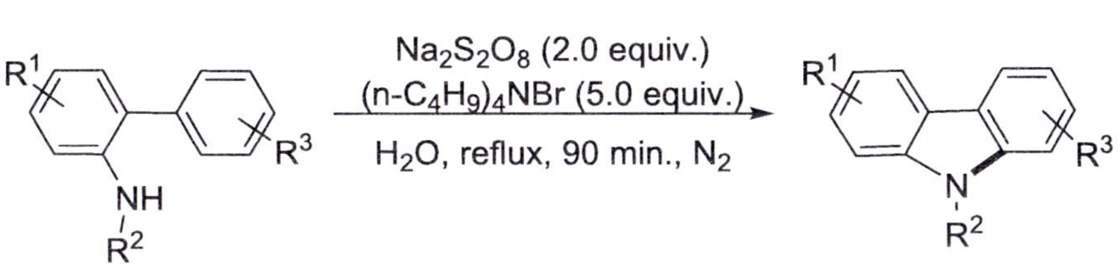

Scheme 7 Peroxodisulphate catalysed synthesis of N-substituted carbazoles

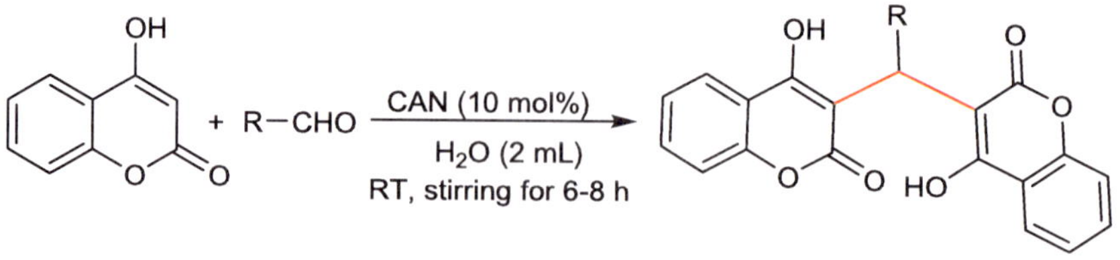

Scheme 8 Synthesis of 3,3′-(arylmethylene)bis(4-hydroxy-2*H*-chromen-2-one) in water

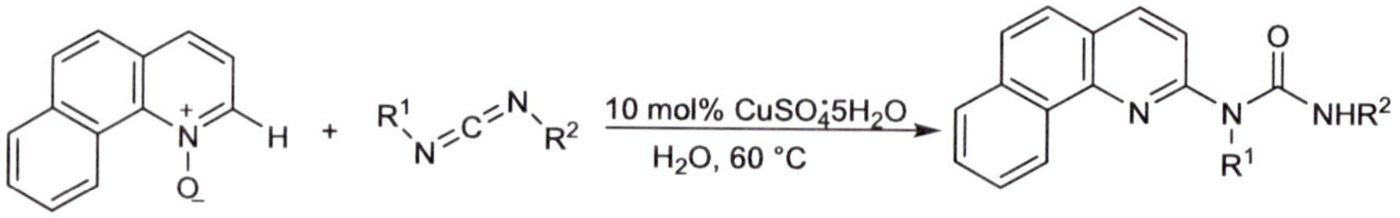

Scheme 9 Synthesis of quinolin-2-yl substituted ureas

the products in good yields. The proposed mechanism suggests that both Knoevenagel and Michael-addition reactions are responsible for the conversion of the reactant.

A hundred percentage atom economic and eco-friendly synthesis of different quinolin-2-yl substituted ureas utilizing water as solvent from carbodiimides and quinoline *N*-oxide was disclosed by He et al. [20] (Scheme 9). This base-free condition yielded products in good percentage with excellent regioselectivity via filtration, followed by ethanol washing. Mild reaction conditions, non-toxic raw materials and absence of organic solvents are the other peculiarities of this reaction.

1.3 Pericyclic Reactions

Yellappa [21] put forward a one-pot synthesis of indole-spiro(indene-pyrrolidine) via a 1,3-cycloaddition of unsymmetrical dipolarophiles with azomethineylide. The dipolarophile was synthesized from 1-acetyl-1*H*-indol-3-yl derivative and indol-3-yl. The decarboxylative addition of sarcosine, an amino acid, with ninhydrin resulted in the azomethineylide. The protocol exhibited wide functional group tolerance and synthesized different substituted indole-spiro(indene-pyrrolidine) in moderate yields with high atom economy.

A new synthetic avenue towards *trans* β-lactams via an "On water" asymmetric Kinugasa reaction between nitrones and alkynes (Scheme 10) was discussed by Feng and co-workers in 2013 [22]. In contradiction to the ordinary Kinugasa reaction (which carry [3 + 2] cycloaddition reaction of a nitrone with an olefin), high percentage yield of desired products with amazing diastereoselectivity and enantioselectivity were acquired.

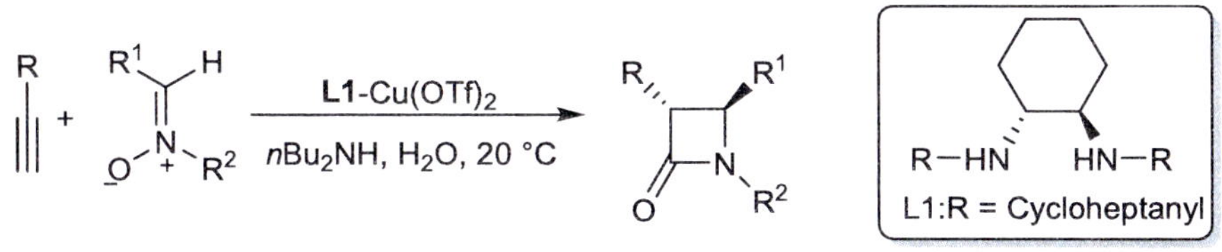

Scheme 10 On-water asymmetric Kinugasa reaction

Aqua-mediated stereoselective one-pot synthesis of spiro[acenaphthylene-1,2′-thiapyrrolizidine] substituents using NaCl via 1,3-dipolar addition of 1,3-thiazole-4-carboxylic acid, acenaphthenequinone and Knoevenagel adduct was established by Dandia [23]. The incorporation of NaCl increased the hydrophobicity of the reaction, which in turn improved the yield of the desired product. The main attraction of this strategy is the formation of four chiral centres consisting of one spiro centre having two C–C and one C–N bonds in a one pot condition, which opens a new area in drug synthesis. Product separation without column chromatography, high atom economy and waste-free isolation of products are the main benefits of this reaction.

Shao and co-workers reported the synthesis of pyrazoles via a Bu_4NI-catalysed strategy [24]. The major steps involved in this protocol were sequential [3 + 2] cycloaddition and oxidative dehydrogenation reactions (Scheme 11). The reaction exhibited wide substrate scope with easily accessible starting reagents and projected itself as an environmentally innocuous protocol rendering high atom economy.

Quinazolines are pharmaceutically relevant heterocyclic systems exhibiting analgesic, anti-hypertensive, anti-convulsant and anti-histaminic properties [25]. A microwave-assisted green approach towards their synthesis was achieved by a Diels–Alder reaction of 2*H*-benzo[*d*][1,3]oxazine-2,4(1*H*)-dione (isatoic anhydride) with azomethines rendering high atom economy (Scheme 12) [26]. This protocol was more efficient than the conventional synthesis and proceeded at a faster rate providing good to excellent yields of quinazolines.

Mo et al. succeeded in developing a [5 + 1] annulation in water for the synthesis of pyrazino[1,2-*a*]indole-2-oxides including 2-carbonyl-1-propargylindoles in the presence of hydroxylamine in water with nickel(II) as the catalyst [27]. The oxime intermediate generated from 2-carbonyl-1-propargylindoles and hydroxylamine, underwent a 6-*exo-dig* cyclization catalysed by nickel. The used nickel catalyst was recycled through seven cycles without losing the catalytic efficiency.

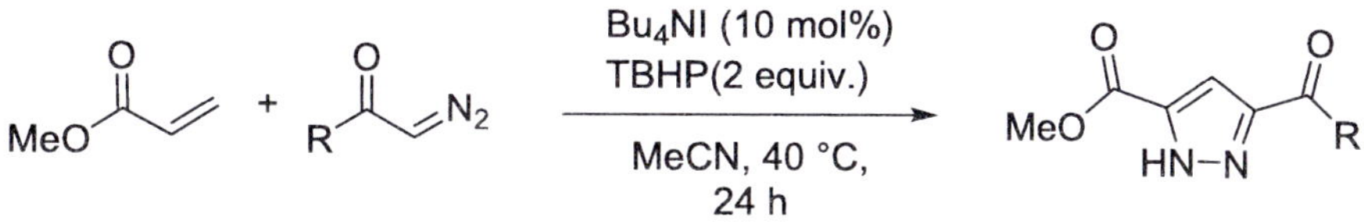

Scheme 11 Bu_4NI-catalysed synthesis of pyrazoles

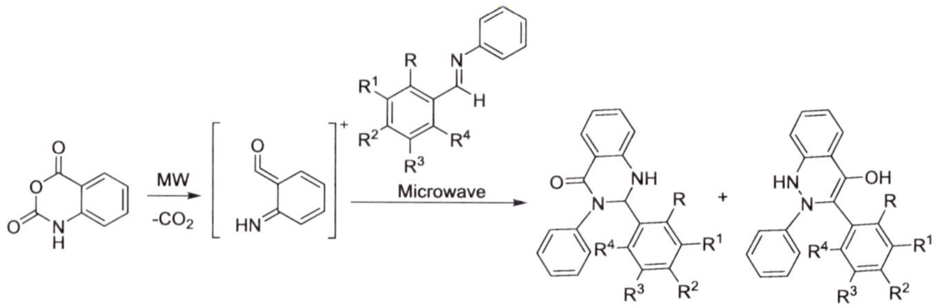

Scheme 12 Synthesis of substituted quinazoline products

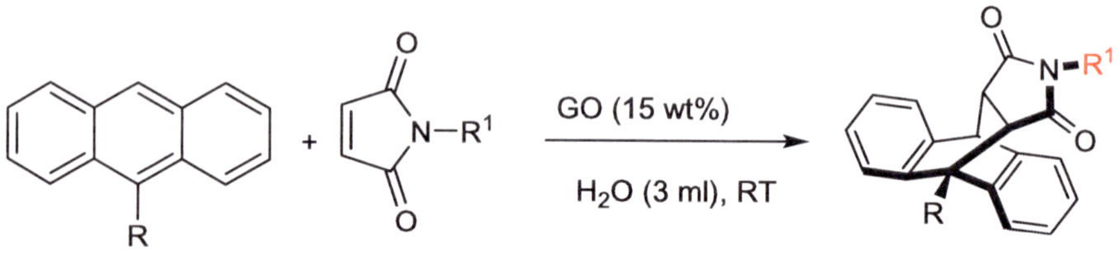

Scheme 13 Diels–Alder reaction using GO as catalyst at room temperature

Graphene oxide (GO)-based Diels–Alder reaction in water was explored by De et al. [28]. This [4 + 2] cycloaddition between 9-hydroxymethylanthracene and *N*-substituted maleimides was feasible at room temperature, when GO is used as carbocatalyst (Scheme 13). The strategy showed many advantages like cost effectiveness, absence of metal catalyst, being a better catalyst than graphene and extensive substrate scope.

1.4 Solvent-Free Reactions

Tufail disclosed a novel malic acid promoted solvent-free Friedlander approach for the construction of polysubstituted quinolones (Scheme 14) [29]. Malic acid as a promoter provided several advantages like high atom economy, broad range of substrate scope and excellent yields.

Pinxterhuis and co-workers explored a solvent-free green Pd-catalysed coupling of organolithium compounds with organic halides [30]. This approach overcomes several challenges that already existed in Pd-catalysed organolithium coupling like removal of humidity, dilution, long reaction period, etc. Even though metal catalysis enhanced atom economy of the reaction, such reactions along with solvent-free conditions significantly improved atom economy furthermore.

A new easy approach to benzothiophene synthesis by the cyclization between thiophenols and alkynes under solvent-free condition was explored (Scheme 15) [31].

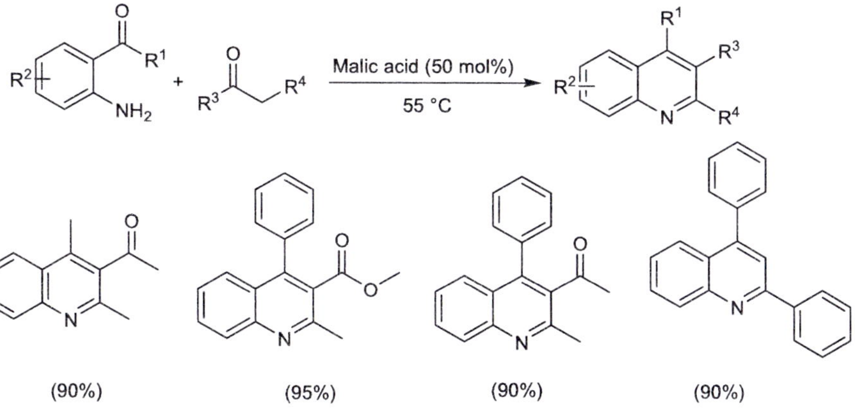

Scheme 14 Substrate scope for acyclic α-methylene compounds

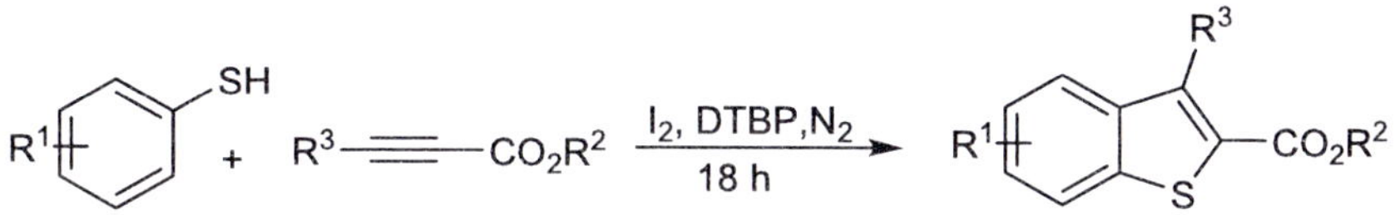

Scheme 15 Synthesis of benzothiophenes through iodine catalysis

This cascade iodine-mediated scheme is inexpensive, efficient and showed reactivity with many functional groups. They could also obtain several benzothiophene frameworks, which are significant in medicinal chemistry in excellent yields. Alkynes having electron-withdrawing substituents resulted in more yield. But thiophenols having electron-withdrawing, releasing and neutral substituents resulted in products without much difference in yields.

Pyrazole-fused 4*H*-pyran and coumarin-fused 4*H*-pyran, namely dihydropyrano[2,3-*c*]pyrazole pyrano[3,2-*c*]chromenone were synthesized [32]. Reaction between aromatic aldehydes, (*E*)-*N*-methyl-1-(methylthio)-2-nitroethenamine (NMSM) with 3-methyl-1-phenyl-5-pyrazolone or 4-hydroxycoumarin under catalyst-free neat conditions accomplished the corresponding products from good to excellent yields. Wide range of substrate study and simple isolation of product are other characteristics of this reaction.

Mirjalili used $TiCl_2$/Nano-γ-Al_2O_3, a new Lewis acid catalyst, along with aldehydes, 1,3-dicarbonyls and ammonium acetate for the one-pot 1,4-dihydropyridines synthesis (Scheme 16) [33]. The condensation in the presence of nano-Lewis acid takes place at 90 °C without any solvent. This greener fruitful reaction is atom economic and superior in selectivity.

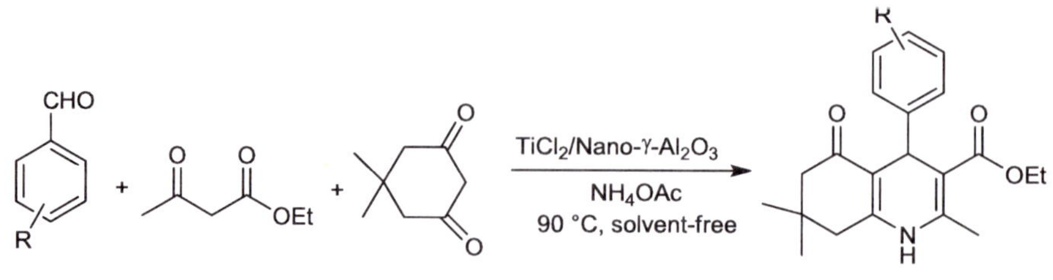

Scheme 16 One-pot 1,4-dihydropyridines synthesis

Wells–Dawson diphosphooctadecatungstic acid ($H_6P_2W_{18}O_{62}{\cdot}24H_2O$) catalysed green synthesis of bis(indolyl)methane analogues under neat conditions was established by Taybee and co-workers [34]. The synthesis involved the electrophilic substitution of a range of carbonyl compounds with indole, providing an excellent atom economic pathway. The reaction of indole with carbonyl compound proceeded via an azafulvenium salt, which then underwent an addition reaction with a second molecule of indole resulting in the bis(indolyl)methane analogues in good to excellent yields.

1.5 *Ionic Liquid-Mediated Reactions*

Sakhalkar et al. explored a novel method for the synthesis of chloroaluminate ionic liquids (ILs), an important catalyst in homogeneous catalysis [35]. Easily accessible starting materials like aluminium chloride and tributylamine provided the desired ionic liquid through an amine–aluminium chloride adduction. This strategy overcomes the problem associated with the conventional method and provided a better atom economy with minimal waste production. The synthesized chloroaluminate ionic liquids were utilized as catalysts for green Friedel–Crafts alkylation reactions.

H_2O_2/[TEBSA][BF_4]/$CO_4HP_2MO_{15}V_3O_{62}$ catalysed oxidation of various ketones and aldehydes was established by Hu and co-workers [36]. Good to excellent yields of the esters and carboxylic acids were obtained by this mild and facile protocol (Scheme 17). Simple workup procedure and improved catalytic ability with efficient atom economy are the highlights of this reaction.

Brønsted acid ionic liquid (BAIL) catalysed synthesis of 6-amino-4-substituted-3-methyl-2,4-dihydropyrano[2,3-*c*]pyrazole-5-carbonitriles was designed by

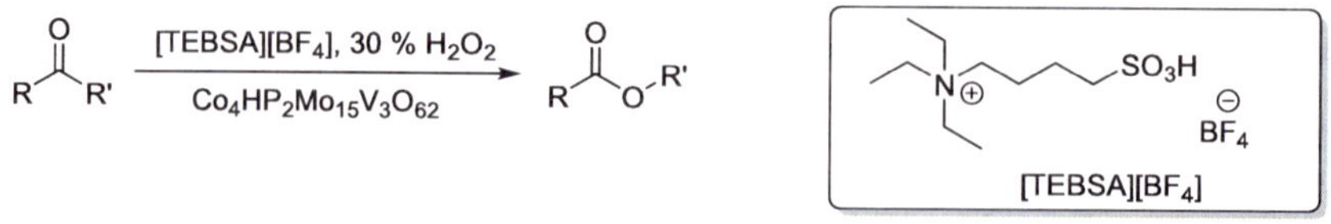

Scheme 17 H_2O_2/[TEBSA][BF_4]/$CO_4HP_2MO_{15}V_3O_{62}$ catalysed oxidation of aldehydes and ketones

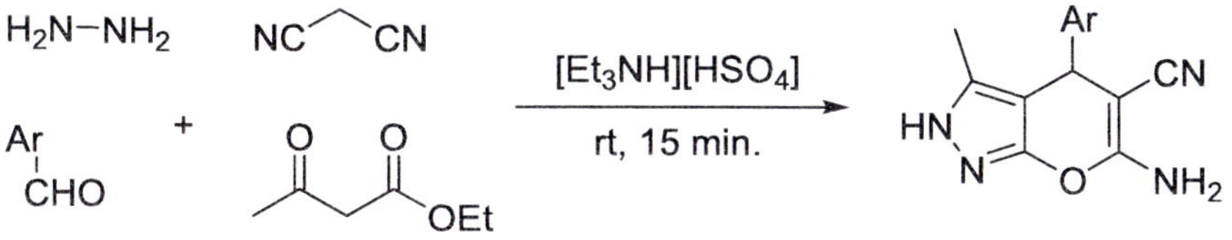

Scheme 18 Synthesis of 6-amino-4-substituted-3-methyl-2,4-dihydropyrano[2,3-*c*]pyrazole-5-carbonitriles

Nimbalkar et al. [37]. Triethylammonium hydrogen sulphate [Et_3NH][HSO_4] was chosen as the BAIL for this strategy and could catalyse the one-pot reaction between aryl aldehydes, hydrazine hydrate, propanedinitrile and ethylacetoacetate under neat conditions (Scheme 18). The reaction proceeded at a faster rate imparting excellent yields with high catalytic recyclability. The synthesized products were active against various cancer cell lines.

Carbon dioxide is very toxic to the environment as well as to the humans. It increases the global temperature of earth significantly. Conversion of carbon dioxide into valuable organic compounds has always gained much attention. Carboxylative cycloaddition of CO_2 to propargylic alcohols is one of the methods for CO_2 fixation [38]. This highly atom economic protocol synthesizes biologically relevant five-membered cyclic carbonates in a single step using metal-, organo-, electrochemical- and ionic liquid-catalysed strategies.

1.6 Metal-Free Dehydrogenative Coupling Reactions

Nowadays, metal-free method for cross-dehydrogenative couplings (CDC) is much significant than metal involved CDC reactions. These types of reactions gained attention since they are important in the construction of C–C, C–N, C–O, C–S and C–Se bonds. But molecular iodine catalysed CDC reactions has been noticed as more atom economic and greener with wide range of substrate scope and regioselectivity [39].

An iodine catalysed cyclopropane ring formation was reported by Itoh from aromatic olefins, using active methylene compounds in the presence of visible light (Scheme 19) [40]. Various substituted cyclopropane derivatives were obtained through a carbon–iodine bond breakage mechanism, in moderate to good yields. Usually, styrenes with electron-deficient groups yielded more product yields than those with electron-rich systems.

Scheme 19 Iodine catalysed cyclopropanation

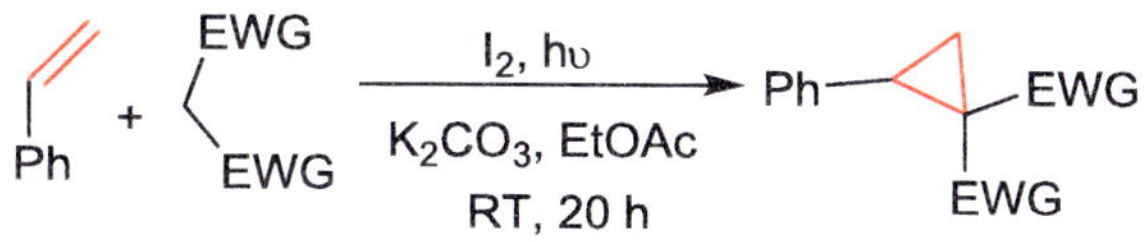

A tandem approach for the stereospecific formation of E and Z isomer of 2-thio-1,4-enediones with TMSOTf and $BF_3{\cdot}Et_2O$, respectively, by self-coupling of terminal alkynes was developed by Shah [41]. This iodine catalysed reaction also provided an opportunity for the synthesis of β-thio-γ-keto-α,β-unsaturated esters via cross-coupling with ethyl glyoxylate. A wide variety of terminal alkynes yielded the desired diones in excellent yields.

Iodine-catalysed β-hydroxysulphide synthesis via the formation of C-O and C-S bonds in one step was reported by Peddinti and co-workers [42]. Readily available styrenes and thiophenols reacted in DMSO affording the corresponding products in good yields (Scheme 20). Starting materials containing halogens and electron-releasing groups afforded better yields of the products. This method is very simple, cost-effective, greener and safe.

Introduction of several aromatic thiols on imidazo[1,2-*a*]-pyridines, -pyrimidines and [1,2-*b*]pyridazines in PEG-400, using molecular iodine as catalyst and H_2O_2 as oxidizing agent was discovered by Hiebel [43]. This sulphenylation method with different thiophenols are effective, facile, regioselective and tolerated many functional groups.

A metal-free molecular iodine-mediated formation of α-carbonyl selenocyanates, which are important starting materials for biologically active molecules, from aromatic methyl ketones via selenocyanation was reported (Scheme 21) [44]. This method avoids the usage of previously prepared α-halo ketones. This convenient strategy provided a broad substrate scope study and supplied desired selenocyanates in good yields. Mechanistic study found that, the generation of iodine radical is essential for the progress of this reaction.

Benzamides can be smoothly prepared by domino approach of ethylarenes with aqueous ammonia taking iodine as the catalyst (Scheme 22) [45]. This novel procedure progressed via a triiodomethyl ketone transition state, which is formed from

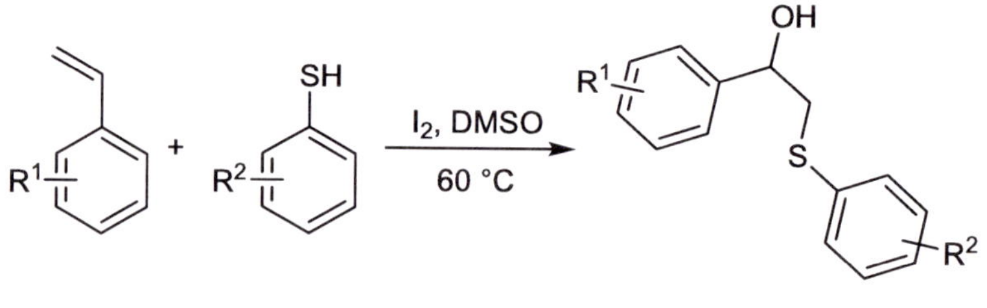

Scheme 20 Synthesis of β-hydroxysulphides through iodine catalysis

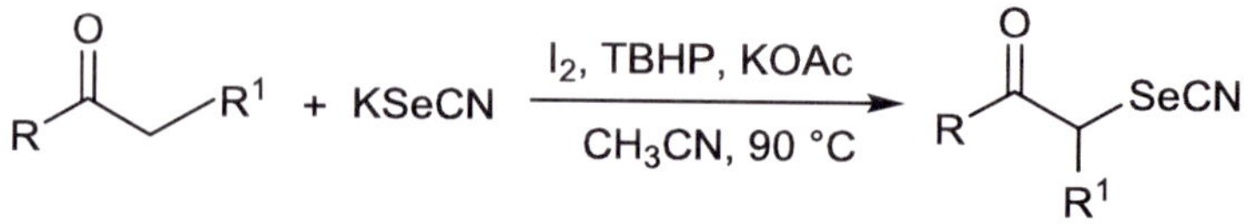

Scheme 21 Formation of α-carbonyl selenocyanates via iodine catalysis

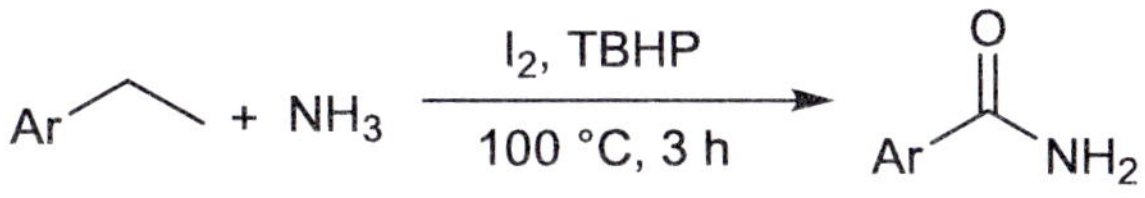

Scheme 22 Metal-free synthesis of benzamides in aqueous ammonia

the oxidant TBHP and iodine. An amide is then formed by the nucleophilic substitution of the intermediate with ammonia. The efficiency of the reaction was confirmed through 17 substrates, which yielded the respective products ranging from 73 to 94%.

A straightforward indoline synthesis from intramolecular amination of anilines was established by Qui (Scheme 23) [46]. This reaction created a novel pathway for the formation of (sp^3)C—N bonds by the selective breakage of (sp^3)C—H bonds over (sp^2)C—H bonds. This iodine-catalysed methodology furnished varieties of heterocycles having nitrogen atom. The proposed mechanism in the study suggested the cleavage of N–I bond.

The single-step oxidative coupling of alkenes with β-keto esters or 2-pyridinyl-β-esters, described by Lei et al., is one of the best avenues for the synthesis of dihydrofurans and indolizines, respectively (Scheme 24) [47]. This iodine-mediated annulation was believed to proceed via a radical addition or cyclization mechanism.

A photo-induced selective synthetic route to substituted piperidines using (sp^3)C—H amination was explored (Scheme 25) [48]. This intramolecular amination path operates through two catalytic cycles, in which the first one is for light-mediated radical C–H abstraction and the other for iodine catalysed C–N bond formation. This protocol also promoted the synthesis of pyrrolidine derivatives, which are commonly prepared through Hofmann–Löffler reaction.

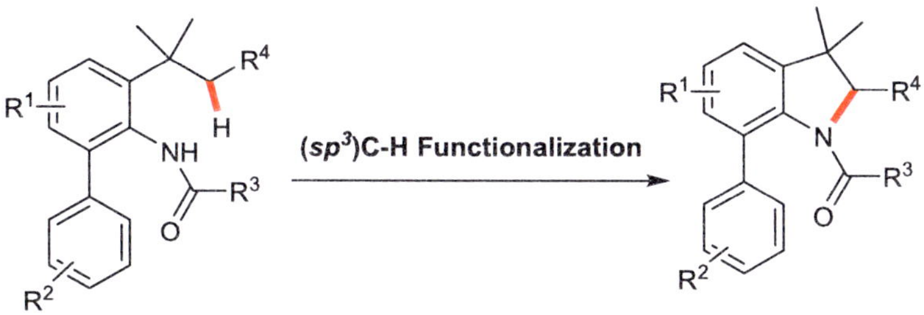

Scheme 23 Selective (sp^3)C—H functionalization

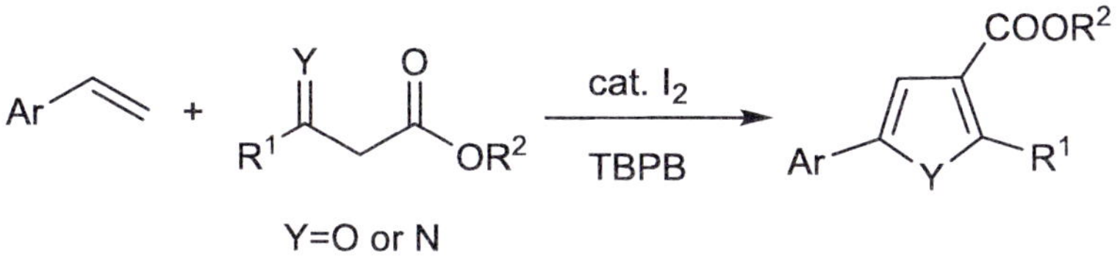

Scheme 24 Synthesis of dihydrofurans and indolizines

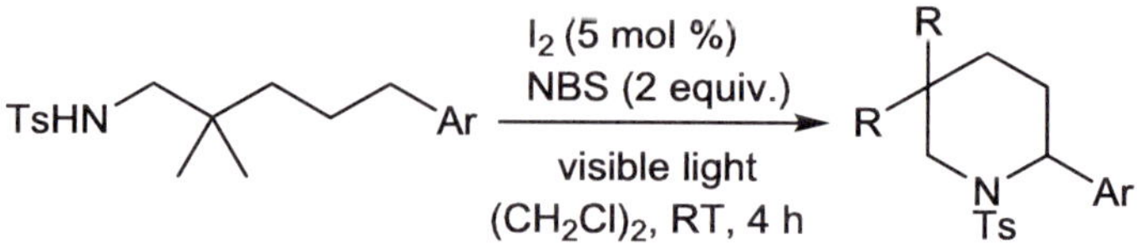

Scheme 25 Iodine-catalysed piperidine formation

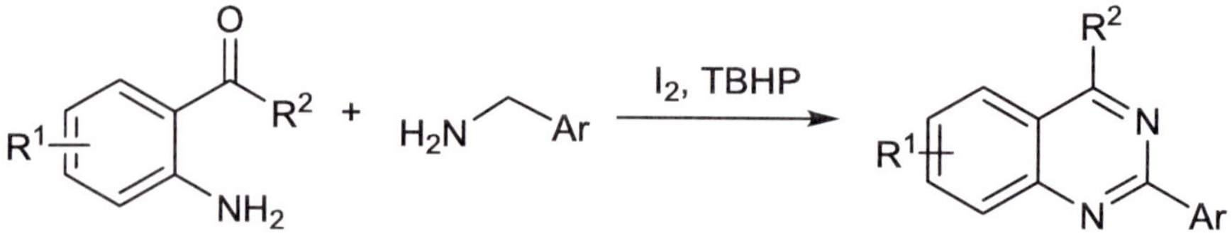

Scheme 26 Synthesis of 2-phenylquinazolines

Taking 2-aminobenzophenones and benzylic amines as starting materials, Wang and co-workers described the formation of 2-phenylquinazolines via using iodine as catalyst (Scheme 26) [49]. This reaction is achieved through a tandem pathway, followed by sp^3 C–H functionalization. Electron-releasing groups on the phenyl ring disgraced the reaction yield. *Ortho* substituents disfavour the reaction due to steric hindrance. This novel iodine-catalysed approach has different merits such as metal-free, elimination of toxic reagents, easy availability of reactants; moreover, it is facile and efficient.

1.7 Miscellaneous Reactions

A mild and efficient green method towards the fixation of CO_2 was manifested by Garai et al. [50]. Covalent organic polymer (COP-213), a zwitterionic π-conjugated catalyst showcased improved selectivity for the cycloaddition of CO_2 to epoxide and could bring about effective conversion of carbon dioxide to cyclic carbonates in the absence of solvents and co-catalysts.

Singh et al. [51] designed a novel approach for the Rose Bengal catalysed cross-coupling of phenylene, 1,2–dicarbonyls and 1,2–diamines in quinoxaline derivative synthesis (Scheme 27). This protocol was found to take place at room temperature in the presence of visible light via a radical route.

(BMOPs) polybismaleimide-based microporous organic polymers were designed via a green, atom economic initiative [52]. The monomer bismaleimide underwent a one-step thermal-initiation homopolymerization in diphenyl sulphone. The highlights of this reaction include exclusion of any initiator and thereby reducing side product formation. These green specialities pave way to a large-scale production of these polymers for industrial applications.

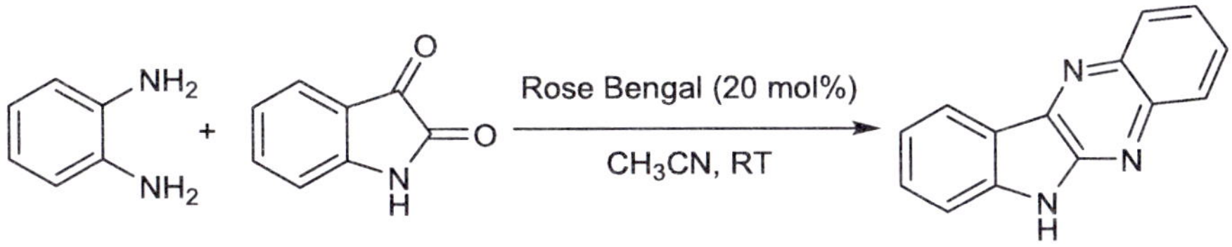

Scheme 27 Synthesis of quinoxalines via Rose Bengal catalysis

Ji and co-workers designed a procedure towards the synthesis of indoles and pyrrolo[3,2,1-*ij*]quinolones [53]. The metal-free protocol utilized NH_4PF_6, a less toxic, economic and safe inorganic salt to promote the synthesis (Scheme 28). The synthesized products obtained after a cyclodehydration were projected as relevant pharmacophores. The process was carried out in HFIP (1,1,1,3,3,3-hexafluoropropan-2-ol) solvent and provided excellent atom-economy.

Pradhan mentioned the preparation of support-free Pd_3CO nanocomposites, found as a good catalytic system for iodonium ylides with acrylates, boronic acids and arylalkynes, in Heck, Suzuki and Sonogashira cross-coupling reactions, respectively [54]. This coupling is progressed by the movement of phenyl ring together with the formation of α-iodoenones, as the intermediate and yielded the coupled product with high atom economy. This coupling accesses a new path in the synthesis of diverse organic compounds.

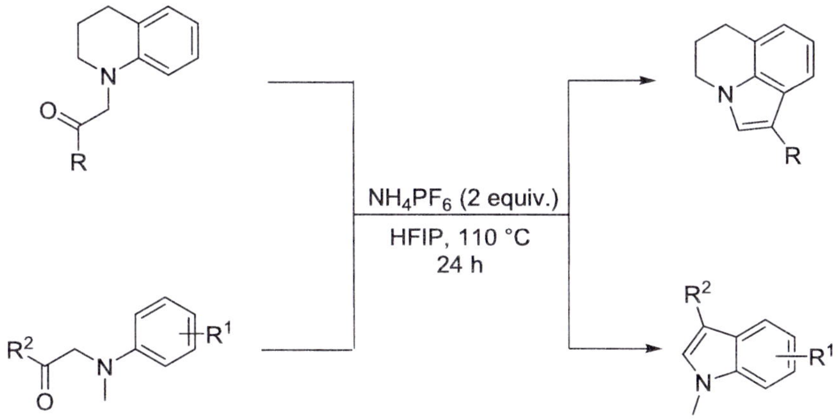

Scheme 28 Synthesis of pyrrolo[3,2,1-*ij*]quinolines and indoles

2 Conclusion

Atom economic reactions play vital roles in the sustainable development, as it goals to bring down the amount of waste in a chemical reaction to a molecular level. High atom economic reactions, obeying greener principles, are safe and efficient. The notion of atom economy basically exhibits the percentage of atoms that is used up in the major product, comparing with the actual no. of atoms initially present in the reaction. So far from the above discussion, it is clear that various modern green organic reactions are highly atom economic and a few of them are hundred percentage atom economic. Scientists of all organic fields are still attempting their best to bring out all reactions to an ideal atom economic situation.

Acknowledgements MN and SR thank the Kerala State Council for Science Technology and Environment (KSCSTE), Trivandrum research fellowships. GA thanks the KSCSTE, Trivandrum, India for a research grant (Order No.341/2013/KSCSTE dated 15.03.2013). The authors are also thankful to EVONIK Industries, Germany for financial support (ECRP 2016 dated 6.10.2016).

References

1. Trost BM (1995) Atom economy—a challenge for organic synthesis: homogeneous catalysis leads the way. Angew Chem Int Ed Engl 34(3):259–281
2. Li C-J, Trost BM (2008) Green chemistry for chemical synthesis. Proc Natl Acad Sci 105(36):13197–13202
3. Trost BM (1991) The atom economy–a search for synthetic efficiency. Science 254(5037):1471–1477
4. Jiménez-González C, Constable DJC, Ponder CS (2012) Evaluating the "Greenness" of chemical processes and products in the pharmaceutical industry—a green metrics primer. Chem Soc Rev 41(4):1485–1498
5. Clarke PA, Santosa S, Martin WHC (2007) Combining pot, atom and step economy (PASE) in organic synthesis. Synthesis of tetrahydropyran-4-ones. Green Chem 9(5):438–440
6. Bai L-G, Zhou Y, Zhuang X et al (2020) Base-promoted aerobic oxidation of N-alkyl iminium salts derived from isoquinolines and related heterocycles. Green Chem 22(1):197–203
7. Mohamadpour F (2019) Four-component synthesis of Dihydropyrano[2,3-*c*]pyrazole scaffolds using glycerol as green reaction media under catalyst-free conditions. Polycyclic Aromat Compd. https://doi.org/10.1080/10406638.2019.1709084
8. Zhao L-Q, Zhou B, Li Y-Q (2011) An efficient one-pot three-component reaction for synthesis of spirooxindole derivatives in water media under catalyst-free condition. Heteroat Chem 22(5):673–677
9. Elinson MN, Ilovaisky AI, Merkulova VM et al (2012) Non-catalytic thermal multicomponent assembling of Isatin, cyclic CH-acids and Malononitrile: an efficient approach to Spirooxindole scaffold. Mendeleev Commun 22(3):143–144
10. Németh AG, Keserű GM, Ábrányi-Balogh P (2019) A novel three-component reaction between isocyanides, alcohols or thiols and elemental sulfur: a mild, catalyst-free approach towards O-thiocarbamates and dithiocarbamates. Beilstein J Org Chem 15:1523–1533
11. Dong Y, Zhang H, Yang J et al (2019) $B(C_6F_5)_3$-catalyzed C–C coupling of 1,4-Naphthoquinones with the C-3 position of Indole derivatives in water. ACS Omega 4(25):21567–21577

12. Chate AV, Shaikh BA, Bondle GM, Sangle SM (2019) Efficient atom-economic one-pot multicomponent synthesis of benzylpyrazolyl coumarins and novel pyrano[2,3-*c*]pyrazoles catalysed by 2-aminoethanesulfonic acid (taurine) as a bio-organic catalyst. Synth Commun 49(17):2244–2257
13. Arbad BR, Khillare SL, Lande MK et al (2017) Synthesis, characterization and catalytic application of V_2O_5/Fe_3O_4 as heterogeneous catalyst for the synthesis of Quinoline-4-carboxylic acid derivatives. Der Pharma Chemica 9(6):30–36
14. Karamthulla S, Pal S, Choudhury LH et al (2013) Synthesis of novel spiro[indoline-3,7′-pyrrolo[1,2-*c*]imidazole]-6′-carbonitrile derivatives in water using a regioselective sequential three component reaction. RSC Adv 3(36):15576–15581
15. Chai S-J, Lai Y-F, Xu J-C et al (2011) One-pot synthesis of Spirooxindole derivatives catalyzed by Lipase in the presence of water. Adv Synth Catal 353(2–3):371–375
16. Chen M, Luo Y, Zhang C et al (2019) Graphene oxide mediated thiolation of indoles in water: a green and sustainable approach to synthesize 3-sulfenylindoles. Org Chem Front 6(1):116–120
17. Nasr-Esfahani M, Mohammadpoor-Baltork I, Khosropour AR et al (2014) Copper immobilized on nanosilica triazine dendrimer (Cu(II)-TD@nSiO_2)-catalyzed regioselective synthesis of 1,4-disubstituted 1,2,3-triazoles and bis- and tris-triazoles via a one-pot multicomponent click reaction. J Org Chem 79(3):1437–1443
18. Natarajan P, Priya CD (2017) Transition-metal-free and organic solvent-free conversion of N-substituted 2-aminobiaryls into corresponding carbazoles via intramolecular oxidative radical cyclization induced by peroxodisulfate. Green Chem 19(24):5854–5861
19. Brahmachari G, Begam S (2019) Ceric ammonium nitrate (CAN): an efficient and eco-friendly catalyst for one-pot synthesis of diversely functionalized biscoumarins in aqueous medium under ambient conditions. ChemistrySelect 4(19):5415–5420
20. Jiang S, Cao Z, He W-M et al (2019) Clean preparation of Quinolin-2-yl substituted ureas in water. ACS Sustain Chem Eng 7(7):7193–7199
21. Yellappa S (2019) An anti-Michael route for the synthesis of indole-spiro(indene-pyrrolidine) by 1,3-cycloaddition of azomethineylide with indole-derivatised olefins. J Heterocycl Chem. https://doi.org/10.1002/jhet.3843
22. Liu X, Feng X, Chen Z et al (2013) Asymmetric synthesis of trans-β-Lactams by a Kinugasa reaction on water. Chem Eur J 19(23):7561–7567
23. Dandia A, Parewa V, Kumar S et al (2016) Imposed hydrophobic interactions by NaCl: accountable attribute for the synthesis of spiro[acenaphthylene-1,5′-pyrrolo[1,2-c]thiazole] derivatives via 1,3-dipolar cycloaddition reaction in aqueous medium. Green Chem 18(8):2488–2499
24. Shao Y, Tong J, Zhao Y et al (2016) [3 + 2] cycloaddition and subsequent oxidative dehydrogenation between alkenes and diazo compounds: a simple and direct approach to pyrazoles using TBAI/TBHP. Org Biomol Chem 14(36):8486–8492
25. Alagarsamy V, Chitra K, Saravanan G (2018) An overview of quinazolines: pharmacological significance and recent developments. Eur J Med Chem 151:628–685
26. Mullasseril A (2013) Substituent effects on solvent-free synthesis (MWAOS) of dihydroquinazolinones by the addition of isatoic anhydride to a series of phenyl-substituted n-(phenylmethylidene) anilines. J Chem 2013:1–4
27. Xiao Y, Su G-F, Mo D-L et al (2019) Nickel(II)-catalyzed [5 + 1] annulation of 2-carbonyl-1-propargylindoles with hydroxylamine to synthesize pyrazino[1,2-a]indole-2-oxides in water. J Org Chem 84(16):9859–9868
28. De M, Girish YR, Pandit S et al (2017) Graphene oxide as a carbo-catalyst for Diels-Alder reaction in aqueous media. Chem Asian J 12(18):2393–2398
29. Singh J, Singh J, Tufail F et al (2017) Bioorganopromoted green friedländer: a versatile new Malic acid promoted solvent free approach to multisubstituted quinolones. New J Chem 41(4):1618–1624
30. Pinxterhuis EB, Giannerini M, Hornillos V et al (2016) Fast, greener and scalable direct coupling of organolithium compounds with no additional solvents. Nat Commun 7:11698
31. Yan K, Yang D, Wang H et al (2015) Facile access to benzothiophenes through metal-free iodine-catalyzed intermolecular cyclization of thiophenols and alkynes. Synlett 26(13):1890–1894

32. Khan MM, Shareef S, Saigala et al (2019) A catalyst- and solvent-free protocol for the sustainable synthesis of fused 4H-pyran derivatives. RSC Adv 9(45):26393–26401
33. Mirjalili BBF, Bamoniri A, Salmanpoor LA (2018) $TiCl_2$/Nano-γ-Al_2O_3 as a novel Lewis acid catalyst for promotion of one-pot synthesis of 1,4-dihydropyridines. J Nanostruct 8(3):276–287
34. Tayebee R, Nehzat F, Rafiee E et al (2011) An efficient and green synthetic protocol for the preparation of bis(indolyl)methanes catalyzed by $H_6P_2W_{18}O_{62}\cdot 24H_2O$, with emphasis on the catalytic proficiency of Wells-Dawson versus Keggin heteropolyacids. J Mol Catal A-Chem 351:154–164
35. Sakhalkar M, Aduri P, Lande S, Chandra S (2020) Single-step synthesis of novel chloroaluminate ionic liquid for green Friedel-Crafts alkylation reaction. Clean Techn Environ Policy 22:59–71
36. Hu Y-L, Li D-J, Li D-S (2015) Efficient and convenient oxidation of aldehydes and ketones to carboxylic acids and esters with H_2O_2 catalyzed by $Co_4HP_2Mo_{15}V_3O_{62}$ in ionic liquid [TEBSA][BF_4]. RSC Adv 5(32):24936–24943
37. Nimbalkar UD, Seijas JA, Nikalje APG et al (2017) Ionic liquid-catalyzed green protocol for multi-component synthesis of dihydropyrano[2,3-c]pyrazoles as potential anticancer scaffolds. Molecules 22(10):1628–1644
38. Li M, Abdolmohammadi S, Vessally E et al (2020) Carboxylative cyclization of propargylic alcohols with carbon dioxide: A facile and Green route to α-methylene cyclic carbonates. J CO2 Util 38:220–231
39. Parvatkar PT, Manetsch R, Banik BK (2019) Metal-Free cross-dehydrogenative coupling (CDC): molecular iodine as a versatile catalyst/reagent for CDC reactions. Chem Asian J 14(1):6–30
40. Itoh A, Tada N, Usami K et al (2016) Intermolecular cyclopropanation of styrenes using iodine and visible light via carbon-iodine bond cleavage. Org Lett 18(1):8–11
41. Devari S, Kumar A, Shah BA et al (2015) C-H functionalization of terminal alkynes towards stereospecific synthesis of (E) or (Z) 2-methylthio1,4-ene-diones. Chem Commun 51(24):5013–5016
42. Tehri P, Aegurula B, Peddinti RK (2017) Iodine-catalysed regioselective synthesis of β-hydroxysulfides. Tetrahedron Lett 58(21):2062–2065
43. Hiebel M-A, Berteina-Raboin S (2015) Iodine-catalyzed regioselective sulfenylation of imidazoheterocycles in PEG_{400}. Green Chem 17(2):937–944
44. Sun K, Lv Y, Wang X et al (2017) A novel metal-free method for the selenocyanation of aromatic ketones to afford α-carbonyl selenocyanates. Org Biomol Chem 15(20):4464–4467
45. Vadagaonkara KS, Prakashb S, Chaskar CA et al (2015) Iodine-mediated domino protocol for the synthesis of benzamides from ethylarenes via sp3 C–H functionalization. Synlett 26(12):1677–1682
46. Qiu R, Au C-T, Yin S-F et al (2017) Intramolecular, site-selective, iodine-mediated, amination of unactivated (sp^3)C—H bonds for the synthesis of indoline derivatives. Org Lett 19(11):2793–2796
47. Gao X, Gao M, Lei A et al (2015) Iodine-catalyzed radical oxidative annulation for the construction of dihydrofurans and indolizines. Org Lett 17(10):2404–2407
48. Zhang H, Muñiz K (2017) Selective piperidine synthesis exploiting iodine-catalyzed Csp^3-H amination under visible light. ACS Catal 7(6):4122–4125
49. Zhang J, Yu C, Wang Z (2010) A simple and efficient approach to the synthesis of 2-Phenylquinazolines via sp^3 C–H functionalization. Org Lett 12(12):2841–2843
50. Garai M, Rozyyev V, Ullah Z (2019) Zwitterion π–conjugated network polymer based on guanidinium and β-ketoenol as a heterogeneous organo-catalyst for chemical fixation of CO_2 into cyclic carbonates. APL Mater 7:111102–111109
51. Sharma AK, Singh J, Singh J et al (2019) Rose Bengal catalyzed coupling of 1,2-dicarbonyls and phenylene 1, 2-diamines: visible-light mediated synthesis of quinoxalines. ChemistrySelect 4(29):8713–8718
52. Zhu T, Yu Q, Ding L (2019) Atom-economical Preparation of polybismaleimide-based microporous organic polymers. Green Chem 21(9):2326–2333

53. Ji X-M, Zhou S-J, Deng C-L et al (2014) NH_4PF_6-promoted cyclodehydration of a-amino carbonyl compounds: efficient synthesis of pyrrolo-[3,2,1-*ij*]quinoline and indole derivatives. RSC Adv 4(96):53837–53841
54. Pradhan S, Mishra K, Lee YR (2019) Support-free Pd_3CO NCs as an efficient heterogeneous nanocatalyst for new organic transformations of C–C coupling reactions. Chem Eur J 25(46):10886–10894

Chapter 13
Green Nanocatalysts in Organic Synthesis

Rossella Santonocito and Giuseppe Trusso Sfrazzetto

1 Introduction

Definition of catalyst and nanocatalyst: We define "catalyst" a chemical specie able to increase the rate of a chemical reaction, without modifications or consumption during the reaction. For this reasons, the stoichiometry of the reaction does not include the catalyst. However, it appears in the kinetic equation. Considering a generic chemical reaction, where A and B are the reagents, C the product and Cat the catalyst,

$$A + B \xrightarrow{\text{Cat}} C$$

the kinetic equation is:

$$V = \frac{\mathrm{d}[C]}{\mathrm{d}t} = k[A]^m[B]^n[\text{Cat}]^\alpha$$

where m, n, and α are the reaction orders relative to A, B, and Cat, respectively, and k is the kinetic constant.

Reaction rate is dependent on the energy barrier required to form the transition state of the reaction, thus leading to the product (Fig. 1).

The energy gap between reagents and the transition state is activation energy. The catalyst increases the reaction rate by decreasing this energy barrier. This phenomenon can be achieved by three different mechanisms:

R. Santonocito · G. Trusso Sfrazzetto (✉)
Department of Chemical Sciences, University of Catania, Viale A. Doria 6, 95100 Catania, Italy
e-mail: giuseppe.trusso@unict.it

R. Santonocito
e-mail: santonocito.rossella@studium.unict.it

G. Anilkumar and S. Saranya (eds.), *Green Organic Reactions*, Materials Horizons: From Nature to Nanomaterials, https://doi.org/10.1007/978-981-33-6897-2_13

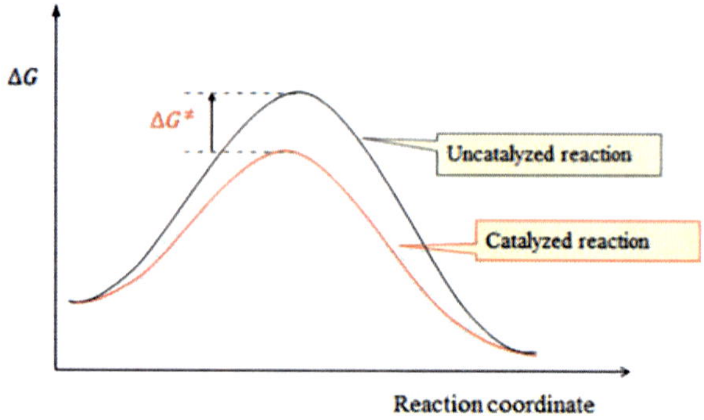

Fig. 1 Energy diagrams of a generic uncatalyzed (black) and catalyzed (red) reaction

(a) reaction can be evolved with a different, but similar, mechanism with respect to the uncatalyzed reaction, and the catalyst transforms the reagents in a less stable species (Fig. 2a);
(b) catalyzed reaction occurs with a similar mechanism with respect to the uncatalyzed one, and the catalyst stabilizes the transition state (Fig. 2b);

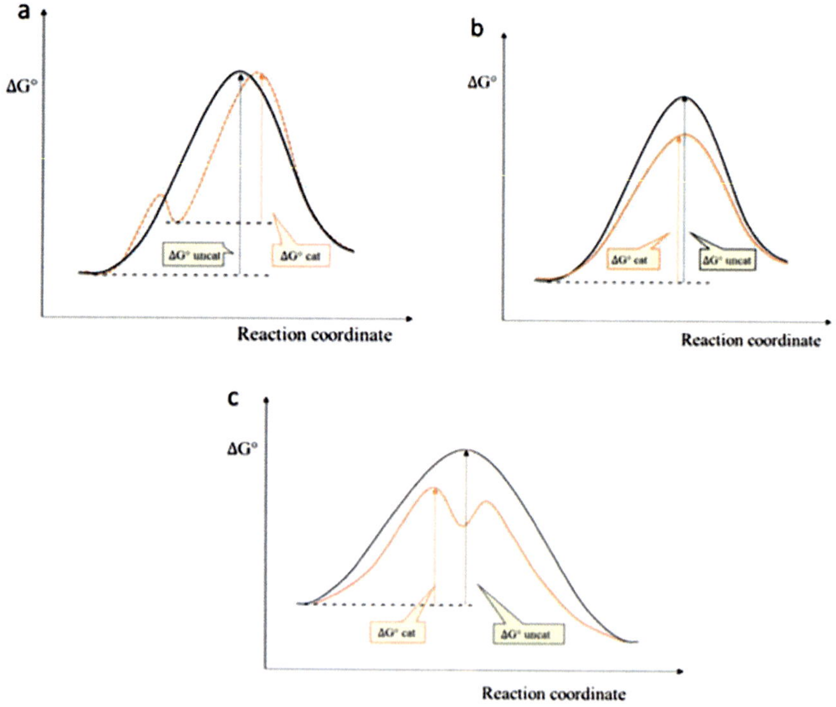

Fig. 2 Different mechanisms of catalysis

(c) the catalyst changes the mechanism of the reaction, leading to an alternative pathway with an activation energy less with respect to the uncatalyzed reaction (Fig. 2c).

As previously described, catalyst can be recovered without modifications or consumption at the end of the reaction. For this reason, in general catalyst is added to the reaction in "catalytic amount," i.e., less than 5% with respect to the starting reagents.

Catalysts can be classified into "homogeneous" and "heterogeneous," depending on the physical conditions with respect to the starting reagents: in paricular, if the catalyst and the reagents are in the same aggregation state (e.g., in solutions), the catalyst is homogeneous; while if the catalyst and the reagents are in different aggregation states (in general, reagents are in solution and the catalyst is in solid state), the catalyst is heterogeneous. Both classes of catalysis show positive and negative aspects: in particular, homogeneous catalysis is characterized by high efficiency in terms of fast reaction rate and high conversion values, but the catalyst is not easily recovered at the end of the reaction. This leads to problems related to the ecological impact of the catalytic process. On the contrary, heterogeneous catalysts can be easily recovered, due to the very low solubility in the reaction media, thus increasing the eco-friendliness of the reaction. The limits of this catalysts are due to the relative low conversion values, if compared to the homogeneous catalysts [1].

Homogeneous catalysts can be recovered and reused by using biphasic liquid–liquid reaction systems. In general, the liquids should be not miscible, for instance water and an organic solvent. The main problem of this strategy is the stability of the catalyst to the aqueous phase [2]. This problem can be resolved by confining the catalyst into a "shielded space," by using an appropriate nanostructure [3]. The possibility to confine the catalyst in a restricted nanometric space leads to many interesting properties:

- one-pot tandem reactions are allowed, due to the possibility to include in the nanospace more than one catalyst;
- the local concentration of the substrate is higher with respect to the bulky solution, thus leading to an increase of the activity and reactivity, with higher conversion values and reaction rates;
- the catalyst is shielded and segregated into a confined space, thus the recovery is easier with respect to a catalyst in solution;
- if the catalyst is not soluble in water, but the nanostructure is water-soluble, catalysis can be performed also in water.

In addition, the goal of nanocatalysts can be found in the possibility to overcome the limits of traditional homogeneous and heterogeneous catalysts (Fig. 3). A nanocatalyst allows an easy purification procedure, typical of homogeneous catalysis, and increases the reactivity with respect to a traditional heterogeneous catalyst.

A nanocatalyst can be defined as a catalyst having nanometer size (10^{-9} m). In this class of compounds, today, we can include mainly:

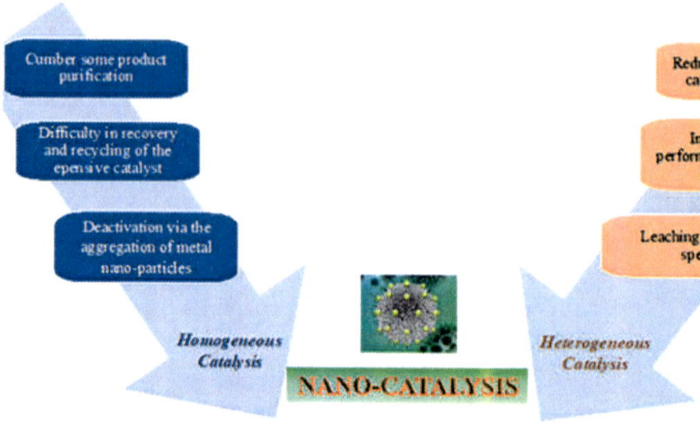

Fig. 3 Goal of nanocatalysis

- self-assembled nanoreactors (micelles, polymeric vescicles, hexameric capsules),
- covalent nanoreactors (dendrimers, metal nanoparticles, and organic nanoparticles).

In this chapter, we focused our attention on these classes of nanocatalysts, paying particular attention to the green aspect related to the catalysis.

2 Green Nanocatalysis

The basic principle of "green nanocatalysis" is to perform a chemical reaction under catalysis, by using "green conditions." In particular, with the term "green condition" we mean reaction with green solvents, reagents, room pressure and temperature, and catalyst.

Green solvents are eco-friendly solvents, also called "bio-solvents," derived from an agricultural process. They are, in general, low molecular weight ketones, alcohols, and esters. Although water is not produced from a similar process, obviously it can be considered as green solvent.

Green reagents can be considered chemical reagents able to carry out the reaction with high efficiency, selectivity, high conversion values, possibly avoiding long purification protocols (e.g., liquid chromatography). In addition, these reagents should generate the desired product and easily removable by-products, such as CO_2, gases, alcohols, or other water-soluble molecules.

3 Self-Assembled Nanoreactors

A self-assembled nanoreactor is a macro- or nanomolecular ordered structure assembled exploiting non-covalent interactions.[4] This confined nanospace creates an environment totally different with respect to the bulk solution.

The key points of the use of a nanoreactor as catalyst in organic synthesis are:

- The catalyst is "shielded" from the solvent molecules, leading to higher stability of the catalyst itself;
- Due to the absence of solvent molecules into the confined space created by the nanoreactor, the reagent species are "desolvated" thus the activity/reactivity is higher with respect to the normal conditions;
- The reagent molecules, confined into a restricted space, are forced to be close to each other, in according to the "proximity effect" [5, 6] thus leading to an increased reaction rate;
- The inner space of a nanoreactor can be compared to the catalytic site of an enzyme, which stabilizes transition state by non-covalent interaction, leading to the decrease of the activation energy and faster reaction rate.

3.1 *Micelles*

Micelles are supramolecular nanostructures assembled exploiting the hydrophobic effect. In particular, the building blocks are amphiphilic molecules, having a hydrophobic chain and a hydrophilic head [7, 8]. In concentration higher than the critical micellar concentration (CMC), the amphiphilic molecules assemble into a micellar nanostructure, that can be spherical or cylindrical (Fig. 4), depending on the "packing parameter":

$$p = v/a_o \cdot \text{lc}$$

Fig. 4 Self-assembly of spherical or cylindrical micelles

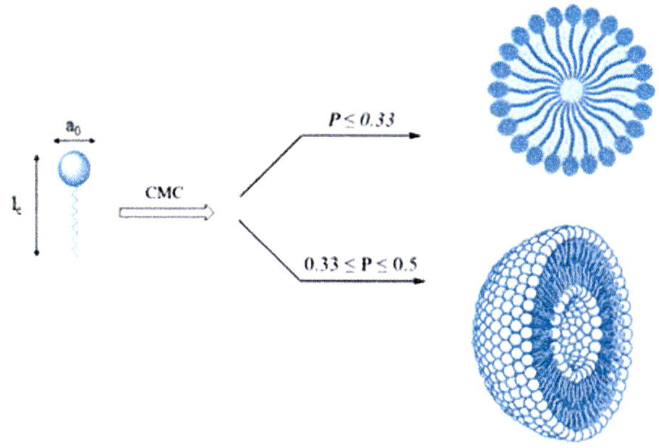

where v is the volume, lc is the length of the hydrophobic chain, and a_o is the area of the head groups [9]. In general, if $p \leq 0.33$ spherical micelles are obtained, while cylindrical (or worm-like) micelles, are preferred when $0.33 \leq p \leq 0.5$.

Many examples of catalysis with micellar systems are reported [8]. The main goal of this strategy is to perform the catalytic reaction, often conducted in organic solvent, in water, [10, 11] with high yields and mainly with easy catalyst recover [12].

The catalytic role of the micelle can be performed by the micellar structure, creating the nanospace described above, or by a catalyst covalently (or not) linked to the amphiphilic molecule.

These nanoreactors are used in a wide range of catalytic application in water [13–18]:

- C–N cross-coupling reactions;
- C–C cross-coupling reactions;
- Amination of allylic ethers;
- Amination of aromatic halide (Buchwald–Hartwig reaction);
- Suzuki–Myaura reaction;
- Aldol reactions;
- Oxidation of alkenes;
- Hydrogenation of alkenes;
- Oxidation of α-hydroxy ketones;
- Click reaction.

In addition, a recent example of micellar nanoreactors has been used for the oxidative degradation of organic contaminants in both liquid and gaseous system [19].

3.2 *Polymeric Vesicles*

Also, polymeric vesicles can be assembled by amphiphilic molecules; however, they consist of a bilayer structure, similar to those of liposomes, in which the inner space of the bilayer structure is hydrophobic, while the external is hydrophilic.

The nature of the structure of these polymeric vesicles leads to higher sizes with respect to those observed with micelle, excellent membrane permeability and stability [20]. This nanostructures can host a catalyst both in hydrophobic (metal catalysts) and in hydrophilic regions (enzymes) [21]. In water, the hydrophobic substrate can be accommodated in the hydrophobic region of the bilayer, closed to the metal catalyst also included into the vesicle, leading also in this case to the proximity effect and higher reaction rate.

Furthermore, the possibility to have two different environments in the same nanostructure (hydrophobic and hydrophilic) can be exploited in order to bring near reagents and catalysts having different solvent solubility.

These nanoreactors have been used as biocatalysts by encapsulation of enzymes:

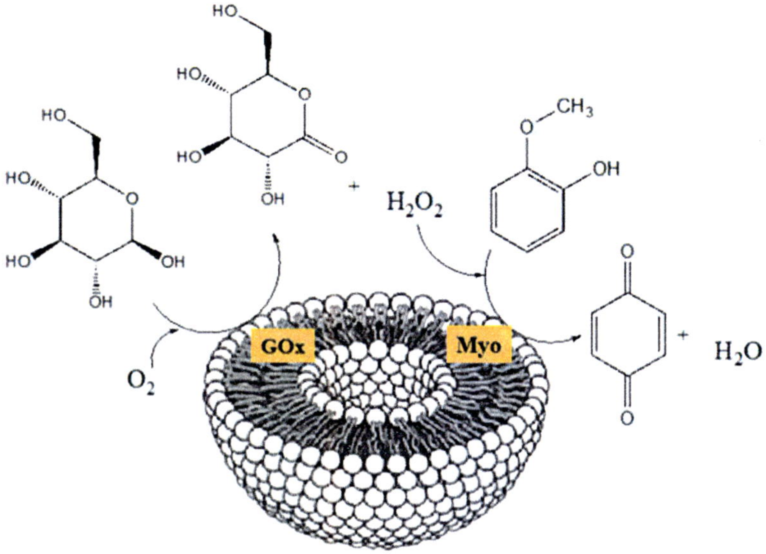

Fig. 5 Multistep conversion of glucose catalyzed by multifunctional vesicle

- Esterification of 1-hexanol with hexanoic acid was performed by including *Candida antarctica* lipase B (CalB), finding higher performance respect to the CalB in water [22];
- CalB encapsulated into a vescicle was also used for ring-opening polymerization (ROP) in water, in particular for the polymerization of lactones [23];
- Multistep conversion of glucose into *D*-glucono-δ-lactone and hydrogen peroxide has been performed by GOx (glucose oxidase), then hydrogen peroxide was employed to oxidize guaiacol to quinone and water by using Myo (myoglobin) hosted in the same nanostructure (Fig. 5) [24];
- Suzuki, Sonogashira, and Heck cross-coupling reactions have been explored into nanostructured vesicles encapsulating a Pd-catalyst [25].

3.3 Resorcin[4]arene Hexameric Capsules

Resorcin[4]arenes are macrocycles, obtained by acid condensation of resorcinol and an aliphatic/aromatic aldehyde, with a hemispherical shape of ca. 10 Å of diameter. They form a hydrophobic cavity, which can be further functionalized increasing the cavity dimensions [26–28]. In the presence of 8 water molecules, 6 units of resorcin[4]arenes assembly into the hexameric capsule (Fig. 6) via hydrogen bonds formation.

The internal volume reached by this self-assembled capsule is of ca. 350 Å^3, thus it can host two or more organic compounds, acting as nanoreactor. The "green

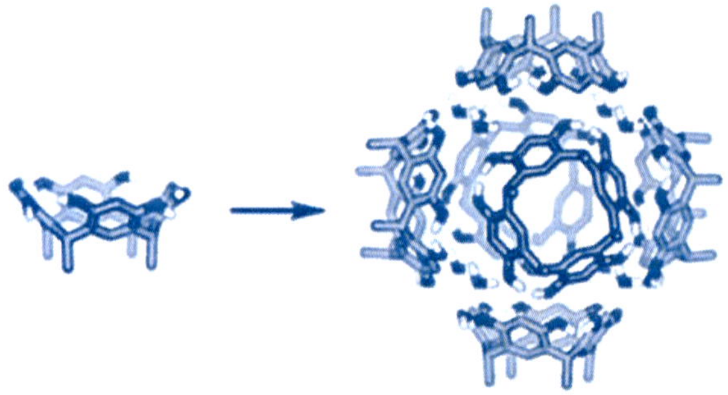

Fig. 6 3D structures of resorcin[4]arene and the self-assembled hexameric capsule

aspect" of this nanocatalyst consists in the possibility to perform reactions in mild conditions and, in most cases, in the absence of a metal catalyst [29].

The possibility to use mild conditions is to ascribe to the proximity effect of the substrates, enclosed into a confined volume. In addition, the catalysis can be conducted in the absence of a metal catalyst due to the property of the hexameric capsule to act as catalyst itself, via hydrogen bond formation or by acid catalysis ascribed to the hydroxylic groups of the resorcin[4]arene units.

Reactions catalyzed by this nanocapsule are:

- hydration of alkynes with a gold catalyst [30];
- cycloaddition reactions conducted in absence of catalyst [31];
- synthesis of amides, by using activating agent [32];
- synthesis of tetrazole [33];
- sulfoxidation of thioethers with hydrogen peroxide [34];
- Friedel–Crafts benzylation of arenes with benzyl chloride [35];
- 1,3-dipolar cycloadditions [31, 36];
- degradation of cellulose [37];
- degradation of organic dyes [38].

4 Covalent Nanoreactors

A covalent approach to create a nanoreactor cannot lead to a nanocapsule, with an internal space isolated from the bulk solution, but creates different systems with respect to the self-assembly process leading to the formation of dendrimers or nanoparticles.

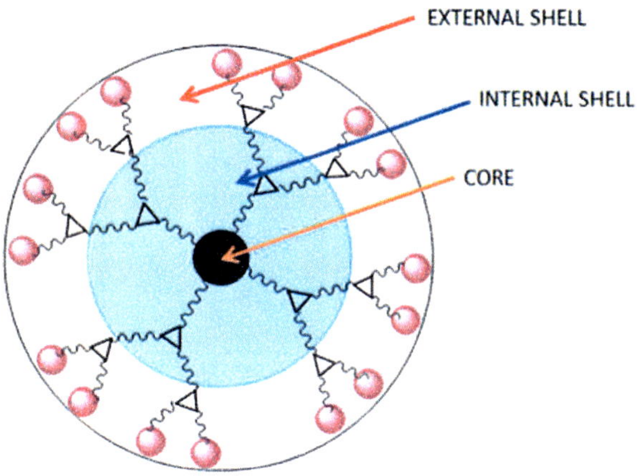

Fig. 7 Schematic representation of a dendrimer

4.1 Dendrimers

A dendrimer can be defined as nanometric highly branched molecule, having high symmetry. We can identify three regions: (i) the core, (ii) the inner shell, and (iii) the external shell (Fig. 7) [39].

These nanostructures can lead to three dimensional structures, with hydrophobic or hydrophilic characteristics, depending on the building blocks used during the synthesis. For these reasons, many catalysts and substrates can be included in the shell, covalently linked or simply adsorbed.

The main goals of dendrimer catalysis are the easy recovery of the catalyst and the possibility to perform reactions in water.

The main reactions performed with these nanosystems are:

- hydrogenation of 1,3-cyclooctadiene into cyclooctene in water [40];
- hydrogenation of olefin in water [41];
- Suzuki reaction [42];
- hydrolysis and aldolase reaction in water [43].

4.2 Nanoparticles

Silica nanoparticles are one of the most used support to create nanocatalysts for a wide range of applications [44–47]. In particular, they have been used as catalysts for the synthesis of thioethers, thioesters, vinylthioethers, and thio-Michael adducts. Furthermore, these catalysts are found to be very efficient in the microwave-assisted homocoupling of terminal alkynes.

Gold/titania nanoparticles are used to remove water pollutants by photocatalytic oxidation [48]. Photocatalytic activity is also used for hydrogen generation, dye degradation, phenol decomposition, and carboxylic acid degradation [49].

Palladium nanoparticles have been used as green electrocatalysts in the hydrogen evolution reaction, in the hydrogenation of alkenes and in the reduction of aromatic compounds into relative amino derivatives [50]. Due to the presence of palladium metal atom, Pd-nanoparticles are successfully used also in the C–C coupling reaction, showing good results in Suzuki, Heck, Stille, and Sonogashira coupling [51–55].

Platinum nanoparticles are also used in green catalysis, due to the presence of ionic liquids as stabilizer of the external shell. These nanocatalysts have been used in the oxidation and reduction of several substrates, including the reduction of molecular oxygen to water [56].

Gold nanoparticles cover one of the main roles in the green nanocatalysis. These nanostructures are synthesized by green methods, by using ionic liquids or other green reduction agents [57]. The success of these nanocatalysts is due to the high stability and the possibility to recover and reuse the nanoparticles. The main applications are the alkane oxidation [58] and the reduction of organic dyes [59, 60].

Recently, also carbon nanoparticles have been used as catalysts for different eco-friendly reactions [61, 62]. In particular, in the photocatalytic degradation of dyes, in the regioselective hydroxylation of aromatic compounds, in the photo-oxidation of benzylalcohol to the benzaldehyde and in the photogeneration of H_2.

All these systems show the typical features of a heterogeneous catalyst, as described before, thus finding important applications in green industrial processes (Scheme 1).

The main industrial reactions are:

- Reduction reactions of *p*-nitro phenol into the *p*-amino phenol in the presence of nickel nanocatalysts [63];
- Oxidation reactions, in particular:

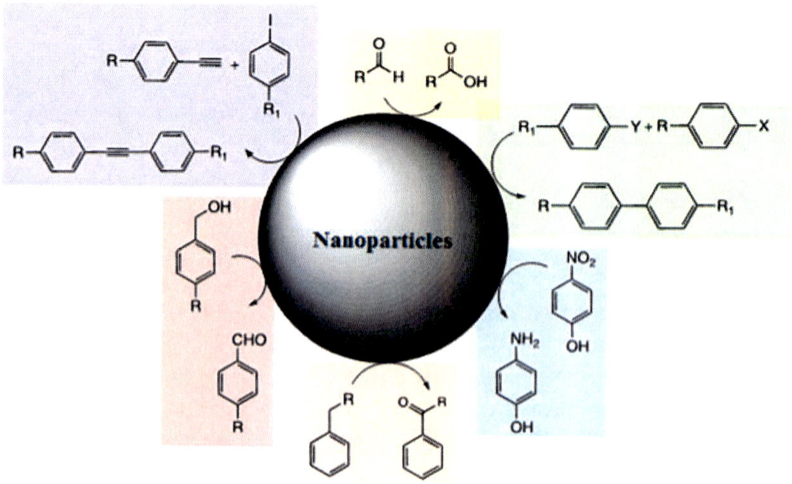

Scheme 1 Summary of reactions catalyzed by nanoparticles, with applications in green industrial processes

(a) starting from benzyl alcohol to benzaldehyde in absence of solvent, by using several metal oxide nanoparticles [64],
(b) starting from the benzylic compounds leading to the corresponding ketones [65],
(c) also aldehyde is easily oxidized into the corresponding carboxylic acids [66].

- Suzuki cross-coupling reactions, performed in water as solvent, using Pd-nanoparticles adsorbed onto diatomite or grapheme [67, 68];
- Heck cross-coupling reaction can be easily driven by palladium nanocatalysts in water, obtaining excellent results in terms of reaction rate and yield [69];
- Sonogashira cross-coupling reactions, by using ethanol as solvent, Pd-nanoparticles as catalyst and potassium carbonate as base [70];
- Hydrogenation of arene compounds in water is performed by rhodium nanoparticles [71], in addition, naphthalene and cyclic polyenes can be hydrogenated in solvent-free conditions, by using hybrid ruthenium–palladium nanostructured;
- Ullmann reaction leads to biaryl compounds starting from aryl chloride, by using Pd-nanoparticles as catalyst, in a mixture of ionic liquid/supercritical carbon dioxide as green solvent [72]. In similar conditions, also graphene oxide has been used as support for the Pd-nanoparticles [73].

5 Other Nanocatalysts

Other examples of nanocatalysts with a wide range of applications are recently prepared. $MoSe_2$ nanosheets show catalytic activity for the hydrogen evolution reaction in water [74], for oxygen reduction [75] and for Knoevenagel condensation [76].

Nanofibers, as well as carbon nanotubes are also used as green catalysts in hydrogenation/dehydrogenation, ammonia synthesis, and oxidation reactions [77].

Noble metal nanoclusters have shown interesting catalytic activity and selectivity in different reactions [78]. These characteristics depend on the ultrasmall size and the high number of active sites.

Other less diffuse nanostructured catalysts are core–shell particles [79], metal nanowires [80], nanolamellae [81], nanorods [82], nanoflowers [83], and nanostars [84].

6 Conclusion

In the last few years, many examples demonstrated the advantages of nanoreactor systems, self-assembled, or covalently built systems with respect to the classic catalysts. In particular, nanoparticles show high surface area and for this reason, high catalytic activity. In addition, nanoreactors are characterized by formation of a confined space in which the reactants are close to each other, thus increasing the reactivity.

Furthermore, the possibility to protect the catalyst from undesired interactions with the environment, in particular with the aqueous solvent, preserves the stability of the catalyst itself, leading to higher conversion values. For this reason, reactions are conducted in water and in normal conditions, with excellent yields and selectivities, which normally can only be obtained by performing reactions in organic media.

In addition, most of these catalysts are still large enough to be isolated from the reaction media by using standard filtration protocols. This aspect leads to important positive applications also for the industrial field.

References

1. Bitter H (2017) Heterogeneous catalyst. In: Kamer PCJ, Dieter V, Thybaut TW (Eds.) Contemporary catalysis, pp 177–188
2. Pagliaro M, Pandaru V, Ciriminna R, Béland F, Demma Carà P (2012) Heterogeneous versus homogeneous palladium catalysts for cross-coupling reactions. Chem Cat Chem 4:432–445
3. Lu Y, Spyra P, Mei Y et al (2007) Composite hydrogels: robust carriers for catalytic nanoparticles. Macromol Chem Phys 208:254–261
4. Gangemi CMA, Pappalardo A, Trusso Sfrazzetto G (2015) Assembling of supramolecular capsules with Resorcin[4]arene and Calix[n]arene building blocks. Curr Org Chem 19:2281–2308
5. Kinghorn MJ, Valdivia-Berroeta GA, Chantry DR et al (2017) Proximity-induced reactivity and product selectivity with a rationally designed bifunctional peptide catalyst. ACS Catal 7:7704–7708
6. Pappalardo A, Puglisi R, Sfrazzetto GT (2019) Catalysis inside supramolecular capsules: recent developments. Catalysts 9:630–650
7. Khullar P, Singh V, Mahal A et al (2013) Block Copolymer Micelles as nanoreactors for self-assembled morphologies of gold nanoparticles. J Phys Chem B 117:3028–3039
8. Dwars T, Paetzold E, Oehme G (2005) Reactions in Micellar systems. Angew Chem Int Ed 44:7174–7199
9. Jain S, Bates FS (2003) On the origins of morphological complexity in block copolymer surfactants. Science 300:460–464
10. Duplais C, Krasovskiy A, Wattenberg A et al (2010) Cross-couplings between benzylic and aryl halides "on water": synthesis of diarylmethanes. Chem Commun 46:562–564
11. Ballistreri FP, Gangemi CMA, Pappalardo A et al (2016) (Salen)Mn(III) catalyzed asymmetric epoxidation reactions by hydrogen peroxide in water: a green protocol. Int J Mol Sci 17:1112–1120
12. Lipshutz BH, Ghorai S (2012) Organocatalysis in water at room temperature with in-flask catalyst recycling. Org Lett 14:422–425
13. Ansari TN, Taussat A, Clark AH et al (2019) Insights on bimetallic micellar nanocatalysis for Buchwald–Hartwig aminations. ACS Catal 9:10389–10397
14. Xiang F, Li B, Zhao P, Tan J, Yu Y, Zhanga S (2019) Copper(I)-chelated cross-linked cyclen micelles as a nanocatalyst for azide-alkyne cycloaddition in both water and cells. Adv Synth Catal 361:5057–5062
15. Mayanka SA, Kaur N, Singh N, Jang DO (2017) A carbon quantum dot-encapsulated micellar reactor for the synthesis of chromene derivatives in water. Mol Catal 439:100–107
16. Bahrami K, Khodaei MM, Meibodi FD (2016) Suzuki and heck cross-coupling reactions using ferromagnetic nanoparticle-supported palladium complex as an efficient and recyclable heterogeneous nanocatalyst in sodium dodecylsulfate micelles. Appl Organometal Chem 1–9

17. Hou S, Xie C, Yu F, Yuan B, Yu S (2016) Selective hydrogenation of α-pinene to cis-pinane over Ru nanocatalysts in aqueous micellar nanoreactors. RSC Adv 6:54806–54811
18. Yu Y, Lin C, Li B et al (2016) Dendrimer-like core cross-linked micelle stabilized ultra-small gold nanoclusters as a robust catalyst for aerobic oxidation of α-hydroxy ketones in water. Green Chem 18:3647–3655
19. Yang F, Gao S, Ding Y et al (2019) Excellent porous environmental nanocatalyst: tactically integrating size-confined highly active MnOx in nanospaces of mesopores enables the promotive catalytic degradation efficiency of organic contaminants. New J Chem 43:19020–19034
20. Sanson C, Schatz C, Le Meins JF et al (2010) Biocompatible and biodegradable poly(trimethylene carbonate)-b-poly(l-glutamic acid) polymersomes: size control and stability. Langmuir 26:2751–2760
21. Wang Z, van Oers MCM, Rutjes FPJT et al (2012) Polymersome colloidosomes for enzyme catalysis in a biphasic system. Angew Chem Int Ed 51:10746–10750
22. van Oers MCM, Veldmate WS, van Hest JCM et al (2015) Aqueous asymmetric aldol reactions in polymersome membranes. Polym Chem 6:5358–5361
23. Nallani M, de Hoog HPM, Cornelissen JJLM et al (2007) Polymersome nanoreactors for enzymatic ring-opening polymerization. Biomacromol 8:3723–3728
24. Gräfe D, Gaitzsch J, Appelhans D et al (2014) Cross-linked polymersomes as nanoreactors for controlled and stabilized single and cascade enzymatic reactions. Nanoscale 6:10752–10761
25. Dergunov SA, Khabiyev AT, Shmakov SN et al (2016) Encapsulation of homogeneous catalysts in porous polymer nanocapsules produces fast-acting selective nanoreactors. ACS Nano 10:11397–11406
26. Amato ME, Ballistreri FP, D'Agata S et al (2011) Enantioselective molecular recognition of chiral organic ammonium ions and amino acids using Cavitand-Salen based receptors. Eur J Org Chem 2011:5674–5680
27. Pappalardo A, Amato ME, Ballistreri FP et al (2012) Pair of diastereomeric uranyl salen cavitands displaying opposite enantiodiscrimination of α-amino acid ammonium salts. J Org Chem 77:7684–7687
28. D'Urso A, Tudisco C, Ballistreri FP et al (2014) Enantioselective extraction mediated by a chiral cavitand–salen covalently assembled on a porous silicon surface. Chem Commun 50:4993–4996
29. Zhu Y, Rebek J Jr, Yu Y (2019) Cyclizations catalyzed inside a hexameric resorcinarene capsule. Chem Commun 55:3573–3577
30. Cavarzan A, Reek JNH, Trentin F, Scarso A, Strukul G (2013) Substrate selectivity in the alkyne hydration mediated by NHC–Au(I) controlled by encapsulation of the catalyst within a hydrogen bonded hexameric host. Catal Sci Technol 3:2898–2901
31. La Sorella G, Sperni L, Strukul G et al (2015) Supramolecular encapsulation of neutral diazoacetate esters and catalyzed 1,3-dipolar cycloaddition reaction by a self-assembled hexameric capsule. Chem Cat Chem 7:291–296
32. Giust S, La Sorella G, Sperni L (2015) Substrate selective amide coupling driven by encapsulation of a coupling agent within a self-assembled hexameric capsule. Chem Commun 51:1658–1661
33. Giust S, La Sorella G, Sperni L et al (2015) Supramolecular catalysis in the synthesis of substituted 1HTetrazoles from isonitriles by a self-assembled hexameric capsule. Asian J Org Chem 4:217–220
34. La Sorella G, Sperni L, Strukul G et al (2016) Supramolecular activation of hydrogen peroxide in the selective sulfoxidation of thioethers by a self-assembled hexameric capsule. Adv Synth Catal 358:3443–3449
35. La Manna P, Talotta C, Floresta G et al (2018) Mild Friedel-crafts reactions inside a hexameric resorcinarene capsule: C-Cl bond activation through hydrogen bonding to bridging water molecules. Angew Chem Int Ed 57:5423–5428
36. La Manna P, De Rosa M, Talotta C et al (2018) The hexameric resorcinarene capsule as an artificial enzyme: ruling the regio and stereochemistry of a 1.3-dipolar cycloaddition between nitrones and unsaturated aldehydes. Org Chem Front 5:827–837

37. Salunke BK, Sawant SS, Kang TK et al (2015) Potential of biosynthesized silver nanoparticles as nanocatalyst for enhanced degradation of cellulose by cellulase. J Nanomat 2015. Article ID: 289410
38. Pandey S, Do JY, Kim J et al (2020) Fast and highly efficient catalytic degradation of dyes using κ-carrageenan stabilized silver. Carbohyd Polym 230:115597
39. Bronstein LM, Shifrina ZB (2009) Nanoparticles in dendrimers: from synthesis to application. Nanotechnol Russ 4:576–608
40. Chung YM, Rhee HK (2003) Pt-Pd bimetallic nanoparticles encapsulated in dendrimer nanoreactor. Catal Lett 85:159–164
41. Ornelas C, Aranzaes JR, Salmon L et al (2008) "Click" dendrimers: synthesis, redox sensing of pd(oac)2, and remarkable catalytic hydrogenation activity of precise pd nanoparticles stabilized by 1,2,3-triazole-containing dendrimers. Chem Eur J 14:50–64
42. Ornelas C, Salmon L, Aranzaes JR et al (2007) Catalytically efficient palladium nanoparticles stabilized by "click" ferrocenyl dendrimers. Chem Commun 2007:4946–4948
43. Clouet A, Darbre T, Reymond JL (2004) A combinatorial approach to catalytic peptide dendrimers. Angew Chem Int Ed 43:4612–4615
44. Wang J, Zhou Y, Shao Y et al (2019) Chitosan–silica nanoparticles catalyst (M@CS–SiO2) for the degradation of 1,1-dimethylhydrazine. Res Chem Interm 45:1721–1735
45. Ramazani A, Farshadi A, Mahyari A et al (2016) Synthesis of electron-poor N-Vinylimidazole derivatives catalyzed by Silica nanoparticles under solvent-free conditions. Int J Nano Dim 7:41–48
46. Maleki A (2012) Fe_3O_4/SiO_2 nanoparticles: an efficient and magnetically recoverable nanocatalyst for the one-pot multicomponent synthesis of diazepines. Tetrahedron 68:7827–7833
47. Sreedhar B, Radhika P, Neelima B et al (2007) Regioselective ring opening of epoxides with amines using monodispersed silica nanoparticles in water. J Mol Cat A Chem 272:159–163
48. Ayati A, Ahmadpour A, Bamoharram FF et al (2014) A review on catalytic applications of Au/TiO_2 nanoparticles in the removal of water pollutant. Chemosphere 107:163–174
49. Primo A, Corma A, García H (2011) Titania supported gold nanoparticles as photocatalyst. Phys Chem Chem Phys 13:886–910
50. Chang F, Kim H, Lee B et al (2010) Highly efficient solvent-free catalytic hydrogenation of solid alkenes and nitro-aromatics using Pd nanoparticles entrapped in aluminum oxy-hydroxide. Tetrahedron Lett 51:4250–4252
51. Nadagouda MN, Polshettiwar V, Varma RS (2006) Self-assembly of palladium nanoparticles: synthesis of nanobelts, nanoplates and nanotrees using vitamin B1, and their application in carbon–carbon coupling reactions. J Mater Chem 19:2026–2031
52. Pacardo DB, Sethi M, Jones SE et al (2009) Biomimetic synthesis of Pd nanocatalysts for the Stille coupling reaction. ACS Nano 3:1288–1296
53. Sawant AD, Raut DG, Darvatkar NB et al (2010) An ester appending multifunctional ionic liquid for Pd(II) catalyzed Heck reaction. Catal Commun 12:273–276
54. Wang Q, Cheng H, Liu R et al (2010) A green and efficient route for preparation of supported metal colloidal nanoparticles in $scCO_2$. Green Chem 12:1417–1422
55. Zhang G, Zhou H, Hu J et al (2009) Pd nanoparticles catalyzed ligand-free Heck reaction in ionic liquid microemulsion. Green Chem 11:1428–1432
56. Li F, Li F, Song J et al (2009) Green synthesis of highly stable platinum nanoparticles stabilized by amino-terminated ionic liquid and its electrocatalysts for dioxygen reduction and methanol oxidation. Electrochem Commun 11:351–354
57. Dash P, Miller SM, Scott RWJ (2010) Stabilizing nanoparticle catalysts in imidazolium-based ionic liquids: a comparative study. J Mol Catal A Chem 329:86–95
58. Chen L, Hu J, Richards R (2009) Intercalation of aggregation-free and well-dispersed gold nanoparticles into the walls of mesoporous silica as a robust "green" catalyst for n-alkane oxidation. J Am Chem Soc 131:914–915
59. Gupta N, Singh HP, Sharma RK (2010) Single-pot synthesis: plant mediated gold nanoparticles catalyzed reduction of methylene blue in presence of stannous chloride. Colloids Surf A 367:102–107

60. Zhan G, Du M, Huang J et al (2011) Green synthesis of Au/TS-1 catalysts via two novel modes and their surprising performance for propylene epoxidation. Catal Commun 12:830–833
61. Testa C, Zammataro A, Pappalardo A et al (2019) Catalysis with carbon nanoparticles. RSC Adv 9:27659–27664
62. Zammataro A, Gangemi CMA, Pappalardo A et al (2019) Covalently functionalized carbon nanoparticles with a chiral Mn-Salen: a new nanocatalyst for enantioselective epoxidation of alkenes. Chem Commun 55:5255–5258
63. Abd El Maksod IH, Hegazy EZ et al (2010) an environmentally benign, highly efficient catalytic reduction of p-nitrophenol using a nano-sized nickel catalyst supported on silica-alumina. Adv Synth Catal 352:1169–1178
64. Choudhary VR, Dhar A, Jana P et al (2005) A green process for chlorine-free benzaldehyde from the solvent-free oxidation of benzyl alcohol with molecular oxygen over a supported nano-size gold catalyst. Green Chem 7:768–770
65. Dapurkar SE, Shervani Z, Yokoyama T et al (2009) Supported gold nanoparticles catalysts for solvent-free selective oxidation of benzylic compounds into Ketones at 1 atm O_2. Catal Lett 130:42–47
66. Gupta NK, Nishimura S, Takagaki A et al (2011) Hydrotalcite-supported gold-nanoparticle-catalyzed highly efficient base-free aqueous oxidation of 5-hydroxymethylfurfural into 2, 5-furandicarboxylic acid under atmospheric oxygen pressure. Green Chem 13:824–827
67. Bernini R, Cacchi S, Fabrizi G et al (2010) Perfluoro-tagged, phosphine-free palladium nanoparticles supported on silica gel: application to alkynylation of aryl halides, Suzuki-Miyaura cross-coupling, and Heck reactions under aerobic conditions. Green Chem 12:150–158
68. Hao C, Zhao X (2010) Highly efficient and recyclable diatomite-supported Pd nanoparticles for the Suzuki-Miyaura coupling reaction. Adv Mater Res 113–116:1824–1827
69. Zhang W, Qi H, Li L et al (2009) Hydrothermal Heck reaction catalyzed by Ni nanoparticles. Green Chem 11:1194–1200
70. Li P, Wang L, Li H (2005) Application of recoverable nanosized palladium(0) catalyst in Sonogashira reaction. Tetrahedron 61:8633–8640
71. Hubert C, Denicourt-Nowicki A, Beaunier P et al (2010) TiO2-supported Rh nanoparticles: from green catalyst preparation to application in arene hydrogenation in neat water. Green Chem 12:1167–1170
72. Cheng J, Tang L, Xu J (2010) An economical, green pathway to biaryls: palladium nanoparticles catalyzed ullmann reaction in ionic liquid/supercritical carbon dioxide system. Adv Synth Catal 352:3275–3286
73. Cheng J, Zhang G, Du J et al (2011) New role of graphene oxide as active hydrogen donor in the recyclable palladium nanoparticles catalyzed Ullmann reaction in environment friendly ionic liquid/supercritical carbon dioxide system. J Mater Chem 21:3485–3494
74. Truong QD, Nakayasu Y, Nguyen QT et al (2020) Defect-rich exfoliated $MoSe_2$ nanosheets by supercritical fluid process as an attractive catalyst for hydrogen evolution in water. App Surf Sci 2020. https://doi.org/10.1016/j.apsusc.2019.144537
75. Wang M, Liu Y, Zhang X, Fan Z, Tong Z (2018) Development of sandwich-structured cobalt porphyrin/niobium molybdate nanosheets catalyst for oxygen reduction. J Mat Res 33:4199–4206
76. Jia H, Zhao Y, Niu P, Lu N, Fan B, Li R (2018) Amine-functionalized MgAl LDH nanosheets as efficient solid base catalysts for Knoevenagel condensation. Mol Catal 449:31–37
77. Pérez-Mayoral E, Calvino-Casilda V, Soriano E (2016) Metal-supported carbon-based materials: opportunities and challenges in the synthesis of valuable products. Catal Sci Technol 6:1265–1291
78. Du Y, Sheng H, Astruc D, Zhu M. (2020) Atomically precise noble metal nanoclusters as efficient catalysts: a bridge between structure and properties. Chem Rev. https://doi.org/10.1021/acs.chemrev.8b00726
79. Palavalli ND, Yaghoubi A, Lai CC, Tin CC, Javey A, Chueh YL (2015) Catalyst-dependent morphological evolution by interfacial stress in crystalline-amorphous core-shell germanium nanowires. RSC Adv 5:28454–28459

80. Wang W, Wu Z, Eftekhari E, Huo Z, Li X, Tade MO, Yan C, Yan Z, Li C, Li Q, Zhao D (2018) High performance heterojunction photocatalytic membranes formed by embedding Cu_2O and TiO_2 nanowires in reduced graphene oxide. Catal Sci Technol 8:1704–1711
81. Wang J, Wang Z, Liu CJ (2014) Enhanced Activity for CO Oxidation over WO_3 Nanolamella Supported Pt Catalyst. ACS Appl Mater Interf 6:12860–12867
82. Wu P, Feng L, Liang Y et al (2020) Magnetic Fe-C-O-Mo alloy nano-rods prepared from chemical decomposition of a screw (a top-down approach): An efficient and cheap catalyst for the preparation of dihydropyridine and dihydropyrimidone derivatives. App Cat A: General. https://doi.org/10.1016/j.apcata.2019.117301
83. Zhao J, Zhang X, Liu M et al (2019) Metal-organic-framework-derived porous 3D heterogeneous NiFex/$NiFe_2O_4$@NC nanoflowers as highly stable and efficient electrocatalysts for the oxygen-evolution reaction. J Mat Chem A: Mater Energy Sustain 7:21338–21348
84. Li Y, Geng X, Leng W et al (2017) Gold nanospheres and gold nanostars immobilized onto thiolated eggshell membranes as highly robust and recyclable catalysts. New J Chem 41:9406–9413

Chapter 14
Visible Light-Catalyzed Asymmetric Synthesis: A Green Perspective

Sankuviruthiyil M. Ujwaldev and Gopinathan Anilkumar

1 Introduction

Asymmetric synthesis forms one of the most critical areas in organic synthesis as it provides the simplest way of excluding the unwanted enantiomer which may overshadow the desired physical, chemical or biological properties of the target enantiomer when administrated to a chiral environment. Chiral catalysis provides one of the most efficient and advanced protocols in the area and in consonance with that the Nobel Prize for chemistry in 2001 was awarded to Knowles, Noyori and Sharpless for the "development of catalytic asymmetric synthesis" [1]. The merging of visible light catalysis with the asymmetric synthesis resulted in a more efficient scenario [2]. The basic principle involved here is the visible light-mediated excitation of the substrate or its derivative by providing a chiral environment. Different approaches were emerged including sensitizer-based protocols [3], use of chiral organocatalysts [4], asymmetric induction [5], etc. In this chapter, we have attempted to summarize the recent developments that had happened in visible light-catalyzed asymmetric synthesis. In accordance with the aspect of this book, we have not included protocols that use transition metal-based photocatalysts in this chapter [6].

S. M. Ujwaldev · G. Anilkumar (✉)
School of Chemical Sciences, Mahatma Gandhi University, Priyadarsini Hills, Kottayam, Kerala 686560, India
e-mail: anilgi1@yahoo.com

S. M. Ujwaldev
e-mail: ujwalsreeragam@gmail.com

G. Anilkumar and S. Saranya (eds.), *Green Organic Reactions*, Materials Horizons: From Nature to Nanomaterials, https://doi.org/10.1007/978-981-33-6897-2_14

2 Photocatalyzed Asymmetric Synthesis

2.1 *Sensitizers as Sole Catalysts*

One of the earliest examples which deals with the photocatalyzed asymmetric synthesis is the report by Hammond and co-workers in which chiral N-(1-(naphthalen-1-yl)ethyl)acetamide was employed as the sensitizer for asymmetric cis/trans isomerization of 1,2-diphenylcyclopropanes [7]. Later, Inoue and co-workers came up with photoisomerization of *cis* cyclooctene [8]. In this protocol, the enantiomeric excess was 2–4% when menthylbenzoate- or isophthalate was used as the sensitizer. Later, many protocols using various sensitizers like chiral naphthalene(di)carboxylates [9], 1,1′-bis(2,10-dicyanoanthracene) [10], *m*-methoxybenzoyl-β-cyclodextrin [11], chiral ketones [12], Xanthones [12], Thioxanthones [13], etc., were also reported.

2.2 *Oraganocatalyst-based Dual Catalysis via Intermediate Formation*

Emergence of organocatalyst-based dual catalysis was another major breakthrough in this area. A chiral molecule with an active functional group is used here in catalytic amounts under visible light irradiation. In most of the cases, a photoredox catalyst is also employed [14]. The organocatalyst initially reacts with the substrate to form a chiral intermediate, e.g., imine formation. This intermediate subsequently undergoes excitation by the visible light and goes through an electron transfer process to generate a radical. The radical undergoes the key bond formation step with the other substrate in an enantiospecific manner. The commonly employed chiral organocatalysts include amino alcohols [15], imidazolidinones [16], pyrrolidines [17], quinine derivatives [18], aminoacids [19], etc. Most of these protocols provide excellent enantiomeric excess.

Visible light promoted enantioselective α-alkylation/perfluoroalkylation of aliphatic aldehydes was demonstrated using eosin as the catalyst (Scheme 1) [20]. Imidazolidinone was employed as the organocatalyst in this protocol. It forms an imine with the aldehyde, and enantio selectively controls the approach of the alkyl radical generated from the corresponding halide on visible light irradiation.

Visible light-enabled enantioselective α-alkylation of aldehydes using electron-deficient alkyl bromides was reported (Scheme 2). The protocol has employed an octahydro-1H-indole derivative as the catalyst [21]. It is proposed to form an imine with the aldehyde which undergoes electron donor–acceptor (EDA) complex formation with the halide. The EDA complex on irradiation undergoes a single electron transfer (SET) process (from imine to halide) to generate a chiral radical ion pair. It follows the imine–enamine isomerization and halide elimination steps resulting in

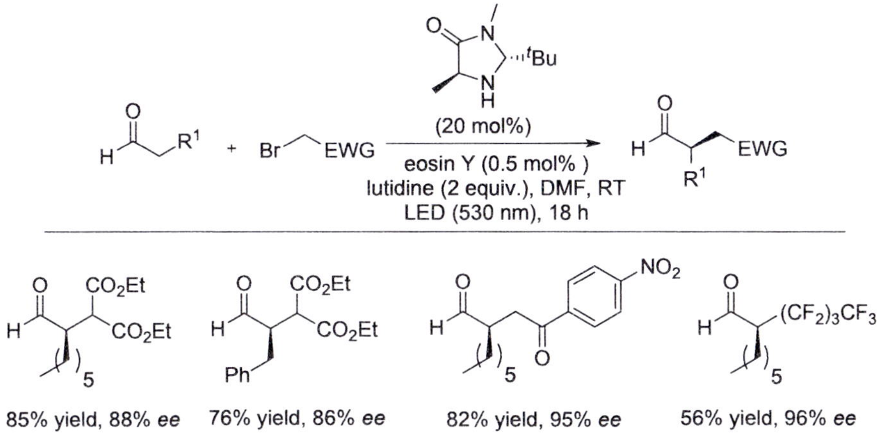

Scheme 1 Visible light promoted enantioselective α-alkylation/perfluoroalkylation of aliphatic aldehydes

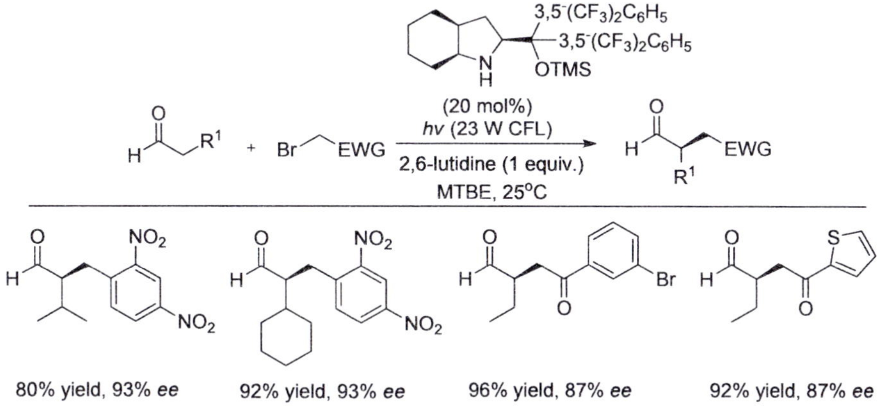

Scheme 2 Enantioselective α-alkylation of aldehydes using electron-deficient alkyl bromides

radical cross-coupling to furnish the product. The substrate scope studies afforded excellent yields of the product with *ee* in the range of 83–94%.

Photochemically induced α-alkylation of aldehydes using electron-deficient organic halides was reported (Scheme 3) [22]. Diarylprolinol silyl ether catalyst was employed as the catalyst. Butanal was alkylated using 2,4-dinitrobenzyl bromide, phenacyl bromide as well as diethyl bromomalonate.

A visible light-driven amine-catalyzed cascade reaction of alkenoic acids/alkenols with cinnamaldehydes was demonstrated to afford butyrolactone/tetrahydrofuran derivatives (Scheme 4) [23]. The protocol involved the simultaneous generation of a carbon–carbon and carbon–oxygen bonds.

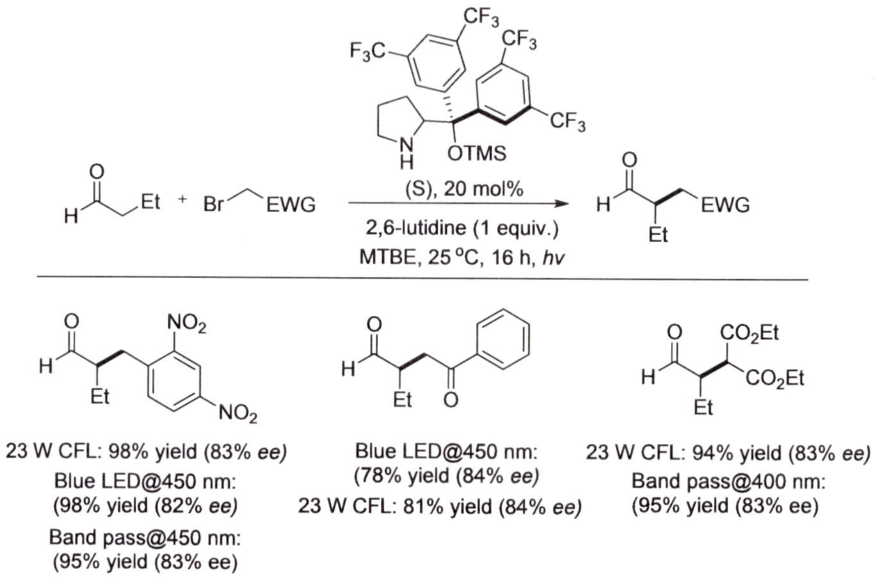

Scheme 3 Visible light-mediated α-alkylation of aldehydes using electron-deficient organic halides

Diastereomeric mixtures of the product were always generated where each diastereomer exhibited excellent *ee*.

Visible light-induced C(sp^3)–H functionalization of various toluene, xylene and benzyl derivatives using cinnamaldehydes was reported (Scheme 5) [24]. The strategy employed a *gem*-difluorinated diaryl prolinol silyl ether as the catalyst along with $Zn(OTf)_2$ as the additive. Enantioselectivities in the range of 60–84% were observed in the case of toluene and xylene while benzylderivatives furnished diastereomeric mixtures.

The proposed mechanism initiates with an imine formation step between cinnamaldehyde and the catalyst. The imine on irradiation with a 420 nm LED gets excited and undergoes SET process with toluene derivative generating the radical. During the process, a benzyl radical is also generated. The radicals react together and lead to the formation of the chiral imine, which undergoes hydrolysis later on to furnish the desired product (Scheme 6).

2.3 *Asymmetric Induction via Non-Covalent Interactions*

Another powerful area in photocatalyzed asymmetric synthesis is the use of a chiral species in catalytic amounts which uses non-covalent interactions-based asymmetric induction leading to enantio enrichment [25]. A photoredox catalyst is often required along with making it a dual-catalyzed protocol. The proposed mechanism begins

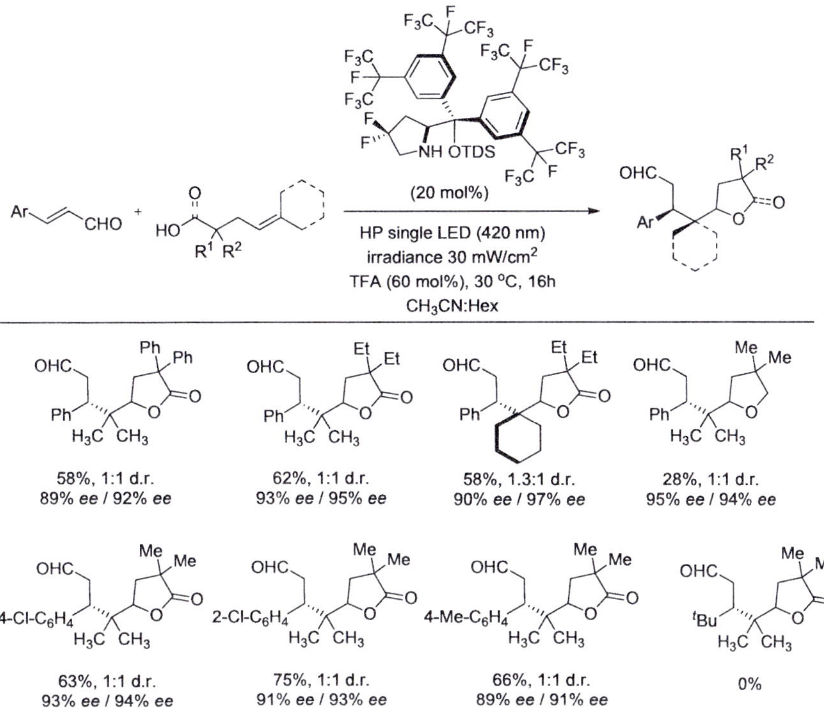

Scheme 4 Visible light-driven amine-catalyzed synthesis of butyrolactone/tetrahydrofuran derivatives

with a SET process between one of the substrates and the photocatalyst to generate a radical species. It is then brought together with the other substrate in a highly specific spacial orientation by the chiral catalyst, leading to the key bond formation step happening in an enantioselective way.

Asymmetric synthesis of α-hydroxy ketones and α-amino ketones was achieved, respectively, from diketones and α-keto ketimines under visible light irradiation. DPZ (i.e., 5,6-Bis(5-methoxythiophen-2-yl)pyrazine-2,3-dicarbonitrile) was employed as the photocatalyst along with **C1/C2** in this protocol (Scheme 7) [26]. DPZ on excitation gets involved in a SET process with diketone/α-keto ketimines to form a ketyl intermediate. Later, **C1/C2** gets hydrogen bonded to the ketyl intermediate and acts as chirality inducer, enabling transfer of hydrogen atom from a terminal reductant THIQ-2 (2-(naphthalen-1-yl)-1,2,3,4-tetrahydroisoquinoline) through the selective face. The generality of the protocol was well established with a variety of diketones and α-keto ketimines. Enantiomeric excess in the range of 80–92% and 64–90% was observed, respectively, in the case of 1,2-diketones and α-keto ketimines.

Asymmetric synthesis of 4-amino-2-methyl tetrahydroquinolines from N-aryl valine derivatives via visible light-mediated decarboxylative Povarov reaction was reported (Scheme 8) [27]. A catalytic duo of DPZ and SPINOL phosphoric acid **C3**

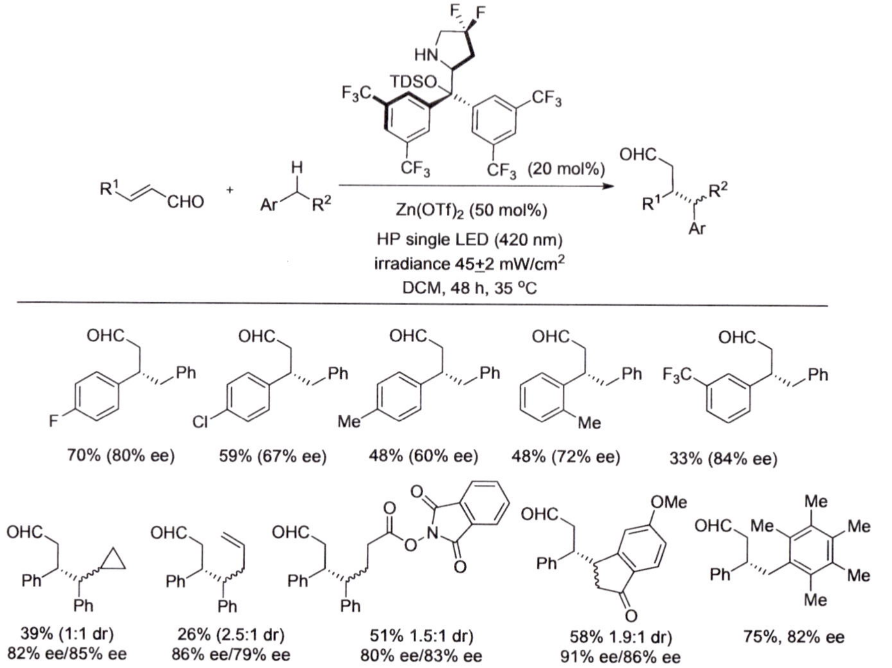

Scheme 5 Visible light-mediated C(sp^3)–H functionalization of benzylic carbon using cinnamaldehydes

was responsible for this conversion. DPZ on irradiation with a 1 W blue LED gets excited and undergoes a SET process with the N-aryl valine derivative to generate a nitrogen-centered radical cation. This gets decarboxylated later to form the corresponding α-amino radical which exists as a mixture of imine and enamine. Formation of the THQ involves a [4 + 2] annulation between these imine (4π system) and enamine (2π system) derivatives during which the chirality is induced by the **C3**. The trans-isomer was formed selectively in all the cases with *ee* in the range of 71–92%.

Enantioselective decarboxylative alkylation of α-aminoacids was demonstrated using α-bromoketones (Scheme 9) [28]. A 1:20 combination of the photocatalyst DPZ and Bronsted acid SPINOL-CPA, **C4** played the synergistic role during conversion.

A mechanism was also proposed which states that DPZ on irradiation gets excited to DPZ* and reacts with α-aminoacids to generate a radical via decarboxylation. The DPZ* gets reduced to DPZ—during the process the bromine from alkyl bromide gets cleaved homolytically furnishing another radical. The radicals couple together enantioselectively to afford the product under the influence of chiral phosphoric acid (Scheme 10).

The protocol was successful in the case of α-bromo-α-fluoroketones to furnish chiral β-fluoro-β-methyl amino ketones at slightly different conditions.

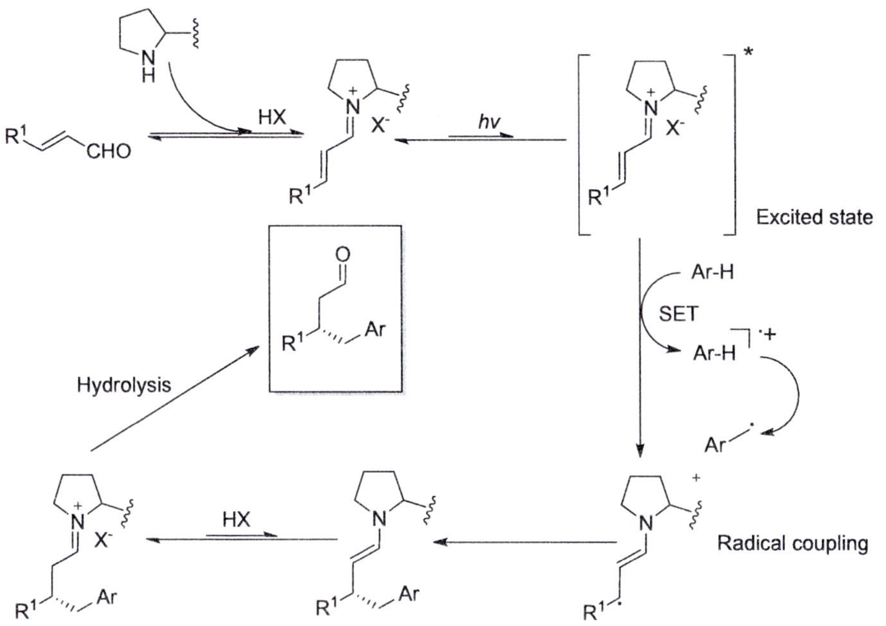

Scheme 6 Plausible mechanism for the visible light-induced functionalization of benzylic carbons using cinnamaldehydes. *Reproduced with permission from* Mazzarella et al. [24]

Visible light-mediated Minisci addition of α-amino benzyl/alkyl radicals to isoquinolines was reported (Scheme 11) [29]. 1,3-Dioxoisoindolin-2-yl 2-((ethoxycarbonyl)amino)alkanoates were acting as the precursors for the α-amino benzyl/alkyl radicals, and enantioselectivities up to 93% were observed. The catalytic system employed here was a duo of DPZ and **C5**. DPZ was the photocatalyst responsible for the radical generation from the precursor. **C5** acts as a chirality inducer during the radical addition step to the isoquinoline via non-covalent interactions. A tertiary radical is being formed which get aromatized subsequently to form isoquinoline via a SET process (Scheme 12).

The substrate scope studies have shown that α-aminobenzyl radicals give better *ee* compared to the α-aminoalkyl radicals.

2-Aryl-3-alkylsubstituted indoles were demonstrated to undergo tandem aerobic oxidation/semipinacol rearrangement sequence to furnish 2,2-disubstituted indolin-3-ones on visible light irradiation (Scheme 13) [30].

The protocol involves a cooperative catalysis composed of photocatalyst DPZ and Brønsted acid catalyst **C6**. Here DPZ generates superoxide from oxygen which reacts with indoles generating 3-hydroperoxy-1H-indole species. Later in the presence of **C6**, cleavage of O–O bond occurs followed by the semipinacole rearrangement in a stereoselective manner.

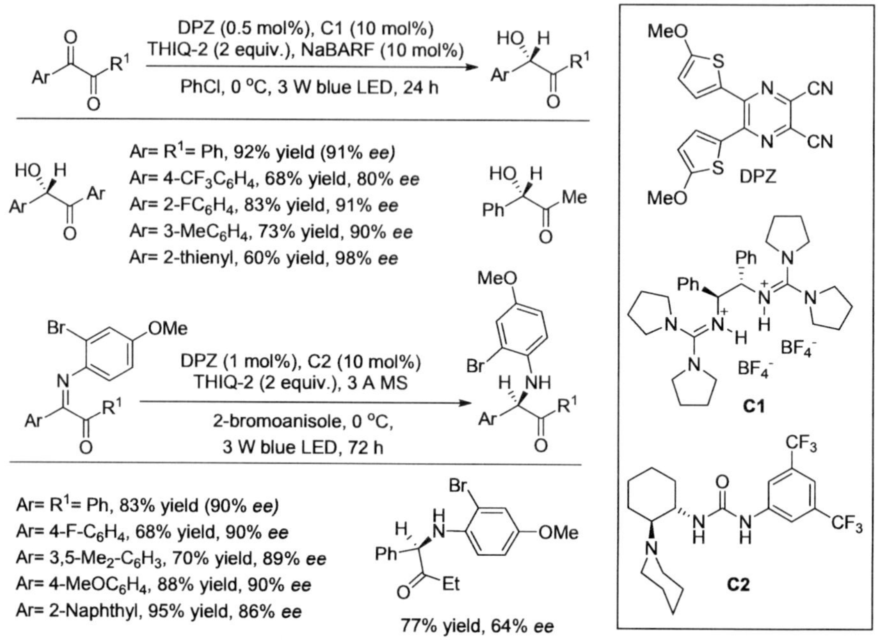

Scheme 7 Asymmetric synthesis of α-hydroxy ketones and α-amino ketones under visible light irradiation

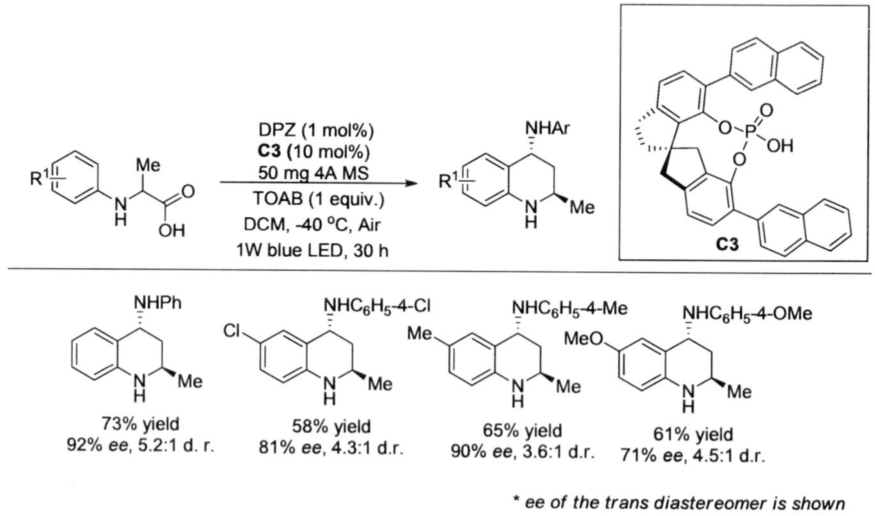

Scheme 8 Visible light-mediated synthesis of 4-amino-2-methyl tetrahydroquinolines from N-aryl valine derivatives via decarboxylative Povarov reaction

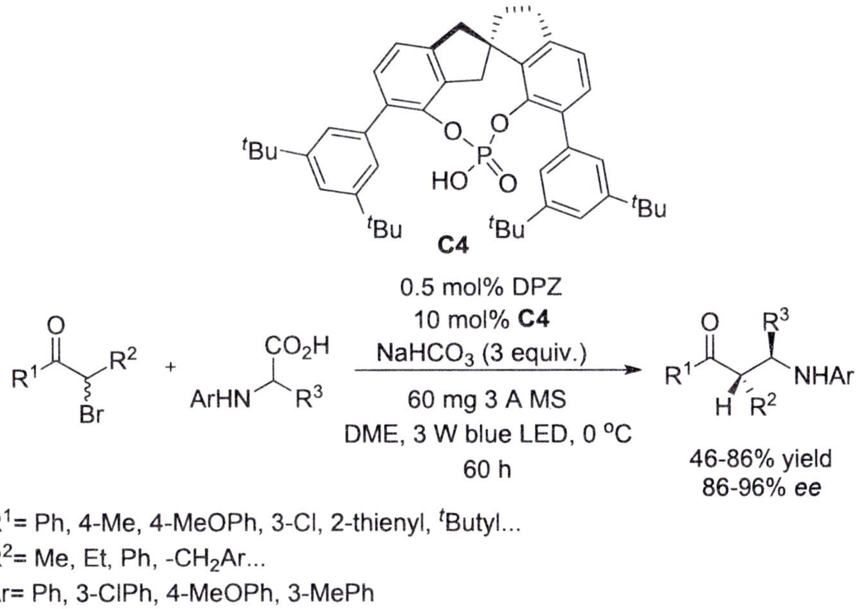

Scheme 9 Decarboxylative alkylation of α-aminoacids under photoredox catalysis

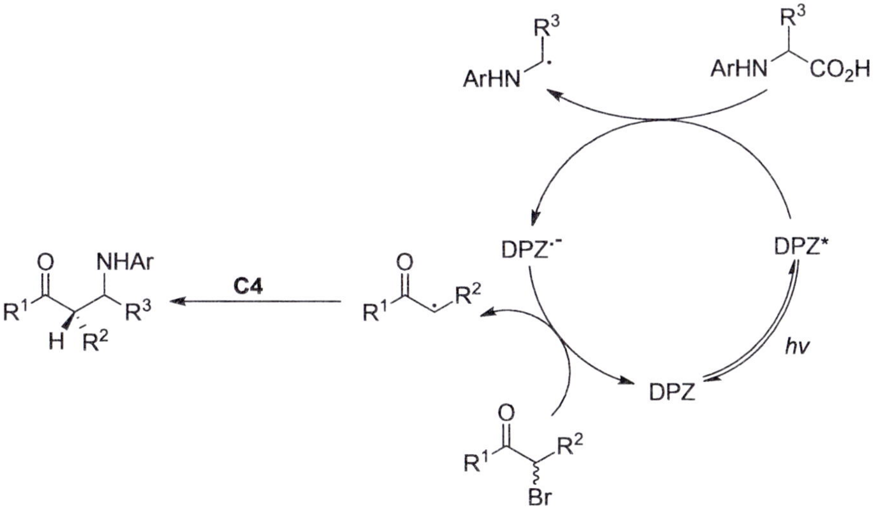

Scheme 10 Plausible mechanism for photoredox-catalyzed decarboxylative alkylation of α-aminoacids. *Reproduced with permission from* Li et al. [28]

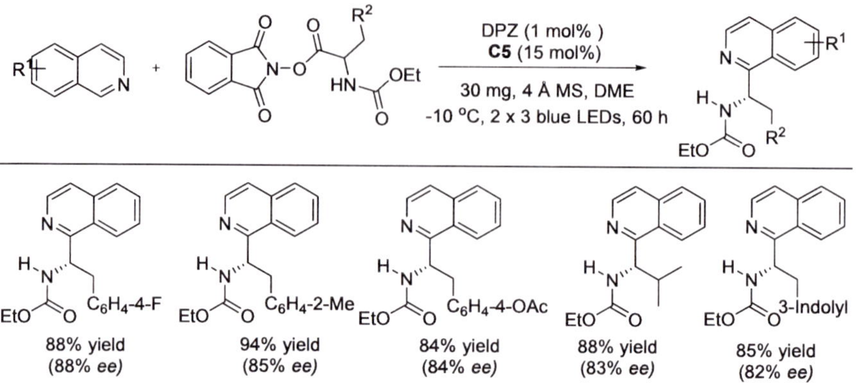

Scheme 11 Visible light-mediated enantioselective Minisci addition of α-amino benzyl/alkyl radicals to isoquinolines. *Reproduced with permission from* Liu et al. [29]

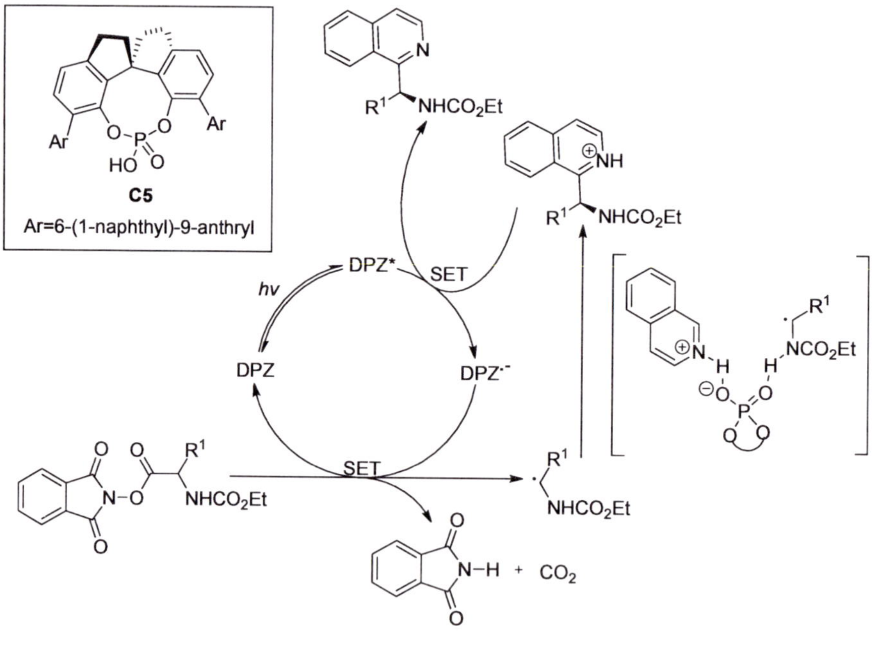

Scheme 12 Mechanism for visible light-mediated enantioselective Minisci addition of α-amino benzyl/alkyl radicals to isoquinolines. (*Reproduced with permission from* Liu et al. [29]

Decarboxylative conjugative addition of α-aminoacids to 2-vinyl pyridines/quinolines followed by enantioselective protonation was reported to furnish α-branched 2-vinylazaarenes (Scheme 14) [31].

DPZ-**C7**/**C8** duo was responsible for this asymmetric transformation, and ee up to 99% was observed. Initially, DPZ* was formed via irradiation with the blue LED,

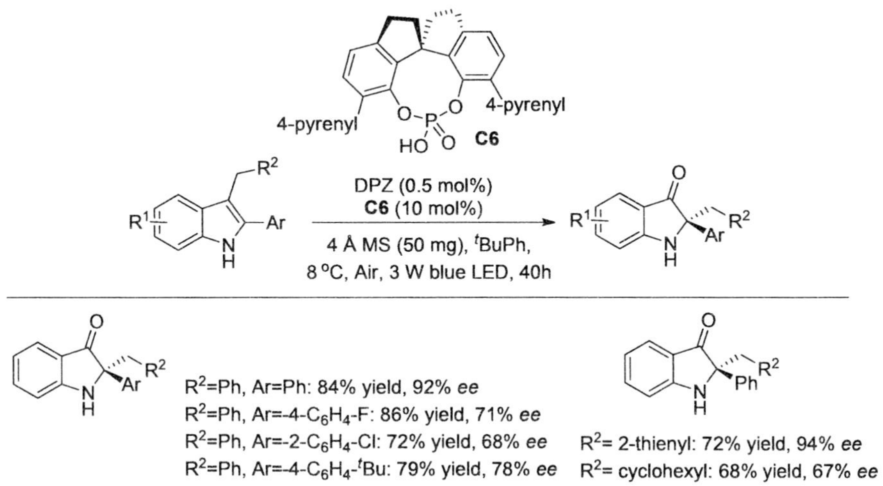

Scheme 13 Oxidation/semipinacol rearrangement sequence leading to the synthesis of 2,2-disubstituted indolin-3-ones under visible light irradiation

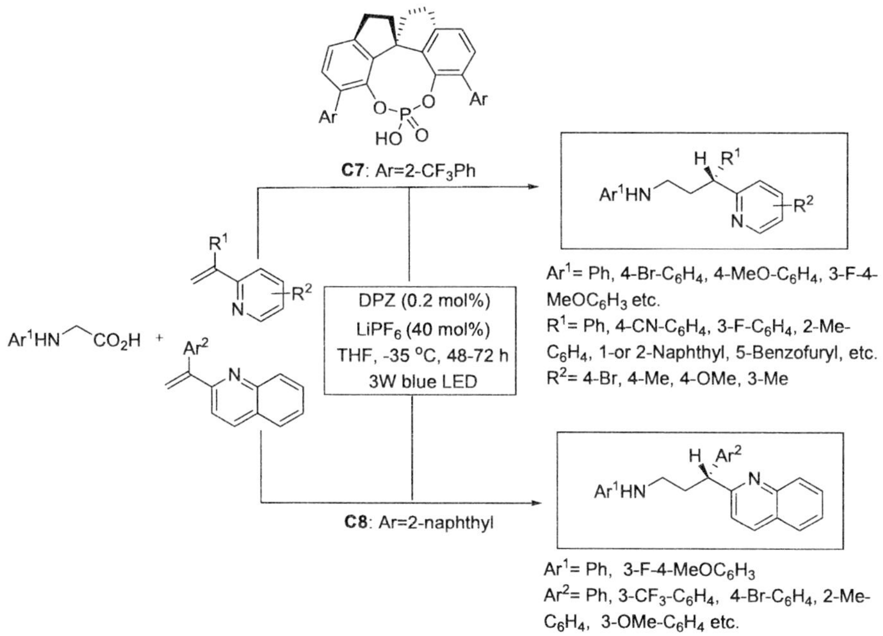

Scheme 14 Visible light-enabled decarboxylative conjugative addition followed by enantioselective protonation

which undergoes SET process with amino acid leading to its decarboxylation and generation of a α-amino radical. The radical makes a 1,4-attack on the CPA coordinated vinyl pyridines/quinolines to furnish a 3o-radical. The radical gets reduced by a SET process later, and the resulting carbanion undergoes an enantioselective protonation by the chiral phosphoric acid to furnish the product.

DPZ-chiral phosphoric acid **C9**-catalyzed enantioselective deuteration of α-chloro-azaarenes was reported (Scheme 15) [32]. The reaction proceeds through a mechanism involving dechlorination followed by **C9**-induced enantioselective introduction of the deuterium atom. Hantzsch ester played the role of electron donor in this protocol by involving in couple of SET processes during the catalytic cycle.

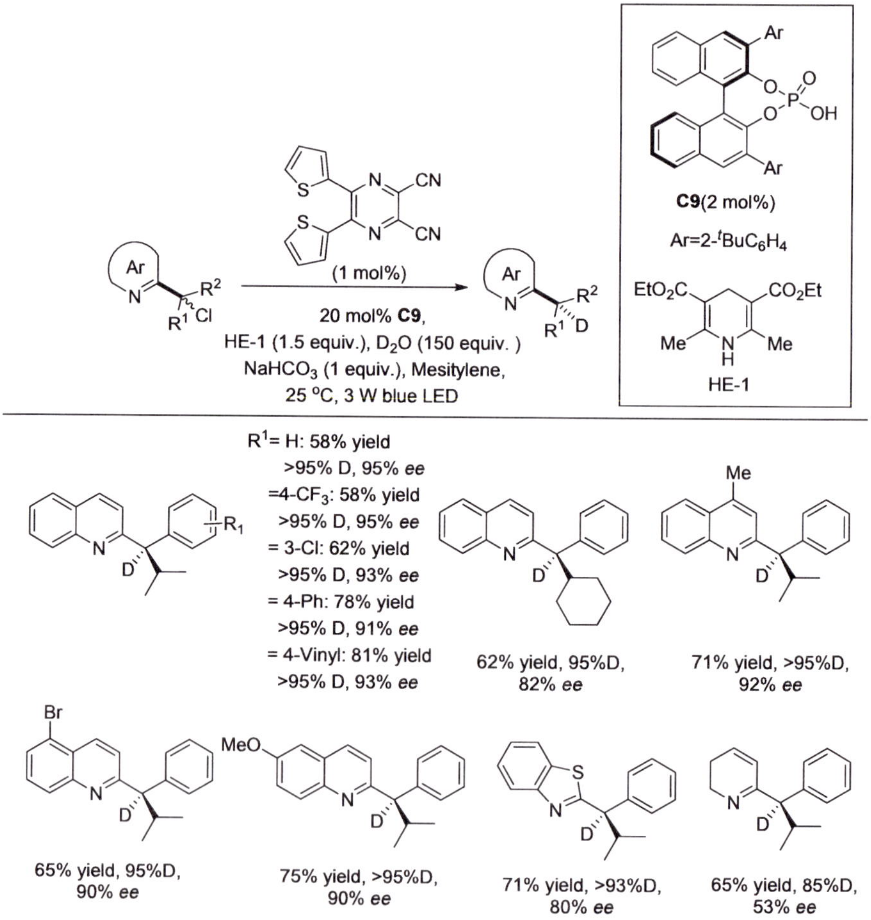

Scheme 15 Visible light-driven dechlorination followed by enantioselective deuteration using photoredox and chiral Brønsted acid catalysis

D_2O provided the deuterium atoms. Enantioselectivities upto 99% were observed with deuterium incorporation ranging between 85 and 95%.

In the same report, authors also demonstrated the enantioselective reductive deuteration of azaarene-substituted ketones to furnish corresponding secondary alcohols. The photocatalyst, electron donor as well as deuterium donor used in this methodology were the same as above while the CPA employed was **C7**.

Tandem visible light-induced cyclization/enantioselective transfer hydrogenation protocol to synthesize 2-substituted tetrahydroquinolines from 2-aminoenones was reported (Scheme 16) [33]. Enantiomeric excess in the range of 81–99% was observed along with moderate to excellent yields. **C10** acts as the catalyst during transfer hydrogenation process where the Hantzsch ester **HE-1** played the role of hydrogen donor.

Blue LEDs irradiated enantioselective addition of aldehydes/imines to vinyl pyridines afforded γ-hydroxy/γ-amino ketones (Scheme 17) [34]. The protocol employed DPZ as the photocatalyst along with a chiral phosphoric acid and a Hantzsch ester. CPA was responsible for the enantioselectivity in the reaction by forming a TS through hydrogen bonds. Once the addition of the aldehyde to the alkene had taken place, the Hantzsch ester provides a hydrogen atom to the resulting radical to furnish the product. During the substrate scope studies, reactivity of a variety of aldehydes and imines against various vinylpyridines were examined. In

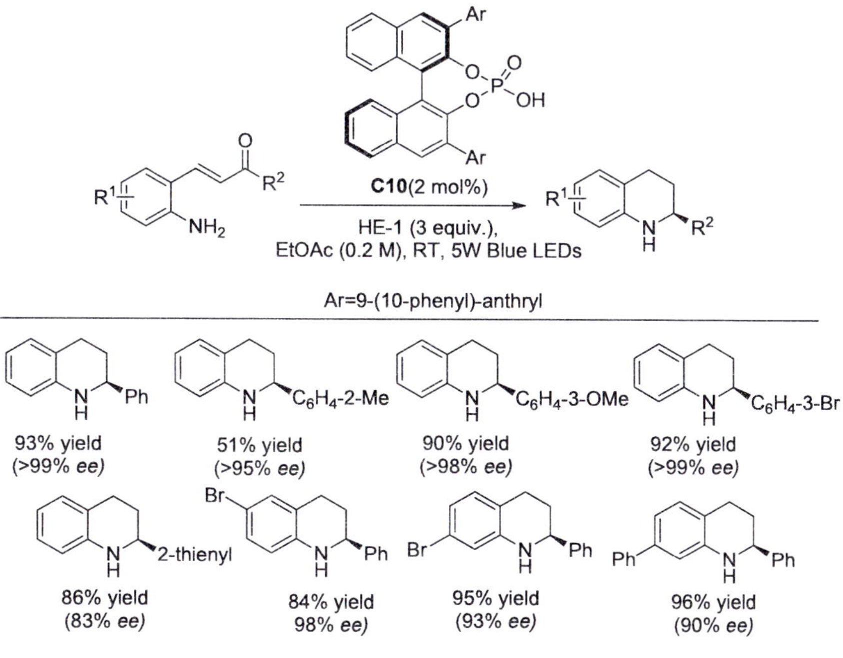

Scheme 16 Visible light-enabled enantioselective synthesize of 2-substituted tetrahydroquinolines from 2-aminoenones

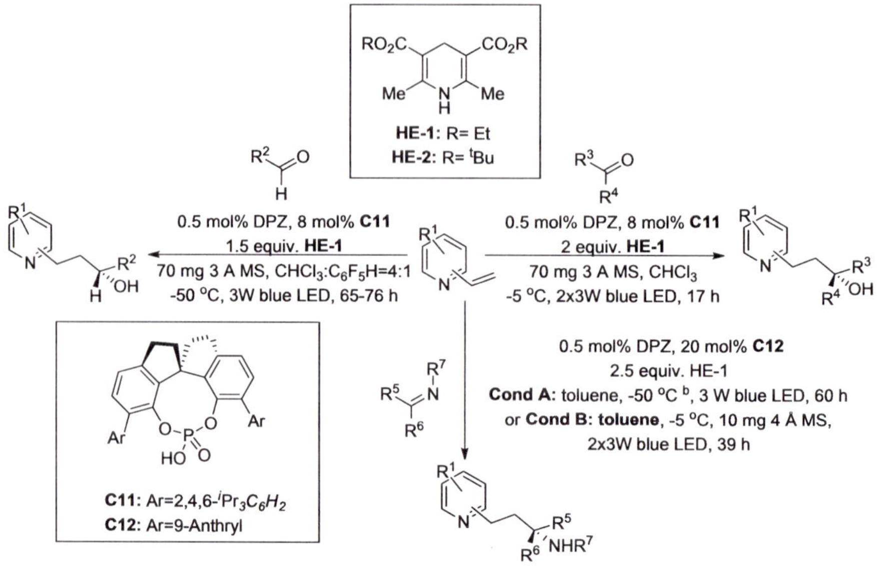

Scheme 17 Photocatalyzed enantioselective addition of aldehydes/imines to vinyl pyridines

the case of aldehydes, γ-secondary alcohols were obtained in 50–92% of the yields along with *ee* of 81–96%. Excellent enantioselectivities were observed with imines also ranging between 80 and >99%.

DPZ-catalyzed asymmetric dechloroprotonation of α-chloro-α-fluoroketones was demonstrated (Scheme 18) [35]. The protocol has employed *L-tert*-leucine-based

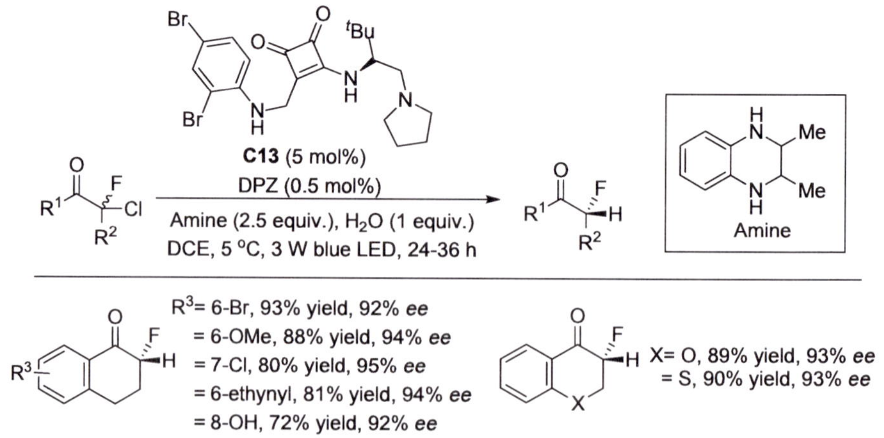

Scheme 18 DPZ-catalyzed enantioselective dechloroprotonation of α-chloro-α-fluoro ketones under visible light irradiation

squaramide-tertiary amine, **C13** as the chirality inducer while 2,3-dimethyl-1,2,3,4-tetrahydroquinoxaline as the hydrogen donor (or terminal reductant). The generality of this protocol was well studied with various ketones including α-chloro-α-fluoro tetralones, indanones, aryl–alkyl ketones, etc. Moderate to excellent yields were observed along with ee in the range of 82–>99%.

2.4 Enzyme-Based Approaches

Enzymes are proven to be suitable co-catalysts in photocatalyzed asymmetric synthesis. They provide suitable binding sites or cavities for substrates as well as possess certain functional groups capable of SET process leading to the generation of active radical species.

Nicotinamide-dependent enzymes, namely RasADH (dehydrogenase derived from ralstonia species) and LKADH (dehydrogenase of the bacterium *Lactobacillus kefiri*) demonstrated to be enantioselectively affecting the dehalogenation of lactones followed by protonation under visible light irradiation (Scheme 19) [36].

Nicotinamide-dependent double-bond reductase, NtDBR, has provided suitable binding sites for the α-acetoxyketone which undergoes asymmetric deacetoxylation under photoredox catalysis (Scheme 20) [37]. The protocol has used 0.5 mol% Rose Bengal (RB) along with 1 mol% $NADP^+$. Once the RB is irradiated to the excited state, it is capable of oxidizing NADPH (derived from NADP+) and gets itself converted to $RB^{\bullet -}$. The proposed mechanism suggests that RB•– goes through a SET process with the enzyme-bound α-acetoxyketone resulting in its deacetoxylation to furnish a α-carbonyl radical. This then undergoes an enantioselective HAT with the NADPH to afford the product. The $NADP^{\cdot}$ generated during this process

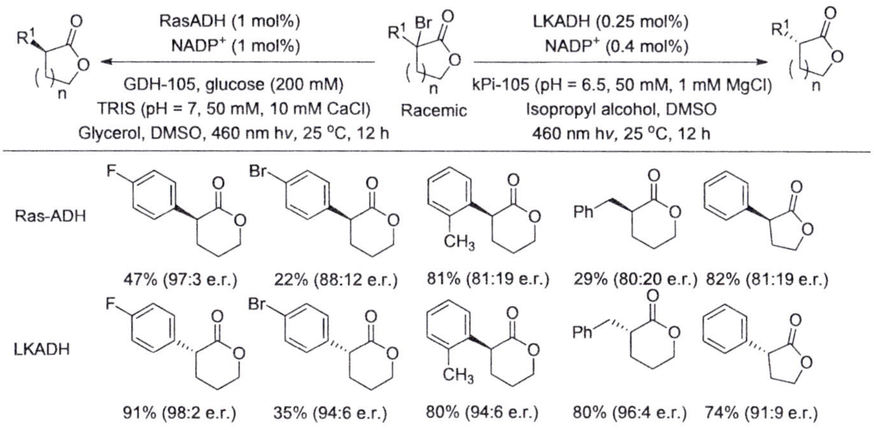

Scheme 19 Enzyme-catalyzed enantioselective hydrodehalogenation of lactones under visible light irradiation

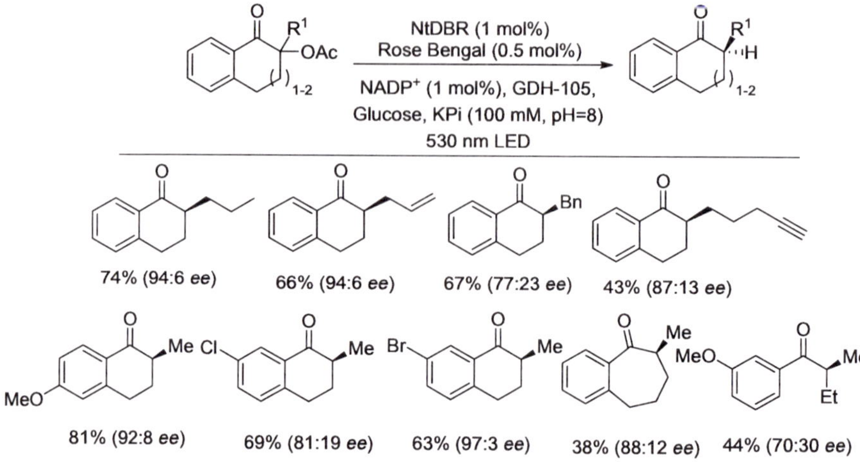

Scheme 20 Photoredox-enabled enantioselective hydrodeacetoxylation of α-acetoxyketones

gets converted back to the $NADP^+$ by RB (Scheme 21). The substrate scope studies were conducted using various 2-alkyl-2-acetoxy tetralones, and enantioselectivities in the range of 77–97% were obtained.

Similarly, the photochemical reaction using the enzyme KRED-P2-D03 (1 mol%) along with eosin Y (0.5 mol%) resulted in dehalogenation of α-bromoamides with good to excellent *ee*.

Flavoenzymes emerged as efficient catalysts for intramolecular asymmetric cyclization of α-chloroamides containing alkene groups under visible light irradiation (Scheme 22) [38]. Under the optimized conditions, five-, six-, seven- and eight-membered lactams were furnished with high enantiomeric purity.

The enzymes employed during this protocol include ene reductases from *Gluconobacter oxydans*, *Lactobacillus casei* and nostoc sp. as well as morpholinone reductase. The active component in the flavoenzyme is the flavin hydroquinone part which forms an electron donor complex with the chloroamide. Photoirradiation of the complex results in electron transfer from flavin hydroquinone to chloroamide. This results in loss of a halide ion from chloroamide and formation of an α-acyl radical. The radical intramolecularly reacts with the alkene part resulting in cyclization which furnishes the product via an enantioselective HAT (Scheme 23).

Anthraquinone sulfate under visible light irradiation was found to catalyze the oxidation of methylene or methyl carbon into the carbonyl group [39]. The methodology accompanied successive asymmetric transformation of the furnished aldehyde/ketone to cyanohydrin, benzyl alcohol, secondary amine, α-chiral ketones, etc., under enzyme catalysis (Scheme 24).

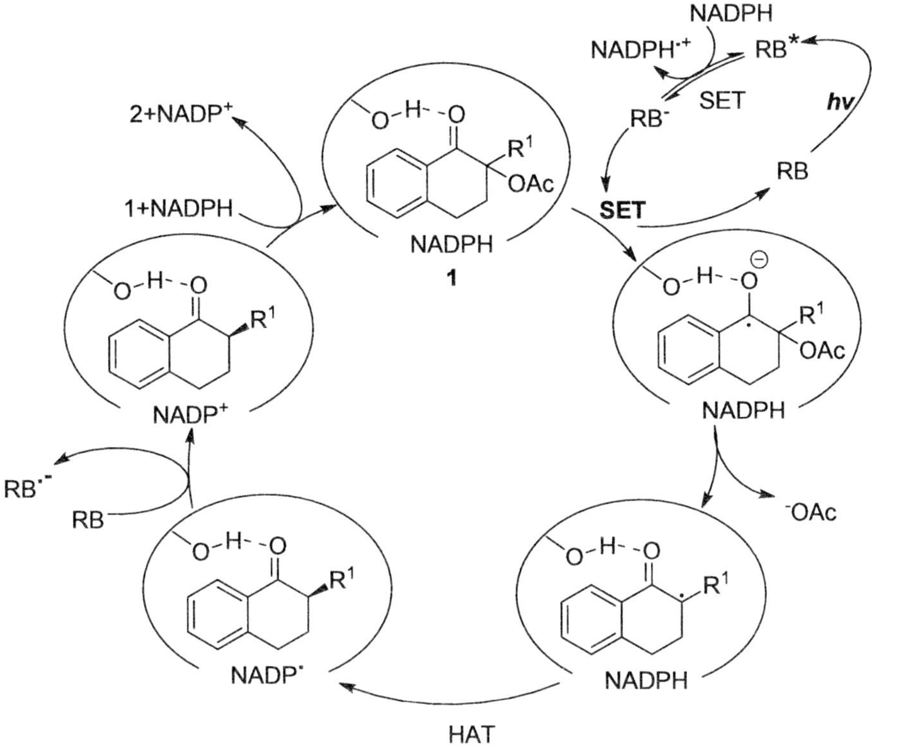

Scheme 21 Plausible mechanism for enantioselective hydrodeacetoxylation of α-acetoxyketones. *Reproduced with permission from* Biegasiewicz et al. [37]

3 Conclusion

Photocatalyzed asymmetric synthesis was proven to be one of the most useful protocols toward chiral synthesis. Starting with sensitizer-based approaches, the advancements ensued in this area are really tremendous. Combination of organocatalysts like amines, amino alcohols, chiral phosphoric acids, thiourea derivatives along with photoredox catalysts vastly improves the efficiency of the area. Enzymes also formed suitable partners for photocatalysts making these protocols greener.

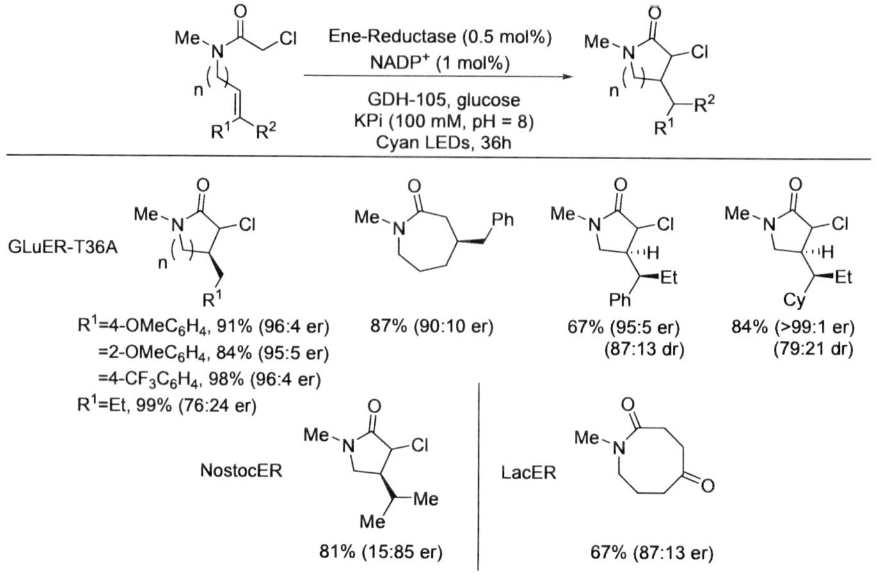

Scheme 22 Flavoenzymes-catalyzed intramolecular asymmetric cyclization of α-chloroamides under visible light irradiation

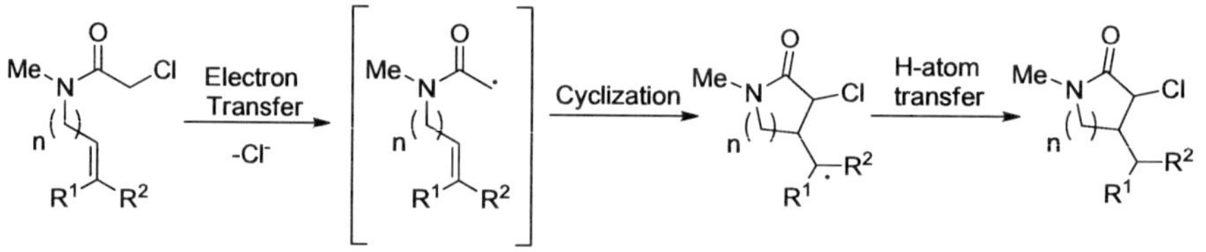

Scheme 23 Plausible mechanism for asymmetric cyclization of α-chloroamides

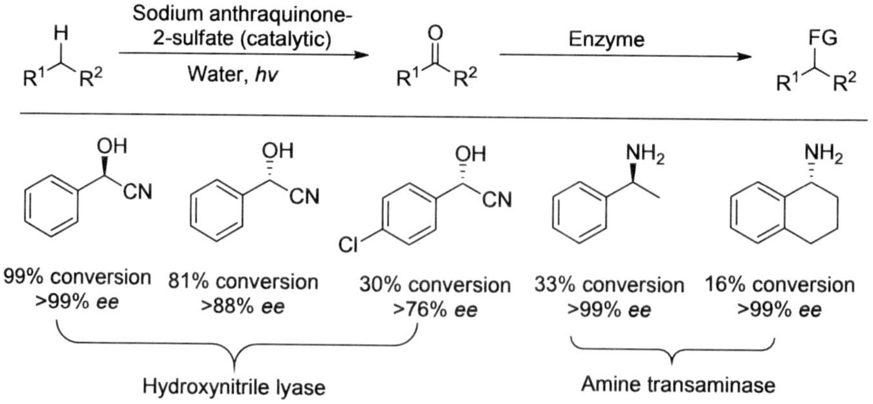

Scheme 24 Asymmetric synthesis involving visible light-mediated sp^3 carbon oxidation followed by enzyme-catalyzed nucleophilic addition

References

1. (a) Noyori R (2002) Asymmetric catalysis: science and opportunities (Nobel lecture). Angew Chem Int Ed 41:2008–2022; (b) Knowles WS (2002) Asymmetric hydrogenations (Nobel lecture). Angew Chem Int Ed 41:1998–2007; (c) Sharpless KB (2002) Searching for new reactivity (Nobel lecture). Angew Chem Int Ed 41:2024–2032
2. Brimioulle R, Lenhart D, Maturi MM, Bach T (2015) Enantioselective catalysis of photochemical reactions. Angew Chem Int Ed 54:3872–3890
3. Kim J-I, Schuster GB (1990) Enantioselective catalysis of the triplex Diels-Alder reaction: addition of *trans*-.beta.-methylstyrene to 1,3-cyclohexadiene photosensitized with (-)-1,1'-bis(2,4-dicyanonaphthalene). J Am Chem Soc 112:9635–9637
4. Ibrahem I, Zhao G-L, Sunden H, Cordova A (2006) A route to 1,2-diols by enantioselective organocatalytic α-oxidation with molecular oxygen. Tetrahedron Lett 47:4659–4663
5. Rono LJ, Yayla HG, Wang DY, Armstrong MF, Knowles RR (2013) Enantioselective photoredox catalysis enabled by proton-coupled electron transfer: development of an asymmetric aza-pinacol cyclization. J Am Chem Soc 135:17735–17738
6. Welin ER, Warkentin AA, Conrad JC, MacMillan DWC (2015) Enantioselective α-alkylation of aldehydes by photoredox organocatalysis: rapid access to pharmacophore fragments from β-cyanoaldehydes. Angew Chem Int Ed 54:9668–9672
7. Hammond GS, Cole RS (1965) Asymmetric induction during energy transfer. J Am Chem Soc 87:3256–3257
8. Inoue Y, Yokoyama T, Yamasaki N, Tai A (1989) An optical yield that increases with temperature in a photochemically induced enantiomeric isomerization. Nature 341:225–226
9. Asaoka S, Kitazawa T, Wada T, Inoue Y (1999) Enantiodifferentiating anti-Markovnikov photoaddition of alcohols to 1,1-diphenylalkenes sensitized by chiral naphthalenecarboxylates. J Am Chem Soc 121:8486–8498
10. Kim J-I, Schuster GB (1992) Enantioselective catalysis of the triplex Diels-Alder reaction: a study of scope and mechanism. J Am Chem Soc 114:9309–9317
11. Lu R, Yang C, Cao Y, Wang Z, Wada T, Jiao W, Mori T, Inoue Y (2008) Supramolecular enantiodifferentiating photoisomerization of cyclooctene with modified β-cyclodextrins: critical control by a host structure. Chem Commun 374–376
12. Müller C, Bauer A, Bach T (2009) Light-driven enantioselective organocatalysis. Angew Chem Int Ed 48:6640–6642
13. Alonso R, Bach T (2014) A chiral thioxanthone as an organocatalyst for enantioselective [2+2] photocycloaddition reactions induced by visible light. Angew Chem Int Ed 53:4368–4371
14. Neumann M, ldner SF, Kçnig B, Zeitler K (2011) Metal-free, cooperative asymmetric organophotoredox catalysis with visible light. Angew Chem Int Ed 50:951–954; Angew Chem 123:981–985
15. Henin F, Muzart J, Pete J-P, M'boungou-M'passi A, Rau H (1991) Enantioselective protonation of a simple enol: aminoalcohol-catalyzed ketonization of a photochemically produced 2-methylinden-3-ol. Angew Chem Int Ed 30:416–418
16. Cecere G, Kçnig CM, Alleva JL, MacMillan DWC (2013) Enantioselective direct α-amination of aldehydes via a photoredox mechanism: a strategy for asymmetric amine fragment coupling. J Am Chem Soc 135:11521–11524
17. Vignola N, List B (2004) Catalytic asymmetric intramolecular α-alkylation of aldehydes. J Am Chem Soc 126:450–451; Hayashi Y, Gotoh H, Hayashi T, Shoji M (2005) Diphenylprolinol silyl ethers as efficient organocatalysts for the asymmetric Michael reaction of aldehydes and nitroalkenes. Angew Chem Int Ed 44:4212–4215; Angew Chem 117:4284–4287
18. Brunner H, Bgler J, Nuber B (1995) Enantioselective catalysis 98. Preparation of 9-amino(9-deoxy)cinchona alkaloids. Tetrahedron Asymmetry 6:1699–1702
19. Córdova A, Sundén H, Engqvist M, Ibrahem I, Casas J (2004) The direct amino acid-catalyzed asymmetric incorporation of molecular oxygen to organic compounds. J Am Chem Soc 126:8914–8915

20. Neumann M, Fuldner S, Konig B, Zeitler K (2011) Metal-free, cooperative asymmetric organophotoredox catalysis with visible light. Angew Chem Int Ed 50(4):951–954
21. Arceo E, Jurberg ID, Alvarez-Fernandez A, Melchiorre P (2013) Photochemical activity of a key donor–acceptor complex can drive stereoselective catalytic α-alkylation of aldehydes. Nat Chem 5:750–756
22. Bahamonde A, Melchiorre P (2016) Mechanism of the stereoselective α-alkylation of aldehydes driven by the photochemical activity of enamines. J Am Chem Soc 138:8019–8030
23. Bonilla P, Rey YP, Holden CM, Melchiorre P (2018) Photo-organocatalytic enantioselective radical cascade reactions of unactivated olefins. Angew Chem Int Ed 57:12819–12823
24. Mazzarella D, Crisenza GEM, Melchiorre P (2018) Asymmetric photocatalytic C–H functionalization of toluene and derivatives. J Am Chem Soc 140:8439–8443
25. Bergonzini G, Schindler CS, Wallentin C-J, Jacobsen EN, Stephenson CRJ (2014) Photoredox activation and anion binding catalysis in the dual catalytic enantioselective synthesis of β-amino esters. Chem Sci 5:112–116
26. Lin L, Bai X, Ye X, Zhao X, Tan C-H, Jiang Z (2017) Organocatalytic enantioselective protonation for photoreduction of activated ketones and ketimines induced by visible light. Angew Chem Int Ed 56:13842–13846
27. Shao T, Yin Y, Lee R, Zhao X, Chai G, Jiang Z (2018) Sequential photoredox catalysis for cascade aerobic decarboxylative povarov and oxidative dehydrogenation reactions of N-aryl a-amino acids. Adv Synth Catal 360:1754–1760
28. Li J, Kong M, Qiao B, Lee R, Zhao X, Jiang Z (2018) Formal enantioconvergent substitution of alkyl halides via catalytic asymmetric photoredox radical coupling. Nat Commun 9:2445–2453
29. Liu X, Liu Y, Chai G, Qiao B, Zhao X, Jiang Z (2018) Organocatalytic enantioselective addition of α-aminoalkyl radicals to isoquinolines. Org Lett 20:6298–6301
30. Bu L, Li J, Yin Y, Qiao B, Chai G, Zhao X, Jiang Z (2018) Organocatalytic asymmetric cascade aerobic oxidation and semipinacol rearrangement reaction: a visible light-induced approach to access chiral 2,2-disubstituted indolin-3-ones. Chem Asian J 13:2382–2387
31. Yin Y, Dai Y, Jia H, Li J, Bu L, Qiao B, Zhao X, Jiang Z (2018) Conjugate addition-enantioselective protonation of n-aryl glycines to α-branched 2-vinylazaarenes via cooperative photoredox and asymmetric catalysis. J Am Chem Soc 140:6083–6087
32. Shao T, Li Y, Ma N, Li C, Chai G, Zhao X, Qiao B, Jiang Z (2019) Photoredox-catalyzed enantioselective α-deuteration of azaarenes with D_2O. Science 16:410–419
33. Xiong W, Li S, Fu B, Wang J, Wang Q-A, Yang W (2019) Visible-light induction/Brønsted acid catalysis in relay for the enantioselective synthesis of tetrahydroquinolines. Org Lett 21:4173–4176
34. Cao K, Tan S, Lee R, Yang S, Jia H, Zhao X, Qiao B, Jiang Z (2019) Catalytic enantioselective addition of prochiral radicals to vinylpyridines. J Am Chem Soc 141:5437–5443
35. Hou M, Lin L, Chai X, Zhao X, Qiao B, Jiang Z (2019) enantioselective photoredox dehalogenative protonation. Chem Sci 10:6629–6634
36. Emmanuel MA, Greenberg NR, Oblinsky DG, Hyster TK Accessing non-natural reactivity by irradiating nicotinamide-dependent enzymes with light. Nature 540:414–417
37. Biegasiewicz KF, Cooper SJ, Emmanuel MA, Miller DC, Hyster TK (2018) Catalytic promiscuity enabled by photoredox catalysis in nicotinamide-dependent oxidoreductases. Nat Chem 10:770–775
38. Biegasiewicz KF, Cooper SJ, Gao X, Oblinsky DG, Kim JH, Garfinkle SE, Joyce LA, Sandoval BA, Scholes GD, Hyster TK, Photoexcitation of flavoenzymes enables a stereoselective radical cyclization. Science 364:1166–1169.
39. Zhang W, Fueyo EF, Hollmann F, Martin LL, Pesic M, Wardenga R, Höhne M, Schmidt S (2015) Combining photo-organo redox- and enzyme catalysis facilitates asymmetric C–H bond functionalization. Eur J Org Chem 2019:80–84

Chapter 15
The Role of Continuous Flow Processing in the Development of Green Chemical Syntheses

Faith M. Akwi and Paul Watts

1 Introduction

The chemical industry's production capacity between the years 2000 and 2017 is said to have almost doubled from 1.2 to 2.3 billion tonnes in 2017 with global sales totalling up to 5.68 trillion US dollars recorded thus positioning the chemical industry as the second largest manufacturing industry in the world [1]. Moreover, these sales are predicted to double by the year 2030. The increased growth in size of this industry does not only mean an increase in returns but also an increased generation of waste which in turn culminates into a number of adverse effects on the environment, health of the human population and aquatic life. According to a United Nation's Environmental Programme (2019) report, it is also predicted that the global goal of minimizing the adverse effects of chemicals and waste by the year 2020 will not be achieved. The authors of the report advise that despite the fact that solutions towards this problem exist, more aggressive worldwide action by stakeholders is urgently required [2].

In the past century however, there has been an increased interest in ensuring that chemical manufacturing processes have minimum effect on the environment and human life. This is mainly driven by the concept of green chemistry that has become a hot topic of discussion in academia and industry. Green chemistry is a concept introduced by Paul Anastas and John Warner in 1998 and is governed by 12 principles [3], which have recently been condensed into a manageable mnemonic [4] described as follows:

P—prevent waste, **R**—renewable materials, **O**—omit derivatization steps, **D**—degradable chemical production, **U**—use safe synthetic methods, **C**—catalytic

F. M. Akwi · P. Watts (✉)
Nelson Mandela University, University Way, Port Elizabeth 6031, South Africa
e-mail: Paul.Watts@mandela.ac.za

F. M. Akwi
e-mail: s211206296@mandela.ac.za

G. Anilkumar and S. Saranya (eds.), *Green Organic Reactions*, Materials Horizons: From Nature to Nanomaterials, https://doi.org/10.1007/978-981-33-6897-2_15

reagents, **T**—temperature, pressure ambient, **I**—in process monitoring, **V**—very few auxiliary substances, **E**—*E*-factor, maximize feed in product, **L**—low toxicity of chemical products and **Y**—yes, it is safe, i.e. **PRODUCTIVELY**.

Ever since its introduction, there has been great debate on how the principles are to be used in defining the greenness of processes. Researchers often claim that the syntheses or chemical processes they develop are green. This is mostly seen in a number of publications in literature; however, the pertinent question to ask is: How can the greenness of a process or synthesis be measured? Must a process tick all the 12 principles such that it can be deemed green? the use of available techniques such as green metrics that can provide a quantitative measure of effect of a process on the environment is one place to start. These green metrics include mass intensity and mass productivity, atom economy, yield, *E*-factor and reaction mass efficiency. These are discussed elsewhere and as such will not be tackled here [5].

Furthermore, in combination with the green metrics, a lifecycle assessment is recommended to evaluate a process in its totality, to elucidate the environmental effects of processes in their life cycle. This is also termed as a cradle to grave assessment [6]. The need to develop green chemical processes has also led to the evolution and use of enabling technologies such as continuous flow processing and microwave technology, mechano-chemistry and reactive extrusion, towards this purpose. In fact, continuous flow processing technology has recently been named by IUPAC as one of the 10 top chemical innovations that will change the world [7].

2 The Benefits of Continuous Flow Processing in Chemical Syntheses

Briefly, continuous flow processing technology entails the use of micro–or meso-structured systems as reactions spaces for chemical syntheses. This approach towards organic synthesis has been instrumental in enabling access to new chemistries that were once thought to be unachievable. Efficient mass and heat transfer in continuous flow reactors enables improved reaction yields, selectivity, efficient atom economy, shortened reaction/processing times which are achieved through excellent diffusion facilitated by the large surface area-to-volume ratio characteristic of the small channel dimensions of the reactors. Additionally, the safe handling of hazardous wastes and minimizing worker personnel exposure is assured with the use of small amounts of reagents [8].

From the above benefits, it can be stated that the use of continuous flow processing is instrumental in the development of green syntheses as it ticks more than half of the 12 principles of green chemistry. In this chapter, a few selected examples of syntheses performed in continuous flow systems are used to illustrate how the use of these systems has facilitated better and greener chemical syntheses compared to traditional reaction systems. The selected examples are neatly categorized and discussed under the benefits of continuous flow processing that correspond to the some of the 12 principles of green chemistry in no particular order.

2.1 Excellent Atom Economy and Yields

The second principle of green chemistry; "Atom economy through synthetic methods designed to maximise the incorporation of all materials used in the process into the final product"

Atom economy and yield are two very common metrics used to evaluate the efficiency of a chemical process. Boros et al. demonstrate how flow synthesis of the final step of Vortioxetine **1** involving piperazine ring formation led to the highest yield and purity ever reported for the API [9]. Vortioxetine is an anti-depressant drug, which belongs to the class of bis-aryl sulphanyl amines. It is used for the treatment of depression by targeting the selective serotonin (5-HT) reuptake inhibitors, 5-HT_{1A} and $5HT_7$. The piperazine ring formation was afforded via the reaction between 2-[(2,4-dimethylphenyl)sulphanyl]aniline **2** with bis 2-(chloroethyl)amine hydrochloride **3**. This reaction in batch mode took up to 27 h to give a 63% isolated yield (99% purity by HPLC) of **1**, however there was a significant amount of side products detected which required laborious purification steps to obtain the pure *L*-(+)-mandelate salt of the compound **1**. Using a specialized reactor system consisting of three independent thermostable modules, i.e. a heated flask, heated high pressure syringe pump (H-ION HI-PD100) and a heater module (H-ION-HI-HM250). The reactor module comprised of three different reactor loops which varied in length and volume. (loop A: reactor volume: 1 mL, i.d: 1 mm, o.d: 1/16″, length: 1.3 m, loop B: reactor volume: 7.85 mL, i.d: 1 mm, o.d: 1/16″, length: 10 m length, loop C: reactor volume: 7.85 mL, i.d: 1.7 mm, o.d: 1/8″, length: 3.46 m). This compact set-up was used to investigate the optimum reaction conditions, i.e. solvent system, temperature, residence time, reactant concentration and coil reactor geometry to give the desired reaction output. The continuous flow process description below summarizes the optimum reaction conditions that were established for this synthesis.

2-[(2,4-Dimethylphenyl)sulphanyl]aniline (0.16 M) **2** and bis 2-(chloroethyl)amine hydrochloride (2 equiv.) **3** were dissolved in NMP:toluene (1:1) solvent system at 90 °C. The substrate solution was thereafter pumped into the prewashed reactor system (coil loop C: reactor volume: 7.85 mL, i.d: 1.7 mm, o.d: 1/8", length: 3.46 m) at a flow rate of 0.28 mL/min (30 min residence time) and a reaction temperature of 190 °C. 79% yield of **1** based on HPLC was recorded (Fig. 1). Compound **1** was transformed into its *L*-(+)-mandelic salt in 99.6% purity.

The efficient control of reaction parameters such as residence time and reaction temperature was very crucial for this chemical transformation. Due to the excellent heat transfer in the continuous flow reactor set-up, the unavoidable side product which was observed in the batch synthesis was minimized in fact a high reaction temperature of 190 °C provided that best reaction output.

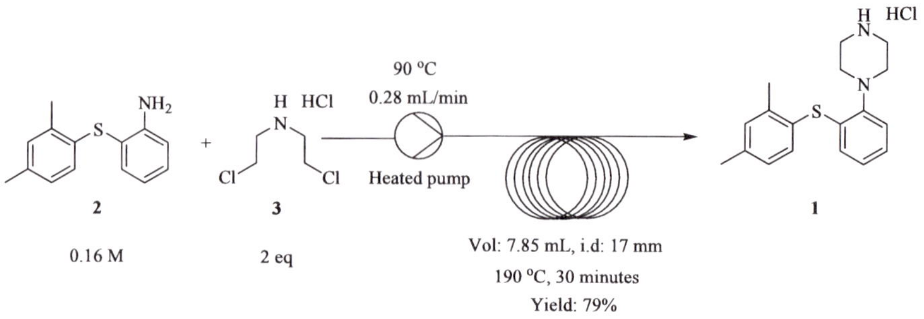

Fig. 1 Final step continuous flow synthesis towards Votioxetine API

Continuous flow synthesis has also been useful in the synthesis of nanoparticles of uniform size and shape distribution moreover in high yield over batch methods [10–12]. Nikam et al. illustrate through the scalable synthesis of CuO nanoparticles achieved through microwave-assisted continuous flow synthesis [13]. At a reaction volume of 25 mL, benzyl alcohol as a solvent, $Cu(Ac)_2$ as a metal precursor (metal precursor concentration: 0.05 M), in the presence of NaOH (30 μL of 0.1 M), 70% yield of CuO nanoparticles were obtained measured by gravimetric estimation after 1 min of irradiation (2.45 GHz, 700 W Ragatech Pvt. Ltd. India). These conditions were optimized in a batch microwave prior to the microwave-assisted continuous flow synthesis. A 25 mL continuous stirred tank reactor was employed for the microwave-assisted continuous flow synthesis (Fig. 2). Using the batch microwave synthesis reaction condition, a solution of $Cu(Ac)_2$ was pumped at a flow rate of 25 mL/min into a CSTR placed in a microwave oven. The formed CuO nanoparticles were collected at the outlet using a peristaltic pump, washed and dried. Characterization of the nanoparticles showed no significant difference between the batch and the flow syntheses (crystal size 3.5 and 3.9 nm, respectively). Spherical-shaped nanoparticles with an average particle size of 4.2 nm were obtained from the microwave-assisted flow synthesis. A high production rate of 5 g/h was successfully achieved with this flow process.

The heat transfer capabilities of CuO nanoparticles synthesized in batch and flow modes were also compared. Using a transient hot wire method, the thermal conductivities of both type of nanoparticles (based on synthesis method) were measured. Due to the similarity in particle shape and size of the nanoparticles, there was no significant difference in thermal conductivity recorded at all volume % of nanoparticles in ethylene glycol that was used for the study. Nonetheless, the flow synthesis was able to afford a high throughput per unit time, therefore providing an efficient method for the synthesis of CuO nanoparticles.

Peris et al. used supported ionic liquid like phases as immobilized catalysts in the multistep and multicatalytic synthesis of enantiopure chiral cyanohydrins in continuous flow [14]. Cyanohydrin synthesis is mainly achieved by two synthetic routes, i.e. the asymmetric addition of a cyanide derivative which requires the use of expensive organometallic systems [15] or via HCN biocatalytic addition, which suffers from

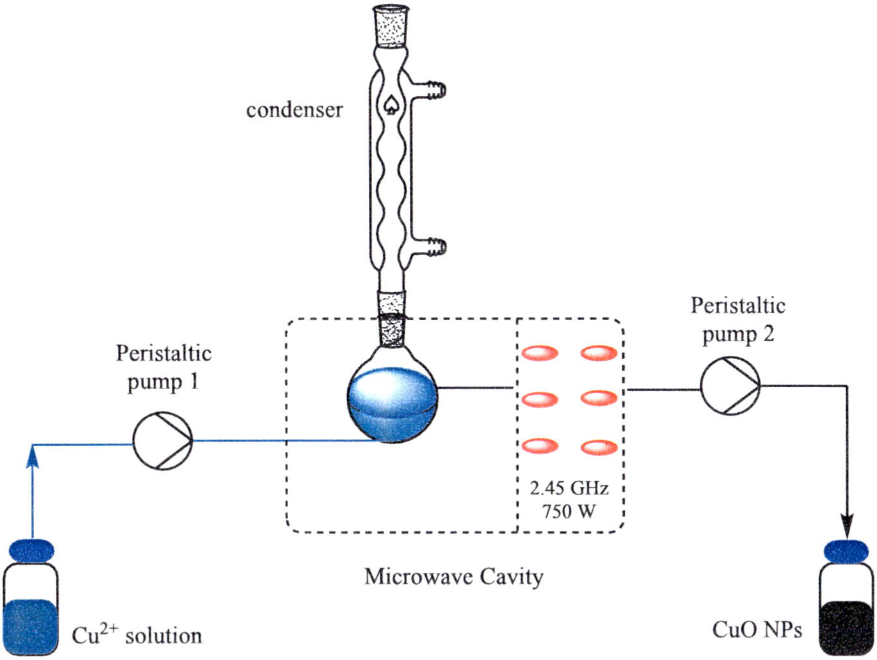

Fig. 2 Microwave-assisted flow synthesis of CuO nanoparticles

reaction reversibility that may cause reduced enantiomeric excess [16]. Four aldehydes, whose corresponding cyanohydrins are key intermediates for the synthesis of different commercial drugs; (*R*)-fluoxetine, (*R*)-atomoxetine, (*R*)-nisoxetine, clopidogel and (*R*)-duloxetine, were used as starting materials for their work. To obtain the desired cyanohydrins, a three-step process, namely organocatalytic cyanosilylation of aldehyde substrate, methyl ester synthesis and kinetic resolution of the formed cyanohydrin ester by transesterification in the presence of alcohol, was employed. All three reactions were successfully optimized in batch mode (Scheme 1). For all four aldehydes used as substrates, high yields for step 1 (95–99%) and step 2 (99%) as well as high enantiomeric excess for all steps were obtained. Step 3 afforded moderate yields (51–57%) of desired products (Table 1).

A single-pot multicatalytic synthesis of chiral cyanohydrins that could not be achieved in batch was easily developed as a multistep multicatalytic continuous flow process. The continuous flow synthesis of cyanohydrin esters was first evaluated in two packed bed reactors using benzaldehyde as a model substrate. The first reactor was packed with catalyst a (1 g) through which a neat mixture of benzaldehyde **4a** and TMSCN (1:1.1) was pumped at a flow rate of 0.01 mL/min at room temperature. Catalyst b was packed into the second reactor (1 g).

Using a T-piece, the outlet stream from the first reactor containing cyanosilyl ether intermediate **5a** was mixed with acetic anhydride (2 equiv., flow rate: 0.0009 mL/min) and flowed into the second catalyst reactor bed to afford hydrolysed intermediate **6a**

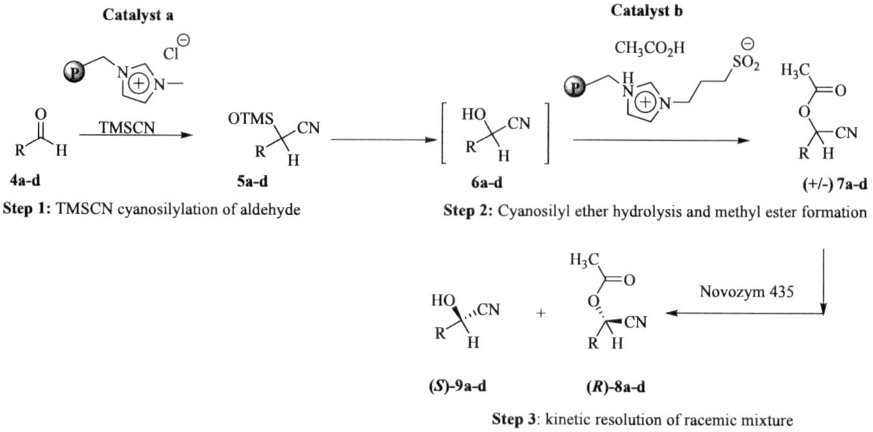

Scheme 1 Multicatalytic batch synthesis of chiral cyanohydrins

Table 1 Tabulated results for the batch synthesis of chiral cyanohydrins

	Substrate	Step 1	Step 2	Step 3	
Entry	R	yield	yield	yield	*ee*
1	C_6H_5–	99	99	56	>99
2	p–ClC_6H_4	95	99	57	>99
3	*o*–ClC_6h_4	95	99	51	>99
4	C_4H_3S	95	99	52	>99

[a] Conditions: 1 equiv. RCHO, 1.2 equiv. TMSCN; rt, 24 h, Cat: 25 mg per mmol of RCHO. [b] Conditions: 1 equiv. substrate, 2 equiv. Ac_2O at 60 °C, 24 h, 50 mg of cat. per mmol of substrate. [c] Conditions: 9.8 mL 2-Me-THF, 0.15 mL *n*-propanol and 100 mg CAL-B (Novozym 435) per mmol of substrate, 60°C, 24 h, yield for 5a–d. [d] *ee* for 14a–d; *ee*>95% for 15a–d. [e] Catalysed by catalyst a.[f] Catalysed by catalyst b. [g] Catalysed by Novozym 435

in *ca* 90%. In a third reactor, catalyst c, Novozym 435 (1 g) was packed to make a bioreactor to which a solution of racemic cyanohydrin ester **7a** in 2-MeTHF (0.1 M) and *n*-propanol (2 equiv., 0.2 M) were fed at a flow rate of 0.35 mL and 40 bar to effect transesterification at 60 °C. 95% *ee* for (*R*)-acetylated mandelonitrile **8a** and hydrolysed (*S*)-mandelonitrile **9a** (> 99% ee) was attained with *ca* 48% conversion of cyanohydrin towards corresponding alcohol recorded. The three-step synthesis was thereafter telescoped in continuous flow after each reactor had reached steady state without separation of intermediates (Fig. 3).

The first reactor reached steady state after 1000 min (>90% yield cyanosilyl ether **5a**) to which the second reactor was connected. The third reactor was joined to the sequence of the first two reactors after 3700 min (> 90% yield of cyanoester derivative **7a**) which selectively hydrolysed the racemic esters to yield the (*S*)-enantiomer of the cyanohydrin **9a** and the corresponding (*R*)-cyanohydrin acetate **8a** (with an enantiomeric excess >99% for a conversion slightly higher than 50%). A space-time

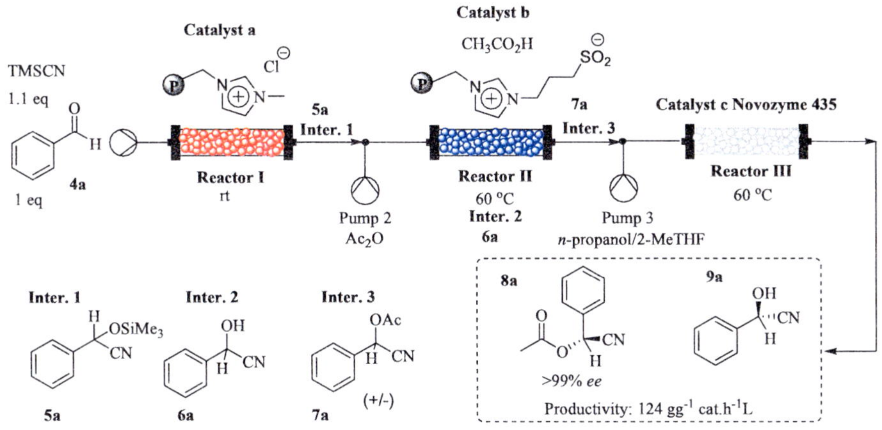

Fig. 3 Continuous flow multicatalytic sequential multistep synthesis of chiral cyanohydrins

yield for this step was reported to be 124 gg^{-1} cat. h^{-1} L. It is of note, that step 1 and step 2 were performed under solventless conditions and had very low *E*-factors (0.1 and 0.93, respectively) thus contributing greatly to the greenness of this flow process. The overall *E*-factor of this continuous flow synthesis was calculated as 21.8 and this was due to the amount of solvent needed in the kinetic resolution of the cyanohydrin esters. A less hazardous reagent, TMSCN was successfully replaced for HCN and the insoluble nature of supported ionic liquid catalysts used allowed for their use in combination with biocatalyst (Novozym 435) in successive reactions. It can be said that the developed continuous flow synthesis of chiral cyanohydrins is superior, greener than the batch process.

The combination of an enabling technology such as continuous flow technology with photochemistry is an avenue with a multitude of opportunities that could drive chemical syntheses towards the development of green processes. For example, in the selective synthesis of azoxybenzenes from nitrobenzenes using visible light irradiation reported by Nishiyama et al., it was found that only corresponding anilines were synthesized as major products, whereas hydroxylamine and nitrosobenzene intermediates were generated when the reaction was performed in a batch set-up (Scheme 2) [17]. In 24 h, 95% conversion was achieved with only 20% yield.

The increase in yield of desired azoxybenzene **12a** in the microreactor synthesis was due to efficient mixing. Vigorous mixing in the batch synthesis of **12b** also confirmed the effect of mixing in the reaction. There was an observed increase in yield of azoxybenzene **12b** from 8% under no mixing conditions to 26% under vigorous stirring while there was an observed decrease in aniline **11b**; 20% to 13% under no mixing and vigorous stirring, respectively. Nonetheless, the microreactor synthesis provided better yield of azoxybenzenes in both substrates. The synthesis of azoxybenzenes was thereafter further studied in the simple microreactor set-up. It was found that at high flow rates (>2 mL/h), i.e. short residence times, high yields of desired azoxybenzenes were generated. High equivalents of triethanolamine were

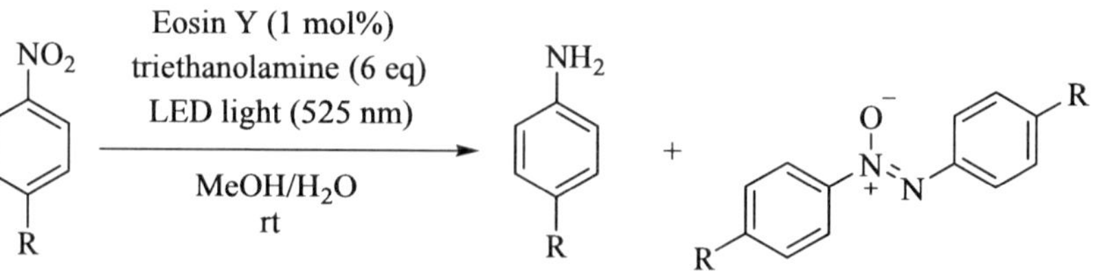

10a **11** **12**

10a; R:H **11a**; R:H **12a**; R:H
10b; R:Cl **11b**; R:Cl **12b**; R:Cl

Scheme 2 Eosin *Y* photocatalysed synthesis of azoxybenzenes

also required to increase the product yields. The photoreaction of *p*-nitrobenzene needed 9 equiv. (2 h, 90% selectivity azoxybenzene) whereas *p*-chloronitrobenzene's photoreaction used up to 12 equiv. (3 h, 87% selectivity azoxybenzene). The desired products were attained in high yields and shorter reaction times compared to the batch synthesis.

A green catalyst-free and simple continuous flow cyclopropenation of alkynes and carbenes from diazoalkanes and rearrangement of sulphides in very short reaction times and high yield has been developed by Hommelsheim et al. [18]. The transposed batch catalyst-free cyclopropenation of alkynes (Scheme 3) was investigated using the diazoester **13** and phenylacetylene **14**. It was found that in the presence of blue LED light (470 nm), using DCM as a reaction solvent, diazoacetate (10 equiv.) **13** and phenylacetylene (1 equiv.) **14** reacted for 16 h gave a yield of 97% cyclopropenated product **15**. The synthesis was extended to a variety of arylacetylenes and diazoalkanes. Excellent yields were reported regardless of substitution pattern (53–99%).

However, heterocyclic alkynes such as thiophene acetylene gave a mixture of products whereas the pyridyl-acetylenes did not provide desired products. The synthesis was also extended to the sigmatropic rearrangement reaction involving the reaction between sulphides with carbenes to yield sulphur ylide via 1,2 or 1,3 sigmatropic

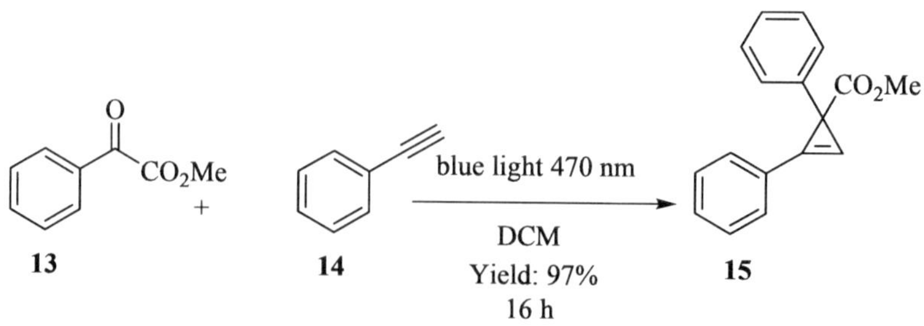

13 **14** **15**

Scheme 3 Batch cyclopropenation of alkynes

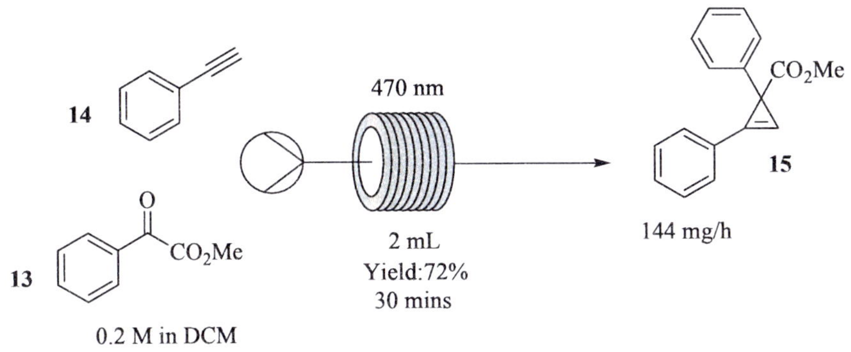

Fig. 4 A photochemical continuous flow cyclopropenation of alkynes

rearrangement. Allylic sulphides reacted well in a Doyle–Kirmse reaction to afford homoallylic sulphides aliphatic and heteroaromatic thioesters in high yields. The flow synthesis of cyclopropene **15** from diazoester **13** and phenylacetylene **14** was thereafter attempted and it was shown that the reaction time could tremendously be shortened to only 30 min. Using two glass microreactors, eight 24 W, 240 Im LEDs, 144 mg of cyclopropene 15 per hour was generated (Fig. 4). It was observed that the established continuous flow cyclopropenation of alkynes process provided 36 times better productivity than the batch protocol thus clearly illustrating the benefit of process intensification using continuous flow systems.

Jang et al. highlights a photocatalysed visible light promoted direct arylation of 2*H*-indazoles in a multicapillary reactor [19]. Using phenyldiazonium salt **16**, the flow synthesis of **18** was performed in a transparent microcapillary reactor (id: 400 μm) with immobilized eosin Y. The reactants, phenyldiazonium salt (1.2 equiv.) **16** and 2*H*-indazole (1 equiv.) **17** in the presence of 1 equiv. DIPEA in DMSO, were irradiated at room temperature using a flexible green LED strip (70 W, 12 V, 5 m SMD5050 chip). Indole **18** was obtained in 69% yield after 3.14 min (Fig. 5). The batch synthesis, on the other hand, took 18 h to give a yield of 65% **18** at room temperature.

This is due to the uniform light distribution and efficient mixing in the microchannel spaces, which is a result of the short diffusive distances and the large surface area-to-volume ratio in microreactors unlike in batch reactors. Additionally, this enables photoredox reactions to occur by engaging in single electron transfer events in their photoexcited states, which is crucial to obtain high yields within short reaction times. A catalyst stability test in the capillary reactor synthesis of **18** showed that no considerable leaching of immobilized eosin Y catalyst was observed up to a period of 1 h with only 0.2 nM eosin Y detected. The yield of **18** was reported to be stable during this period. From 20 min to 60 min residence time, with leaching tests performed at a 10 min interval, product yields of 61, 61, 58, 57 and 57% were recorded.

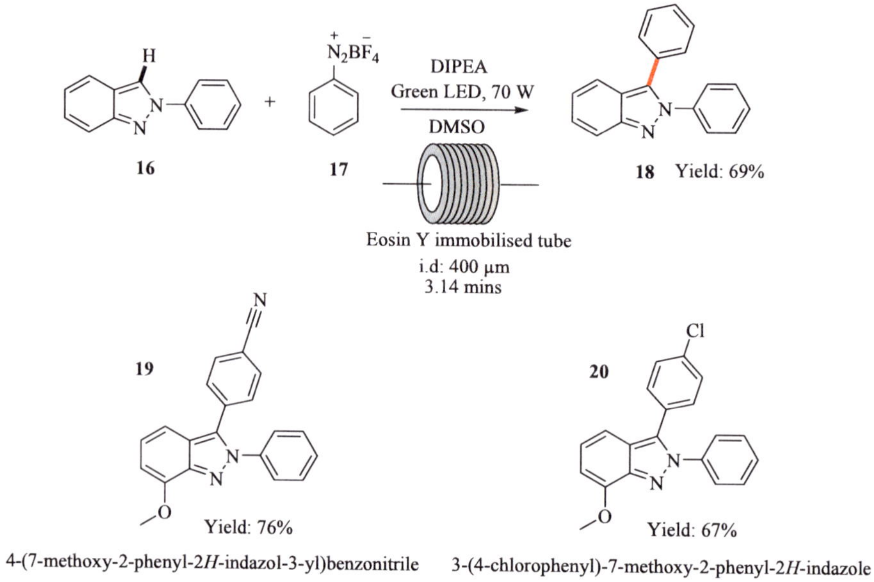

Fig. 5 Photocatalysed visible light-assisted direct arylation of 2*H*-indazoles in continuous flow

The generality of the developed process was thereafter checked using the synthesis of two 2*H*-indazole-based drugs, 4-(7-methoxy-2-phenyl-2*H*-indazol-3-yl)benzonitrile **19** and 3-(4-chlorophenyl)-7-methoxy-2-phenyl-2*H*-indazole **20** which took only 0.63 min to give 76% and 67% yield, respectively. The result of their batch syntheses was no different from the model reaction in this study. It took 18 h to achieve comparable yields: 72% and 65% of **19** and **20,** respectively.

The success of this study encouraged further studies into the multicapillary reactor scale up synthesis of liver *X* receptor agonist, 3-(4-chlorophenyl)-7-methoxy-2-phenyl-2H-indazole **20**. This was performed in a photocatalytic multicapillary assembly reactor constituting of 10 capillary tubes (Length: 1 m) which were vertically aligned. A one-body system was fabricated by cylindrically configuring the capillary tubes connected to two conical 3D printed fixtures at the bottom and top. These fixtures functioned as an inlet and outlet, respectively; 63% overall yield **20** was recorded. Additionally, a work up procedure was successfully integrated into the continuous flow process. DMSO and DIPEA base were removed from the reaction stream via liquid–liquid extraction using a diethyl ether and water system; 63% isolated yield was obtained using a 10 m tube for 1.57 min of flow.

Otake et al. recently published the use of $BrCCl_3$, a rarely used benzylic brominating reagent yet readily available, in a photochemical benzylic bromination reaction in continuous flow [19]. The benzylic bromination of toluene **21** was used to investigate and optimize the reaction. Under different light sources (254 nm UV-lamp (8 W), 365 nm UV lamp (8 W) and a medium pressure Hg lamp 150 W), 365 nm UV lamp (8 W) was found to be ideal for the reaction. Benzophenone **22** as a photosensitizer was also found to be a necessity to increase the rate of reaction and thus yield of the brominated compound. At a reaction temperature of 45 °C, benzophenone (2 mol%) **22**, $BrCCl_3$ (4 equiv.) using a 365 nm UV lamp (8 W) over period of 15 min, toluene was brominated to afford **23** after which morpholine (2 equiv.) **24** was reacted with the reactor output stream to give **25** in 47% isolated yield in batch (Scheme 4). To establish the applicability of the flow synthesis procedure, various substrates were reacted under the reaction conditions above. The products were collected and immediately reacted with morpholine (2 equiv.) **24**, for 16 h at room temperature. *Para, ortho* substitutions, electron-deficient and electron-rich substituents and highly substituted substrates were all tolerated (isolated yield: 28–84%).

The authors were very much interested in the benzylic bromination of *p*-methoxytoluene **32** to afford a protecting group used in the synthesis for an API precursor **35**, cipargamin (KAE609) **36** that is a potent anti-malarial candidate currently under phase II clinical trials. The synthesis of cipargamin **36** usually begins with a challenging PMB protection of isatin **34** that requires potassium iodide. In this study, the authors employed an in situ generation and reactive quench of the unstable *p*-methoxybenzyl bromide **33**.

Using the established benzylic bromination flow procedure, *p*-methoxytoluene **26** (neat), $BrCCl_3$ (4 equiv.), benzophenone (2 mol%) were fed into a photoreactor (365 nm, 8 W, 45 °C) at a flow rate of 0.133 mL/min (residence time: 15 min). The outlet of the photoreactor is fed into a batch reactor containing 5-chloroisatin **28** and potassium carbonate in acetonitrile maintained at 60 °C. PMB protected isatin derivative **29** was isolated in 91% yield on a 0.6 mmol scale in 1 h. The synthesis was scaled up and there was no decrease in yield recorded; 11 g of **35** was isolated in 4 h upon non-stop dosing of *p*-methoxybenzylbromide **27** (Fig. 6). A neat high yielding and productive semi-continuous flow synthesis of API precursor **30** was developed via the efficient flow benzylic bromination of **26** using a $BrCCl_3$, an environmentally benign reagent.

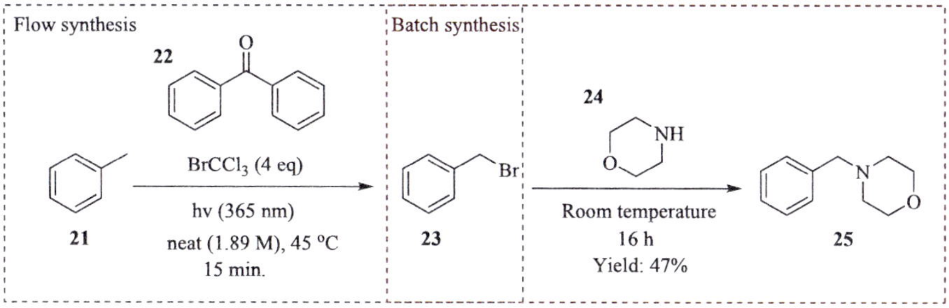

Scheme 4 A flow and sequential batch synthesis of morpholine adduct 25

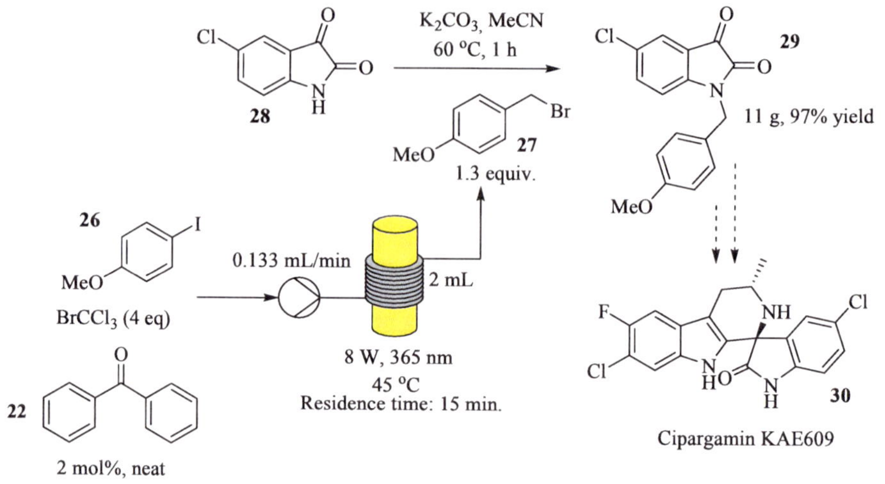

Fig. 6 A semi-continuous flow synthesis of API precursor 30 towards cipargamin

2.2 *Waste Minimization*

The first principle of green chemistry; "Prevention of waste to avoid treating or cleaning up waste after it has been created"

An ideal synthesis is one that has none or few purification steps and most especially avoids column chromatography but rather employs easier techniques such as crystallization or product precipitation.

Indoles are some of the most abundant heterocycles in natural products and are widely found in many pharmaceutical and agrochemical agents. Chiurchiu et al. demonstrated the synthesis of 3-substituted alkylated indole compounds using polymer-supported borohydride under flow conditions [20]. This protocol entailed a sulphonyl indole reduction with only one downstream unit operation, i.e. solvent evaporation, required (Fig. 7). To illustrate, sulphonyl indole **31** (0.25 mmol in solvent: 0.05 M) was flowed at flow rates ranging between 0.025 and 0.05 mL/min

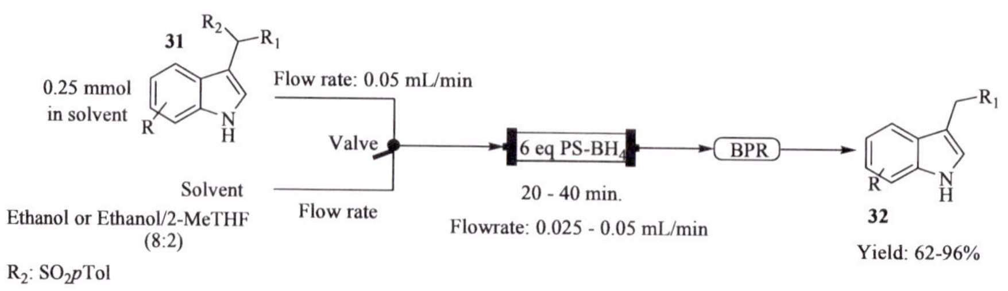

Fig. 7 Continuous flow synthesis of 3-alkylated indoles

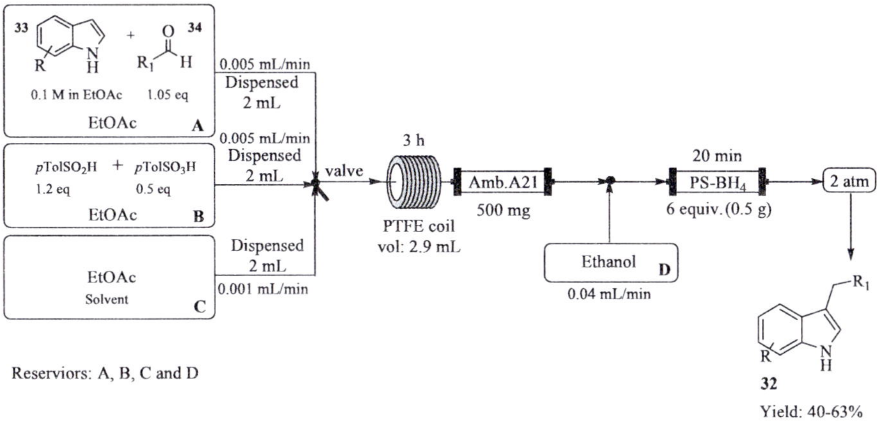

Fig. 8 Two-step continuous flow synthesis of 3-alkylated indoles from aldehydes

through a packed bed reactor filled with PS–BH_4 resin (0.53 g, 1.2 mmol) fitted with a back pressure regulator (2 atm) to obtain desired compounds **32** in 62–96% yield in 20–40 min residence time.

A multistep synthesis of the 3-substituted alkylated indoles was also attempted from coupling selected indoles and aldehydes in the presence of aryl sulphinic acids followed by subsequent reduction of the sulphonyl indoles to generate the desired C-3 alkylated indoles. This multistep approach enabled the preparation of these important molecules using reduced amounts of solvent and fewer purification steps (Fig. 8).

Andrade et al. describe a transposed isocyanide-based Passerini reaction from batch microwave synthesis to a continuous flow synthesis [21]. A reaction usually marred by extremely long reaction times was done in only in 5 min to achieve desired products in very good yields (Fig. 9). At 80 °C, using acetonitrile as a reaction solvent, various α-acyloxy ketones **38** were synthesized from different isocyanides **35**, arylglyoxals **36** and carboxylic acids **37** in excellent reaction yields (89–97%). Furthermore, there was no need for purification of the products. The continuous flow synthesis was easily scaled up to 10-fold which corresponded to a productivity of 0.32 g/min.

Since scale up in batch microwave is not an easy task, the continuous flow synthesis enabled scale-up of the synthesis of an α-acyloxy ketone (substrate substitution; Ar: phenyl, R_1: *t*Bu and R_2:-$CH_2CH_2N_3$) up to 10-fold, which corresponded to a productivity of 0.32 g/min.

A continuous one-pot flow synthesis of trifluoromethylated *N*-fused heterocycles involving only one step (cyclization) has been reported [22]. Trifluoroacetic anhydride **39** was used as a CF_3 source and dehydrative cyclization agent (Fig. 10). The amine (0.133 mmol, 1 equiv.) **40** or **41** and triethylamine (0.43 mmol, 3.2 equiv.) in tetrahydrofuran as a solvent were fed using one pump into a 10 mL Hastelloy coil via a T-mixer while trifluoroacetic anhydride **39** was fed into the system with another pump. At a reaction temperature of 80 °C (6 bar back pressure), residence

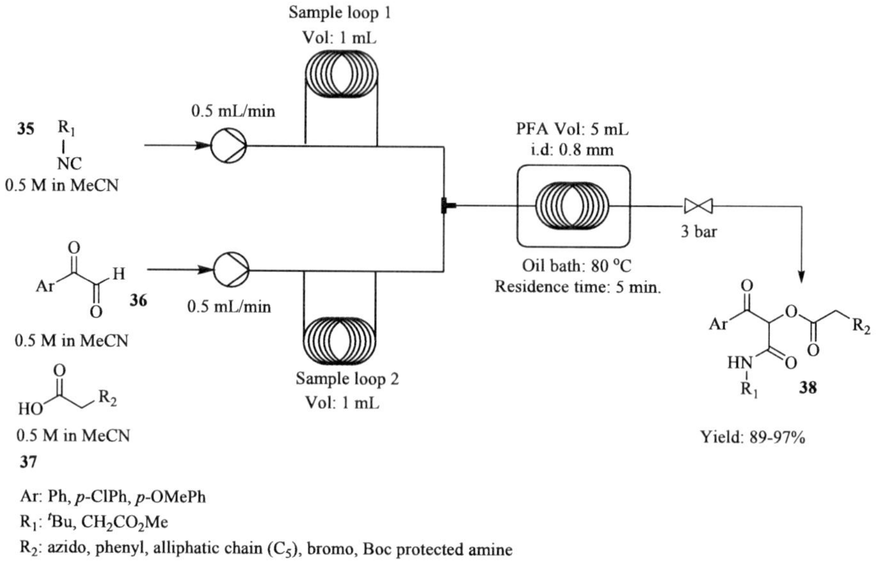

Fig. 9 Flow synthesis of α-acyloxy ketones via isocyanide-based Passerini reaction

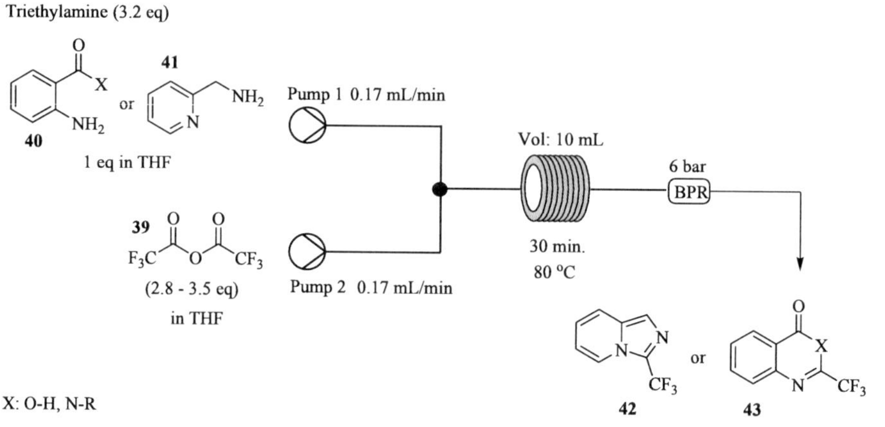

Fig. 10 Flow synthesis of trifluoromethylated heterocycles

time of 30 min, a small library of trifluoromethylated heterocycles such as imidazopyridines (49-81%), imidazopyridazines (56%), benzoxazinones (44–99%) and quinazolinones (6–57%) from various substrates were generated.

Process scalability was demonstrated using the synthesis of an imidazopyridine derivative generated from substrate, (6-methylpyridin-2-yl)methanamine. After an 8-hour continuous operation, 60% *ca* yield was obtained with a productivity of 1 g/h (99% purity by HPLC). This synthesis avoided intermediate work-up and purification steps. The developed flow process had better green metric values, for example overall yield: 57%, PMI: 183 and *E*-factor: 182, RPG%: 14, RPI: 9 and OpI: 5.

Compared to the previously reported two-step batch (overall yield: 13%, PMI: 1728, *E*-factor: 1727) and the two-step flow synthesis (overall yield 51%, PMI: 496, *E*-factor: 495, RPG%: 5, RPI: 4 and OpI: 2), this synthesis is greener, sustainable and cost effective (100 USD per g of product).

2.3 Enhanced Process Safety via Safe Reaction Spaces

The twelfth principle of green chemistry; "Inherently safer chemistry for accident prevention substances and the form of a substance used in a chemical process to be chosen to minimize the potential for chemical accidents, including releases, explosions, and fires"

The recent unfortunate explosion at a chemical plant in Port Neches [23] is an immense indicator of how safety should be at the forefront of any kind of chemical manufacture/production. However, the commonly used conventional batch synthesis is still faced with difficulties in handling hazardous and explosive syntheses. With the use of continuous flow synthesis, the impact of some of these unsafe syntheses is mitigated. For example, the hydrolysis of nitriles to amides using hydrogen peroxide as an oxidant is a highly exothermic and potentially explosive reaction on large scale. However, Zhan et al. show a versatile, safe, robust and scalable continuous flow protocol for the hydrolysis of nitriles to amides in the presence of DMSO [24]. Excellent yields were reported for the hydrolysis of hetero-aromatic nitriles containing pyridine, piperazine and thiophene (92–97%) in 150 s (Fig. 11). Substituted nitriles with different functional groups were also well-tolerated (80–99%) moreover; nitriles substituted with strong electron withdrawing groups required only 480 s to achieve comparably excellent yields. The authors extended the synthesis to aliphatic nitriles, which also behaved exceptionally well under the optimized continuous flow conditions.

Diazoalkanes, due to their high reactivity even under mild conditions, are dangerous chemical species to handle. They are said to be very explosive, sensitive to shock and are toxic as such, they are hardly synthesized on large scale. The upside is that they are important precursors that can be employed in the synthesis of highly reactive carbenes, carbenoids or carbocations. They are used to facilitate C–H, C–C and C–X bond insertions [25] as well as various cycloadditions [26]. It is for this sole reason that safer methods of synthesis are sought after. Rulliere et al. recently reported a safe synthesis of diazoalkanes in continuous flow [27]. It was postulated that the use of continuous flow synthesis would circumvent some of the bottlenecks experienced in the diazoalkane synthesis and would facilitate (i) Precise reaction parameter control. (ii) Easy purification by use of solid reagents. (iii) In situ generation and consumption reactive intermediates avoiding any risk associated with storage. Hydrazones were first prepared to afford substrates for the diazoalkane synthesis in continuous flow. For this purpose, a broad range of ketones and aldehydes

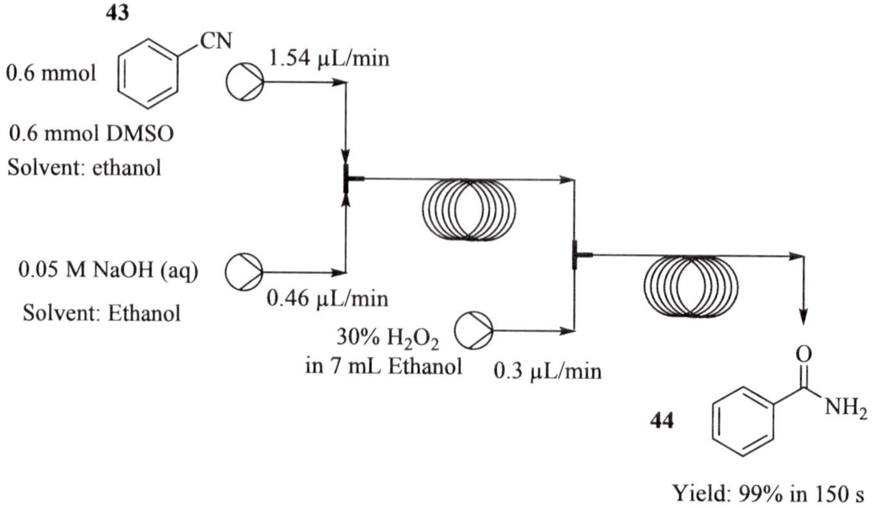

Fig. 11 Hydrolysis of nitriles using hydrogen peroxide as an oxidant

was treated with 1.5–2.4 equiv. of hydrazine monohydrate at reaction temperatures 50–100 °C for 10–20 min. The corresponding hydrazones were generated in high yields (53–99%) (see reference [28] for details).

Dimethylhydrazone **46** was the model substrate of choice for the formation of diazopropane **47**, which was reactively quenched with dimethylacetylenedicarboxylate **48** to afford corresponding [3+2] cycloadduct **49** in continuous flow mode. This was achieved by pumping a solution of substrate (0.1 M in DCM) **46** through a column (1.7 mL) packed with Ag_2O (1.2 equiv.) and K_2CO_3 (2.0 equiv.) at −20 °C to give 80% of [3+2] cycloadduct **49** in 36 s (Fig. 12). The reaction was carried out at a 1.4 mmol scale. The applicability of the process to oxidation of previously synthesized hydrazones was investigated at different reaction temperatures and residence times. Bisalkyl diazoalkanes were generated from cyclic substrates, sterically hindered centers and alkenes in (30–93%) Oxidation of hydrazones generated from aldehydes gave moderate yields of 48–62%, which was explained by their instability. Aryl-alkyl diazoalkanes were also synthesized in excellent yields (61–95%). With this flow synthesis, the electron neutral aryl alkyl diazo compounds were only obtained at higher temperatures (19 °C) in 84–99% yield while the electron rich aryl diazo compounds were furnished at lower temperatures (−15 °C) in 78–88% yield. A safe and easy synthesis of highly non-stabilized diazoalkanes was successfully established in continuous flow systems.

The authors took a step further to investigate the use of diazoalkanes in the esterification of carboxylic acids in a continuous flow mode. Diazoalkane **48** was reacted for 2 h at room temperature with various substituted carboxylic acids, e.g. amide, indole, protected amine, free alcohol and pyridine substituted to give corresponding alkylated products in high yields (83–99%). The use of diazoalkanes as partners in the

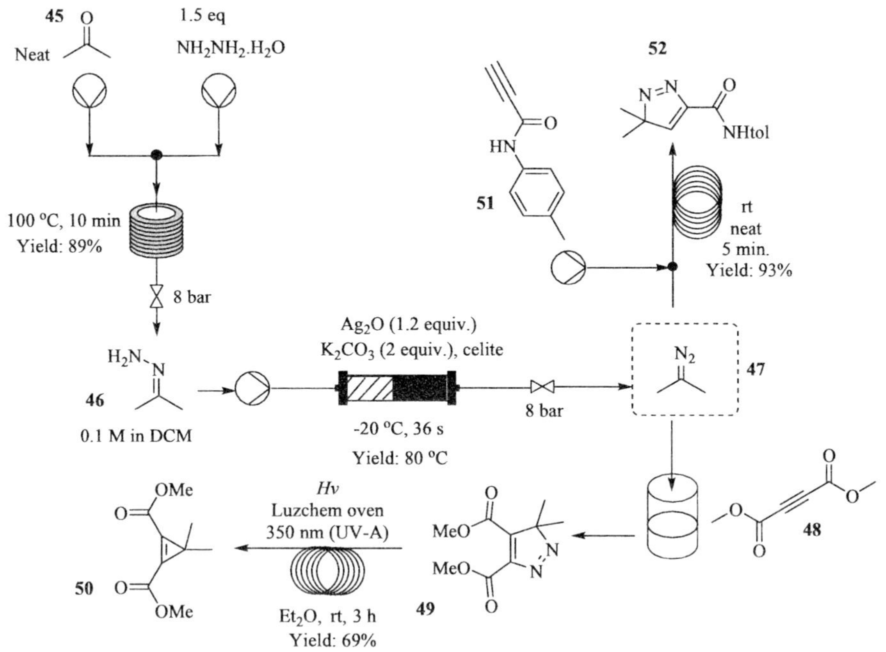

Fig. 12 Schematic for the synthesis of hydrazones, diazoalkanes, [3+2] cycloadducts and cyclopropenes

Michael-induced ring closure reaction with α,β unstaturated carbonyl compounds was also attempted from which good yields of desired products (53–99%) were obtained (see reference for details). The [3+2] cycloaddition of various alkenes and alkynes with non-stabilized diazoalkanes was also studied in continuous flow mode. The reaction was done neat at room temperature with 1 equiv. of alkyne or alkene and 2 equiv. of unstabilized diazoalkane (0.1 M in DCM) to give corresponding pyrazoline and pyrazolenine (isolated yield: 34–99%) in 5 min residence time. The versatility of alkyl diazoalkanes was showcased in this work in addition to the suitability of continuous flow systems to enable safe synthesis and in situ reactive quench of this unstable chemical species to different chemical compounds such as [3+2] cycloaddition products from alkynes and alkenes and cyclopropenes from photochemical arrangement of pyrazolenines etcetera. Finally yet importantly, the large-scale synthesis of furylalkyl diazoalkane **54** was used to finally demonstrate the suitability of the developed continuous flow method (Fig. 13). 8.3 mmol of diazo compound **53** was obtained in 60 min and subsequently consumed in situ in a [3+2] cycloaddition to form adduct **55** in 76% yield (1.7 g in 60 min).

The synthesis of *N,N*-diarylbenzamides and diphenylamines via the Chapman rearrangement usually requires high reaction temperatures ranging between (250–300 °C) to effect the 1,3 aromatic migration of the aryl group from the oxygen to the nitrogen. Imidates derived from acidic phenols will however rearrange at lower

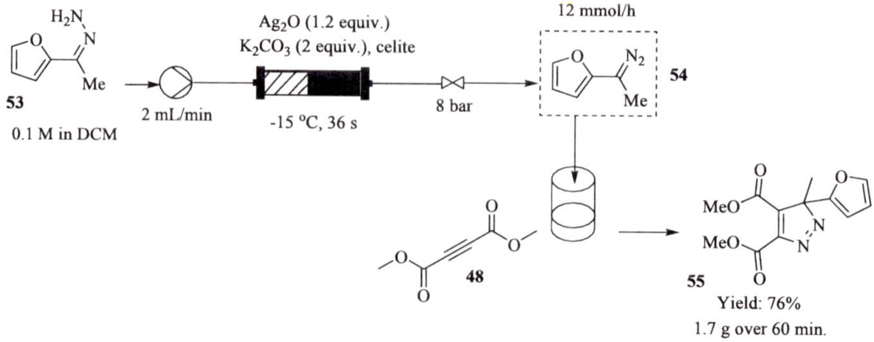

Fig. 13 Scaled up synthesis of furyl-alkyl diazoalkane **54**

temperatures (160–170 °C). Furthermore, precise temperature control is required for this rearrangement. Employing a reaction temperature, above that which provides the required activation energy, leads to product or reactant decomposition. In fact, in some cases where the reaction is in itself exothermic, these high reaction temperatures become very problematic by causing an uncontrollable rapid increase in temperature rendering the reaction unsafe and difficult to handle in conventional reactors [29]. The efficient heat transfer and precise temperature control observed in microreactor synthesis of organic molecules is an ideal approach which can enable safe and non-hazardous synthesis of *N,N*-diarylbenzamides and diphenylamines via the Chapman rearrangement.

Fang et al. illustrates with an investigative study on the chapman rearrangement of 2,6-dichloro-phenyl-*N*-phenyl benzimidate **56** both in batch and a microreactor system [30]. In a microreactor system comprising of a 250 μL piece of stainless steel coil (id: 0.588 mm, l: 920 mm), benzimidate **56** in tetraethylene glycol dimethyl ether was pumped into a reactor coil immersed in an oil bath (Fig. 14). An additional loop (id: 0.588 mm) was added to the set-up and kept at room temperature to efficiently effect a reaction quench.

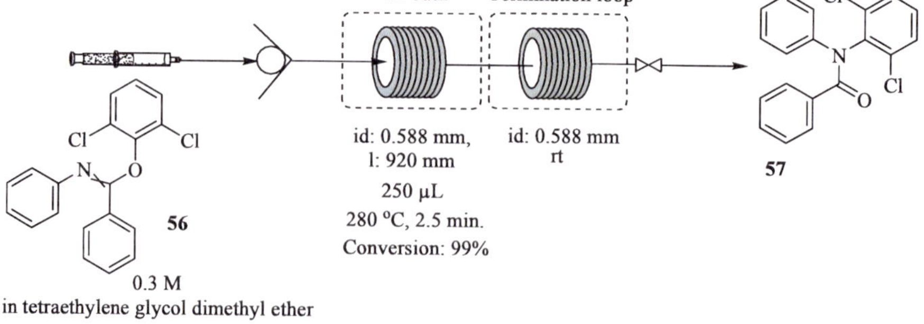

Fig. 14 Chapman rearrangement in a continuous flow system

A temperature of 280 °C was found to be the optimum reaction temperature with 99% conversion of **56** achieved in only 2.5 min; however, 240 °C was chosen as an ideal optimum temperature for this rearrangement. While the microreactor system enabled almost full conversion (94.42%) of **56** to **57** in 20 min at 240 °C, the batch synthesis required almost 3 times more that time to achieve nearly similar reaction conversion. This established synthesis was applied to various substrates with dichloro, fluoro, bromo substitutions, methoxy and ester functionality from which excellent conversions (72.4->99%) were recorded. The authors attempt at increasing throughput by increasing the microreactor channel size (id: 1.755 mm, l: 1650 mm) using 2,6-dichloro-phenyl *N*-phenyl benzimidate **56** as a substrate was successful. The reactor volume was increased up to 16-fold with the residence time kept constant. A throughput of 0.949 gh^{-1} was obtained in the scaled up synthesis equating to 22.776 g of 57 per day.

In 2018, Xue et al. reported a continuous flow the amination of heteroaryl chlorides to generate heteroaryl amines using aqueous ammonia as a NH_3 source [31]. A gas permeable Teflon AF-2400 tube in tube system was employed to safely deliver ammonia gas through separated diffusion from a feed stream of aqueous ammonia subjected to elevated temperatures to effect the amination reaction. Aqueous ammonia (25%) and 2-chloro-8-nitroquinoline **58** (0.4 M in DMSO) were fed into the tube-in-tube system (inner/outer id: 0.04"/0.08") that was kept at a reaction temperature of 150 °C (350 and 300 psi back pressure for the inner and outer tubes, respectively). 93% isolated yield of desired product **59** was furnished in 30 min residence time (Fig. 15). It should be noted that at these reaction conditions, 4.6 wt% and 20.7 wt% ammonia and water concentration were measured.

The highly pressurized system and elevated reaction temperature in this continuous flow system was able to provide satisfactory yield. In contrast, the microwave batch synthesis of **59** at 150 °C for 30 min. afforded only 13% of **59** and was reported to be problematic with pressures higher than 300 psi generated as well as violent escape of ammonia gas observed during venting of the reactor to release the pressurized vial. A broad substrate scope including heteroaryl fluorides, chlorides and highly substituted heteroaryl chlorides were aminated to afford heteroaryl

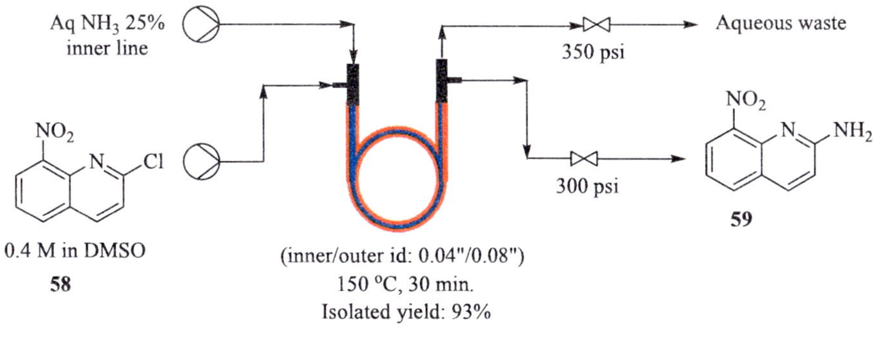

Fig. 15 Amination of heteroaryl chlorides using aqueous ammonia as an NH_3 source

amines in excellent yields (55–97%) and very good productivity in a safe continuous flow process.

Thus far, the use of continuous flow synthesis in enabling safe handling of hazardous chemicals and intermediates, the safe use of high pressure and temperature have been discussed but it should be noted that it also facilitates the use of safer and milder alternative reagents and chemicals for synthesis in turn allowing for the development of green chemical processes. This is associated with the enhanced reaction rates observed in continuous flow reactors due to efficient mixing. This characteristic opens up possibilities of reactions occurring in unexpectedly milder conditions or even using safe and green reagents.

The third principle of green chemistry; "Less hazardous chemical syntheses designed to use and generate substances that possess little or no toxicity to human health and the environment"

To illustrate, Liu et al. discuss the use of a natural ligand, β-cyclodextrin, in the palladium catalysed Suzuki–Miyaura-cross-coupling in continuous flow [32]. In their work, natural β-cyclodextrin was used as a supermolecular host to prepare the Pd-β-CD compound catalyst and was used to study the Suzuki–Miyaura cross-coupling of 4-bromoanisole **60** and boronic acid **61**. Using Pd-β-CD (0.0001 mol%), aryl halide **60** was reacted with **61** (1.2 equiv.) in the presence of K_2CO_3 (1.5 equiv.) in an ethanol:H_2O (1:1) solvent system; 90% yield of 4-methoxybiphenyl **62** was obtained at a reaction temperature of 80 °C in 1 min residence time (Fig. 16). A lower activation energy for the oxidative addition (50.1 ± 4.7 $KJmol^{-1}$) than the previously reported value (60–118 $KJmol^{-1}$) involving phosphine ligands was recorded.

The authors explain that this is due to the presence of two catalyst sites per cyclodextrin molecule used. The continuous flow biphasic liquid–liquid cross-coupling facilitated the separation of the product from the aqueous phase which contained the catalyst thus providing pure product with no traces of metal catalyst present. The droplet formation observed in the flow synthesis also increased the surface area-to-volume ratio of the reaction space thus increasing the reaction rate of the oxidative addition stage, which was found to be the rate-determining step in

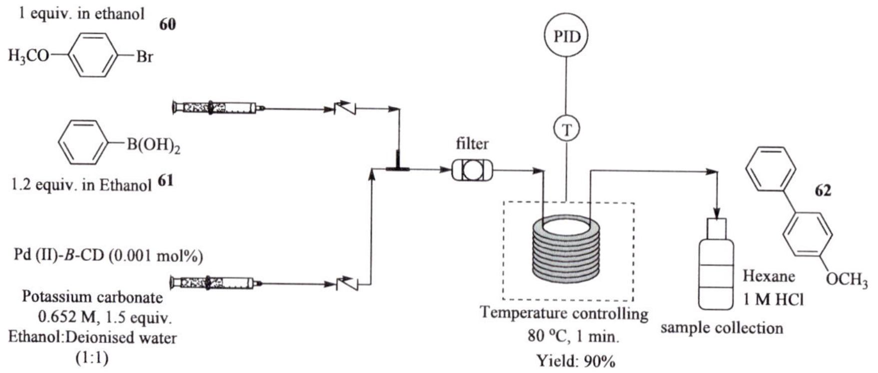

Fig. 16 Palladium catalysed Suzuki–Miyaura cross-coupling using *β*-cyclodextrin as a ligand

this synthesis. It is of note that this was a selective synthesis, with no side products formed moreover; very low catalyst loading was employed.

The seventh principle of green chemistry; "Promotion of the use of renewable raw materials or feedstock instead of depleting ones whenever technically and economically practicable"

Zhang et al. employ a bio-derived green solvent, 2-methyltetrahydrofuran in the continuous flow sequential synthesis of diaryl ketones by coupling aryl Grignard reagents with acyl chlorides [33]. A coupling between phenylmagnesium bromide **63** and benzoyl chloride **64** was chosen to determine the reaction feasibility and optimum reaction conditions in continuous flow reactors. The reaction was performed in a flow system using 2-MeTHF as the ideal solvent. Reactants, phenylmagnesium bromide (0.6 M) **63** and benzoyl chloride (0.4 M) **64** were fed into a T-shaped mixer (Peek i.d: 1/32") and standard PTFE reactor (i.d: 1/6", internal volume: 120 mL) using two standard pressure HPLC pumps at 1 mL/min each. At 25 °C, an 85% isolated yield of **65** was achieved in 60 min. The batch synthesis, on the contrary, was low yielding. At similar reactant concentrations and reaction conditions, only 34% isolated yield of **65** was obtained (Fig. 17).

Using a continuous flow parallel system adjusted by a five-way valve, a substrate scope of various aryl acid chlorides coupled with phenylmagnesium bromide was evaluated. Moderate to good yields were observed in the continuous flow synthesis of diaryl ketones from acyl chlorides. Electron withdrawing groups on acyl chlorides led to better yields in corresponding diaryl ketones (65–82%). Acyl chloride substitution also had an effect on the yield where *ortho* substituted acyl chlorides (79–85%) gave higher yields of corresponding ketones compared to *meta* (57–65%) or *para* (55–64%) substituted acyl chlorides. Additionally, aryl magnesium bromides with electron withdrawing, neutral and donating groups were well tolerated in the coupling with 2-fluorobenzoyl chloride (43–95%). The authors also demonstrated the scalability of their green approach towards diaryl ketones using the synthesis of 2-(3-benzoylphenyl)propionitrile **67**, an intermediate in the synthesis of Ketoprofen

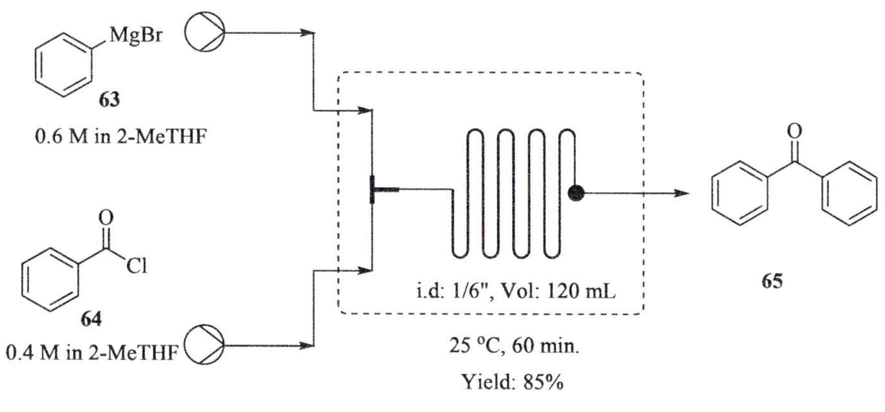

Fig. 17 Continuous flow synthesis of diaryl ketones

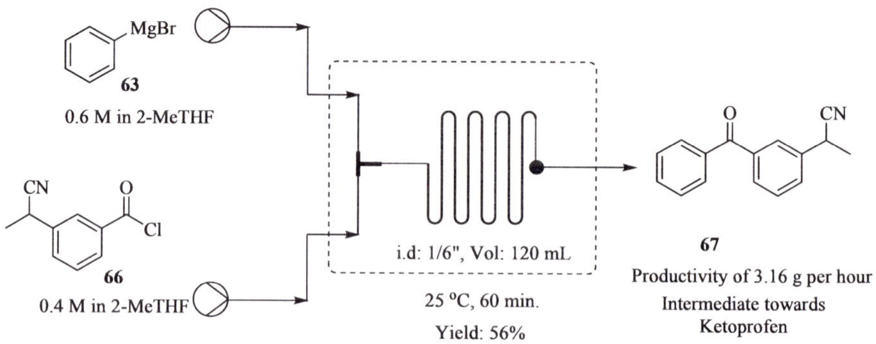

Fig. 18 Scaled up continuous flow synthesis of an intermediate towards ketoprofen

(Fig. 18). Without any further optimization of the reaction parameters, 4.74 g was produced in 1.5 h, which translates to a productivity of 3.16 g per hour.

Ferlin et al. employ cyclopentylmethylether as an environmentally friendly and safe reaction medium in the synthesis of 2-arylbenzoxazoles. Heterogeneous mixed valence manganese octahedral molecular sieves (OMS) using oxygen as an environmentally benign terminal oxidant and cyclopentylmethyl ether as a peroxide safe reaction were used to effect the synthesis [34]. Using benzyl alcohol **68** as a model substrate, the synthetic route chosen was as follows: oxidation of the alcohol followed by imine formation and finally oxidative C-H functionalization/cyclization (Scheme 5). The three reaction steps were fully optimized for relevant reaction parameters in batch mode prior to translation to flow systems. H–OMS and K–OMS (mixed valence manganese octahedral molecular sieves) were identified as ideal catalysts for the alcohol oxidation reaction (>99% conversion in 24 h at 106 °C and 20 mol% H–OMS, 1 mmol 68) and oxidative C–H functionalization reaction (>99% conversion in 60 min at 106 °C and 100 mol% K–OMS, 1 mmol **71**), respectively. The imine **71** synthesis was achieved, by reacting benzaldehyde (0.4 mmol) **4a** with 2-aminophenol (0.2 mmol) **70** at 120 °C in the presence of O_2 (1 atm); >99% conversion to imine **71** was achieved in 10 min. A suitable approach for catalyst regeneration was also investigated, where it was found that washing the catalysts with either cyclopentylmethylether or acetone followed by flushing with oxygen (at least 2 bar) at 100 °C was sufficient to regenerate the catalysts. A catalyst leaching test also showed very low concentrations of manganese in reaction solutions.

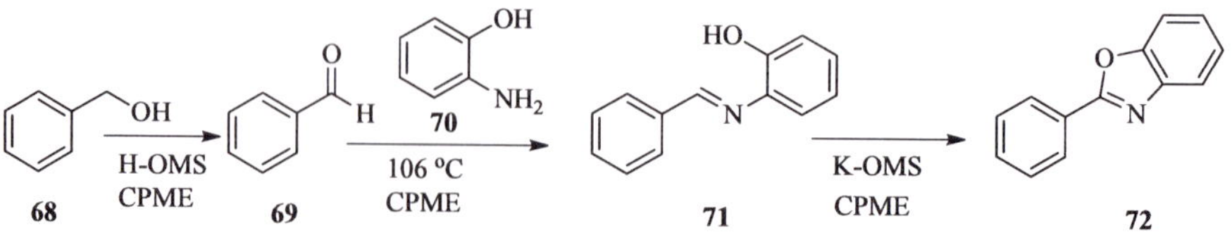

Scheme 5 Developed synthesis route towards benzoxazoles

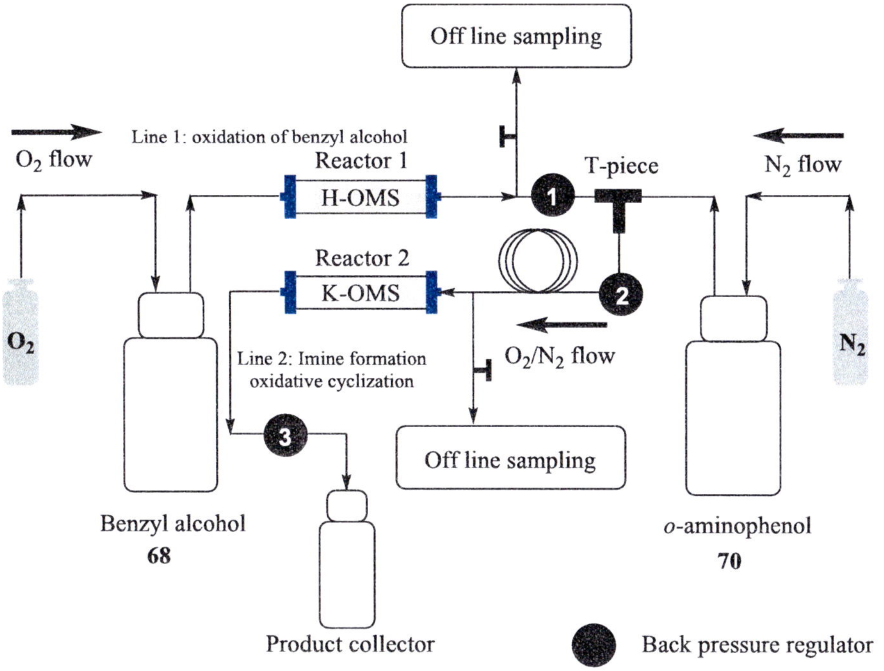

Fig. 19 Schematic diagram of an O_2–N_2 gas pressure-driven continuous flow process for the synthesis of benzoxazoles

A continuous flow protocol that included a non-stop flow continuous flow procedure for the regeneration of catalyst was thereafter developed (Fig. 19). The reagent solutions were fed into the reactor system using pressurized O_2 and N_2. Reactor **1** and **2** were filled with 40 mg H–OMS (62% Mn loading, 0.05 mmol of Mn) and 34 mg K–OMS (62% Mn loading, 0.4 mmol of Mn).

Using 0.4 mmol benzaldehyde **4a**, 0.37 mmol 2-aminophenol **70**, a total back pressure of 12.9 bar, O_2-N_2 pressure of 5 bar each, 98% 2-aryl benzoxazole **72** was achieved in a total residence time of 58 min (alcohol oxidation: 13 min, imine formation: 5 min and benzoxazole formation: 40 min). The applicability of the flow process was demonstrated using substituted benzyl alcohols and 2-aminophenols. Sulphur containing substrates (isolated yield: 98%) as well as halogenated substrates (isolated yield: 95–98%) were tolerated without any detection of disulphides, sulphoxides, sulphones and dehalogenated products, respectively. Electron-rich and poor 2-arylbenzoxazoles were also synthesized in high isolated yields (92–98%). The protocol was also immune to highly substituted substrates and those containing alkyl groups (isolated yield: 98–99%). Tafamidis **77**, an API has been synthesized in 92% isolated yield using this approach with an *E*-factor of 4.4 (Scheme 6). The authors also went ahead to test the durability and stability of the process in an elongated non-stop flow continuous synthesis of 2-phenylbenzoxazole

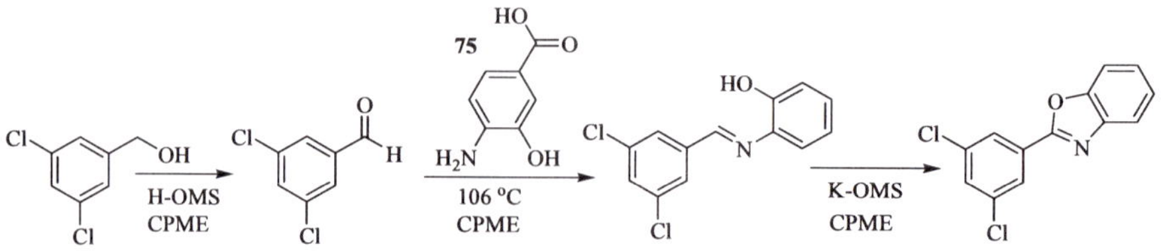

Scheme 6 Continuous flow synthesis of tafamidis 77

72. The system was run for 24 h with 280 mmols of reagents. 2.3 gh^{-1} was obtained after steady state was achieved in 58 min.

Compared to the batch process whose *E*-factor is 42, this flow synthesis boasts of a low *E*-factor of 1.7 for the multigram scale synthesis and 6.4 for smaller-scale procedures. Therefore, a truly waste minimization and sustainable approach towards the synthesis of 2-arylbenzoxazoles was developed in continuous flow mode. Besides successfully developing a non-stop continuous catalyst regeneration protocol, the authors introduce an elegant continuous system devoid of pumps to facilitate the flow of reagents.

Labes et al. reported a high temperature and pressure conversion of alcohols to amines via hydrogen borrowing approach in a continuous flow system. A relatively inexpensive air stable catalyst system of ([Ru(*p*-cymene)Cl_2]$_2$ and DPEPhos) was used for this purpose [35]. With the help of design of experiments, the reaction parameters were optimized. A solution of morpholine **24** and benzyl alcohol **68** (1.3 equiv.) in toluene (2 M) in the presence of [Ru(*p*-cymene)Cl_2]$_2$ (0.5 mol%) and DPEPhos (0.1 mol%) catalyst system was pumped into a Pheonix flow reactor (8 mL stainless steel flow coil) at 250 °C to furnish 100% yield of desired amine in 8 min (Fig. 20).

The flow synthesis protocol seemed robust as is evident from the different substrates used to investigate its applicability. High yields were achieved for benzylic and aliphatic alcohols while allylic alcohols had reduced yield mainly due to the reduction of the double bond. Secondary alcohols on the other hand gave no product. Heterocycles like piperidine (76%), methylenedioxy and piperazine (62–74%), indole (81–90%) and morpholine (81–99%) also generated product in good yield; however, thiazole gave poor yield. Esters (76%) and phenolic (96%) substituted substrates remain unaffected in this synthesis with ester and phenolic functionalities retained. Meanwhile, *tert*-butylamine (78%) and aniline (91%) required higher catalyst loading to give substantial yield. After this success, the direct methylation of amines using methanol (1.1 equiv.) at 0.8 mL/min was investigated under the same reaction conditions and flow system. It was found that substrates with ester functionalities (isolated yield: 86%) and chlorine substitutions (isolated yield: 40%) were smoothly reacted without any occurrence of transesterification and halogen substitution, respectively. Morpholine was also methylated in very good yield (93%). A

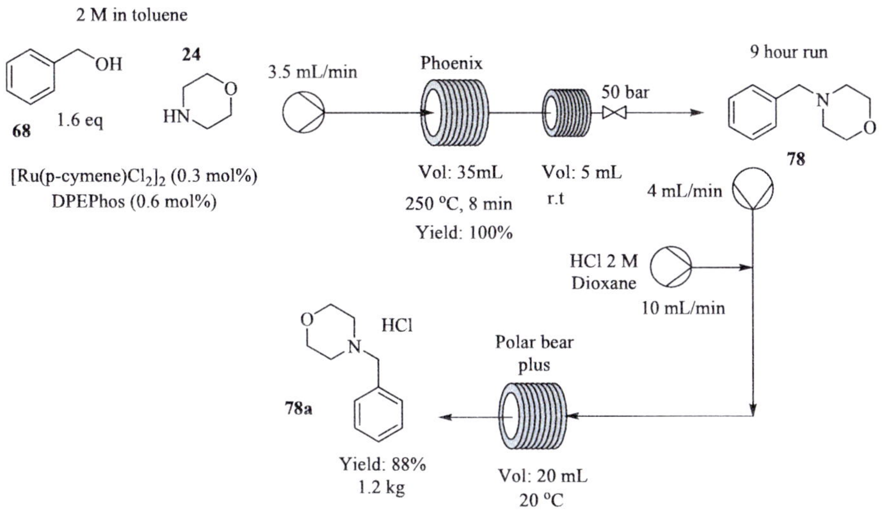

Fig. 20 Scaled up flow synthesis of **78a** hydrochloride

smooth, direct, efficient and safe high temperature continuous flow process was established for the transformation of alcohols to amines in very short reaction times. To demonstrate process scalability, the reaction between morpholine **24** and benzyl alcohol **68** with downstream processing was used. (Figure 20) 1.2 kg of **78a** hydrochloride was obtained from a 500 mL crude material gotten from the first reaction step (1.9 L over a 9-hour period).

The sixth principle of green chemistry; "Design for energy efficiency of chemical processes to minimize their environmental and economic impacts and if possible, to introduce synthetic methods to be conducted at ambient temperature and pressure"

The conventional synthesis of 6-and 1,6-substituted 3-cyano-4-methyl-2-pyridones involves the use of high temperatures and prolonged reaction times. Tadic et al. have reported the use of a microreactor assembly to perform a room temperature synthesis of 6- and 1,6-substituted 3–cyano-4-methyl-2-pyridones from 1,3-dicarbonyl compounds and *N*-substituted cyanoacetamides in the presence of NaOH as a catalyst [36]. Aqueous solutions of NaOH and *N*-substituted cyanoacetamides were premixed in a T-mixer and then flowed to mix in a second T-mixer with a solution of 1,3-dicarbonyl substrate in methanol. The resultant fluid stream was thereafter flown into a PEEK capillary microreactor (i.d: 0.5 mm, volume: 5 ml, length: 25 mL) and quenched off-line using HCl. The reaction was investigated at room temperature for optimum flow rates and reactants concentrations. A 60% yield of 3-cyano-4,6-dimethyl-2-pyridone **81** was obtained in 8.2 min at equimolar substrate concentration of both pentane-2,4-dione **79** and cyanoacetamide **80** in the presence of 0.7 M NaOH (Fig. 21) whereas in batch mode, 60% yield 3-cyano-4,6-dimethyl-2-pyridone **81** was obtained in 60 min under reflux conditions.

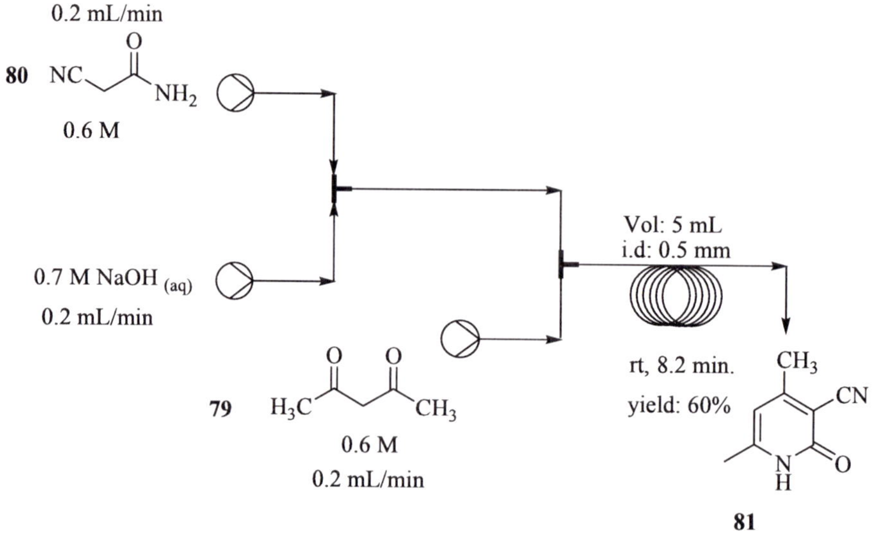

Fig. 21 Flow synthesis of 3-cyano-4,6-dimethyl-2-pyridone **81**

Unfortunately, *N*-substituted 3-cyano-4-methyl-2-pyridones did not yield satisfactory results using this method. At increased cyanoacetamide (1.5 M) 80 and NaOH (2 M) concentrations, no improvement in yield of **81** was observed. However, in the synthesis of 3-cyano-6-hydroxy-4-methyl-2-pyridone **83** from cyanoacetamide **80** and ethylacetoacetate **82** an improvement in reaction yield was recorded (from 22% to 59% yield) in 8.2 min while the batch synthesis took 480 min to afford 3-cyano-6-hydroxy-4-methyl-2-pyridone **83** in 61% yield (Fig. 22). Nonetheless, the use of ambient reaction conditions in flow systems to furnish product usually generated at harsher reaction conditions was demonstrated.

2.4 The Use of Heterogeneous Catalysis

The ninth principle of green chemistry, "Catalytic reagents as selective as possible"

Applications of catalysis, heterogeneous or homogenous for synthesis in continuous flow systems have been discussed in some of the above sections. The use of heterogeneous catalysts, in particular, in flow syntheses has many advantages over their use in batch systems [37]. These include easy recovery and reuse of catalysts. Due to efficient heat transfer in continuous flow systems, thermal catalyst decomposition is avoided thus allowing for longer periods of catalyst activity before regeneration is required. Heterogeneous catalysts are employed in packed bed reactors on supported materials or in monolith tubular reactors [38–40]. This allows for easy or work-up/purification procedures after a synthesis. Additionally, the use of solvents

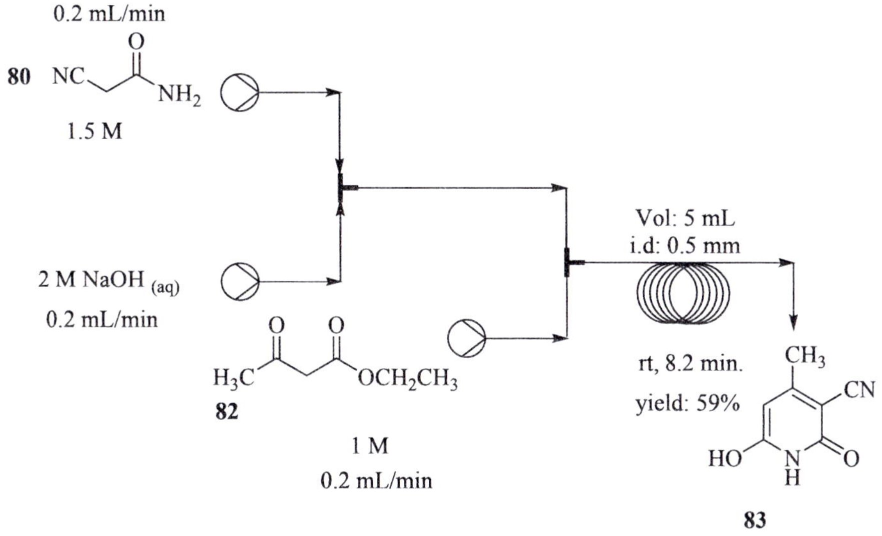

Fig. 22 Flow synthesis of 3-cyano-6-hydroxy-4-methyl-2-pyridone **83**

can also be eliminated in cases where all reagents and starting materials are liquids at the reaction temperature required therefore generally reducing the amount of waste generated. On the other hand, the use of homogeneous catalysts presents numerous challenges in terms of product purification and separation of catalyst from the reaction mixture. This in turn leads to loss of both catalyst and desired products during tedious work-up procedures. In this section, we highlight some of these aspects.

The synthesis of alkenyl chlorides from alkenyl iodides using copper reactor tubing has been reported as a better alternative to the batch synthesis as well as flow synthesis of alkenyl chlorides. In the batch synthesis, Me_4NCl was used as a chloride source, CuI (10 mol%) as a catalyst and *trans-N,N'*-dimethylcyclohexane-1,2-diamine (20 mol%) as the diamine ligand. At a reaction temperature of 110 °C and reaction time of 48 h, yields of 50-92% were reported for various alkenyl chlorides synthesized [41]. The authors transposed this batch synthesis into a PFA tube coil (10 mL) flow synthesis using β-iodo-styrene **84** as a model substrate and Aliquat 336® (2 equiv.) as a replacement for Me_4NCl. At reaction temperature of 110 °C, residence time of 120 min., ethanol as a reaction solvent and a backpressure regulator set at 8 bar, a yield of only 8% **86** was obtained. On switching the PFA tube coil for a copper reactor coil (10 mL), the in situ generation of the copper catalyst without any need of an external source of copper (CuI) as in the previous syntheses, i.e. batch and PFA coil flow syntheses was successfully achieved; 84% yield of **86** was obtained in 40 min at 130 °C (Fig. 23). The optimized reaction conditions were translated into the synthesis of other alkenyl chlorides.

E-Iodo substituted alkenyl chlorides were achieved in 62–91% yield at 110 °C in 60 min residence time. On the other hand, the synthesis of *Z*-iodo-substituted alkenyl

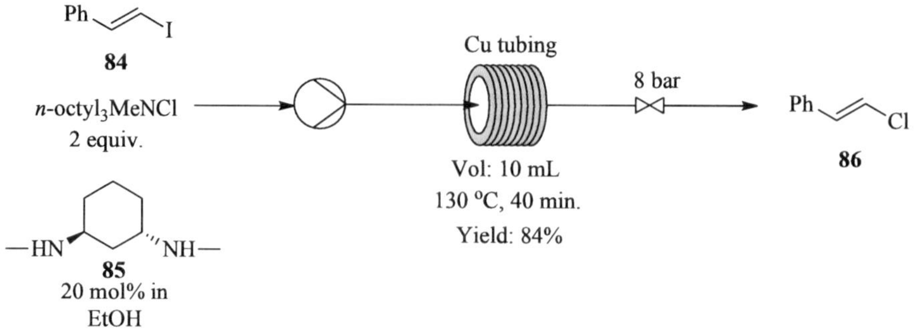

Fig. 23 Continuous flow chlorination of alkenyl iodides promoted by copper tubing

chlorides required an increase in reaction temperature (120 °C) and residence time (120 min.) to achieve yields ranging between 79–88%. In addition, electron donating, withdrawing, protecting groups in addition to silyl ether, acetate, phthalimide were all well tolerated. The most significant observation of this study was the tremendous decrease in the reaction time effected by the flow synthesis in copper tubing (1–2 h) compared to the batch synthesis (24–48 h). The capability of the flow system to be pressurized (8 bar) also enabled the use of the reaction solvent, ethanol above its boiling point (110–130 °C). On scaling up of the synthesis, identical yields were obtained.

The use of unavoidably expensive catalysts in various organic syntheses [42] such as palladium, ruthenium, gold and platinum-based catalysts, that sometimes not only requires high catalyst loading to effect a chemical transformation, but also require harsh conditions, does not offer sustainable synthesis. Catalyst deactivation, due to the high temperatures, which in turn culminates into side reactions and unwanted by-products, is a common occurrence in conventional batch Pd catalysed cross-coupling reactions. Sieber et al. recently reported a first low Pd loading Suzuki–Miyaura cross-coupling catalysed by preformed (BIDIME)Pd, a Buchwald's third-generation palladacycle [43]. Using a Uniqsis flow screening system, the optimized conditions for the synthesis of **90** from 2-chloropyridine **87** and phenolic boronic acid **88** were investigated. A mixed solvent system of IPA/DME/H_2O was identified as ideal for the study. 8.4 wt% phenolic boronic acid (10 volumes of 85/15 H_2O/IPA) 145 and 24 wt% 2-chloropyridine 87 (4 volumes of 50/50 DME/IPA) were reacted in the presence of 0.5 mol% palladacycle **89** and K_2CO_3 (1.67 equiv.) in a 1.8 mL microchip reactor heated at 120 °C to give 98% conversion in 2 min (Fig. 24).

The established continuous flow cross-coupling reaction at these optimum conditions did not suffer catalyst deactivation and side reactions, i.e. dehalogenation of **87** and protodeboronation of **88** unlike in the batch synthesis where only 85% yield of **90** was attained in 24 h at 1 mol% Pd loading. The incomplete reaction towards the desired product **90** was attributed to catalyst deactivation even at a low reaction temperature of 80 °C compared to the 120 °C used in the flow cross-coupling reaction.

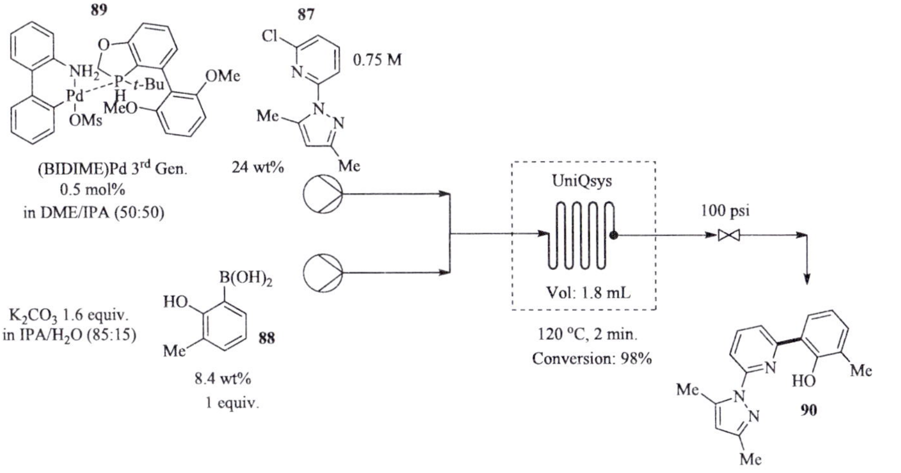

Fig. 24 Suzuki–Miyaura cross-coupling catalysed by preformed (BIDIME)Pd Buchwald's third-generation palladacycle

In comparison to in situ generation of the catalyst from a combination of the ligand and precatalyst, palladacycles have been reported to have better stability and shelf life since they are not sensitive to air [44]. They are also designed to enable facile formation of stoichiometric monoligated Pd(0), which is said to be the most active catalytic species [45]. As a result, these characteristics coupled with the added advantages of flow systems facilitate low Pd loadings and render the use of palladacycles a green approach towards Suzuki–Miyaura cross-couplings.

The synthesis of vinylogous esters in continuous flow is reported by Mohanta et al. Using Amberlyst-15 as a H^+ source and supported catalyst, the synthesis of β-keto-vinylogous esters, transetherification and their reverse reaction was achieved from simple precursors, i.e. diketo compounds, alcohols and water. Cyclohexane-1,3-dione **91** and ethanol **92** were chosen substrates for the β-keto enol ether synthesis in continuous flow [46]. At a flow rate of 0.3 ml/min, a 0.1 M solution of cyclohexane-1,3-dione **91** in ethanol **92** was fed into an Omnifit packed bed reactor (6.6 × 150 mm) containing 628 mg of Amberlyst-15 (4 cm bed height) which was heated at 80 °C. In 4.6 min., 99% yield of β-keto enol ether **93** was furnished (Fig. 25). The stability of the catalyst was demonstrated by continuously running a 40 h reaction of cyclohexane-1,3-dione **91** and ethanol **92**. With 72.12 mmol of reactant pumped continuously at 0.3 mL/min, 57.22 mmol of product **93** was obtained (90% isolated yield) with a TON and TOF of 28.61 and 0.715 h^{-1}, respectively. The robustness of the flow synthesis was evaluated with various alcohols, e.g. linear, secondary, aliphatic, allylic and aryl alcohols gave excellent yields (80–99%). Different diketo compounds such cyclic C6 and C5 cyclic diketo substrates were also reacted with linear and branched alcohols to give the corresponding products in excellent yield (73–99%). In addition to the above, acyclic carbonyl compounds were reacted with methanol, and propanol and also provided excellent yields (88-95%) under similar reaction conditions.

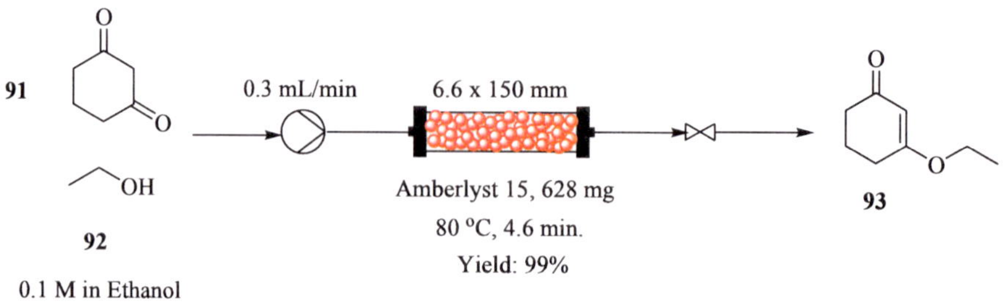

Fig. 25 Continuous flow synthesis of β-vinylogous esters catalysed by amberlyst 15

Alcohols bearing sensitive alkyne functionality as well were tolerated (76% yield) using cyclohexane-1,3-dione **91** as a substrate. On the other hand, C5 cyclic diketo substrate gave β-vinylogous esters in 65% yield when reacted with a similar alcohol under the optimized conditions. Alcohols with longer aliphatic straight chain and aromatic primary alcohols gave β-vinylogous esters in moderate yields (61–65%) from C6 and C5 cyclic diketo substrates. The authors took a step further to develop a selective, milder hydrolysis of vinylogous esters using an environmentally benign continuous flow approach. A solution of vinylogous ester **93** in water (0.1 M) was pumped through an Omnifit packed bed reactor (6.6 × 150 mm) containing Amberlyst-15 (628 mg, 4 cm bed height) at room temperature (Fig. 26). > 98% conversion to ketone **91** was obtained in 13.6 min at these reaction conditions. Selected 1,3-diketones were smoothly obtained using this simple method with water as a solvent (conversion: 96–99%). Transetherification using vinylogous esters was also investigated in continuous flow. Vinylogous ester **94** was used as a model substrate for the purpose. The flow synthesis was optimized to give 92% yield **93** at 80 °C from a solution of vinylogous ester in ethanol (0.1 M) pumped at 0.3 mL/min through an Omnifit reactor bed prepacked with Amberlyst-15 (Fig. 26). Excellent

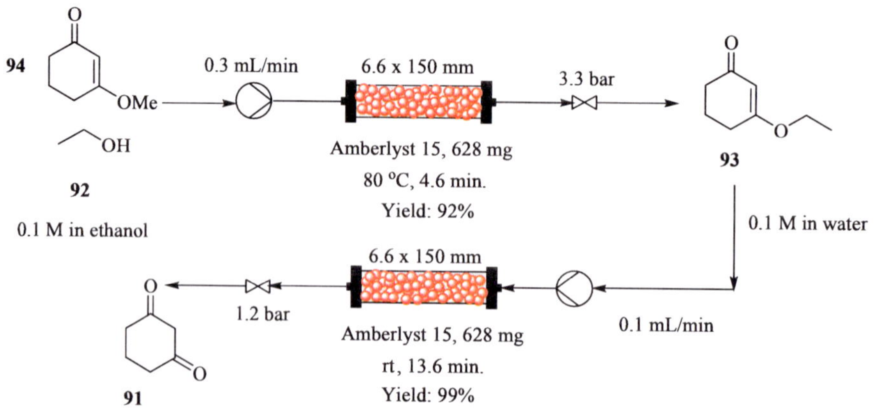

Fig. 26 Vinylogous ester hydrolysis and transetherification reactions in packed bed column reactors

yields were attained from the transetherification of a few vinylogous esters using C4–C5 aliphatic straight chain alcohols (84–93%).

Miyamura et al. developed heterogeneous Au–Pt nanoparticle catalysts immobilized on dimethyl polysilane (Pt–Au/(DMPSi–Al_2O_3)) which were successfully employed in the continuous flow direct synthesis of hydroquinones from quinones in high reactivity and selectivity [47]. The derivatization of electron-rich hydroquinones is problematic in conventional modes of synthesis. Their synthesis requires stronger reductive conditions that can lead to over-reduction and side reactions [48]. They are also difficult to isolate and react with molecular oxygen even without catalysts at ambient conditions [49]. The authors devised a simple, sequential and continuous flow hydrogenation derivatization system to tackle some of these issues. Catalyst activity in the hydrogenation reaction was initially screened in batch where it was found that using Pt-Au (1:1, 0.05 mol%), the room temperature hydrogenation of 2-ethyl anthraquinone **95** in the presence of H_2 and sequential reaction between the formed 2-ethyl anthrahydroquinone **96** with molecular oxygen, afforded 93% H_2O_2 in 23 h. This synthesis was thus translated to a flow reactor to develop a continuous flow process.

Despite the fact that the bimetallic catalyst, Pt-Au/HB, was also found to be effective for the hydrogenation of **95** in good yield as estimated from the yield of H_2O_2 (81%) generated from the reaction of 2-ethyl-anthrahydroquinone **96** and molecular oxygen, moreover with excellent catalyst reusability (75% yield H_2O_2 on second use) in flow, it was prone to swelling. As such, it was replaced with Pt-Au/DMPSi-Al_2O_3 and Pt-Au/PMPSi-Al_2O_3. Using 0.028 and 0.045 mmol for catalysts Pt-Au/PMPSi-Al_2O_3 and Pt-Au/DMPSi-Al_2O_3 respectively, at a $H_{2(g)}$ flow rate of 3.3 mL/min (1 atm), 0.075 M substrate concentration flowed through a column reactor packed with catalyst (Φ 5 mm × 5 cm), H_2O_2 yield of 76% and 91% was obtained for both catalysts, respectively (Fig. 27). Pt-Au/DMPSi-Al_2O_3 was found to be more durable as a stable yield of 95% was recorded for the second and third reuse whereas the Pt–Au/PMPSi-Al_2O_3 reduced in activity as was seen in the decreased

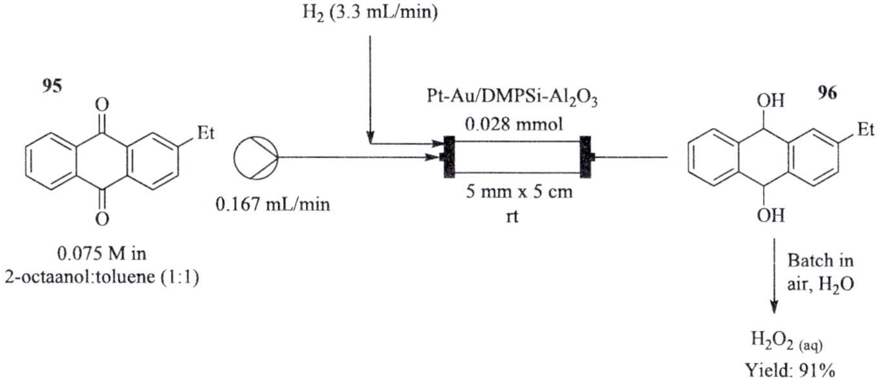

Fig. 27 Pt-Au/DMPSi-Al_2O_3 catalysed flow hydrogenation of anthraquinone **95**

H_2O_2 yield (fresh use: 69%, second use: 67% and third use: 61%) after 3 cycles of use.

A sequential hydrogenation and hydroquinone methylation of **96** was thereafter investigated; $H_{2\ (g)}$ and 2-ethyl anthraquinone **95** (0.075 M in THF) were flowed at flow rates 3.3 mL/min and 0.1 mL/min, respectively, through a packed bed column filled with Pt-Au/DMPSi–Al_2O_3 catalyst (864 mg). The hydrogenation reaction was performed at 30 °C. Excess hydrogen was removed via a phase separating unit and the outlet reaction stream was channelled into the next tubular reactor (Φ 3 mm × 5 cm), dipped in a sonicator for the methylation reaction. At flow rates of 0.3 mL/min and 0.036 mL/min for KOH (0.25 M) and dimethylsulphate (2.1 M in THF) respectively, dimethylated compound **97** was obtained in 93% yield. This sequential reaction was run for 7 days and 80–90% yield of **97** was recorded therefore demonstrating the stability and durability of the catalyst. In residence times ranging between 10–30 s, a variety of dimethylated compounds were generated in good yields. Napthoquinone derivatives were neatly methylated to give products in 76–97% yield. Napthoquinones with electron withdrawing and donating groups were shown to give corresponding products in excellent yields (90–96%). *p*-Benzoquinone derivatives also provided desired products in good yield (90->99%). The protocol was shown to be inclusive of various substrates; moreover, air-sensitive hydrogenated intermediates were also methylated in good yields.

The authors also investigated the acetylation reaction in the sequential continuous flow derivatization of hydroquinones using DMF as a solvent. The same flow set-up that used for the methylation reaction was used for this short study (Fig. 28). The diacetylation product (83% yield) was obtained by flowing acetyl chloride (5 equiv. in DMF, flow rate: 0.053 mL/min) and DMAP (20 mol%) (pyridine/DMF(1/19) solvent system in the presence of flow rate: 0.03 ml/min) through the second tubular reactor in Fig. 28. Benzoylation reaction using benzoyl chloride also afforded desired product in 99% yield. Trifluoromethane sulphonic anhydride and methacrylic anhydride were

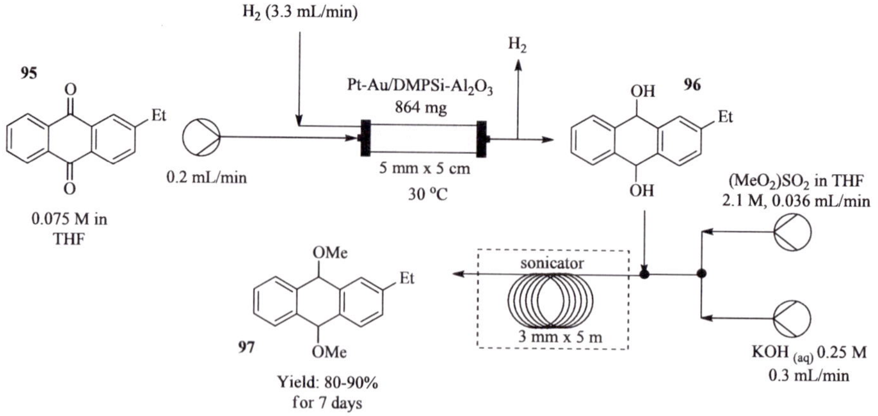

Fig. 28 A continuous flow hydrogenation of anthraquinone **95** and sequential methylation reaction

also employed as substrates to generate trifluoromethanesulphonylated (89%) and methacrylated products (>99%), respectively, in excellent yields. The robustness of the heterogeneous Pt-Au/DMPSi-Al_2O_3 catalyst and the ability of continuous flow reactors to facilitate sequential reactions to give useful compounds in good yields; avoiding the almost impossible isolation of the air-sensitive hydroquinone intermediate is a major plus.

2.5 *Process Automation, in-Line Monitoring, Purification and Separation*

The eleventh principle of green chemistry; "The development of analytical methodologies needed to allow real-time analysis for pollution prevention, in-process monitoring and control prior to the formation of hazardous substances"

The versatility of continuous flow systems allows for integration of in-line microprocessing analytics as well as purification and separation processes. The latter is still challenging but researchers are still making headway by developing innovative platforms to enable these unit operations to be carried out in-line rather than off-line. The plus here is in the rapid reaction screening of processes, minimized waste generation since recycle streams can be incorporated and increased overall productivity of chemical processes. Moreover, this has opened up the possibility of a process operation realm that is either void of or requires very little human intervention and rather applies the use of robotics for the mundane activities. As a result, continuous flow processing greatly adds to the efforts towards the development of green processes.

To elaborate, Gerdy et al. in the synthesis of cyclic organic carbonates under organocatalytic continuous flow conditions demonstrates the use of a low-yield in-line NMR analysis for the quantitative real-time reaction monitoring [50]. The continuous flow process developed is characterized by short reaction times (3–9 min), low catalyst loading (3.5 mol%) and the use of environmentally benign ammonium and

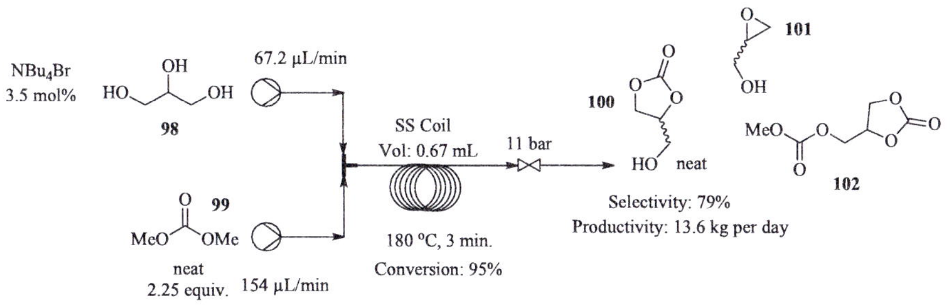

Fig. 29 Flow synthesis of cyclic organic carbonates

phosphonium organic salts as catalyst (Fig. 29). A wide substrate scope of carbonates was synthesized with good conversions (49–95%) (see reference [50] for further details)

Carbonation of glycerol **98** was achieved in 79% selectivity for **100** in 3 min residence time at 180 °C. An induced process failure, where the reaction temperature was decreased to 100 °C from 180 °C, was successfully monitored by in-line NMR inserted downstream of the reactor with the aim of evaluating the potential of real-time qualitative process monitoring. This is an example of how easy it is to integrate the right analytical tool into a continuous flow process to curb and rectify process failures, which could be caused by deviation from set reactions conditions or malfunctioning of other units in the continuous flow system. Ultimately, this leads to reduced waste and a greener approach to organic synthesis.

Poscharny et al. employed an in-line ReactIR spectrometer and a computer-based interface to optimize the synthesis of oxetanes from benzophenone and furan derivatives (Scheme 7), a Parteno–Buchi reaction in a capillary-type microfluidic photoreactor [51]). The self-optimizing reactor assembly comprising of a medium pressure Hg lamp and a Pyrex glass filter coupled with an IR took measurements before and after the reaction. The computer-based IR data analysis was used to control the pump by mainly following the disappearance of the characteristic C=O band of the aldehyde or ketone substrate. The reaction between benzophenone (1.37 mmol) **22** and furan (10 mL) **103** was nicely optimized in a 5 mL volume photoreactor irradiated with an Hg lamp. IR software (iCIR™) was used for IR data interpretation by following the disappearance of benzophenone C=O band at 1662 nm^{-1}. Sample collection was effected by a computer-assisted valve control having passed the IR flow cell. 97% yield of **104** in a total residence time of 83 min was obtained using a one-pump system (flow rate: 360 μL/min).

A substrate scope was evaluated from which it was observed that substrates with electron donating (82–96%) and withdrawing (82–90%) substituents on the aromatic group of the carbonyl compound provided desired products in high yields. In addition, substitution patterns on the aromatic group has negligible effect on the product yields; however, it was noted that aliphatic carbonyl substituents gave the lowest yield (23%). This was attributed to insufficient irradiation (lower than 300 nm). The reaction productivity was also improved using a two-pump system (Fig. 30) to deliver

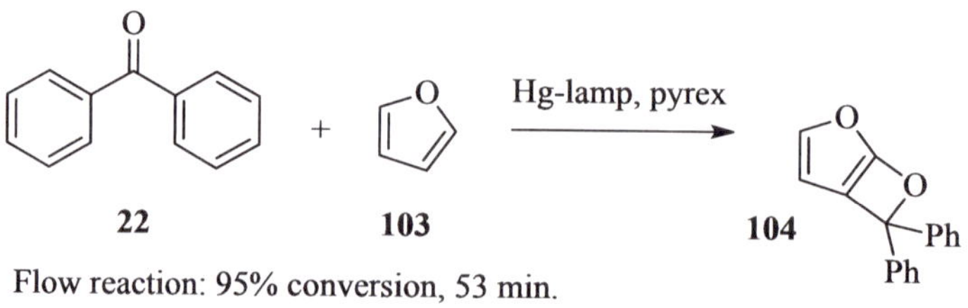

22 103 104

Flow reaction: 95% conversion, 53 min.

Scheme 7 Synthesis of oxetanes

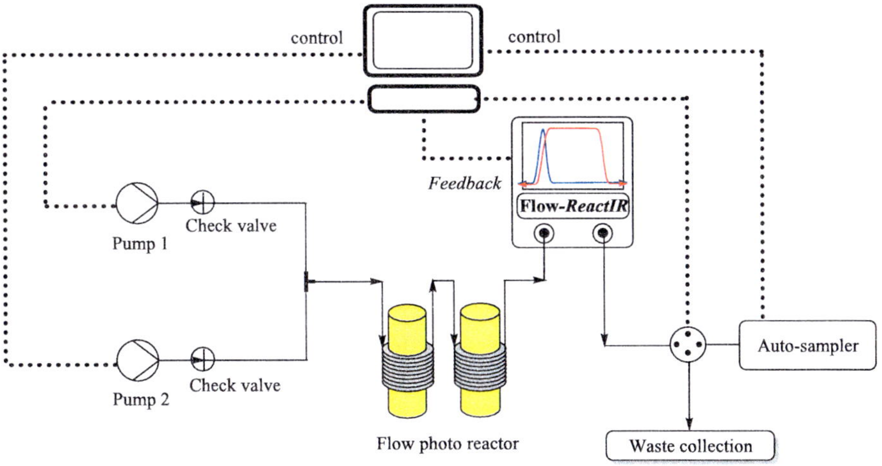

Fig. 30 A schematic representation of the computer-based flow-IR optimization of Parteno–Buchi reaction in a flow photoreactor

the reactants, benzophenone **22** (65 μL/min) and furan **103** (115 μL/min) into the photoreactor (95% conversion in 56 min).

On-line HPLC has also been successfully used in in-line monitoring of reactions [52], for example in a two-step telescoped final bond forming step in the synthesis of AZD9291 (Fig. 31). Osimertinib (AZD9291) is a third-generation irreversible

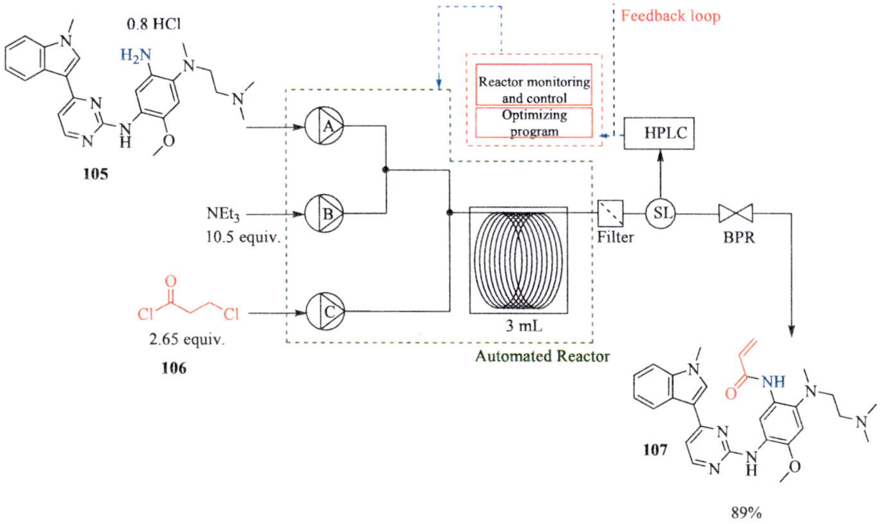

Fig. 31 Continuous flow synthesis of osimertinib (AZD9291) with in-line HPLC reaction monitoring

epidermal growth factor receptor kinase inhibitor [53] used in the treatment of non-small cell lung carcinoma. Using a self-optimizing flow reactor system, Holmes et al. achieved the synthesis using a self-optimizing flow reactor system coupled with reaction monitoring and feedback control (Fig. 31). SNOBFIT and MATLAB algorithms were used to facilitate the reaction optimization [54]. At 0.11 mL/min of **105**, acid chloride **106** (2.65 equiv.) in the presence of NEt_3 (10.5 equiv.) at 123.9 °C, 89% yield **107** was obtained.

With this protocol, the authors were able to determine the possible experimental domain within which the dimer impurity of **107** would have the highest yield. This ensures that a product with high purity is synthesized. Furthermore, the material minimization functions in this process were reported to have improved the efficiency of material use by 22%. The automated optimization was reported to have required 42 experiments and an overall run time of 26 h. Moreover, only 10 g of material was used for this optimization process.

Echtermeyer et al. in the Pd-catalysed C-H activation of aliphatic secondary amines resulting in the formation of aziridines demonstrates the use of model-based design of experiments and self-optimization approach [55]. The combination of model-based design of experiments with continuous flow synthesis was shown to rapidly deliver optimized reaction conditions with minimum effort required. The flow UV cell was used for in-line detection of products in reactions slugs that were formed from reactants that were pumped into a solvent stream, which was continuous, flowed in the system. The flow system was also coupled to an in-line GC for product analysis (Fig. 32). Once again, the versatility of continuous flow is exemplified.

Other examples of in-line monitoring techniques that have been successfully integrated with continuous flow systems include mass spectrometry [56] and size exclusion chromatography [57]. Coupling of in-line separation techniques with continuous flow systems is still in its developing stages. This is because not one size fits all reaction syntheses however, in-line purification techniques such a liquid–liquid extraction [58–60], microdistillation [61], microcrystallization [62] and free-flow electrophoresis [63] have successfully been coupled to continuous flow systems and in turn afforded greener processes.

2.6 *Multistep Syntheses*

Lastly but not least, as can be seen in the examples discussed in the above sections, multistep synthesis is a common feature of continuous flow processing. This is a key characteristic that allows for syntheses to be performed without isolation of intermediates thus ensuring safety, minimizing final product loses caused by sequential off-line purification, minimizing the amount of solvent and reagents used and encourages the efficient use of energy since steps with similar energy requirements can be coupled in the same reactor unit.

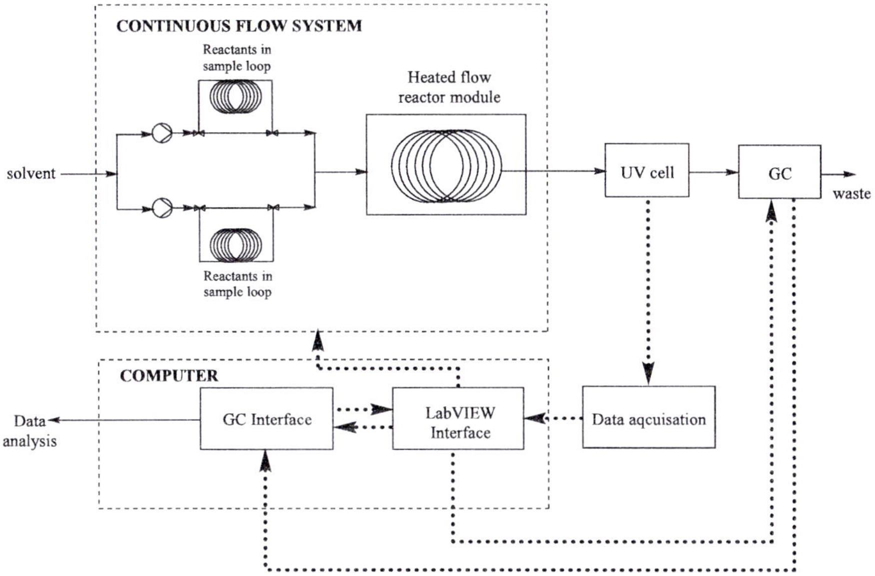

Fig. 32 Schematic representation of the automated continuous flow system used for the model development and black box sequential optimization

The eighth principle of green chemistry; "Reduce derivatives through minimizing or avoiding the use of blocking groups, protection/deprotection, and temporary modification of physical/chemical processes that require additional reagents and can generate waste"

For example, Marsini et al. reported a concise, robust and scalable synthesis of a developmental CCR1 **111** [64]. The three-step reaction process, i.e. azide formation, rearrangement and isocyanate trapping was successfully established within very short residence times (Scheme 8). At a 40 kg scale reaction scale up, a throughput of 0.8 kg/h was achieved with an *E*-factor of approximately 9 associated with the process. What is even more interesting, the authors were able to carry out the Curtius rearrangement and acid-isocyanate coupling in one single protecting group free reaction step in a semi-continuous manner. The process was reported to have a relative process greenness of 166%.

3 Conclusion

Using a few current examples from a multitude found in literature, we exemplify the role of continuous flow processing in the development of green syntheses by enabling excellent yields, selectivity, conversions and product purity in the shortest time possible. With regard to safety concerns in the synthesis and use of unstable, explosive but rather important compounds, continuous flow processing has revealed

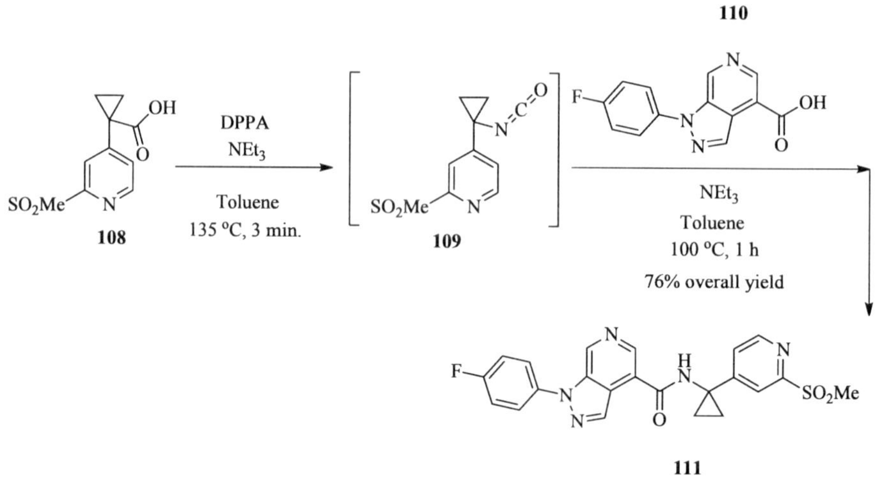

Scheme 8 Direct, semi-continuous curtius synthesis

safe reaction spaces for this purpose. As such, in an era where the long-term adverse effects of human activity such as chemical production on the environment, human health and aquatic life are being experienced, technologies such as these that drive towards cleaner and greener chemical production processes should be embraced whenever possible.

References

1. Economic UN, Commission SI (2019) The evolving chemicals economy : status and trends relevant for sustainability. pp 24–25
2. UNEP promotes environmentally sound practices globally and in its own activities (2019)
3. Anastas P, Eghbali N (2010) Green chemistry: principles and practice. Chem Soc Rev 39(1):301–312
4. Tang S, Bourne R, Smith R et al (2008) The 24 principles of green engineering and green chemistry: improvements productively. Green Chem 10:268–269
5. Constable DJ, Jimenez-Gonzalez C (2012) Evaluating the greeness of synthesis. In: Green processes, pp 35–65
6. Swarr TE, Cespi D, Fava J et al (2017) Application of life cycle assessment to green chemistry objectives. In: Handbook of green chemistry, pp 1–28
7. Gomollón-bel F (2019) Ten chemical innovations that will change our world. Chem Int 12–17
8. Akwi FM, Watts P (2018) Continuous flow chemistry: where are we now? recent applications, challenges and limitations. Chem Commun 54(99):13894–13928
9. Boros Z, Nagy-gy L, Kátai-fadgyas K, Imre K, Ling I, Nagy T (2019) Continuous flow production in the final step of vortioxetine synthesis. piperazine ring formation on a flow platform with a focus on productivity and scalability. J Flow Chem 9:101–113
10. Gioria E, Wisniewski F, Gutierrez L (2019) Journal of environmental chemical engineering microreactors for the continuous and green synthesis of palladium nanoparticles: Enhancement of the catalytic properties. J Environ Chem Eng 7(3):103136

11. Roberts EJ, Habas SE, Wang L et al (2017) High-throughput continuous flow synthesis of nickel nanoparticles for the catalytic hydrodeoxygenation of guaiacol. ACS Sustain Chem Eng 5(1):632–639
12. Larrea A, Sebastian V, Ibarra A et al (2015) Gas slug microfluidics: a unique tool for ultrafast, highly controlled growth of iron oxide nanostructures. Chem Mater 27(12):4254–4260
13. Nikam AV, Dadwal AH (2019) Scalable microwave-assisted continuous flow synthesis of CuO nanoparticles and their thermal conductivity applications as nanofluids. Adv Powder Technol 30(1):13–17
14. Peris E, Porcar R, Burguete I et al (2019) Supported ionic liquid-like phases (SILLPs) as immobilised catalysts for the multistep and multicatalytic continuous flow synthesis of chiral cyanohydrins. Chem Cat Chem 11:1–9
15. Torrelo G, Van Midden N, Stloukal R et al (2014) Immobilized hydroxynitrile lyase: a comparative study of recyclability. Chem Cat Chem 6(4):1096–1102
16. Delville MME, Koch K, Van Hest JCM, Rutjes FPJT (2015) Chemoenzymatic flow cascade for the synthesis of protected mandelonitrile derivatives. Org Biomol Chem 13(6):1634–1638
17. Nishiyama Y, Fujii A, Mori H (2019) Selective synthesis of azoxybenzenes from nitrobenzenes by visible light irradiation under continuous flow conditions. React Chem Eng 4:2055–2059
18. Hommelsheim R, Guo Y, Yang Z et al (2019) Photochemical synthesis blue-light-induced carbene-transfer reactions of diazoalkanes. Angew Commun 58:1203–1207
19. Jang S, Vidyacharan S, Gyak K et al (2019) Photocatalysis in a multi-capillary assembly microreactor: toward up-scaling the synthesis of *2H*-indazoles as drug scaffolds. React Chem Eng 4:1466–1471
20. Chiurchiù E, Palmieri A, Petrini M (2019) 3-Alkylated indoles by reduction of sulfonyl indoles under flow chemical conditions Arkivoc iv:69–79
21. Salvador CEM, Andrade CKZ (2019) A mild, fast, and scalable synthesis of substituted α -acyloxy ketones via multicomponent reaction using a continuous flow approach. Front Chem 7:2–11
22. Amini-rentsch L, Vanoli E, Richard-bildstein S et al (2019) A novel and e ffi cient continuous-flow route to prepare trifluoromethylated n - fused heterocycles for drug discovery and pharmaceutical manufacturing. Ind Eng Chem Res 58:10164–10171
23. Port Neches (2019): Texas chemical plant blasts lead to evacuation order. BBC News
24. Zhan W, Ji L, Ge Z et al (2018) A continuous flow synthesis of primary amides from hydrolysis of nitriles using hydrogen peroxide as oxidant. Tetrahedron 74(13):1527–1532
25. Guttenberger N, Breinbauer R (2017) C-H and C-C bond insertion reactions of diazo compounds into aldehydes. Tetrahedron 73(49):6815–6829
26. Breugst M, Huisgen R, Reissig HU (2018) Regioselective 1,3-Dipolar cycloadditions of diazoalkanes with heteroatom-substituted alkynes: theory and experiment. Europ J Org Chem 20:2477–2485
27. Rulli P, Benoit G, Allouche EMD et al (2018) Diazo compounds safe and facile access to nonstabilized diazoalkanes using continuous flow technology. Angew Chemie Int Ed 57:5777–5782
28. Rullière P, Cyr P, Charette BA (2016) Difluorocarbene addition to alkenes and alkynes in continuous flow. Org Lett 18(9):1988–1991
29. Schulenberg JW, Archer S (2011) The chapman rearrangement. In: Organic reactionsp, pp 1–51
30. Fang J, Ke M, Huang G et al (2019) The Chapman rearrangement in a continuous-flow microreactor. RSC Adv 9:9270–9280
31. Xue C, Li J, Lee JP et al (2018) Continuous amination of aryl/heteroaryl halides using aqueous ammonia in a Teflon AF-2400 tube-in-tube micro-flow reactor. React Chem Eng 4(2):346–350
32. Hartman RL, Liu Y (2019) Reaction kinetics of a water-soluble palladium–β-cyclodextrin catalyst for a suzuki-Miyaura cross- coupling in continuous flow. React Chem Eng 4:1341–1346
33. Zhang C, Zhu R, Wang Z et al (2019) Continuous flow synthesis of diaryl ketones by coupling of aryl Grignard reagents with acyl chlorides under mild conditions in the ecofriendly solvent 2-methyltetrahydrofuran. RSC Adv 9:2199–2204

34. Ferlin F, Hulst MK, Der Van et al (2019) Continuous flow/waste-minimized synthesis of benzoxazoles catalysed by heterogeneous manganese systems. Green Chem 21:5298–5305
35. Labes R, Mateos C, Battilocchio C et al (2019) Fast continuous alcohol amination employing a hydrogen borrowing protocol. Green Chem 21:59–63
36. Tadić J, Mihajlović M, Jovanović M et al (2019) Continuous flow synthesis of some 6- and 1, 6-substituted. J Serb Chem Soc 84(6):531–538
37. Masuda K, Ichitsuka T, Koumura N et al (2018) Flow fine synthesis with heterogeneous catalysts. Tetrahedron 74(15):1705–1730
38. Yue J (2018) Multiphase flow processing in microreactors combined with heterogeneous catalysis for efficient and sustainable chemical synthesis. Catal Today 308:3–19
39. Munirathinam R, Huskens J, Verboom W (2015) Supported catalysis in continuous-flow microreactors. Adv Synth Catal 357:1093–1123
40. Haas CP, Müllner T, Kohns R (2017) High-performance monoliths in heterogeneous catalysis with single-phase liquid flow. React Chem Eng 2:498–511
41. Nitelet A, Kairouz V, Lebel H et al (2019) Continuous flow chlorination of alkenyl iodides promoted by copper tubing. Synthesis (Stuttg) 51:251–257
42. Monguchi Y, Ichikawa T, Yamada T (2019) Continuous-flow suzuki-miyaura and mizoroki-heck reactions under microwave heating conditions. Chem Rec 19:3–14
43. Sieber JD, Buono F, Brusoe A et al (2019) Application of a preformed Pd-BIDIME precatalyst to suzuki–miyaura cross-coupling reaction in flow. J Org Chem 84:4926–31
44. Li H, Seechurn CCCJ, Colacot TJ (2012) Development of preformed Pd catalysts for cross-coupling reactions, beyond the 2010 nobel prize. ACS Catal 2:1147–1164
45. Gildner PG, Colacot TJ (2015) Reactions of the 21st century: two decades of innovative catalyst design for palladium-catalyzed cross-couplings. Organometallics 34(23):5497–5508
46. Mohanta N, Chaudhari MB, Digrawal NK et al (2019) Rapid and multigram synthesis of vinylogous esters under continuous flow: an access to transetherification and reverse reaction of vinylogous esters. Org Process Res Dev 23:1034–1045
47. Miyamura H, Tobita F, Suzuki A (2019) Direct synthesis of hydroquinones from quinones through sequential and continuous-flow hydrogenation-derivatization using heterogeneous Au–Pt nanoparticles as catalysts. Angew Chem Int Ed 58:9220–9224
48. Greeves N (1991) Reduction of C=O to CHOH by metal Hydrides. In: Comprehensive organic synthesis, pp 1–22
49. Roginsky V, Barsukova T (2000) Kinetics of oxidation of hydroquinones by molecular oxygen. Effect of superoxide dismutase. J Chem Soc Perkin Trans 2 (7):1575–1582
50. Gérardy R, Estager J, Luis P et al (2019) Versatile and scalable synthesis of cyclic organic carbonates under organocatalytic continuous flow conditions. Catal Sci Technol 9(24):6841–6851
51. Poscharny K, Fabry DC, Heddrich S et al (2018) Machine assisted reaction optimization: a self-optimizing reactor system for continuous- fl ow photochemical reactions. Tetrahedron 74(25):3171–3175
52. Heiland JJ, Warias R, Lotter C et al (2017) On-chip integration of organic synthesis and HPLC/MS analysis for monitoring stereoselective transformations at the micro-scale. Lab Chip 17:76–81
53. Greig SL, Osimertinib (2016) First global approval. Drugs 76(2):263–273
54. Holmes N, Akien GR, Blacker JA et al (2016) Self-optimisation of the final stage in the synthesis of EGFR kinase inhibitor AZD9291 using an automated flow reactor. React Chem Eng 1(4):366–371
55. Echtermeyer A, Amar Y, Zakrzewski J (2017) Self-optimisation and model-based design of experiments for developing a C–H activation flow process. Beilstein J Org Chem 13:150–163
56. Bristow TWT, Ray AD, O'Kearney-Mcmullan A et al (2014) On-line monitoring of continuous flow chemical synthesis using a portable, small footprint mass spectrometer. J Am Soc Mass Spectrom 25(10):1794–1802
57. Rubens M, Vrijsen JH, Laun J et al (2019) Ultra-precise polymer synthesis by autonomous self-optimizing flow reactors. Angew Chemie Int Ed 58(10):3183–3187

58. Vural Gürsel I, Kockmann N, Hessel V (2017) Fluidic separation in microstructured devices-concepts and their integration into process flow networks. Chem Eng Sci 169:3–17
59. Herath A, Cosford NDP (2017) Continuous-flow synthesis of highly functionalized imidazo-oxadiazoles facilitated by microfluidic extraction. Beilstein J Org Chem 13:239–246
60. Sagandira RC, Watts P (2017) Synthesis of amines, carbamates and amides by multi-step continuous flow synthesis. Eur J Chem 44:6554–6565
61. Yang R, Liu C, Wang Y et al (2016) A comprehensive review of micro-distillation methods. Chem Eng J 313:1509–1520
62. Escriba M, Hessel V, Maier MC et al (2018) Continuous-flow in-line solvent-swap crystallization of vitamin D 3. 22(2):178–189
63. Agostino FJ, Krylov SN (2015) Advances in steady-state continuous-flow purification by small-scale free-flow electrophoresis. Trends Anal Chem 72:68–79
64. Marsini MA, Buono FG, Lorenz JC et al (2017) Development of a concise, scalable synthesis Curtius rearrangement 1454–1461

Chapter 16
Green Chemistry of Recoverable Catalysts

Rasmi P. Bhaskaran and Beneesh P. Babu

1 Introduction

The recovery of catalysts after the completion of a reaction is of great importance in green chemistry [1]. Green chemistry addresses the impact of any process on the environment, and it is primarily concerned about preventing pollution [2]. Catalysis plays a vital role in the chemical industry [3]. The waste generated in chemical synthesis is mainly due to the usage of stoichiometric reagents, especially the metal-based reagents [4]. For example, stoichiometric reducing agents, both metals (Na, Mg, Zn, Fe) and metal hydrides ($LiAlH_4$, $NaBH_4$, etc.), oxidizing agents such as permanganate, manganese dioxide, and chromium(VI) reagents are the commonly used ones. Reactions such as sulfonation, nitration, halogenation, diazotization, and Friedel–Crafts acylation employ stoichiometric amounts of mineral acids (H_2SO_4, HF, H_3PO_4) and Lewis acids ($AlCl_3$, $ZnCl_2$, BF_3) [5].

The broad classification of catalysis is heterogeneous and homogeneous catalysis. In heterogeneous reactions, the phase of the catalyst is different from that of the reactants, and hence, the separation of the catalyst from the reaction medium is relatively easy. On the other hand, in homogeneous reactions, the catalyst and other reactants are in the same phase and are miscible with one another [6]. Solvent extraction, [7] filtration, nanofiltration, [8] catalyst leaching, chemical precipitation, [9] magnetic recovery, [10], and adsorption are catalyst recovery techniques typically employed in industrial and pharmaceutical processes. Nanofiltration is a relatively new membrane filtration technique used only in the homogenous process, and the catalysts are immediately reused. Materials such as activated carbon and ion exchange resins are used

R. P. Bhaskaran · B. P. Babu (✉)
Department of Chemistry, National Institute of Technology Karnataka (NITK), Surathkal, Mangalore, Karnataka, India
e-mail: pbbeneesh@nitk.edu.in

R. P. Bhaskaran
e-mail: rasmibhaskaranp93@gmail.com

G. Anilkumar and S. Saranya (eds.), *Green Organic Reactions*, Materials Horizons: From Nature to Nanomaterials, https://doi.org/10.1007/978-981-33-6897-2_16

in the adsorptive recovery techniques and can include the recovery of either homogeneous catalysts or homogeneous ligands/catalysts. Magnetic regeneration of the catalysts allows the easy retrieval of catalysts with minimal loss.

2 Classification

Catalyst recovery can be broadly classified based on the recovery techniques used, and the choice of method for the catalyst separation is dependent on the nature of the catalysts. The separation techniques are time-consuming and have the limitations of cost, efficiency, generation of secondary wastes, etc. [11].

2.1 Metallodendrimers-Dendrimers Supported Catalysts

A widely used method for the easy recovery of the catalyst is the attachment of the metal catalysts to a solid support [12]. Generally used supports include the organic, inorganic, and hybrid, which work through heterogeneous catalysis. The solid support silica is the universal inorganic support used because of its chemically inert nature and physical strength. But the catalyst efficiency in this type of heterogeneous catalysis is inferior due to the very slow diffusion of the catalyst into the reaction medium, which decreases the rate of mass transfer and also the generally less activity of heterogeneous catalyst compared to the homogeneous ones. These limitations focus on the development of soluble catalyst support, which makes the reaction medium seamless. Dendrimers are such types of soluble catalyst supports used. The globular macrostructure of such supports enables the catalyst recovery effortless, and it is suited for membrane filtration [13]. The first example of a recoverable metallodendritic catalyst was reported in 1997 by Reetz for the Heck reaction using Pd dendrimers derived from the dendritic phosphines [14]. Potentially speaking, a dendrimer can be considered as a homogenous catalytic system as it is a combination of a large number of monomeric units with active sites. The total catalytic features such as selectivity, stability, and recyclability depend on the dendritic architecture of the metallodendrimers [15] (Fig. 1).

A copper-free recoverable dendritic palladium catalyst for the Sonogoshira reaction was reported in 2003 by Heuze and co-workers [16]. The catalyst prepared bears two cyclohexyl substitutions on the phosphorous atom, and the catalyst is quickly recovered and reused by a simple precipitation method. They have developed three generations of the dendritic palladium complexes by treating $Pd(OAc)_2$ with aminophosphines. The catalytic activity of the synthesized compounds was tested on the coupling between phenylacetylene and iodobenzene or bromobenzene in a copper-free Sonogashira procedure (Scheme 1). The first two generations of the complex showed excellent catalytic activity. Still, for the third generation complex, the activity was less because of the increased steric effect of the dendrimer branch.

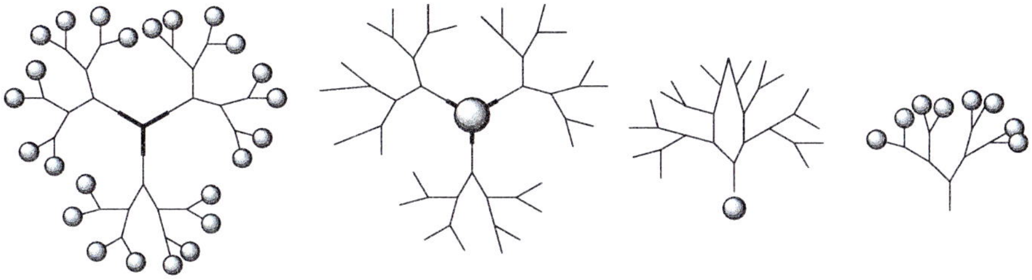

Fig. 1 Dendrimer architectures

Scheme 1 Dendritic Pd catalyzed Sonogoshira reaction

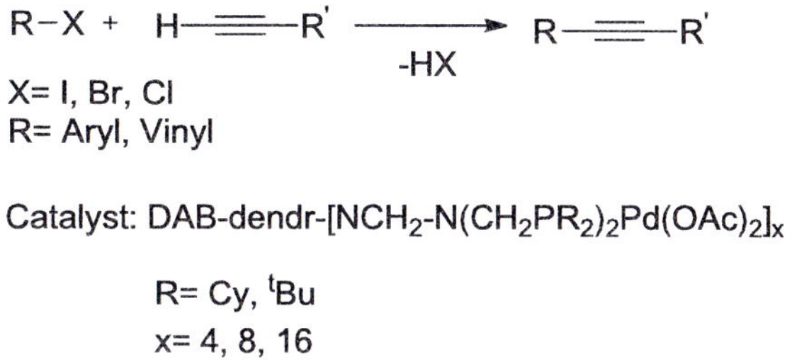

In the year of 2004, Nlate and co-workers synthesized a stable polyoxometallate (POM) centered dendrimer and studied its catalytic activity in different oxidation reactions [17]. The core present in the dendritic catalyst was peroxophosphotungstate and was developed by the ionic bonding approach (Scheme 2). The application of this catalyst lies in the oxidation reactions using peroxides, and they were found to be highly air-stable and recoverable. The catalyst was efficient for performing the epoxidation of the alkene, selective oxidation of sulfide to sulfone, and also on the oxidation of cyclohexanol to cyclohexanone. The bulkier branches surrounding

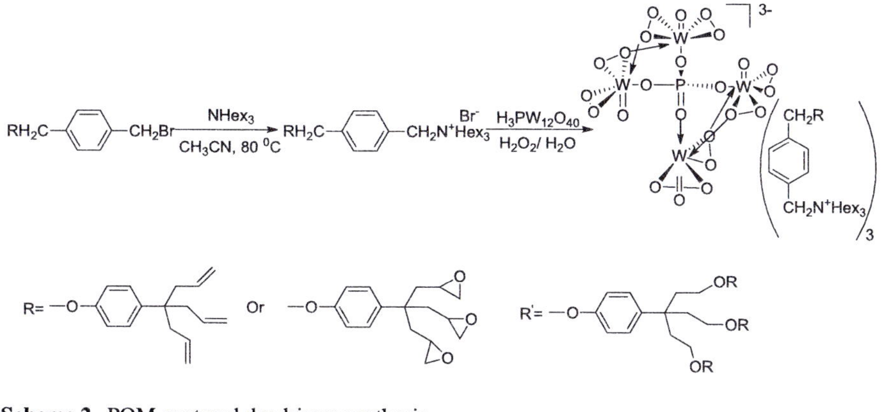

Scheme 2 POM-centered dendrimer synthesis

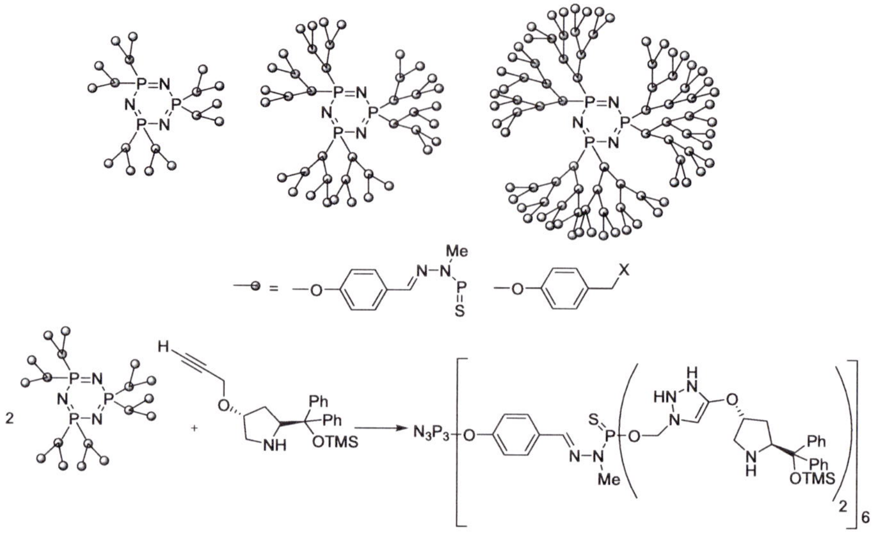

Scheme 3 Preparation of Azido terminated phosphorous dendrimers

the POM center increase the stability of the catalyst, which prevents it from the degradation.

An efficient catalyst supported on phosphorous dendrimers and nanobeads of Co/C for the Michael addition of aldehydes to nitro olefins was developed by Keller and co-workers in 2013 [18]. The dendritic catalyst was a combination of Jorgensen-Hayeshi organocatalyst grafted via triazole linkers on phosphorus dendrimers (Scheme 3). A comparative study on the catalytic performances of dendrimer supported and magnetic nanoparticle supported organocatalysts on Michael addition was carried out. The dendrimer supported catalyst was efficient with maximum conversion at a reasonable time. It showed excellent recycling ability via precipitation method and were reused up to seven consecutive catalytic cycles.

2.2 *Fluorous Biphasic Catalyst-PTC*

Phase transfer catalyst (PTC) allows the transfer of reagents from one phase and creates suitable conditions for the reaction to happen. Generally, organic salts with fluorinated anions act as phase transfer catalysts in organic transformations [19].

In the year of 2004, several cinchona derived alkaloids were anchored to different polystyrene supports by Chinchilla et al., resulting in the formation of chiral polymeric ammonium salts (Scheme 4) [20]. The catalyst prepared was used for the asymmetric alkylation of *N*-(diphenyl methylene) glycine esters. The best *S*-enantiomer

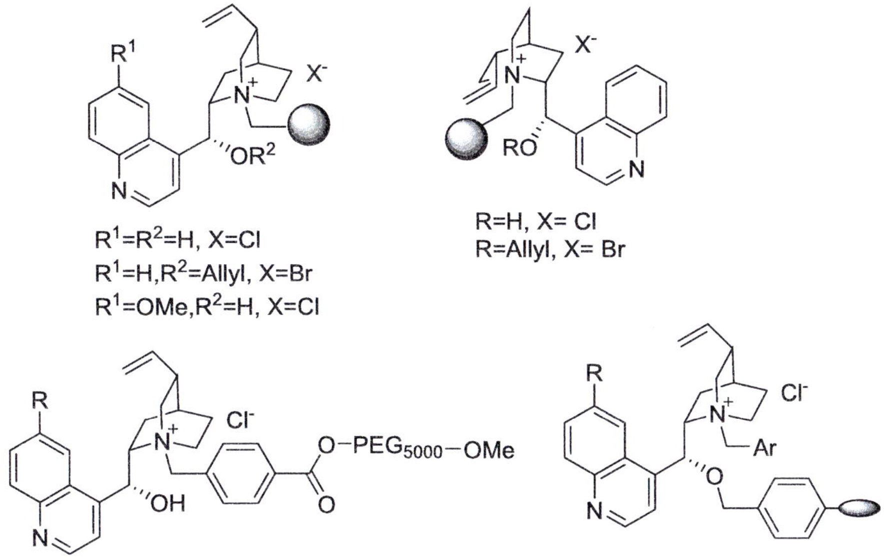

Scheme 4 Polystyrene anchored cinchona ammonium salts

selectivity was found for the Merrifield resin supported cinchonidium salt, and the *R*-selectivity was observed for the cinchonine derived salts. Simple filtration techniques were used to separate the catalysts after the reaction and were reused.

Tetraalkyl ammonium salts containing PF_6- and BF_4- anions were developed by Kryshtal and co-workers and have used as efficient PTC for the trichloromethylation and dichlorocyclopropanation reactions of alkenes and aldehydes (Scheme 5) [21]. The fluorinated PTCs were readily recovered from the reaction medium and washed with H_2O and Et_2O. The catalyst retained its activity over several cycles of reactions.

A silica nanoparticle supported phosphonium-palladium phase transfer catalyst matrix was reported by Hajipour et al. The application of this catalyst in the Heck reaction in neat aqueous media was studied [22]. The palladium complexes along with the quarternary phosphonium cations were covalently immobilized on a silica support. The catalyst was synthesized by a complex formation of quaternary phosphonium iodide with palladium chloride, which resulted in the formation of $[PdCl_2I_2]^{2-}$, and it was reacted with SiO_2 nanoparticles (Scheme 6). The activity of Pd/PTC was tested in the coupling reaction of iodobenzene and methyl acrylate, using pure water as the

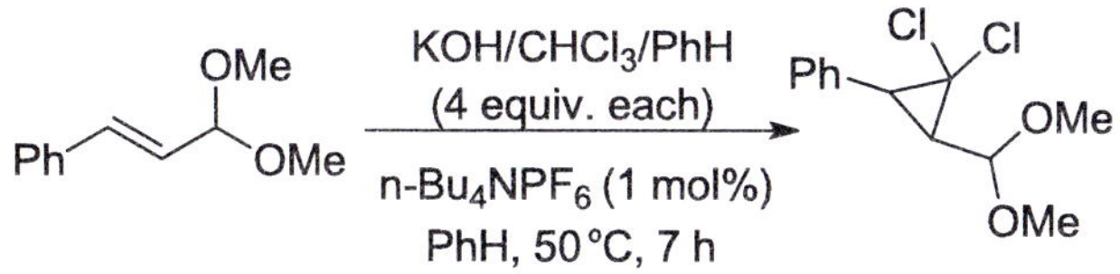

Scheme 5 Cyclopropane synthesis using fluorinated PTCs

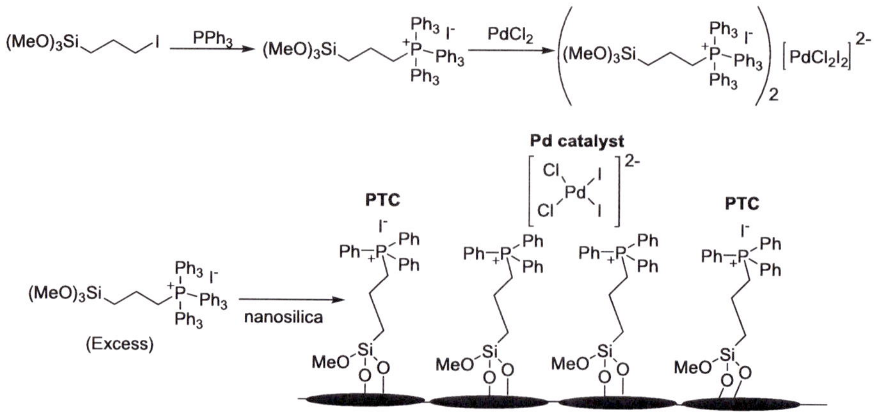

Scheme 6 Immobilization of Pd/PTC on silica

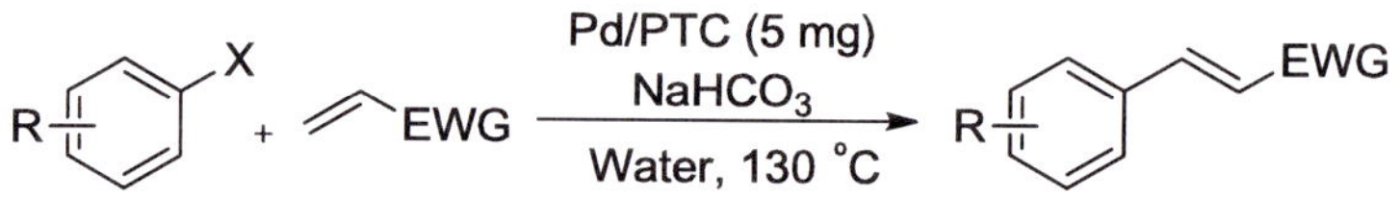

Scheme 7 Heck reaction catalyzed by immobilized Pd/PTC on silica

solvent (Scheme 7). The catalytic recyclability was also checked in the same reaction conditions. Catalyst recovery from the medium was done expertly by centrifugation method and reused after washing it with ethanol and acetone. The synthesized catalyst exhibited better reactivity in terms of turnover numbers.

Commercially available cinchona alkaloids were used for the synthesis of different dendrimeric phase transfer catalysts by Rull et al. The quaternization of the quinuclidine nitrogen resulted in the formation of PTC, and they catalyzed asymmetric organic transformations [23] (Scheme 8). The catalytic activity was checked on the asymmetric alkylation of the glycinate Schiff base with benzyl bromide. The recovery and reusability of the catalyst were valid for five consecutive cycles, maintaining the activity. The enantioselectivity of the catalyst was observed to reduce slightly during the runs.

2.3 *Magnetically Recoverable Metal Nanoparticles*

Magnetically recoverable catalysts are green chemistry aspects as they enable the environmentally friendly catalytic approach. Various strategies, like grafting, surface modification, and self-assembly, etc., provide a wide range of scope for the preparation of metal nanoparticles (MNPs). The separation of the catalyst from the reaction

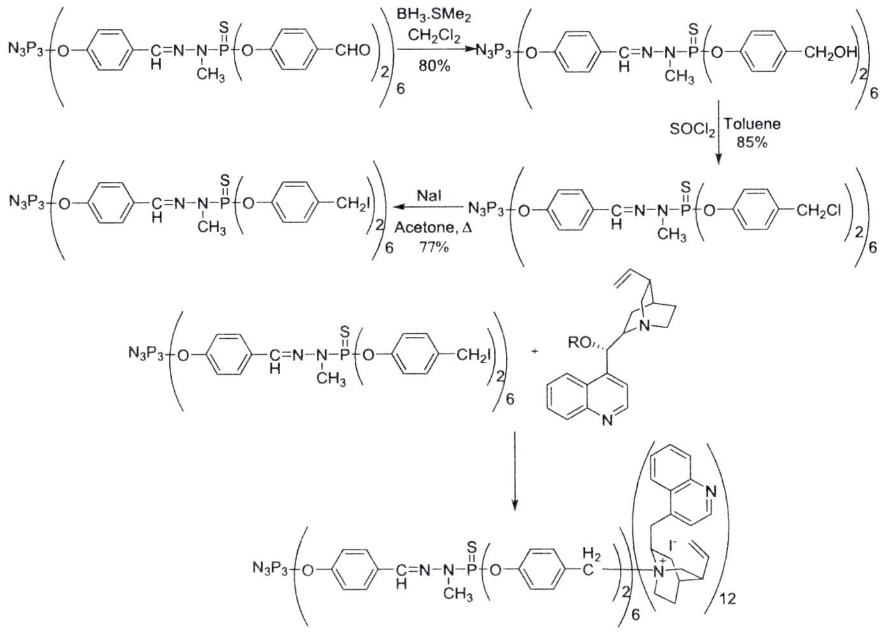

Scheme 8 Synthesis of benzyl chloride/iodide functional phosphorous dendrimers

medium is also cheap, feasible, and straightforward. Most of the Fe-, Ni-, Co-based MNPs enhance the reaction in both homogenous and heterogeneous media [24].

Considering the photocatalytic application of magnetically recoverable catalysts, Boruah and co-workers synthesized Fe_3O_4/reduced graphene oxide catalysts [25]. The synthesis was carried out following eco-friendly solution chemistry. The performance of the prepared nanoparticle was checked on the dye degradation. Carcinogenic and mutagenic dyes such as methyl green, methyl blue, and rhodamine B were effectively degraded by Fe_3O_4/rGO (Scheme 9). The catalytic application was

$$Fe_3O_4 + h\nu \longrightarrow Fe_3O_4\,(e^- + h^+)$$

$$\text{Dyes} + h\nu \longrightarrow (\text{Dyes})^*$$

$$Fe_3O_4/\,rGO + h\nu \longrightarrow Fe_3O_4\,(h^+) + rGO\,(e^-)$$

$$rGO\,(e^-) + O_2 \longrightarrow rGO + O_2^{\bullet -}$$

$$Fe_3O_4\,(h^+) + OH^- \longrightarrow Fe_3O_4 + \dot{O}H$$

$$\text{Dye Molecules} + \dot{O}H \longrightarrow \begin{array}{c}\text{Degraded products}\\ (Cl^-,\ NO_3^-,\ SO_4^-,\ CO_2 \text{ and } H_2O)\end{array}$$

Scheme 9 Photo dye degradation using Fe_3O_4/reduced graphene oxide extended to the photoreduction of mutagenic aqueous Cr (VI) to nontoxic Cr (III) solution under sunlight. The activity was found to be valid up to ten reaction cycles and was stable enough.

Vavasori et al. synthesized a new magnetically recoverable Pd catalyst by depositing 1% palladium on Poly-(1-oxo-trimethylene) with a 15% magnetite [26]. Alkoxycarbonylation of iodobenzene to benzoic esters was catalyzed using synthesized metal nanoparticles under phosphine free conditions (Scheme 10). The conversion was found to reach up to 100% within 3 h. The catalyst was recovered and recycled for about five times.

Later, in 2017, a Pd(0)-based magnetically recoverable nanocatalyst supported on nano Fe_3O_4–SiO_2 was developed by Khazaei et al. (Scheme 11) [27]. The prepared catalyst was used for the Suzuki coupling reaction (Scheme 12). Eggshell was utilized as a low-cost solid base for the reaction. The catalyst was characterized using SEM, EDS, XRD, TEM, etc. and confirmed the formation of Fe_3O_4@SiO_2–Pd. An external

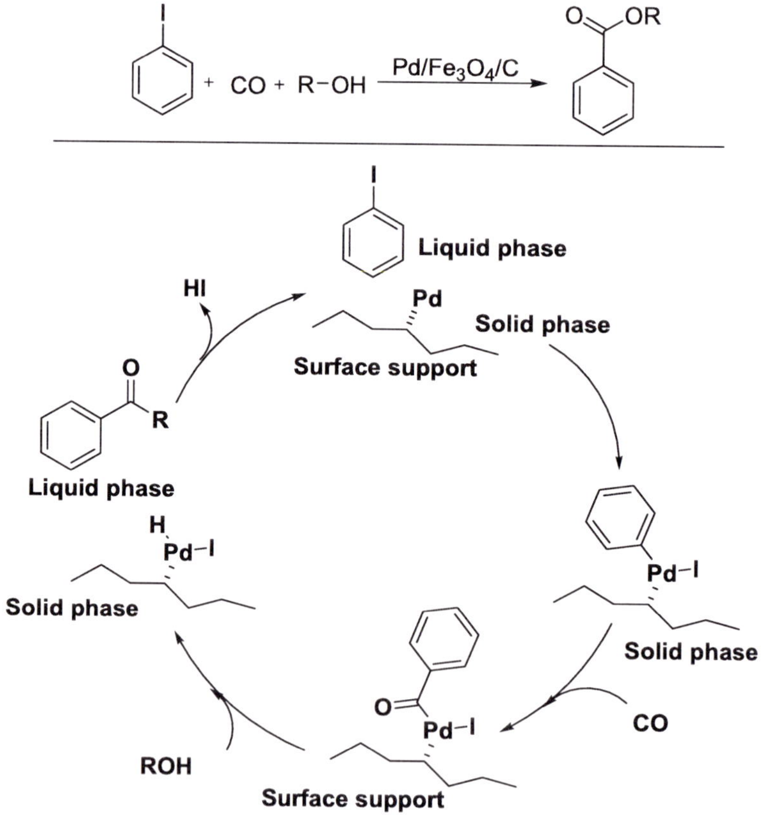

Scheme 10 Alkoxy carbonylation catalyzed by Pd/Fe_3O_4 with the mechanism

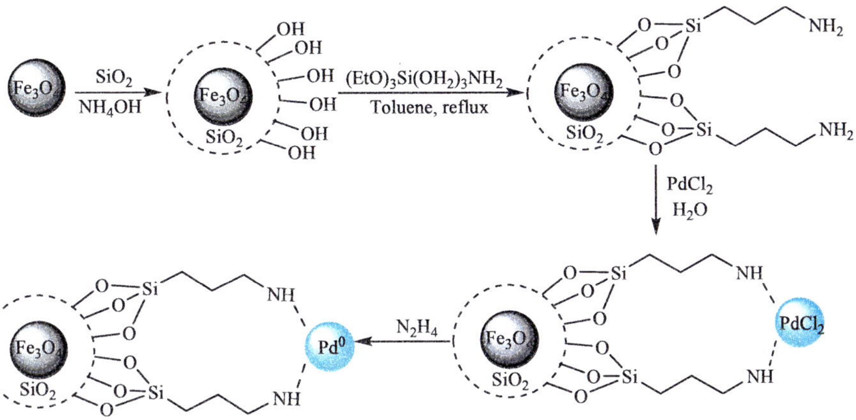

Scheme 11 Preparation of Pd-magnetic nanoparticles

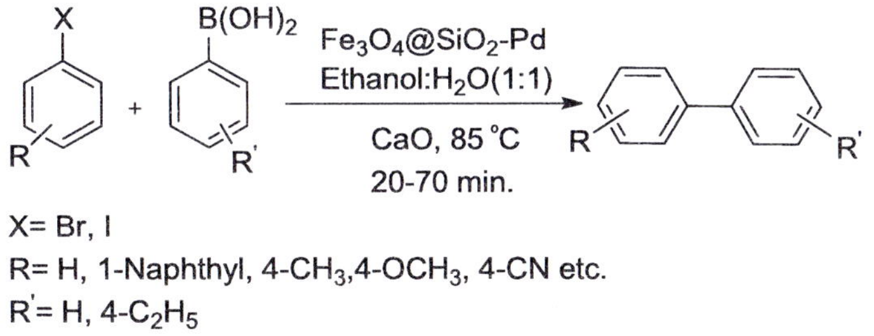

Scheme 12 Suzuki coupling in the presence of Fe_3O_4@SiO_2-Pd

magnet recovered the nanoparticles and was further used certain times without losing the catalytic activity.

A catalytic reduction of nitroaromatic compounds using magnetically separable nanoparticles was reported by Hong et al. in 2018 [28]. The catalyst synthesized was a combination of 13 nm Au nanoparticles enclosed in a hollow mesoporous carbon. The nanocatalyst Au@hmc-FeCo/GC was prepared from the thermal decomposition of Fe and Co on silica support containing Au nanoparticles (Scheme 13). Au@hmc-FeCo/GC showed superparamagnetism at room temperature. Catalytic reduction of 4-nitrophenol, 4-nitrotoluene, etc. was carried out, and the results were confirmed to

Scheme 13 Synthesis of Au@hmc-FeCo/GC

Scheme 14 Au@hmc-FeCo/GC catalyzed nitroarene reduction

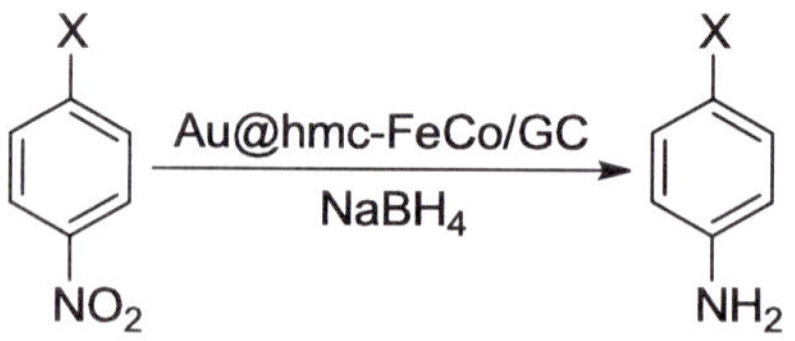

be better (Scheme 14). The same reaction cycle can be repeated at least five times without losing the initial catalytic properties.

2.4 *Organic Resin and Polymer-Supported Recoverable Catalysts*

Insoluble organic resins and polymers such as polystyrene-based cross-linkers have been used as supports in heterogeneous catalysis, which allows the active recovery and reuse of many metal catalysts by simple methods. Organic polymer supports provide improved catalytic stability, surface activity, and high selectivity for the reaction.

Angelini et al. reported the conjugate addition of an N-nucleophile for the synthesis of β-amino carbonyl compounds catalyzed by polystyryl-DABCOF$_2$ (Scheme 15) [29]. The reaction was carried out in an aqueous medium. The porous polymeric materials were found to be more compatible with the conditions. DABCO, a diamine moiety carries high loading fluoride as the counter ions and efficient than the commercially available fluoride catalysts. The catalyst was used in the β-azidation of enones (Scheme 16). The reaction was carried out in a continuous flow method, and it helps for the recovery of the catalyst used.

Merrifield resin supported peroxomolybdenum (VI) compounds for the selective oxidation of organic sulfides was developed by Boruah et al. in 2013 [30]. The catalyst was prepared by the immobilization of dioxomonoperoxomolybdenum (VI) on amino acid-functionalized Merrifield resins (Scheme 17). The catalyst showed excellent activity and selectivity for the oxidation of thioethers and dibenzothiophene to sulfoxides/sulfones by peroxides (Scheme 18). H_2MoO_4 reacted with 30%

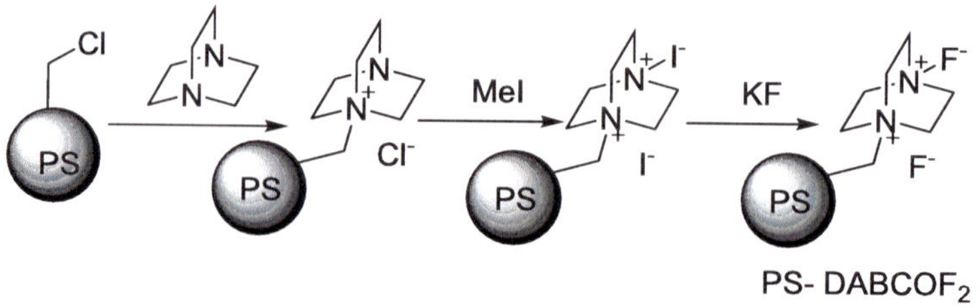

Scheme 15 Preparation of PS-DABCOF$_2$

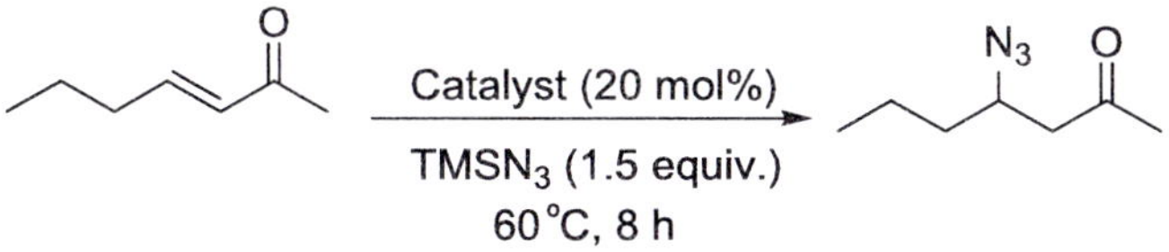

Scheme 16 β-Azidation in the presence of PS-DABCOF$_2$ based catalysts

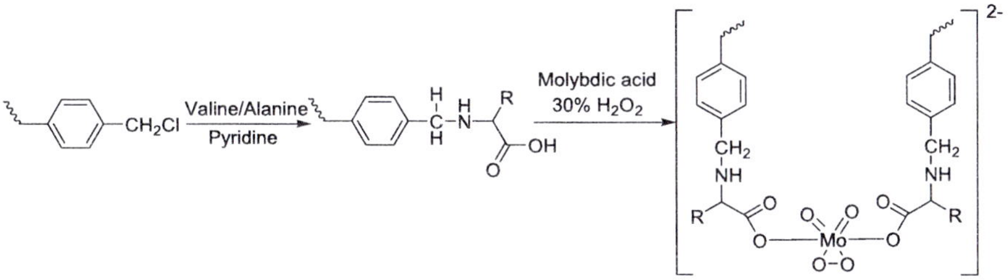

Scheme 17 Synthesis of Merrifield resin supported PMO complexes

Scheme 18 Selective oxidation of methyl phenyl sulfide

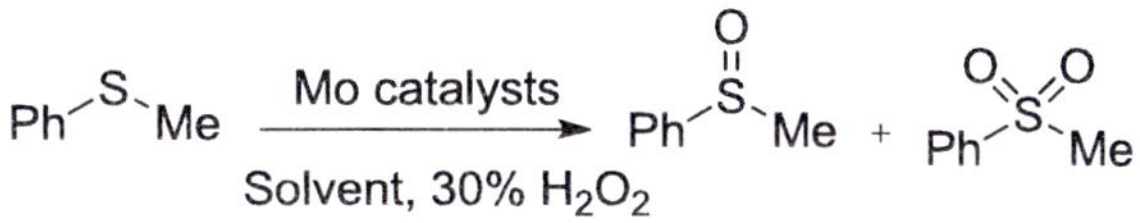

H_2O_2, and amino acid-functionalized resin at neutral pH resulted in the catalyst $[MoO_2(O_2)L_2]^{2-}$MR, L = Valine/Alanine; MR = Merrifield resin. The straightforward synthetic route, clean and safe reaction conditions, commercially cheap starting materials, easy recovery and reusability of the catalyst for several cycles, etc. are the compelling features of the catalyst.

Liang and co-workers studied soluble polymer supports for homogenous continuous flow reactions [31]. The transition metal entrapped polyisobutylene, and poly(4-dodecylstyrene) catalyzes the cyclopropanation, ring-closing metathesis, and nucleophilic catalysis. Reactions were carried out in the heptane phase, saturated with acetonitrile (Scheme 19a). The catalyst was finally separated via gravity-based liquid/liquid separation. Leaching helps the recovery of the catalyst after the reaction. The desired products were obtained in moderate to good yields, and the reaction outputs were compatible with the prior results obtained with the same phase insoluble catalyst. The utility of heptane soluble polymer-supported catalyst was first studied in a batch reaction of rhodium-catalyzed cyclopropanation. Heptane soluble PIB (Polyisobutylene)-bound Rh (II) carboxylate catalyzes cyclopropanation (Scheme 20) was reported. A commercially available PIB-bound iodide was used as the starting material. In the same way, a ruthenium-based PIB-bound Hoveyda-Grubbs catalyst affected in the ring-closing metathesis of diene substrates

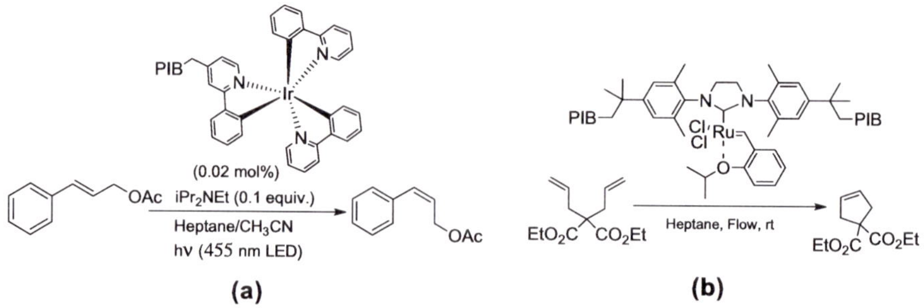

Scheme 19 **a**) Photo *E/Z* isomerization using PIB-Ir complex, **b**) reaction catalyzed by PIB-bound Hoveyda-Grubb's catalyst

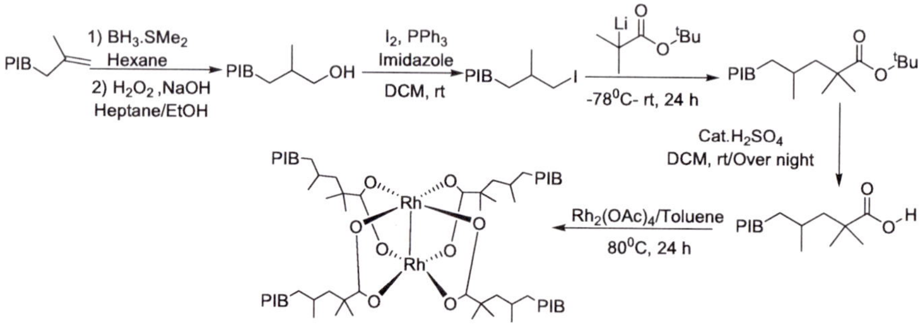

Scheme 20 Synthesis of PIB-Rh (II) Carboxylate

(Scheme 19b). The constant recovery and reuse of the catalyst were achieved because of the phase selectivity of PIB.

An efficient, recyclable catalyst based on ruthenium complexes immobilized on hyper cross-linked polymers for organic transformations was reported by Jia et al. (Scheme 21) [32]. The synthesized catalyst was characterized by employing physicochemical methods. It was observed to possess high thermal stability, chemical stability, and low synthetic cost. Organic synthesis of 2,4-diaryl substituted pyridine

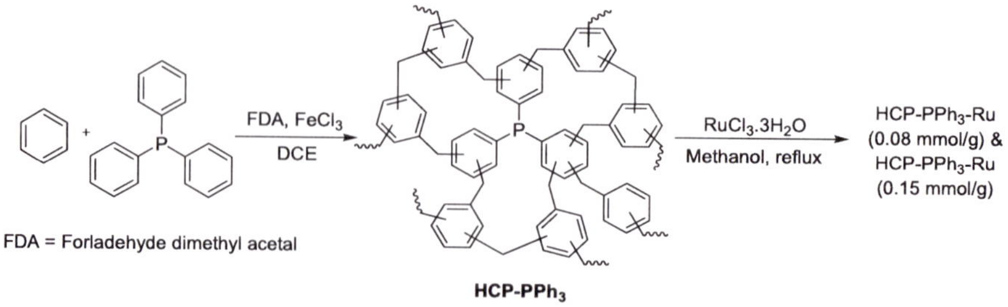

Scheme 21 Preparation of $HCP\text{-}PPh_3\text{-}Ru$ catalyst

was carried out from acetophenones, NH_4OH, and DMF using ruthenium catalysts, and it also catalyzed the formation of dihydrofurans via cycloaddition reaction of diazodicarbonyl compounds with olefins. After the reaction, the HCP-PPh_3-Ru catalyst was recovered and reused without significant loss in reactivity. The study indicates that the PPh_3-Ru catalyst immobilized on external cross-linked HCP materials are quite robust for the organic transformations.

A highly recoverable polymer-supported ionic salen-Pd complex (PS-Pd-salen) was developed by Balinge et al. for the Suzuki- Miyaura cross-coupling reactions (Scheme 22) [33]. The catalytic activity was checked on the coupling of phenylboronic acid with different aryl halides using water as a solvent, under 20% catalyst loading. The PS-Pd-salen complex was found to be thermally stable, and it showed excellent reusability for five consecutive runs.

Very recently, Yang et al. developed a heterogeneous system in which chloride quinones supported on an organic resin was used for the synthesis of trisubstituted pyridines (Scheme 23) [34]. Recoverable Merrifield resins were used as organic support. The synthesized catalyst was characterized using FTIR, XPS, EDX techniques. The chemoselective synthesis of 2, 4, 6-trisubstituted pyridines offered good yields under mild conditions. The proposed mechanism was well established by the successful capture of the intermediate during the reaction pathway. The catalyst was

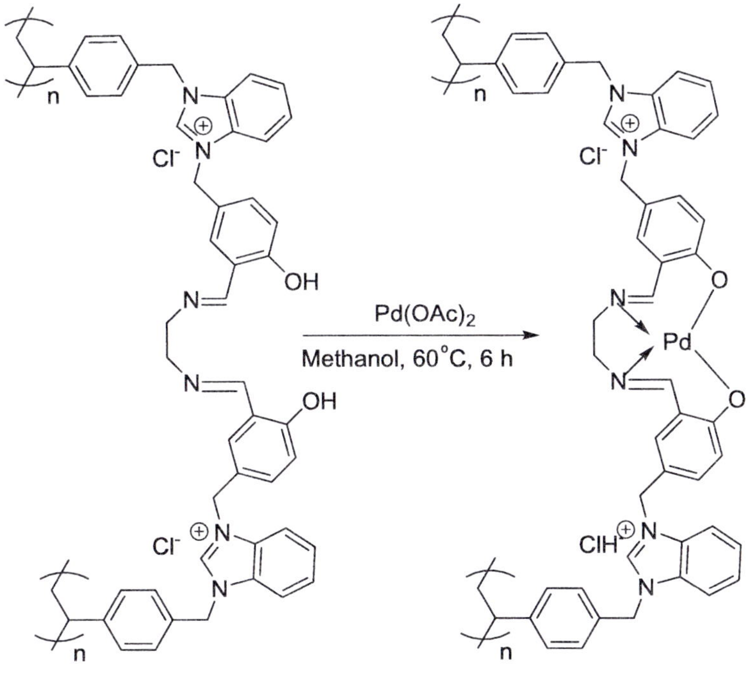

Scheme 22 Synthesis of polymer-supported Pd-Salen complexes

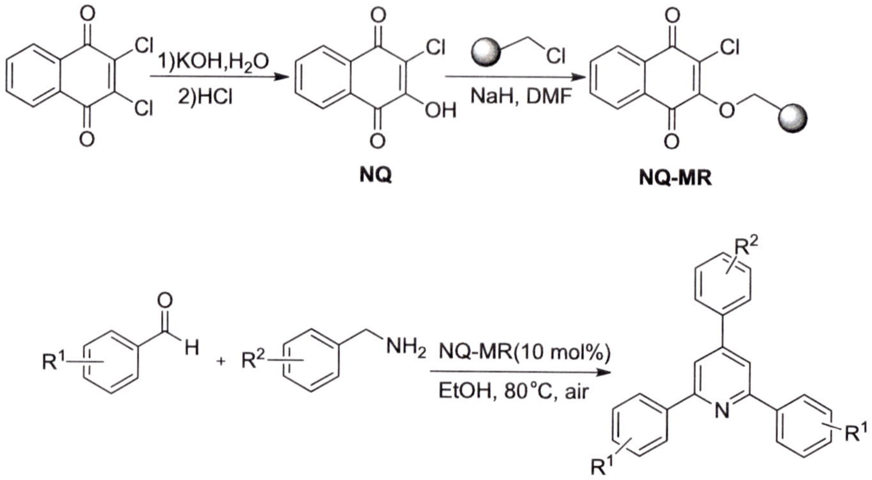

Scheme 23 Supported NQ-MR catalyzed pyridine synthesis

recovered by simple centrifugation techniques, washed with water, dried, and used further for the synthesis.

Bartacek reported a heterogeneous catalyst for enantioselective Sharpless epoxidation in 2019 (Scheme 24) [35]. The metal present in the catalyst was Ti (IV). The titanium complexes were supported on cross-linked polystyrene beads with ethyl-(4-vinyl benzyl)-*L*-tartrate/ethyl-(2*R*,3*R*)-2,3-dihydroxy-4,5-oxo-5-(4-vinyl phenyl) pentanoate using suspension polymerization. The catalytic activity was checked on the cinnamyl alcohol epoxidation, and high conversion and enantioselectivity were observed for the reaction.

Pisk and co-workers synthesized adipic acid via oxidation reaction using Molybdenum/Tungsten-based Keggin polyoxometallates (Scheme 25) [36]. The POM was surrounded by organic cations or functionalized Merrifield resins. Aqueous H_2O_2 (30%) was used as the oxidizing agent, and the starting materials used were cyclohexene, epoxy cyclohexane/cyclohexane diol. The catalyst loading was very low, about 0.001%- 0.007% for the POM, and adipic acid was obtained in excellent yields. The resin was finally isolated at the end of the catalytic cycle, washed with diethyl ether, dried, and used further.

Very recently, Chen et al. demonstrated the fabrication of thermoresponsive polymer-based TEMPO nanoreactors [37]. The application of the catalyst lies in the selective oxidation of alcohols in aqueous media. Catalyst synthesis was carried out initially using an amphiphilic co-polymer NHS-P(MMA_{25}-b-$OEGMA_{75}$) consisting of PMMA and poly(oligo ethylene glycol) methyl ether methacrylate with terminal NHS (Scheme 26). Then, TEMPO was introduced to the terminal of the hydrophobic part. The self-assembly of the amphiphilic co-polymer in water is resulted in the

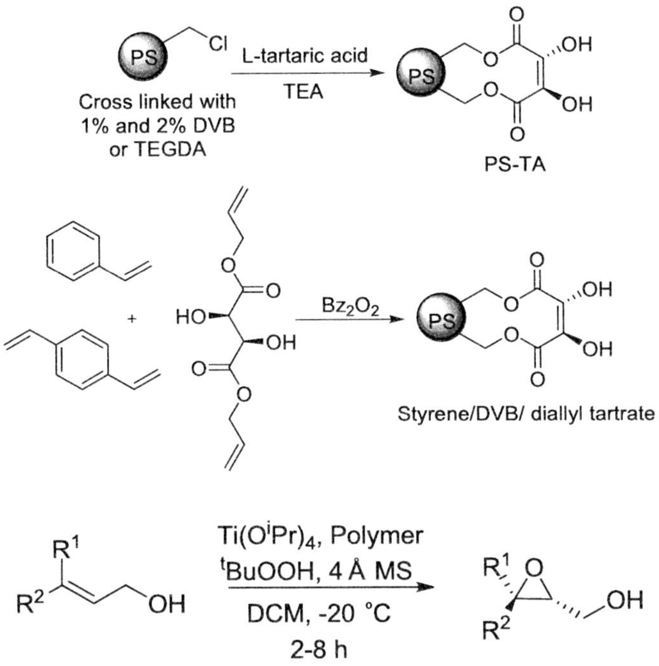

Scheme 24 Synthesis and application of *L*-(+)-tartaric acid esters immobilized on styrene on Sharpless epoxidation

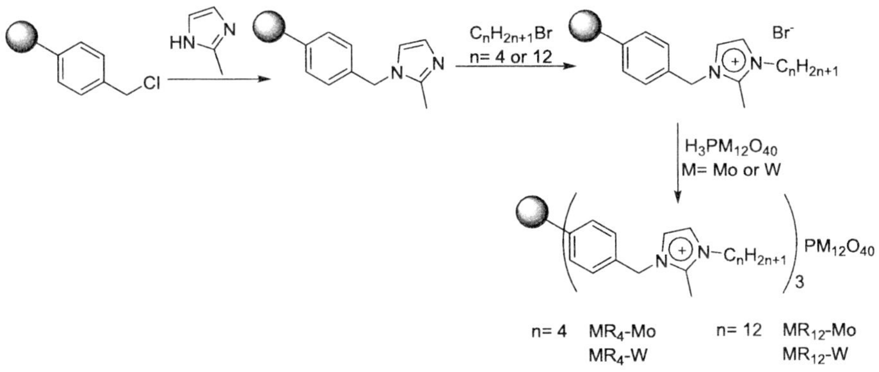

Scheme 25 Merrifield supported heterogeneous catalysts- synthetic route

core-shell type of polymeric nanoreactors, which catalyzed the selective oxidation of the alcohol and yielded excellent amounts of aldehydes and ketones. The thermoresponsive character of the catalyst helps for its easy recovery by precipitation.

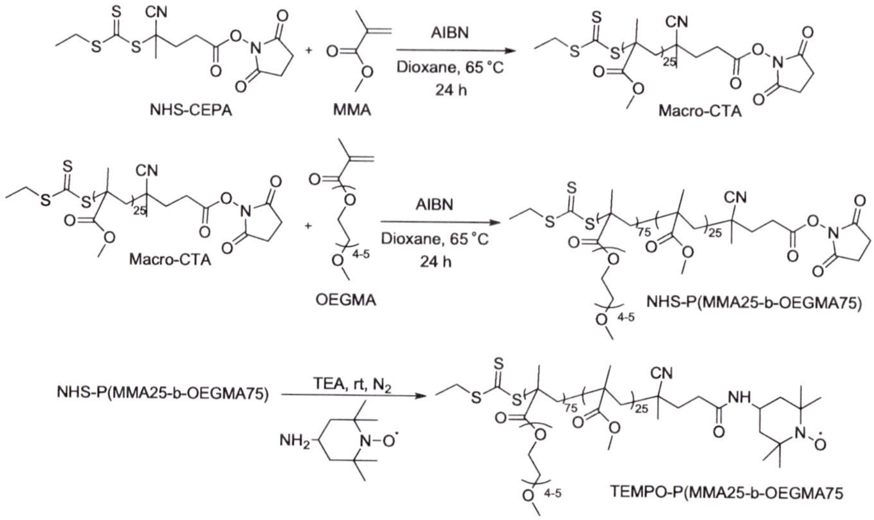

Scheme 26 Synthesis of TEMPO-(PMMA$_{25}$-b-OEGMA$_{75}$)

2.5 *Silica-Supported Organocatalysts*

An efficient synthesis of silica-supported gold (I) complexes were developed by Shu et al. [38]. The catalyst was used for the regio- and enantio-selective lactonization reactions. The preparation of the catalyst was carried out by stirring silica materials in a CH_2Cl_2 solution of Ph_3PAuBF_4 at room temperature for 4 h (Scheme 27). Catalytic activity was observed on the nucleophilic addition reaction of allenes. High enantioselectivities for the lactonization were obtained using heterogeneous catalysts. The catalysts were quickly recovered and reused up to 11 consecutive cycles without loss in the enantioselectivity.

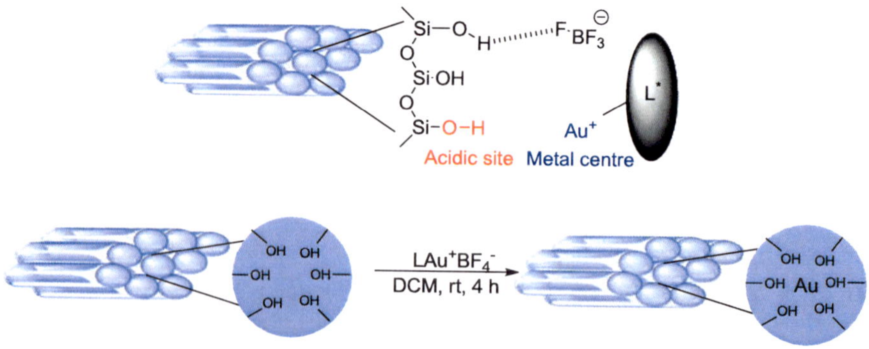

Scheme 27 One-step synthesis of heterogenous gold@SiO_2 catalysts

A novel and efficient boronic acid substituted silica-supported Pt was synthesized by Li et al. in 2019 [39]. The boronic acid-functionalized silica gel complex (SiO_2- BA) was prepared by grafting silica gel and 3-aminophenyl boronic acid using 3-chloropropyltrimethosilane, γ-glycidoxypropyltrimethoxysilane, and 3-aminopropyl tri methoxysilane as complexing agents. Further, Pt nanoparticles were dispersed on the complex using a solution of $H_2PtCl_6.6H_2O$ in ethanol. The catalytic activity of the synthesized compounds was investigated by the hydrosilylation of olefins with methyl dichlorosilane (Scheme 28). The substrate scope was explored for a wide range of terminal aliphatic olefins and methyl dichlorosilane. The recycled catalyst was found to be useful up to 10 catalytic cycles without any noticeable loss in catalytic activity.

Mesoporous recoverable silica-based nanomaterials for the synthesis of substituted pyridodipyrimidines were reported by Kohzadian and co-workers [40]. The catalyst preparation was straightforward and cost-effective (Scheme 29). The nanocatalyst [SiO_2-R-NMe_2SO_3H]Cl was detached from the reaction medium via centrifugation and decantation. Further, it was washed with methanol and reused. Synthetic protocol carried out for the preparation of pyrido[2,3-*d*: 2,3-*d*']dipyrimidine was a one-pot multicomponent reaction of aryl aldehydes, 2-thiobarbituric acid, and NH_4OAc in the presence of the synthesized catalyst. The reaction time, yield, as well as the condition were better compared to the reported methods.

Post-transition metal triflates as solid catalysts are being used in heterogeneous green catalysis. An indium (III) triflate [SISAln(OTf)$_2$] supported on sulfoacetate modified silica for the synthesis of tetrasubstituted imidazoles was established by Vaid et al. (Scheme 30) [41]. 1,2,4,5-Tetrasubstituted imidazoles were obtained by the

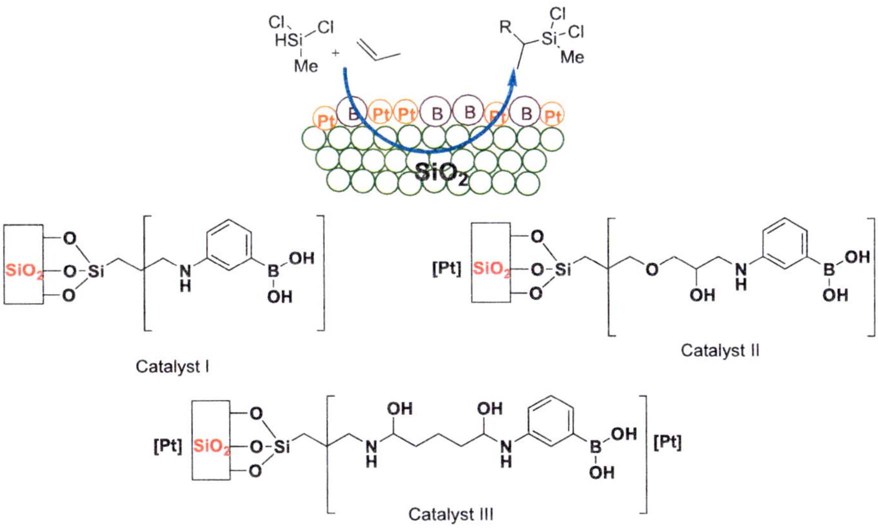

Scheme 28 Boronic acid substituted Pt@SiO_2 catalyzed reaction

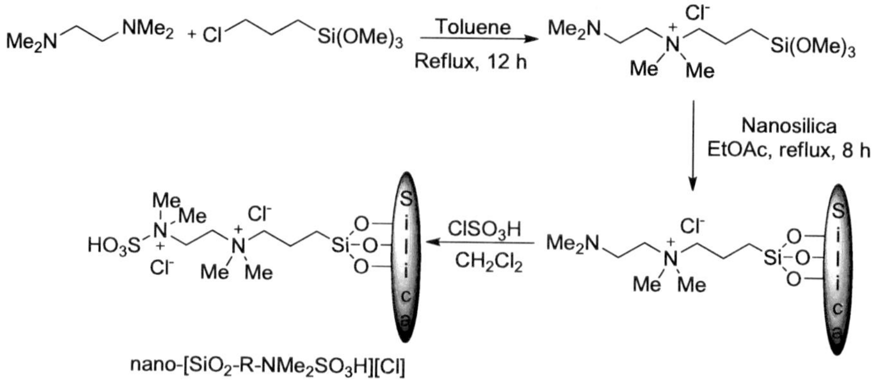

Scheme 29 Preparation of nano-[SiO_2-R-NMe_2SO_3H][Cl]

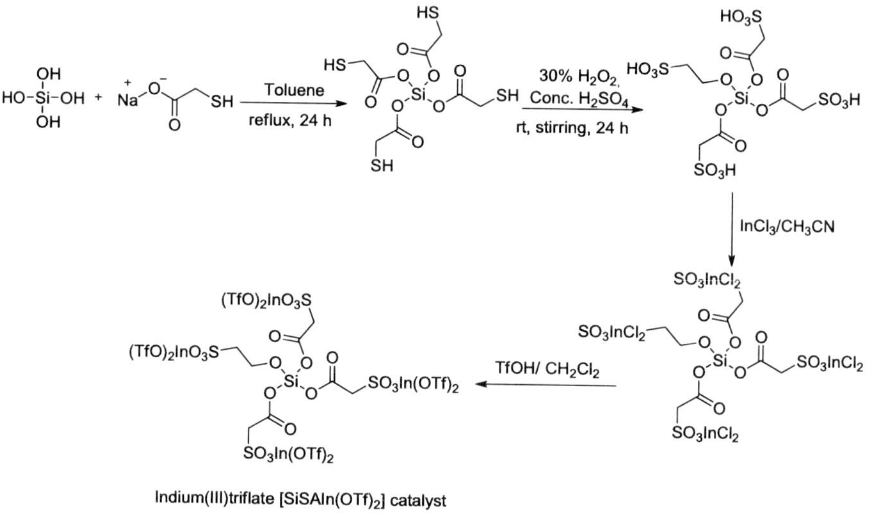

Scheme 30 Multi-step synthesis of Sulfoacetate modified silica-supported indium (III) triflate [SiSAIn$(OTf)_2$]

reaction of aromatic aldehydes, benzyl, aromatic amines, and ammonium acetate. To investigate the stability and recyclability, the catalyst was separated from the reaction mixture by simple filtration. The catalytic activity was found to be stable up to five consecutive runs.

2.6 Thermomorphic Catalysts

Thermomorphic catalysts are separated from the reaction mixture by varying the temperature of the system. This helps in the phase separation based on their temperature-dependent solubility behavior. The separation strategies include (i) cooling (for the catalysts which are soluble under elevated temperatures)—catalysts separate by lowering the temperature, (ii) temperature-dependent miscibility of solvent mixtures.

A novel thermomorphic system for the electrolytic Diels-Alder reaction was reported by Imada et al. [42]. They replaced the polar $LiClO_4/CH_2NO_2$ solution with a less polar LiTFSI/1-nitropropane solution (Scheme 31). During the electrochemical response, methylcyclohexane forms a monophasic condition with LiTFSI/1-nitropropane at rt. After the completion of the reaction, at −50 °C, it turns into a biphasic system. Cycloadducts move into the upper methylcyclohexane phase, and in the lower phase, $LiTFSI/PrNO_2$ gets separated, which can be used for further cycles.

Heijazifar et al. proposed the use of thermomorphic ionic liquid microemulsion applied for the Ru-catalyzed asymmetric transfer hydrogenation of ketones (Scheme 32) [43]. $[C_{12}mim][(^iC_8)_2PO_2]$ and $[C_{12}dmim][(^iC_8)_2PO_2]$ were found to be the best thermomorphic ionic liquids for the reaction (Fig. 2). Use of water-soluble ligands improve the immobilization of the catalyst, and it minimizes the chances of leaching into the organic layer during extraction. The catalytic system was found to be more effective for aromatic ketones than the aliphatic ones in terms of conversion

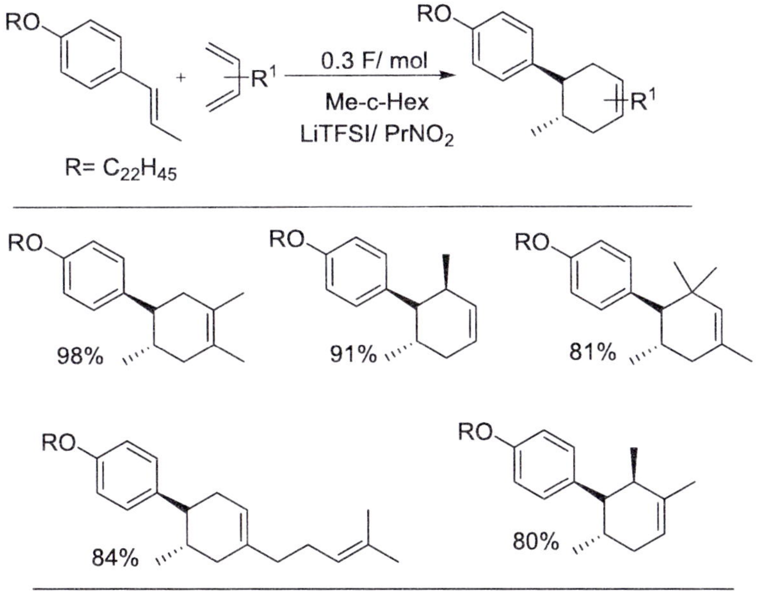

Scheme 31 Electrocatalytic Diels-Alder reaction under thermomorphic conditions

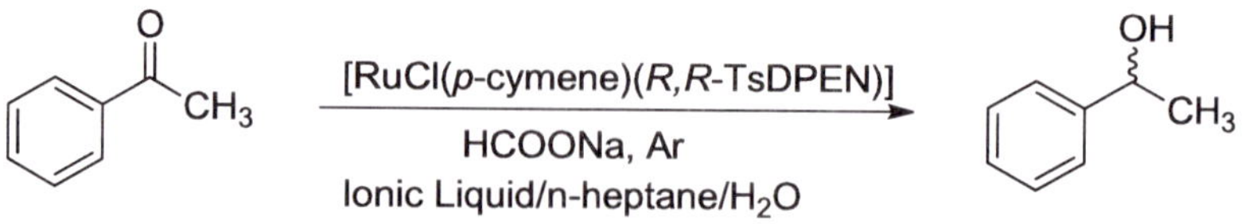

Scheme 32 Asymmetric *trans* hydrogenation of ionic liquid microemulsion

Fig. 2 Surface active ionic liquids

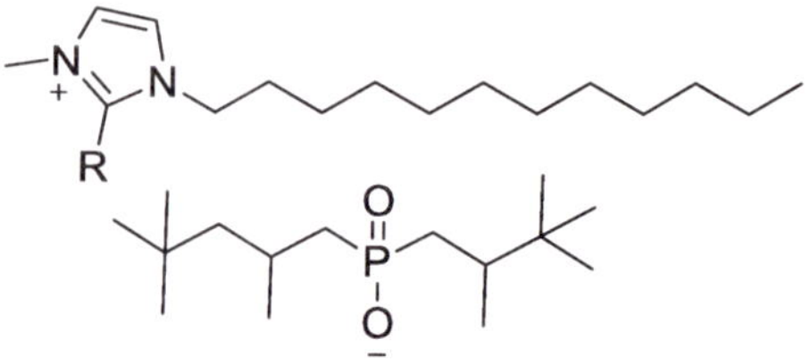

and enantioselectivity. Moderately good yields were obtained during the recyclability study, and the reuse of the catalyst was effective.

Homogenously catalyzed reductive amination in a thermomorphic system was first reported very recently by Künnemann and co-workers (Scheme 33) [44]. Thermomorphic multiphase system was established for the recovery and reuse of the catalyst (Scheme 34). Commercially available PPh_3 ligand and Rh(acac)(cod) catalyst precursor were used for the reaction. The researchers have scaled up the reaction

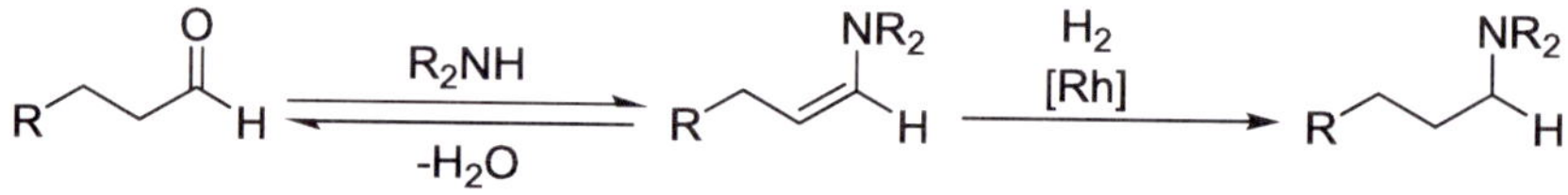

Scheme 33 Thermomorphic Rh complex catalyzed reductive amination

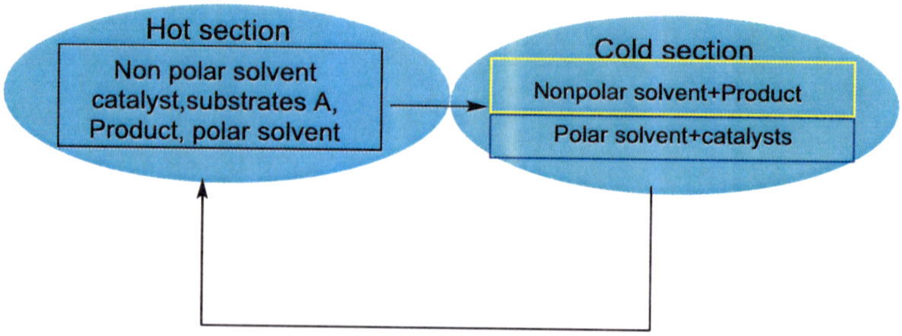

Scheme 34 Representation of thermo multiphase system

into a continuous system that recycles the catalyst inflow over 70 h. The overall reaction possessed above 90% yield and meager percentage of catalyst leaching, 1%/h.

3 Conclusion

The development of a recoverable catalyst is a rapidly growing area in catalysis. Effective recycling of the catalyst from the reaction medium can be performed using various methodologies. Depending on the nature of the catalyst used, different recovery methods are applied. Significant progress has been observed in the preparation of solid-supported catalysts—both dendritic and silica-supported, magnetically supported, and also organic resin supported catalysts. In most of the methods, the effective recovery of the catalysts is done without altering its efficiency. Moderate to high reaction rates can be sustained over multiple consecutive cycles.

Magnetically recoverable metal nanoparticles enhance the catalytic activity because of its increased surface area. The advantages of metallodendritic catalysts extend to both homogenous and heterogeneous catalysis, and these can be separated employing simple techniques such as precipitation, filtration, and centrifugation. These modified recoverable catalysts could benefit from the additional active supports (dendrimers, organic resins, silica, polymers, etc.), which helps in the catalyst formation and reaction. Ease of synthesis, the serenity of the reaction, recyclability, and cost-effectiveness make these methods more sustainable toward green chemistry approach.

Acknowledgements B.P.B. acknowledges the Department of Science and Technology (DST), Government of India, for an INSPIRE Faculty Award (IFA-13 CH-98). B.P.B. and R.P.B. thank NITK, Surathkal, for providing laboratory facility and research fellowship for R.P.B.

References

1. a) Gladysz JA (2002) Introduction: recoverable catalysts and reagents—perspective and prospective. Chem Rev 102:3215–3216; b) Kirchhoff MM (2003) Promoting green engineering through green chemistry. Environ Sci Technol 37:5349–5353
2. a) Glaser JA (2012) Green chemistry with nanocatalysts. Clean Techn Environ Policy 14:513–520; b) Sheldon RA (2016) Green chemistry and resource efficiency: towards a green economy. Green Chem 18:3180–3183. www.rsc.org/greenchem. 23 Feb 2020
3. Schmidt F (2004) The importance of catalysis in the chemical and non-chemical industries. Springer, Berlin, Heidelberg
4. Sheldon RA, Arends IWCE, Hanefeld U (2007) Green Chemistry and Catalysis Wiley, Germany
5. Liu F, Hung K, Zheng A et al (2018) Hydrophobic solid acids and their catalytic applications in green and sustainable chemistry. ACS Catal 8:372–391

6. Cole-Hamilton DJ (2003) Homogeneous catalysis—new approaches to catalyst separation, recovery, and recycling. Science 299:1702–1706
7. a) Sun PP, Lee MS (2011) Separation of Pt from hydrochloric acid leaching solution of spent catalysts by solvent extraction and ion exchange. Hydrometallurgy 110: 91–98; b) Lee JY, Raju B, Kumar BN et al (2010) Solvent extraction separation and recovery of palladium and platinum from chloride leach liquors of spent automobile catalyst. Separ Purif Tech 73:213–218; c) Gürsel IV, Noël T, Wang Q et al (2015) Separation/recycling methods for homogeneous transition metal catalysts in continuous flow. Green Chem 17: 2012–2026
8. a) Nair D, Luthra SS, Scarpello JT et al (2002) Homogeneous catalyst separation and re-use through nanofiltration of organic solvents. Desalination 147:301–306; b) Dijkstra HP, Klink GPMV, Koten GV (2002) The use of ultra- and nanofiltration techniques in homogeneous catalyst recycling. Acc Chem Res 35:798–810
9. Dloumaev VK, Bullock RM (2003) A recyclable catalyst that precipitates at the end of the reaction. Nature 424:530–532
10. a) Polshettiwar V, Luque R, Fihri A et al (2011) Magnetically recoverable nanocatalysts. Chem Rev 111:3036–3075; b) Shylesh S, Schünemann V, Thiel WR (2010) Magnetically separable nanocatalysts: bridges between homogeneous and heterogeneous catalysis. Angew Chem Int Ed 49:3428–59; c) Shifrina ZB, Bronstein LM (2018) Magnetically recoverable catalysts: beyond magnetic separation. Front Chem 6:1–6; d) Rossi LM, Costa NJS, Silva FP et al (2014) Magnetic nanomaterials in catalysis: advanced catalysts for magnetic separation and beyond. Green Chem 16:2906–2933
11. Vono LLR, Damasceno CC, Matos JR et al (2018) Separation technology meets green chemistry: development of magnetically recoverable catalyst supports containing silica, ceria, and titania. Pure Appl Chem 90:133–141
12. a) Munnik P, de Jongh PE, de Jong KP (2015) Recent developments in the synthesis of supported catalysts. Chem Rev 115:6687–6718; b) Wang D, Astruc D (2013) Dendritic catalysis-basic concepts and recent trends. Coord Chem Rev 257:2317–2334
13. Van Heerbeek R, Kamer PCJ, Van Leeuwen PWNM et al (2002) Dendrimers as support for recoverable catalysts and reagents. Chem Rev 102:3717–3756
14. Reetz MT, Lohmer G, Schwickardi R (1997) Synthesis and catalytic activity of dendritic diphosphane metal complexes. Angew Chem Int Ed 36:1526–1529
15. Méry D, Astruc D (2006) Dendritic catalysis: major concepts and recent progress. Coord Chem Rev 250:1965–1979
16. Heuzé K, Mery D, Gauss D et al (2003) Copper-free, recoverable dendritic pd catalysts for the sonogashira reaction. Tetrahedron 22:74–75
17. Nlate S, Astruc D, Neumann R (2004) Synthesis catalytic activity in oxidation reactions, and recyclability of stable polyoxometalate-centred dendrimers. Adv Synth Catal 346:1445–1448
18. Keller M, Perrier A, Linhardt R et al (2013) Dendrimers or nanoparticles as supports for the design of efficient and recoverable organocatalysts? Adv Synth Catal 355:1748–1754
19. a) Dobbs AP, Kimberley MR (2002) Fluorous phase chemistry: a new industrial technology. J Fluor Chem 118:3–17; Pozzi G, Shepperson I (2003) Fluorous chiral ligands for novel catalytic systems. b) Coord Chem Rev 242:115–124; c) Hejazifar M, Pálvölgyi ÁM, Bitai J, Lanaridi O, Bica-Schröder K (2019) Asymmetric transfer hydrogenation in thermomorphic microemulsions based on ionic liquids. Org Process Res Dev 23(9):1841–1851
20. Chinchilla R, Mazón P, Nájera C (2004) Polystyrene-anchored cinchona ammonium salts: easily recoverable phase-transfer catalysts for the asymmetric synthesis of α-amino acids. Adv Synth Catal 346:1186–1194
21. Kryshtal GV, Zhdankina GM, Zlotin SG (2008) Recoverable phase-transfer catalysts with fluorinated anions: generation and reactions of dichlorocarbene and CCl_3 anion in the heterogeneous system $KOH(s)/CHCl_3/nBu_4NPF_6$. Eur J Org Chem 1777–1782
22. Hajipour AR, Azizi G (2014) Simultaneous immobilization of a matrix containing palladium and phase transfer catalyst on silica nanoparticles: application as a recoverable catalyst for the heck reaction in neat water. RSC Adv 4:20704–20708

23. Rull J, Jara JJ, Sebastian RM et al (2016) Recoverable dendritic phase-transfer catalysts that contain (+)-cinchonine-derived ammonium salts. Chem Cat Chem 8:2049–2056
24. Baig RBN, Varma RS (2013) Magnetically retrievable catalysts for organic synthesis. Chem Commun 49:752–770
25. Boruah PK, Borthakur P, Darabdhara G et al (2016) Sunlight assisted degradation of dye molecules and reduction of toxic Cr(VI) in aqueous medium using magnetically recoverable Fe_3O_4/reduced graphene oxide nanocomposite. RSC Adv 6:11049–11063
26. Vavasori A, Calgaro L, Quartarone G et al (2016) New magnetically recoverable palladium-based catalysts active in the alkoxycarbonylation of iodobenzene. Pure Appl Chem 88:445–455
27. Khazaei A, Khazaei M, Nasrollahzadeh M (2017) Nano-Fe_3O_4@SiO_2 supported Pd(0) as a magnetically recoverable nanocatalyst for suzuki coupling reaction in the presence of waste eggshell as low-cost natural base. Tetrahedron 73:5624–5633
28. Hong Y, Choi IA, Seo WS (2018) Au Nanoparticle@hollow mesoporous carbon with FeCo/graphitic shell nanoparticls as a magnetically recyclable yolk-shell nanocatalyst for catalytic reduction of nitroaromatics. Sci Rep 8(1)-7469
29. Angelini T, Lanari D, Maggi R et al (2012) Preparation and use of polystyryl-DABCOF2: an efficient recoverable and reusable catalyst for β-azidation ofα, β- unsaturated ketones in water. Adv Synth Catal 354:908–916
30. Boruah JJ, Das SP, Ankireddy SR et al (2013) Merrifield resin supported peroxomolybdenum(vi) compounds: recoverable heterogeneous catalysts for the efficient, selective and mild oxidation of organic sulfides with H_2O_2. Green Chem 15:2944–2959
31. Liang Y, Watson C, Malinski T et al (2016) Soluble polymer supports for homogeneous catalysis in flow reactions. Pure Appl Chem 88:953–960
32. Jia Z, Wang K, Tan B et al (2017) Ruthenium complexes immobilized on functionalized knitted hypercrosslinked polymers as efficient and recyclable catalysts for organic transformations. Adv Synth Catal 359:78–88
33. Balinge KR, Khiratkar AG, Bhagat PR (2018) a highly recoverable polymer-supported ionic salen-palladium complex as a catalyst for the suzuki-miyaura cross coupling in neat water. J Organomet Chem 854:131–139
34. Yang Q, Zhang Y, Zheng W et al (2019) Merrifield resin-supported quinone as an efficient biomimetic catalyst for metal-free, base-free, chemoselective synthesis of 2,4,6-trisubstituted pyridines. Green Chem 21:5683–5690
35. Bartáček J, Drabina P, Váňa J et al (2019) Recoverable polystyrene-supported catalysts for sharpless allylic alcohols epoxidations. React Funct Polym 137:123–132
36. Pisk J, Agustin D, Poli R (2019) Organic salts and merrifield resin supported $[PM_{12}O_{40}]_3$ − (M = Mo or W) as catalysts for adipic acid synthesis. Molecules 24:783
37. Chen T, Xu Z, Zhou L et al (2019) Highly efficient polymer-based nanoreactors for selective oxidation of alcohols in water. Mol Catal 474:110422
38. Shu XZ, Nguyen SC, He Y et al (2015) Silica-supported cationic gold(I) complexes as heterogeneous catalysts for regio- and enantioselective lactonization reactions. J Am Chem Soc 137:7083–7086
39. Li L, Li Y, Wang H et al (2019) Preparation and evaluation of a novel and high efficient boronic acid-substituted silica supported Pt catalyst colloids. Surf A Physicochem Eng Asp 570:322–330
40. Kohzadian A, Zare A (2019) Effective and rapid synthesis of pyrido[2,3-d:6,5-d′]dipyrimidines catalyzed by a mesoporous recoverable silica-based nanomaterial. Silicon 12:1407–1415
41. Vaid R, Gupta M, Kour G et al (2019) Sulfoacetate modified silica supported indium(III) triflate [SiSAIn$(OTf)_2$]: a novel solid acid nano-catalyst and investigation of its catalytic potential for one-pot synthesis of 1,2,4,5-tetrasubstituted imidazole derivatives. Chem Sel 4:9179–9184
42. Imada Y, Shida N, Okada Y (2019) A novel thermomorphic system for electrocatalytic diels-alder reactions. Chin J Chem 37:557–560
43. Hejazifar M, Pálvölgyi AM, Bitai J et al (2019) Asymmetric transfer hydrogenation in thermomorphic microemulsions based on ionic liquids. Org Process Res Dev 23:1841–1851

44. Künnemann KU, Bianga J, Scheel R et al (2020) Process development for the rhodium-catalyzed reductive amination in a thermomorphic multiphase system. Org Process Res Dev 24:41–49

Chapter 17
Supercritical Fluids as Green Solvents

Ani Deepthi and Vidya Sathi

1 Introduction

Development of alternate solvents to combat and replace conventional organic solvents has been a major focus ever since the implementation of green chemistry principles put forward by Paul Anastas and John Warner in 1998. [1] These alternate solvents mainly include water, [2] ionic liquids, [3] renewable solvents from biomass, [4] fluorous solvents [5], and supercritical fluids. [6] Among these, supercritical fluids have received much attention as evidenced by the large number of reviews, books, and book chapters on the topic. Theoretically any substance above its critical point can be called as a supercritical fluid; critical point referring to the point at the end of liquid-vapor equilibrium line. Above this point there is no individual existence of gas or liquid—it is a fluid which exists that has a liquid-like density and a gas-like viscosity. This fluid can diffuse through solids like a gas and can dissolve materials like a liquid. Conventional solvents require large variation in pressure to cause alterations in density, whereas under the critical conditions the fluid density becomes very sensitive to even small changes in pressure and temperature. As the solvent power is approximately proportionate to its density, the solvent power can be fine-tuned by altering the temperature and pressure; therefore, supercritical fluids are considered as tunable solvents.

The critical pressure and temperature of various substances differ significantly. Generally more polar compounds have larger value of critical temperature and pressure. Table 1 shows the critical temperature and pressure for various substances.

A. Deepthi (✉) · V. Sathi
Department of Chemistry, University of Kerala, Kariavattom Campus, Thiruvananthapuram, Kerala 695581, India
e-mail: anideepthi@gmail.com

V. Sathi
e-mail: 92vidyanathan@gmail.com

G. Anilkumar and S. Saranya (eds.), *Green Organic Reactions*, Materials Horizons: From Nature to Nanomaterials, https://doi.org/10.1007/978-981-33-6897-2_17

Table 1 Critical temperature and pressure of various substances

Substance	Temperature (°C)	Pressure (bar)
Ammonia	132.5	112.8
Benzene	289.0	48.9
Chlorotrifluoromethane	28.9	38.7
CO_2	32.1	73.8
Cyclohexane	280.3	40.7
Ethane	32.2	48.8
Ethylene	9.3	50.4
Isopropanol	235.2	47.6
n-Pentane	196.6	33.7
Propane	96.7	42.5
Propylene	91.9	46.2
Toluene	318.6	41.1
Trichlorofluoromethane	198.1	44.1
Trifluoromethane	26.2	48.6
Water	374.2	220.5

Of the various supercritical fluids studied supercritical carbon dioxide ($scCO_2$) and water (scH_2O) are found to be most promising. The inert, non-toxic and non-flammable nature of $scCO_2$ along with its moderate critical constants makes it an economic and viable alternative to conventional solvents for extraction and organic synthesis. In contrast scH_2O, due to its high temperature requirement (374.2 °C), is unsuitable for organic compound synthesis but is used for synthesis of inorganic solids and in waste treatment applications. The growth of supercritical fluid technology has led to its use in power generation, biodiesel production, dry cleaning, refrigeration, impregnation, and dyeing. The main areas where supercritical fluids find application as green solvents can be categorized as follows:

1. Supercritical fluid extraction (SFE)
2. Supercritical fluid chromatography (SFC)
3. Supercritical fluids as reaction media
4. Supercritical fluid deposition (SFD)

A brief overview of these applications with emphasis on the developments which has happened from 2018 to 2019 is given in the following sections.

1.1 *Supercritical Fluid Extraction (SFE)*

The low viscosity and high diffusivity associated with supercritical fluids have long made them ideal solvents for extraction of various substances. Of the various supercritical fluids, supercritical carbon dioxide is most preferred due to its moderate critical temperature which is suitable for temperature sensitive compounds as well [7]. SFE technology is being widely used for extraction of caffeine from tea and coffee (decaffeination of coffee is the first industrial application of $scCO_2$ extraction), flavors used in brewing, flavors from spices and for high precision cleaning of machine parts. The extraction of food and natural products using SFE technology mainly consists of two steps: (i) extraction of supercritical fluid soluble components from the crude and (ii) separation of the components from supercritical fluids either by varying the pressure or via adsorption by addition of an external agent. Worldwide there are more than 150 facilities which use the SFE technology for large scale extractions and purifications. Of late SFE is being mainly used for preparation of high value products such as flavors and fragrances, food supplements and nutraceuticals and for the purification of materials that are used for the production of medical devices. Apart from the latter, $scCO_2$ is used for the removal of organic contaminants from machined parts, electronic assemblies, optical and LASER components and computer parts. After the extraction process, CO_2 is depressurized and the residue is trapped on liquid or solid traps for further determination. SFE is often done using a modifier/co-solvent along with the supercritical fluid which has the role to enhance the solubility of polar components. Ethanol is the most common co-solvent and is added either by pre-mixing with the supercritical fluid or by impregnation in the matrix followed by extraction. Coupling of SFE with supercritical fluid chromatography (SFC) has gained great importance as well [8]. Several details of extraction of lipophilic molecules including natural products, bioactive compounds, and lipids are available in literature [6, 9] and some specific examples are presented here.

A research group from Brazil reported the consecutive use of two green technologies—SFE for non-polar fractions and pressurized liquid extraction (PLE) for polar components from araçá fruits, a native of Brazilian Atlantic forest [10]. Antioxidant and antibacterial activities and total phenolic content of the extracts were determined using the combined technique. SFE using neat CO_2 and CO_2-ethanol was used for the recovery of phenolic antioxidants from the seeds of papaya and it was found that use of the co-solvent ethanol was found to be useful for the recovery of phenolic acids from the seeds (considered as the agriwaste of papaya products) [11]. SFE technology has also been used for e-waste treatment and this topic has been reviewed [12]. The ability of supercritical fluids to extract heavy metals and to remove hazardous organic materials (used as flame retardants or decolorants) from e-waste through various steps like debromination, dechlorination, liquefaction, and degradation renders them as ideal solvents for the purpose. Extraction of seed oils is another extensively studied area in supercritical CO_2 extraction probably due to the greater solubility of oils in high pressure CO_2. Supercritical extraction of high value seed oils for production of nutraceutical, cosmeceutical, dietary supplements, and animal health products has been widely explored [13–15]. Passion fruit seed oil, a rich source of value-added compounds such as sterols, phenolic compounds, tocols, and carotenes have been extracted and their phase equilibria have been studied using $scCO_2$ [16]. Extraction of oregano oil obtained from *Oregano Vulgare L.* showed that α-linolenic, palmitic, oleic, linoleic acids contribute to almost 75% of the total fatty acids present in the oil and that the extracts obtained showed good antimicrobial and antioxidant activities [17]. Bioactive residues with antioxidant activities were extracted from soybean oil residues using $scCO_2$ after optimizing the temperature and pressure conditions. [18] Extraction of bioactive compounds from *Hibiscus Sabdariffa* [19] and *Cannabis Sativa L.* [20] are also reported.

As mentioned earlier many of the $scCO_2$-based extractions involve the use of an organic co-solvent which decreases the green character of the technique. Therefore, coupling of SFE with supercritical fluid chromatography (SFC) has become quite popular [21, 22]. The main advantages associated with this sort of coupling include minimal sample requirement, decrease in analysis time, and avoidance of co-solvent. The next section details about SFC and SFE-SFC techniques.

1.2 Supercritical Fluid Chromatography (SFC)

Supercritical fluid chromatography has emerged as a powerful analytical tool due to its inherent flexibility and high throughput. Here carbon dioxide is used as the mobile phase and other polar solvents such as methanol or isopropanol are used as co-solvents or modifiers [23]. Use of eco-friendly co-solvents along with shorter analysis time and lesser waste generation have made SFC a green and popular technique for a wide variety of separations. Details regarding certain applications of SFC will be presented here rather than the instrumentation for which the readers may refer to other sources [6]. SFC is most popularly used in combination with SFE but other hyphenated techniques like SFE-MS, [24] using APCI or ESI modes of mass analysis, have also gained popularity in recent times. In addition, two-dimensional chromatographic analysis using SFC has gained great attention of late. 2D chromatography refers to integration of two different separation modes on the basis of compatibility and orthogonality [25]. Certain examples of some separations done using 1D and 2D SFC are given below.

1.2.1 Separation of Complex Lipid Mixtures

Complex lipid mixtures contain a large range of fatty acids and triglycerides which cannot be separated using a single column. Hirata et al. used 2D SFC x SFC chromatographic technique for such separations [26, 27]. The separations can be done either by using the same packing material in the two columns (ODS—octadecyl silica—is one material that is widely used) or by using ODS in one column and silica gel in another. The effluent from the first column containing the solute can be trapped and is subsequently fed into the second column after depressurizing; both columns being maintained at different temperatures. The ODS column separates the fatty acid methyl esters according to their chain length while the silica gel column separates the components according to the polarity.

1.2.2 Profiling of Gangliosides (GLS)

Gangliosides are a group of molecules composed of glycosphingolipids and one or more sialic acids linked on a sugar chain. These are largely present in the plasma membrane of nervous systems and modulate cell signal transmission mechanisms. Gangliosides are continuously synthesized and degraded in cells, catalyzed by a set of lysosomal enzymes [28]. Mutations in gene coding for these enzymes lead to partial breakdown of gangliosides which results in various diseases such as Alzheimer's, Parkinson's, and cancer. The diversity and complexity of gangliosides result in difficulties in their separation and analysis which are very important in order to study them. Recent studies have used an off-line two-dimensional (2D-SFC) x reversed phase liquid chromatography (RPLC)/QToF system for fast screening and identification of gangliosides. A total of 153 gangliosides were separated and 79 were identified using this method [29].

1.2.3 Profiling of Natural Products

Supercritical fluid chromatography is a well-established analytical technique for natural product separations. Various natural products such as lipids, carbohydrates, polyphenols, alkaloids, flavonoids, fatty acids, coumarins, terpenes, and triacylglycerols have been successfully separated using SFC [30–32] Very recently, ultrahigh performance SFC (UHPSFC) has been combined with RPLC, HRMS, and MN (molecular networking) for the profiling of *Venenum Bufonis* (*VB*), a product of secretion of the Asiatic toad, which is widely used as a traditional chinese medicine. The various types of bufadienolides present in *VB* and two new compounds were isolated and characterized using this method (Fig. 1) [33]. SFC coupled with quadrupole-time-of-flight mass spectrometry (SFC-Q-ToF/MS) was used for the chiral separation and residue determination for diniconazole enantiomers (Fig. 2), a class of fungicides used in plants and fruit bearing trees [34]. The residues of tea and fruit samples were extracted using acetonitrile/water and were purified by solid phase extraction column and were analyzed by SFC-Q-ToF/MS. Optimal separation of the *R*- and *S*- enantiomers were obtained using a chiral CCA column by employing carbon dioxide/isopropyl alcohol as the mobile phase and with a flow rate of 2 mL/min.

1.2.4 Chiral Separations

Just as in the example given above, SFC has been used for bioanalytical enantioseparations over the years and many reviews have appeared in this topic [35, 36]. The same technique was used for the analytical and preparative separation of (*R*,*S*)-goitrin (Fig. 3), an active ingredient present in a traditional Chinese herbal medicine. (*R*)-Goitrin is known for its antiviral property whereas (*S*)-goitrin has antithyroid activity; so the separation of these enantiomers is most essential to develop new

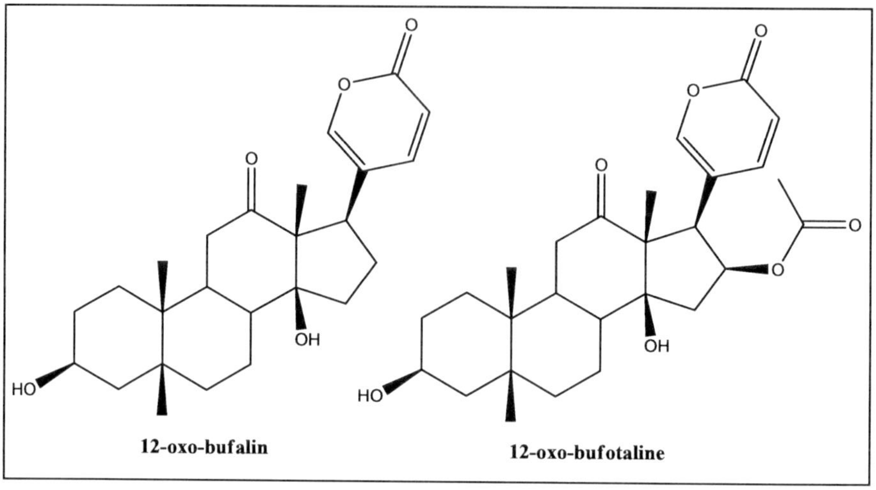

Fig. 1 New bufadienolides isolated

Fig. 2 Structure of Diniconazole

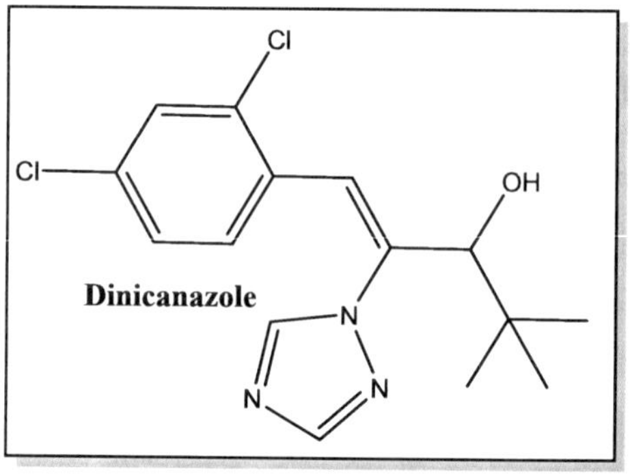

Fig. 3 (*R*,*S*)-goitrin

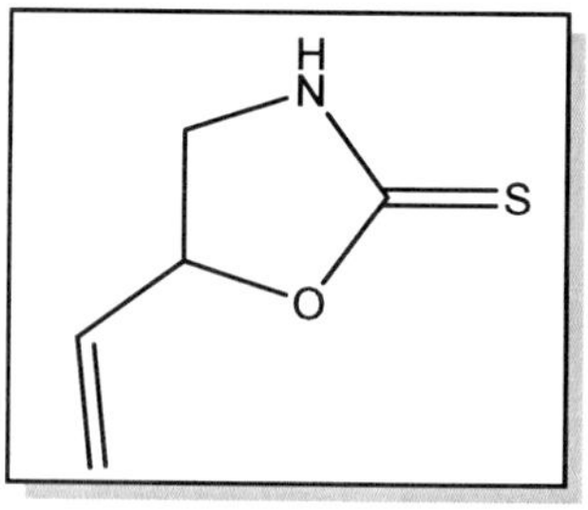

antiviral agents. Scale-up purification of (*R*)-goitrin and (*S*)-goitrin was done on an SFC column with >90% total recovery [37].

1.2.5 Polymer Separations

SFC is a highly accurate tool for measuring the molecular weight distributions of polymers [38]. Molecular weight distributions of polystyrene, polyethylene glycol, and polyprenol have been done using SFC. Quite recently racemic ephedrine enantiomers (Fig. 4) were separated by SFC using (−) ephedrine molecularly imprinted polymers (MIPs) as the chiral stationary phase. The separations obtained with SFC were found to be much superior to that obtained with chiral HPLC. The authors of the paper have reported that 21 different (−) ephedrine imprinted polymers were used for the enantioseparation of the racemic ephedrine [39].

1.3 Supercritical Fluids as Reaction Media

The peculiar properties of $ScCO_2$ like diffusivity combined with solvent power make it suitable for use in many chemical reactions. Supercritical fluids have been used as solvents for biocatalysis, biomass conversion, homogeneous and heterogeneous catalysis, polymerization, material synthesis, and chemical synthesis [40–42]. Certain examples of reactions done in $scCO_2$ are highlighted in Table 2. High miscibility of hydrogen in supercritical carbon dioxide compared to that in other organic solvents makes hydrogenation reactions very feasible in supercritical media. As a consequence of this many homogeneous and heterogeneous-catalyzed hydrogenation reactions have been found to be effective in $scCO_2$ [43, 44]. In addition several enantioselective hydrogenations have also been reported in $scCO_2$ using chiral catalysts (entries 1, 2 in Table 2) [45–47]. As in the case of hydrogenation and hydroformylation, the high solubility of oxygen in $scCO_2$ makes oxidation reactions in this media also very feasible. Oxidation of alcohols to carbonyl compounds, [48] sulfides to sulphoxides and epoxidations are the widely studied oxidation reactions in $scCO_2$. Quantitative conversion of *cis*-cyclooctene into epoxycyclooctane was achieved using an aldehyde as the co-oxidant and the reactor material, stainless steel (which facilitated formation of acylperoxy radicals) provided a non-catalytic radical pathway for the oxidation (entry 5, Table 2). Lower viscosity leads to higher rates of diffusion in $scCO_2$ which facilitates radical reactions like cyclizations and additions

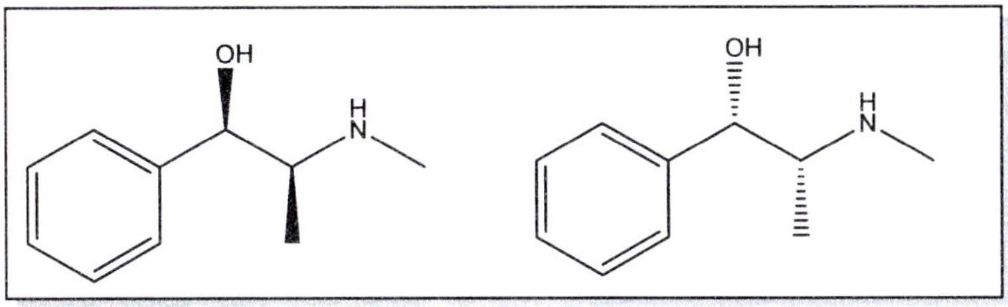

Fig. 4 Ephedrine enantiomers

Table 2 Some examples of reactions in supercritical media

S. No	Type of reaction	Reaction
1	Hydrogenation	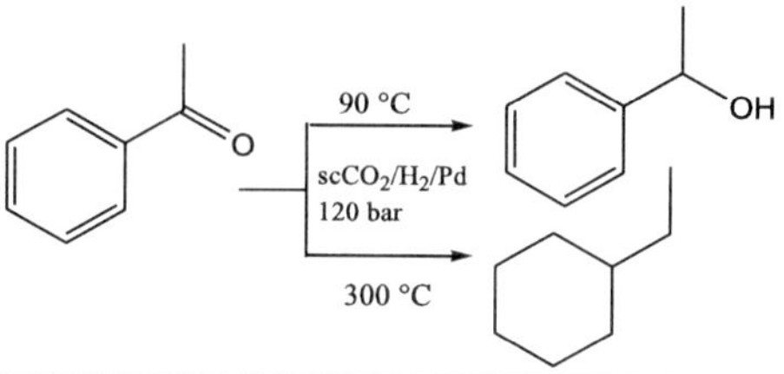
2	Asymmetric hydrogenation	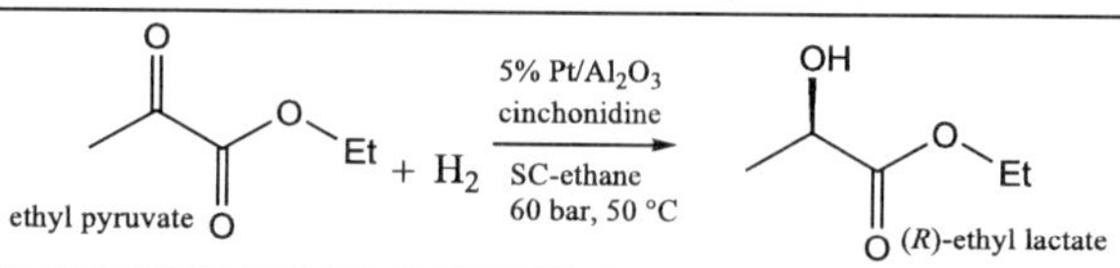
3	CO_2 as reactant	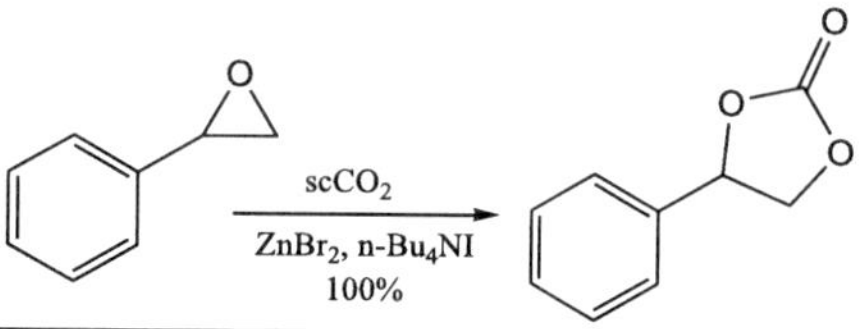
4	Hydroformylation	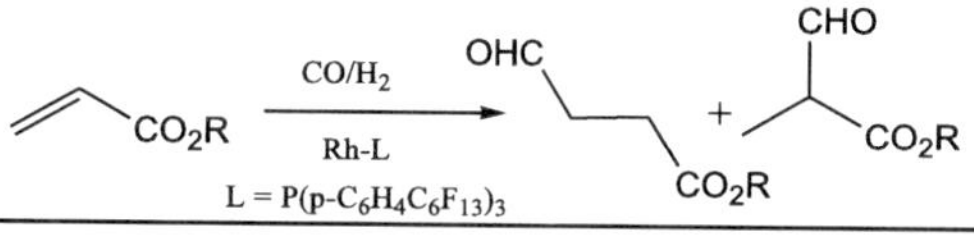
5	Oxidation	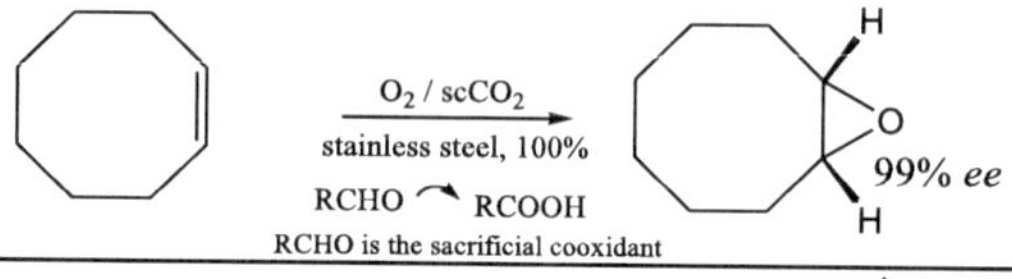
6	Free-radical reaction	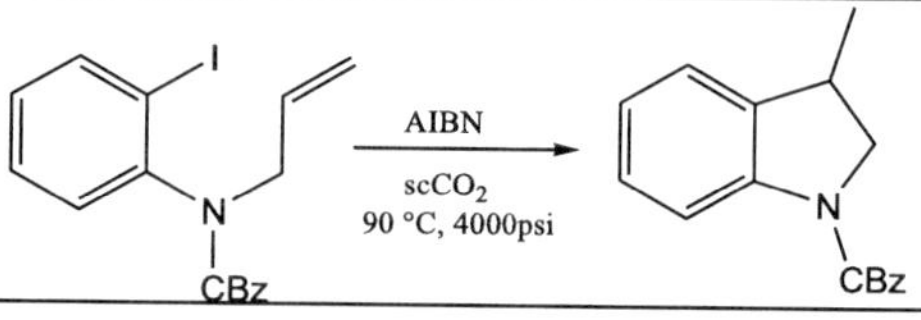
7	Coupling reactions	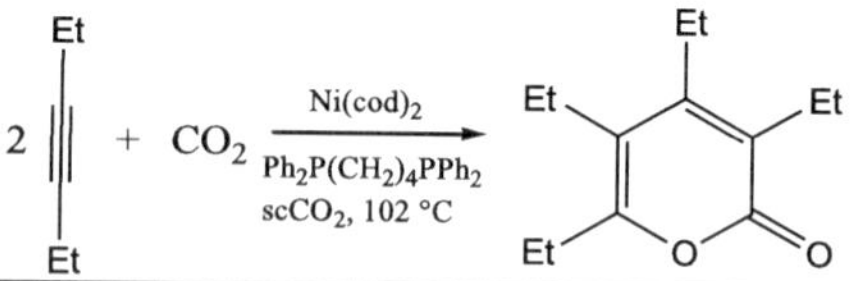
8	Hydroboration	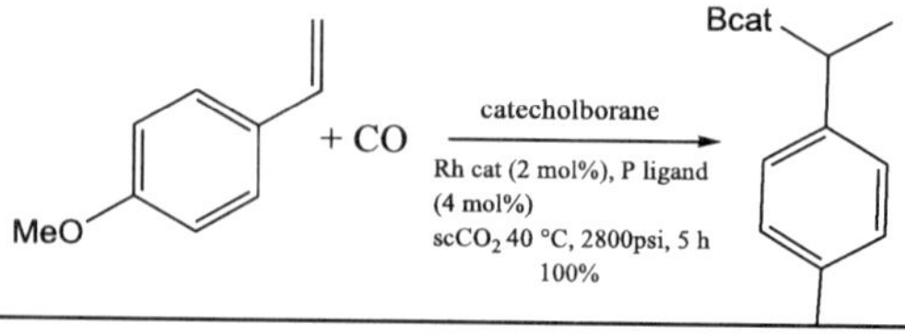

in presence of AIBN in $scCO_2$ (entry 6, Table 2). Apart from those listed in the table, other well-studied reactions in $scCO_2$ include Diels-Alder reaction, [49, 50] Pauson Khand reaction, [51] Baylis-Hillman reaction, [52] alkylation, acetylation, esterification, and palladium catalyzed coupling reactions [2].

However, the high cost associated with the requirement of maintaining high pressure vessels to carry out continuous reactions in $scCO_2$ along with the difficulty to reproduce reactions on a commercial scale requiring precise control of pressure and temperature are the major hurdles in the use of supercritical fluids as reaction solvents. Nonetheless several commercial plants were set-up for doing reactions in supercritical fluids. An overview of the reactions in $scCO_2$ and their advantages compared to conventional solvents has been published [53]. A process for the production of biodiesel from soy sauce residue (a waste product) under supercritical carbon dioxide was reported to occur most efficiently at 100 °C and at a pressure of 16 MPa over a time of 180 min [54].

1.3.1 Supercritical Polymerizations

It is well-known that density of the supercritical medium can be altered through simple variations of pressure which makes these fluids suitable for carrying out polymerization reactions leading to the synthesis of polymers of different molecular weights. Additionally the low viscosity of the supercritical medium and its inertness for chain-transfer terminations renders the supercritical medium very attractive for polymerizations. Polymerization of acrylate **1** and propylene oxide **3** have been successfully carried out in supercritical carbon dioxide (Scheme 1) [55]. In the former case, the initiator AIBN is found to be more efficient in $scCO_2$ than in conventional solvents and in the latter case CO_2 also acts as the reagent in presence of heterogeneous zinc or aluminum catalyst to bring about the polymerization reaction. In

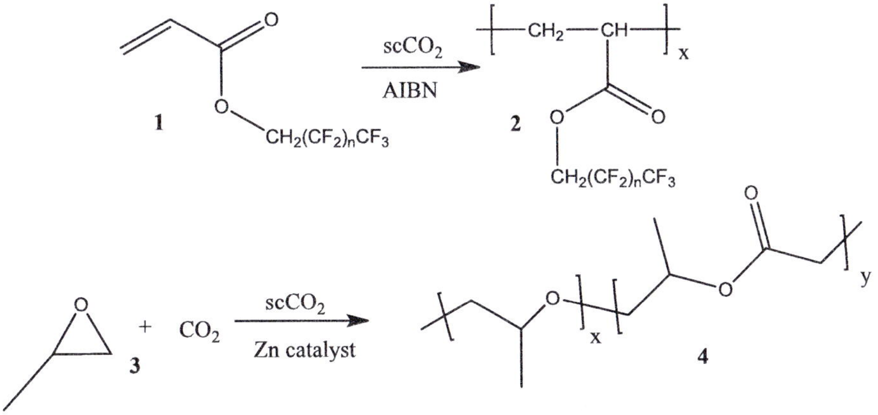

Scheme 1 Polymerization of acrylate and propylene oxide in $scCO_2$

addition to these, polymerization of ethylene terephthalate, ring-opening metathesis polymerization of norbornenes, and cationic polymerization of isobutenes are also known to proceed well in supercritical carbon dioxide medium [56].

Polymer synthesis and processing using supercritical fluids is an area which has received much attention. The uniqueness of supercritical fluids such as high diffusivity, sustainability, and less-residue interactions have been utilized to synthesize polymethylmethacrylate and its fluorinated analogues, polystyrenes, and various other cross-linked polymeric systems [57–60]. Belomoina et al. synthesized a series of poly(naphthoylenebenzimidazole)s of the type **7** in $scCO_2$ in absence of other hazardous solvents. The catalyst used in the reaction was a mixture of benzoic acid and benzimidazole. The synthesized polymers displayed a yellow-orange light emission property. Conventional method for the synthesis of these polymers involved use of phosphoric acid-based solvents and high temperatures. The $scCO_2$ based method worked well at 90 °C temperature and involved no toxic chemicals and thereby proved to be advantageous (Scheme 2) [61].

Using sequential reactions in order to avoid intermediate depressurization steps proved to be advantageous in devising new reactions in $scCO_2$. Development of such continuous flow reactors has helped in enhancing profitability and reducing reaction time [62]. For instance, Friedel-Crafts alkylation of *m*-cresol with different alkylating agents using supercritical carbon dioxide was carried out for the selective synthesis of thymol, the isomers of which are all having a similar boiling point which makes their separation otherwise difficult [63].

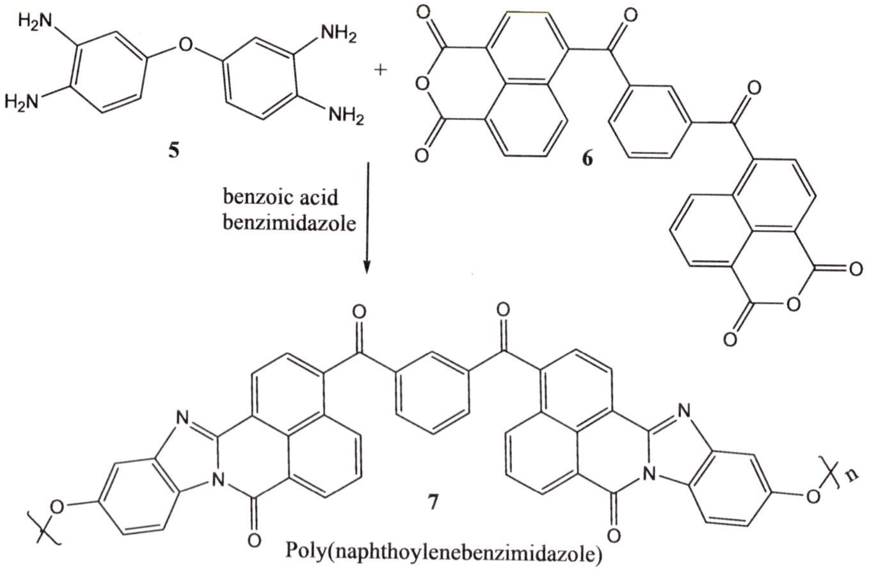

Scheme 2 Synthesis of poly(naphthoylenebenzimidazole) in $scCO_2$

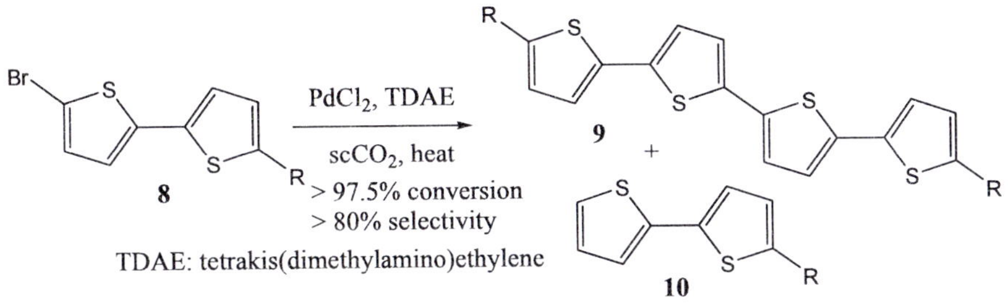

Scheme 3 Synthesis of an organic n-type semiconductor material in $scCO_2$

A recent research has also shown that zeolite-imidazole framework (a subclass of MOF's) has been synthesized by crystallization in $scCO_2$ [64]. An organic n-type semiconductor material 5,5‴-bis(tridecafluorohexyl)-2,2′:5″,2″:5″, 2‴-quaterthiophene **9** was synthesized and purified by $scCO_2$ (Scheme 3) [65]. Due to the high affinity of $scCO_2$ to fluorine, [66] fluoroalkyl group could be easily introduced to the quaterthiophene core.

Thus supercritical CO_2 is one of the most promising alternative solvents for organic reactions. Research teams have devised various methods to reduce the cost of reactions based on $scCO_2$ like use of gas-expanded liquids (GXLs) and controlled phase separation using ionic liquid/CO_2 combination. Several reviews on catalytic including enzymatic reactions in supercritical fluids can be found in the literature and is therefore not detailed here [67–70].

1.4 Supercritical Fluid Deposition

Supercritical fluid chemical deposition (SFCD) was first introduced by Sievers et al. in the year 1991 [71]. In supercritical fluid chemical deposition (SFCD or SCFD), the precursor is dissolved in the supercritical fluid, the latter acting as a medium. Using this method, Watkins deposited organometallic precursors into porous carriers by impregnating the carriers with the mixture of $scCO_2$ and the precursors and this was one of the early reports using this technique. [72] Subsequently the technique was used for preparing supported nanoparticles, [73] thinfilms [74], and controlled-release drug delivery systems [75]. Owing to the good diffusivity associated with the supercritical fluid, conformal deposition of Pd, Cu, Ni, Pt, Co, and Ru were also achieved using SFCD technology [76]. The lack of surface tension enables efficient electrodeposition of supercritical fluids onto fragile substrates and a number of metals including silver, copper, and cobalt and many conducting polymers have been electrodeposited from supercritical fluids [2].

The ability of supercritical fluids to assist solutes for permeating into small pores was utilized for modifying ionic liquids (ILs). Due to the high viscosity of ionic liquids, the gas diffusion co-efficient in ILs is very small and this has

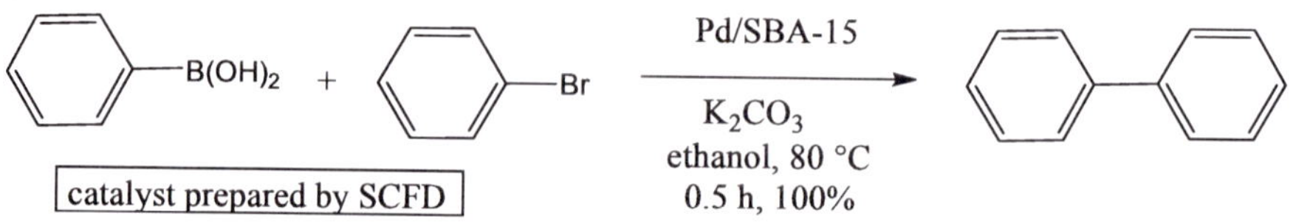

Scheme 4 Suzuki coupling using Pd/SBA-15 nanoparticle catalyst

led to the development of supported ionic liquid membranes (SILM). Xu et al. [77] deposited [bmim][BF_4] into SBA-15 using SCFD with methanol as the co-solvent. The prepared samples exhibited larger pore volume and specific surface area compared to those prepared using impregnation method. SCFD can also be used to prepare SILMs on asymmetric γ-Al_2O_3 membranes. With the assistance from scCO_2 and co-solvent (ethanol), the ILs could penetrate into the small and large pores in γ-Al_2O_3 efficiently and exhibited a larger CO_2 permeance and improved CO_2/N_2 selectivity [78].

Supercritical fluid deposition of metals and supercritical impregnation of drugs into aerogels have been studied extensively and has been reviewed [79]. Such methods are very promising for incorporating drugs within porous carriers and the enhancement in the interactions between the compound and the aerogel matrix depends largely on the supercritical impregnation conditions. Supercritical CO_2 deposition method has been used to prepare solid material supported metal nanoparticles. Here the metal-organic precursor is first dissolved in scCO_2 followed by adsorption of the organometallic complex to the solid surface and final reduction by thermal or chemical means. An improved deposition of Pd nanoparticles on SBA-15 templated mesoporous silica was successfully carried out using Pd(II) complex of 2,2′-bipyridyl as the precursor in scCO_2. The prepared nanoparticle catalyst proved to be efficient in Suzuki reactions as shown in the scheme below (Scheme 4) [80].

2 Conclusion

In summary, use of supercritical fluid as a green solvent is highly beneficial in terms of product recovery and tuning. The technology has already reached a state of "adulthood" and is routinely being used for various industrial processes such as decaffeination of coffee beans, flavor/aroma extraction, dyeing, polymer processing, cork cleaning, sterilization, fractionation of fish oil omega-3 fatty acids, and leather tanning. Development of new processes with supercritical fluids is a major focus of many research groups and is sure to culminate in innovative techniques suited to the industry and academia.

References

1. Anastas PT, Warner JC (1988) Green chemistry—theory and practice. Oxford University Press, New York
2. Ahluwalia VK (2013) Green chemistry. Narosa Publishing House, New Delhi
3. Macfarlane DR, Kar M, Pringle JM (2017) Fundamentals of ionic liquids: from chemistry to applications. Wiley-VCH Verlag GmbH & Co, KGaA, Weinheim
4. Klass DL (1988) Biomass for renewable energy, fuels and chemicals. Elsevier Inc, USA
5. Gladysz JA, Curran DP, Horvath IT (eds) (2004) Handbook of fluorous chemistry. Wiley-VCH Verlag GmbH & Co.KGaA, Weinheim
6. Hunt AJ, Attard AM (eds) (2018) Supercritical and other high pressure solvent systems for extraction, reaction and material processing. Royal Society of Chemistry, UK
7. Edward R, Qiubai S, Zhiqiang Z et al (2009) Mini review: green sustainable processes using supercritical fluid carbon dioxide. J Envir Sci 21:720–726
8. Zoccali M, Donati P, Mondello L (2019) Recent advances in the coupling of carbon dioxide-based extraction and separation techniques. Tr Analy Chem 116:158–165
9. Catchpole O, Moreno T, Montanes F et al (2018) Perspectives on processing of high value lipids using supercritical fluids. J Supercrit Fluids 134:260–268
10. Bittencourt GM, Firmiano DM, Fachini RP et al (2019) Application of green technology for the acquisition of extracts of araçá (Psidiumgrandifolium Mart. ex DC.) using supercritical CO_2 and pressurized ethanol: characterization and analysis of activity. J Food Sci 84:1297–1307
11. Castro-Vargas HI, Baumann W, Ferreiro SRS et al (2019) Valorization of papaya (Carica papaya L.) agroindustrial waste through the recovery of phenolic antioxidants by supercritical fluid extraction. J Food Sci Technol 56:3055–3066
12. Li K, Xu Z (2019) A review of current progress of supercritical fluid technologies for e-waste treatment. J Cleaner Prod 227:794–809
13. King JW, List GR (eds) (1997) Supercritical fluid technology in oil and lipid chemistry. AOCS Press, Illinois
14. Valli JMD, Fuente JCD, Uquiche E (2012) A refined equation for predicting the solubility of vegetable oils in high-pressure CO_2. J Supercrit Fluids 67:60–70
15. Bourgou S, Rebsy IB, Dakhlaoui S et al (2019) Green extraction of oil from Carum carvi seeds using bio-based solvent and supercritical fluid: Evaluation of its antioxidant and anti-inflammatory activities. Phytochem Anal. https://doi.org/10.1002/pca.2864
16. Santes LC, Bitencourt RG, Santos P et al (2019) Solubility of passion fruit (passiflora edulis sims) seed oil in supercritical CO_2. Fluid Phase Equilib 493:174–180
17. Garcia-Perez JS, Cuellar-Bermidez SP, Cruz-Quiroz R et al (2019) Supercritical CO_2-based tailor made valorization of origanum vulgare L extracts: a green approach to extract high-value compounds with applied perspectives. J Environ management 232:796–902
18. Alvarez MV, Cabred S, Ramirez CL et al (2019) Valorization of an agroindustrial soybean residue by supercritical fluid extraction of phytochemical compounds. J Supercrit Fluids 143:90–96
19. Pimentel-Moral S, Borrás-Linares I, Lozano-Sánchez J et al (2019) Supercritical CO_2 extraction of bioactive compounds from Hibiscus sabdariffa. J Supercrit Fluids 147:213–221
20. Gallo-Molina AC, Castro-Vargas HI, Garzon-Mendez WF et al (2019) Extraction, isolation and purification of tetrahydrocannabinol from the Cannabis sativa L. plant using supercritical fluid extraction and solid phase extraction. J Supercrit Fluids 146:208–216
21. Sánchez-Camargo AP, Parada-Alfonso F, Ibáñez E et al (2019) Recent applications of on-line supercritical fluid extraction coupled to advanced analytical techniques for compounds extraction and identification. J Sep Sci 42:243–257
22. Sánchez-Camargo AP, Parada-Alfonso F, Ibáñez E et al (2017) On-line coupling of supercritical fluid extraction and chromatographic techniques. J Sep Sci 40:213–227
23. Brondz I, Sedunov B, Sivaraman N (2017) Influence of modifiers on supercritical fluid chromatography (SFC) and supercritical fluid extraction (SFE) part I. Int J Anal Mass Sp Chr 5:17–39

24. Pilařová V, Plachká K, Khalikova MA et al (2019) Recent developments in supercritical fluid chromatography—mass spectrometry: is it a viable option for analysis of complex samples? Tr Anal Chem 112:212–225
25. Pirok BWJ, Gargano AFG, Schoenmakers PJ (2018) Optimizing separations in online comprehensive two-dimensional liquid chromatography. J Sep Sci 41:68–98
26. Hirata Y, Sogabe I (2004) Separation of fatty acid methyl esters by comprehensive two dimensional supercritical fluid chromatography with packed columns and programming of sampling duration. Anal Bioanal Chem 378:1999–2003
27. Hirata Y, Hashiguchi T, Kawata E (2003) Development of comprehensive two-dimensional packed column supercritical fluid chromatography. J Sep Sci 26:531–535
28. Yu RK, Tsai YT, Ariga T et al (2011) Structures, biosynthesis, and functions of gangliosides–an overview. J Oleo Sci 60:537–544
29. Si W, Liu Y, Xiao Y et al (2020) An offline two-dimensional supercritical fluid chromatography × reversed phase liquid chromatography tandem quadrupole time-of-flight mass spectrometry system for comprehensive gangliosides profiling in swine brain extract. Talanta 208:120366
30. Gibitzeisath N, Sturn S, Stuppner H (2018) Supercritical fluid chromatography in natural product analysis—an update. Planta Med 84:361–371
31. He P, Zhang Y, Zhou Y et al (2019) Supercritical fluid chromatography-a technical overview and its applications in medicinal plant analysis: an update covering 2012–2018. The Analyst 144:5324–5352
32. Mentañés F, Tallon S (2018) Supercritical fluid chromatography as a technique to fractionate high-valued compounds from lipids. Separations. Separations 5(3):38–50
33. Wei W, Hou J, Yao C et al (2019) A high-efficiency strategy integrating offline two dimensional separation and data post-processing with dereplication: characterization of bufadienolides in Venenum Bufonis as a case study. J Chr A 1603:179–189
34. Xinzhong Z, Yuechen Z, Xinyi C et al (2018) Application and enantiomeric residue determination of diniconazole in tea and grape and apple by supercritical fluid chromatography coupled with quadrupole-time-of-flight mass spectrometry. J Chr A 1581–1582:144–155
35. Speybrouck D, Emmanuelle L (2016) Preparative supercritical fluid chromatography: a powerful tool for chiral separations. J Chr A 1467:33–55
36. Harps LC, Joseph JF, Parr MK (2019) SFC for chiral separations in bioanalysis. J Pharm Biomed Anal 162:47–59
37. Nie L, Dai Z, Shuangcheng M (2016) Improved chiral separation of (R,S)-goitrin by SFC: an application in traditional chinese medicine. J Anal Meth Chem, S)-Goitrin by SFC. https://doi.org/10.1155/2016/5782942
38. Berger TA (1995) Packed column supercritical fluid chromatography. Royal Society of Chemistry, UK
39. Ansell RJ, Kuah JKL, Wang D et al (2012) Imprinted polymers for chiral resolution of (±)-ephedrine, 4: packed column supercritical fluid chromatography using molecularly imprinted chiral stationary phases. J Chr A 1264:117–123
40. Knez Ž, Markočič E, Leitgeb M et al (2014) Industrial applications of supercritical fluids: a review. Energy 77:235–243
41. Olmos A, Asensio G, Pérez PJ (2016) Homogeneous metal-based catalysis in supercritical carbon dioxide as reaction medium. ACS Catal 6:4265–4280
42. Mayadevi S (2012) Reactions in supercritical carbon dioxide. Ind J Chem 51A:1298–1305
43. Jessop PG, Ikariya T, Noyori R (1999) Homogeneous catalysis in supercritical fluids. Chem Rev 99:475–494
44. Baiker A (1999) Supercritical fluids in heterogeneous catalysis. Chem Rev 99:453–474
45. Burk MJ, Feng S, Gross MF et al (1995) Asymmetric catalytic hydrogenation reactions in supercritical carbon dioxide. J Am Chem Soc 117:8277–8278
46. Xiao J, Nefkens SCA, Jessop PG et al (1996) Asymmetric hydrogenation of α, β-unsaturated carboxylic acids in supercritical carbon dioxide. Tetrahedron Lett 37:2813–2816
47. Kainz S, Brinkmann A, Leitner W, Pfaltz A (1999) Iridium-catalyzed enantioselective hydrogenation of imines in supercritical carbon dioxide. J Am Chem Soc 121:6421–6429

48. Jerner G, Sueur D, Mallat T et al (2000) Partial oxidation of alcohols in supercritical carbon dioxide. Chem Commun 2247–2248
49. Clifford AA, Pople K, Gaskill WJ et al (1998) Potential tuning and reaction control in the Diels-Alder reaction between cyclopentadiene and methyl acrylate in supercritical carbon dioxide. J Chem Soc, Faraday Trans 94:1451–1456
50. Renslo AR, Weinstein RD, Tester JW et al (1997) Concerning the regiochemical course of the Diels–Alder reaction in supercritical carbon dioxide. J Org Chem 62:4530–4533
51. Jeong N, Hwang SH, Lee YW et al (1997) Catalytic Pauson–Khand reaction in super critical fluids. J Am Chem Soc 119:10549–10550
52. Rose PM, Clifford AA, Rayner CM (2002) The Baylis–Hillman reaction in supercritical carbon dioxide: enhanced reaction rates, unprecedented ether formation, and a novel phase-dependent 3-component coupling. Chem Comm 968–969
53. Knez Ž, Pantić M, Cör D et al (2019) Are supercritical fluids solvents for the future? Chem Engg Proc 141:107532
54. Xiang C, Liu SY, Fu Y et al (2019) A quick method for producing biodiesel from soy sauce residue under supercritical carbon dioxide. Renew Energy 134:739–744
55. DeSimone M, Guan Z, Elsbernd CS (1992) Synthesis of fluoropolymers in supercritical carbon dioxide 257:945–947
56. Beckman EJ (2004) Supercritical and near-critical CO_2 in green chemical synthesis and processing. 28:121–191
57. Kiran E (2016) Supercritical fluids and polymers—the year in review—2014. J Supercrit Fluids 110:126–153
58. Belomoina NM, Bulycheva EG, Nikitin LN et al (2016) Study of the process of poly(phenylquinoxaline)s formation in supercritical carbon dioxide. J Supercrit Fluids 113:66–71
59. Ngo MT, Dickmann JS, Hassler JC et al (2016) A new experimental system for combinatorial exploration of foaming of polymers in carbon dioxide: the gradient foaming of PMMA. J Supercrit Fluids 109:1–19
60. Cooper AI, Hems WP, Holmes A (1999) Synthesis of highly cross-linked polymers in supercritical carbon dioxide by heterogeneous polymerization. Macromolecules 32:2156–2166
61. Belomoina NM, Bulycheva EG, Elmanovich IV et al (2019) Supercritical carbon dioxide as an effective medium for poly(naphthoylenebenzimidazole)'s synthesis. J Supercrit Fluids 148:148–154
62. Han X, Poliakoff M (2012) Continuous reactions in supercritical carbon dioxide: problems, solutions and possible ways forward. Chem Soc Rev 41:1428–1436
63. Amandi R, Hyde JR, Ross SK et al (2005) Continuous reactions in supercritical fluids; a cleaner, more selective synthesis of thymol in supercritical CO_2. Green Chem 7:288–293
64. Wang S, Tang S (2019) Rapid and facile synthesis of metal organic framework materials by reaction crystallization in supercritical CO_2. Mat Lett 251:65–68
65. Hirase R, Honda K, Ishihara M et al (2018) A new method of using supercritical carbon dioxide as a green solvent for synthesis and purification of 5,5‴-bis(tridecafluorohexyl)-2,2′:5′,2″:5″,2‴-quaterthiophene, which is one of n-type organic semiconducting materials. Tetrahedron Lett 59:469–472
66. Ganapathy HS, Yuvaraj H, Hwang HS (2006) CO_2-soluble semiconducting polymers synthesized in supercritical carbon dioxide. Synth Met 156:576–581
67. Habulin M, Primozic M, Knez Z (2007) Supercritical fluids as solvents for enzymatic reactions. Acta Chim Slov 54:667–677
68. Lozano P, de Diego T, Iborra JL (2010) Supercritical solvents. In: Anastas PT (ed) Handbook of green chem. Wiley-VCH Verlag GmbH & Co. KGaA, New York, Part 4
69. Kamat SV, Beckman EJ, Russell AJ (2008) Enzyme activity in supercritical fluids. Cr Rev Biotech 15:41–71
70. Hyde J, Leitner W, Poliakoff M (2002) Catalytic reactions in supercritical fluids. In: van Eldik R, Klärner F-G (eds) High pressure chemistry: synthetic, mechanistic, and supercritical applications. Wiley-VCH Verlag GmbH, New York, pp 371–397

71. Hybertson BM, Hansen BN, Barkley RM et al (1991) Deposition of palladium films by a novel supercritical fluid transport-chemical deposition process. Mater Res Bull 26:1127–1133
72. Watkins JJ, McCarthy TJ (1995) Polymer/metal nanocomposite synthesis in supercritical carbon dioxide. Chem Mater 7:1991–1994
73. Xu Q, Zhao X, Yin et al (2016) Direct growth of highly-dispersed $MnCl_2$ centerdot $4H_2O$ nanostructures with different morphologies on graphene in supercritical CO_2. Mater Res Expr 3:65012
74. Hwang J, Min D, Yoon D et al (2016) Liquid carbon dioxide-based coating of auniform carbon layer on hierarchical porous MoO_2 microspheres and assessment of their electrochemical performance. Chem Eng J 290:335–345
75. Jain A, Ong V, Jayaraman S et al (2016) Supercritical fluid immobilization of horseradish peroxidase on high surface area mesoporous activated carbon. J Supercrit Fluids 107:513–518
76. Kondoh E (2018) Cu wiring fabrication by supercritical fluid deposition for MEMS devices. In: Sone M, Masu K (eds) Novel metal electrodeposition and recent applications. Intech Open. https://doi.org/10.5772/intechopen.81636
77. Xu Q, Yin J, Zhou X et al (2016) Impregnation of ionic liquids in mesoporous silica using supercritical carbon dioxide and co-solvent. RSC Adv 6:101079–101086
78. Liu YF, Xu QQ, Wang YQ et al (2018) Preparation of supported ionic liquid membranes using supercritical fluid deposition based on γ-alumina membrane and imidazolium ionic liquids. J Supercrit Fluids 139:88–96
79. Tkalee G, Pantić M, Novak Z et al (2015) Supercritical impregnation of drugs and supercritical fluid deposition of metals into aerogels. J Mater Sci 50:1–12
80. Ulusal F, Erünal E, Güzel B (2018) Green preparation of Pd nanoparticles on SBA-15 via supercritical fluid deposition and application on Suzuki-Miyaura cross-coupling reaction. J Nanopart Res 20:219. https://doi.org/10.1007/s11051-018-4325-0